Andreas Niederberger
Uchenna Okeja • Johanna Gördemann
Editors

Handbook of Migration Ethics

Springer

Editors
Andreas Niederberger
Universität Duisburg-Essen
Institut für Philosophie
Essen, Nordrhein-Westfalen, Germany

Johanna Gördemann
Universität Duisburg-Essen
Institut für Philosophie
Essen, Nordrhein-Westfalen, Germany

Uchenna Okeja
Nelson Mandela University
Gqeberha, South Africa

ISBN 978-3-031-89876-1 ISBN 978-3-031-89877-8 (eBook)
https://doi.org/10.1007/978-3-031-89877-8

This Springer imprint is published by the registered company Springer Nature Switzerland AG
The registered company address is: Gewerbestrasse 11, 6330 Cham, Switzerland

Introduction

In many parts of the world, we are currently seeing heated arguments about refugees and migration, and increasingly drastic measures are being adopted to counteract the supposedly disruptive consequences of migration for societies and states. Yet, the history of mankind has always been a history of settlement and migration; human existence cannot be limited to one or the other. Even nowadays, without migration, societies become encrusted and economies and social systems weakened. People are not deterred from migrating if this is an effective means of significantly improving their own situation and escaping miserable circumstances. This certainly does not apply to all people in all places to the same extent, but it does apply to enough people in a sufficient number of places for us to be able to say that the history of humanity remains and will remain a history of migration.

Given the profound importance of migration to humanity, it is quite surprising how little attention it has received in the history of philosophy. Today, the ethics of migration is one of the fast-growing fields of research, not only in the West, but in almost all parts of the world. Of course, migration has also been thought about at other times. However, if we consider the significance that migration has had for human history in the truest sense of the word, that is, for example, for contact between people which has significantly changed and brought forward communities, we must ask why philosophy has not always been much more centrally concerned with migration in all its different forms and with all its different effects. This question becomes even more pertinent when we realise that the topic of encountering the other, the stranger, understood as those who do not already belong to the circle of family and friends, has always been central to ethics. Yet in ethics in particular, matters relating to duties, responsibilities and goals have been thought of primarily in terms of strangers who are simply there and are not characterised by a history of migration.

As a first step in exploring the big question of what migration and refugees should mean in and for philosophy in general, it is necessary to gain an overview of the current state of philosophical research and discussion on these topics. This is what this handbook provides. Studies on the ethics of migration in the broadest sense exist not only in ethics, but also in political philosophy, applied ethics, political theory, legal theory, social philosophy and the study of migration in the social sciences with their philosophical dimensions. Approaches in this now broad field have already taken a wide variety of forms as theorists attempt to make sense of different aspects of migration and refugees. Some approaches have developed a phenomenology of the migrant. Others have worked out possible rights, duties and responsibilities of migrants, host societies and other agents that are relevant to emigration, immigration or the migration process itself. A number of approaches have debated the role of borders and admissible criteria for immigration policies, the selective access to citizenship or the differences between claims refugees and other migrants can raise. Some of these approaches focus directly on migrants or the emigration and immigration nexus, while others look at migrants in the context of questions regarding the nature of a successful life, social or, more precisely, global justice or the cultural and political conditions for the constitution and maintenance of communities and polities.

Migration often involves crossing borders. These do not necessarily have to be national borders, even if definitions of migration often link the phenomenon to the fact that migrants move their centre of life more or less permanently from one country to another. Such definitions seem plausible to many because without the crossing of state borders, it is unclear when a move on a territory becomes migration. From the perspective of migrants, crossing borders is certainly important: they see themselves as migrants when they cross the border of what they know and are used to. However, such a border is often also crossed when moving from the countryside to the city or from a small town to the capital of a country. In many cases, migration across national borders does not represent anything qualitatively new beyond this internal migration. The idea of a homeland, i.e. an entire country in which everything is familiar to us, in contrast to other countries in which most things are foreign to us, is a fiction.

Philosophy can do much to clarify contingent definitions and to show how they contribute to distinguishing the supposedly normal from the supposedly problematic issues in migration. For those who restrict migration conceptually to the crossing of state borders, their understanding of migration is that it is a movement from one's own order to another order, as the change of state is the presumed purpose of migration. In contrast, those who understand migration as leaving behind the familiar and turning towards the unknown tend to see migrants as the incarnation of curiosity, openness and a willingness to become involved in a new context. As many contributions to this handbook show, there is no shared definition of migration in philosophy and, accordingly, no shared interpretation of what migration is about. In short, it makes sense to question the meaningfulness of subsuming all migrants or all refugees under a single definition of what a "migrant" or a "refugee" is, thereby obscuring the diverse reasons and contexts of migration and flight. Notwithstanding, it must be conceded that in this respect there are tensions in the philosophical method between the analytical aim of differentiation, on the one hand, and the normative aim of justifying claims or rights, on the other. Differentiation makes it difficult to grant the differentiated the same claims.

Beyond offering conceptual clarifications and rigorous analysis of the questions we pose about migration, there are other ways philosophical analysis is crucial for the study of migration, especially the ethics of migration. One such use of philosophy is the questioning of assumptions and the expansion of the boundaries we arbitrarily impose on the limits of the categories by which we co-exist with the so-called compatriots and strangers. Take, for instance, the assumption that people classified as refugees are clearly distinguishable from those classified as economic migrants. When this assumption of categorical difference between different kinds of immigrants is critically analysed and evaluated, it could become evident that we are mistaken in assuming that certain factors count more than others in our definitional approaches to the phenomenon of migration. Furthermore, the questioning of assumptions could lead to the recognition of the need to expand the scope of the categories by which we live. Such an expansion could improve our perception of our duties to strangers and other complex questions discussed by theorists of immigration. All this should not be taken to mean that improving our concepts and assumptions about migration through philosophical analysis implies that philosophical study of migration always and everywhere leads to better outcomes of concrete governance of migration. Our point is that philosophy has a very close bearing on migration which should not be ignored by philosophers. Besides, the critical reflection philosophers offer on questions about immigration can help us to improve policies and institutions.

Even if philosophy seems to relate intrinsically to migration, the recent popularity of the ethics of migration can certainly also be explained by the political controversies mentioned at the beginning. In its various disciplines, philosophy reacts to global mass migration, increasing numbers of refugees and the globalisation of societies and seeks to put forward original and innovative normative perspectives. Philosophy is therefore not only discovering migration and taking a fresh look at its own history and ideas. It sees migration and, most recently, the question of how to deal with refugees at least also as subjects and challenges that it wants to address to prove the political-practical relevance of philosophy. But several factors must be acknowledged as they pose challenges to the theoretical development of ethics and the philosophy of migration. Some of these factors include uncertainty regarding future patterns of migration, the impact of political rhetoric on civic and political cultures and the overall resistance to normative and fact-based theorising of migration due to political and other convictions.

The field of the ethics of migration is growing for diverse reasons in many disciplines and prominent among these reasons is the political transformations we can currently observe. Thus, the ethics of migration deals with key aspects of standard or classical positions in the disciplines involved in the debate. This is the case, for instance, with regard to the discussion of the role of the nation, the state and the relationship between the Global North and Global South. The ethics of migration therefore contributes to the reconsideration and revision of central concepts and assumptions of modern political and social theory.

Although the discourse on the ethics of migration is multifaceted, a significant challenge of the theoretical development of the field is the protracted initial debate about open borders. Given the prominence of this debate, it became easy for many to reduce the entire spectrum of concerns encapsulated in the ethics of migration to the debate about borders. This is especially the case because the focus on borders led to further derivative question, such as the role of the state in immigration governance, the claims of sovereignty and the limits of cosmopolitan justice. Important as it is, delimiting the problems of justice implicated in the open borders debate is definitely neither the most important challenge of practical action nor the most crucial philosophical problem to consider in the discourse on the ethics of migration. This is the case because most of the crucial concerns about migration arise in relation to people who are already within the jurisdiction of a state.

Against this background, what is lacking so far is a comprehensive review of the state-of-the-art in migration ethics, which serves both academics and general readers interested in normative questions regarding current political and social developments. This handbook presents structured analyses of key topics and existing research and highlights important complex problems that should be explored in future research. By including subjects, theories and authors from most regions of the world, the handbook attempts to move the discourse on the ethics of migration beyond the Euro-Atlantic focus of most of the debate. The goal of doing this is to initiate an exchange between approaches, positions and theoretical paradigms. The handbook extends the scope of the ethics of migration to include the historical evolution of normative positions on migration. This provides a robust grounding of current debates against their often neglected historical situatedness.

The first part of the handbook takes a look at the foundations of migration ethics. It presents important perspectives on the development of migration ethics in global history with particular emphasis on crucial periods and authors. The chapters in this part of the handbook also examine the place of migration ethics in philosophy. The second part reconstructs the different philosophical disciplines and methodologies that have shaped the theorising of migration ethics. In part three, the handbook provides analyses of important topics in migration ethics, providing thereby a robust overview of the state of the debate regarding the topics, while indicating critical shortcomings that merit further investigation. The fourth part of the handbook introduces readers to the specific debate on refugee ethics and the issues that are of particular importance to this group of migrants. Part five is dedicated to analyses of the relevance of the normative discussion of migration and asylum in disciplines beyond philosophy.

In part 1, articles by Phillip Cole, Oliver Eberl, Naoto Higuchi, Gilberto Rodrigues and Jose Blanes Sala and Uchenna Okeja address foundational issues about migration ethics. In his chapter on the normative foundations of migration ethics, Cole argues that the foundation of migration ethics is derived from liberal democratic theory. He considers the implications of the key principles of moral equality and rationality for migration ethics as a way to map the arguments about the ethics of migration. He recommends, among other things, that theorists should pay closer attention to empirical work on migration as well as issues that manifest themselves outside the context of the Global North. Eberl explores basic issues regarding the rights of strangers, particularly in the context of Europe's historical experience up to 1800, noting how ideas about the territorial state and territorial sovereignty shaped the migration of both citizens and strangers. He further considers the evolution of the migration discourse in European social and political theory to show how cosmopolitanism bridges the conceptual dichotomy between citizens and strangers. Higuchi takes on the topic of migration ethics in East Asia. In this regard, he examines the three-layered structure of migration in Japan, South Korea, Hong Kong and Taiwan, namely, the legacy of colonialism, the impact of the developmental states and the crucial role of feminisation of migration through marriage. Underlying this consideration is the key claim that ethnic nationalism is not the main driver of migration restrictions in this region. Rodrigues and Sala explore migration ethics in South America by focusing on its contribution to identity formation and other key facts shaping the history of migration in the region. They show that different facets of migration produced unique ethical challenges in the region and so migration ethics in this region must be considered in a manner that takes history seriously. Focusing on one of the key ethical challenges related to migration, namely, nationalism, Okeja shows in his chapter how migration and nationalism are related. He argues that nationalism-inspired arguments about migration are limited, hence the need to consider in future research the role of the nation and borders as factors that have histories.

Part two has as overarching goal the exploration of migration ethics discourse in philosophy. The chapters by Julian Müller, Iseult Honohan, François Boucher, Daniel Kersting and Julian Culp address the question of how migration is treated differently in different areas of philosophy, both methodologically and in terms of content. Focusing on applied ethics, Müller considers how the methods of applied ethics are applied to migration ethics. By this means, he highlights the importance for theorists of migration ethics to pay closer attention to their methods and what role the inclusion of methods from political philosophy can play in the treatment of migration issues in applied ethics. Honohan considers migration ethics in relation to political philosophy. Underscoring that so far political philosophers viewed issues about justice, freedom, democracy and citizenship mainly through the reference point of a bounded state, Honohan points out how recognition of the importance of migration changes this approach and the nature of the controversies this realisation occasions for political philosophy. In his chapter in this section, Boucher considers how theories of rights are relevant to migration ethics by discussing the crucial issue of entry. Legal concepts that are often invoked in migration ethics, such as rights and obligations, are discussed within the context of their development in law. This enables him in the end to show how these developments shape debates in migration ethics and reflect some of the tasks for future research on migration ethics from the standpoint of legal philosophy. Kersting extends the discussion in this section in his consideration of the often tensed relationship between migration and democracy due to migrants' claims to moral equality and democratic citizens' de facto right to self-determination. He considers ways of thinking about this challenge through the lens of different models of democracy and the need to take more seriously the theorising of the relationship between migration and democracy from a cosmopolitan perspective. Culp rounds off the section with his analysis of the meaning and relevance of the distinction between ideal and non-ideal theories to ethics of migration. He points to specific issues in the debates between proponents and opponents of open borders to show how ideal and non-ideal theorising may be mobilised to conceptualise and justify different arguments.

The third part of this handbook is the most comprehensive and maps the major controversies in migration ethics as a specific field of philosophy dating back to the 1980s at the latest. Alex Sager considers humanity and mobility against the background of the problems that assumptions about sedentarism and mobility raise for conceptual categories and theories of migration ethics. Andreas Niederberger asks how claims and rights that are justified in the field of migration ethics relate to expectations of justice and, in particular, global justice. James Souter discusses the implications of historical injustice for debates that currently shape migration ethics. Sarah-Lea Effert explores the implications of borders for migration ethics, while Andreas Cassee examines the issue of international freedom of movement. The other controversies include the challenge of migration policies discussed by Desiree Lim, the moral issues raised by integration of immigrants which is the core of Sophie Guérard de Latour's chapter, xenophobia, gender, expatriation and denationalisation, reverse migration and securitisation considered by Matolino, Higgins and Reed-Sandoval, Lenard, Adogame and Hermann, respectively. This part presents not only a robust mapping of the debates that currently shape the landscape of the ethics of migration, but it also demonstrates the growing recognition of ethical dimensions of emerging patterns of international migration.

Next, questions surrounding the ethics of refugees and asylum are taken up in part 4. In Georgiana Turculet's chapter, the nexus between migrants, refugees and tourists is considered. David Owen explores the contentious question of refugee ethics in his chapter, while Christopher Boom tackles the core ethical questions about refugees discussed in international law. The other chapters are by Max Gabriel Cherem, on humanitarianism; Mareike Gebhardt on refugee agency; Eilidh Beaton and Matthew Lister on refugee and family unification; Gottfried Schweiger on children refugeehood; Silja Klepp and Moritz Riemann on climate mobility; Andreas Niederberger and Uchenna Okeja on economic refugees and Mollie Gerver on refugee repatriation. Evident in the chapters that make up this part is that the conceptual distinction between refugees and other categories of immigrants has far-reaching implications for philosophical and other normative theorising of the ethics of migration.

The final part is devoted to exploring the framing and theorising of migration ethics beyond philosophy. Benjamin Boudou examines the way the ethics of migration is imagined and debated in political science. Anja Weiß discusses sociological positions on migration ethics. Johan Rochel's chapter tackles the ethical dimensions of the economics of migration. Brigitte Suter analyses the ethical questions that arise in the context of migration studies itself and Nausikaa Schirilla rounds the section off with her chapter that unpacks the relationship between normative positions in cultural studies and migration ethics. Overall, the chapters follow the format of laying out the context of the problem, before proceeding with the examination of concepts, approaches and central arguments. Each chapter concludes with a brief overview of desiderata for future research. This is a way of underscoring the open-ended nature of the migration ethics as a field of critical inquiry.

The articles in this handbook are not simply juxtaposed but are often interlinked and offer valuable insights by complementing each other. For example, the chapters by Cole, Honohan and Kersting examine the interplay between liberal democratic values and the normative foundations of migration. They explore how the principles of liberal democracy shape and complicate debates on migration ethics. Similarly, the chapters by Niederberger, Rodrigues and Sala and Matolino are linked by tackling the intricate issues of justice and labour migration policy. These contributions highlight the conceptual and policy implications of paying close attention to historical contexts and dynamics. As these chapters illustrate, a deeper understanding of the conceptual complexity and historical evolution of specific ethical questions regarding immigration enables scholars in the field of migration ethics to identify the limitations of existing theories and explore the potential for new perspectives. The handbook also provides an intersectional view on migration, particularly in the exploration of the relationship between economically motivated migration and gender. This is exemplified in the contributions by Higgins and Reed-Sandoval, Higuchi, and Beaton and Lister, which examine how gender shapes immigration policies and frames the ethics of migration. While the chapters address diverse issues, they share a focus on the complexity of gender and its role in structuring migration ethics. For all the breadth and diversity of the chapters in this handbook and despite the debates among migration ethics theorists on many particular points, including the importance of key concepts and the best framing of the issues they examine, their links demonstrate a shared understanding that migration is a dynamic feature of societies. This handbook brings together leading philosophers and scholars from other disciplines working on migration and refugee ethics to analyse important topics in the ethics of migration. The contributions are accessible and highly relevant to issues studied in the field and publicly debated. An assumption that informs this handbook is that the chapters provide insights that are useful to scholars and general readers interested in understanding ethical problems raised by migration against the background of their social and political implications. With the clarity it provides about arguments, concepts and the historical and theoretical evolution of debates and approaches to migration ethics, the handbook is instrumental in deepening the interest of philosophers in refining their perspectives and contributions to resolving, or at least better framing, established and emerging ethical questions about human mobility.

Andreas Niederberger
Uchenna Okeja

Contents

About the Editors and Contributors

About the Editors

Andreas Niederberger is Professor of Political, Social and Legal Philosophy at the University of Duisburg-Essen (Germany), where he is also Deputy Chair of the Board of the Centre for Global Cooperation Research. In 2018–2021 he was coordinator of the Horizon 2020 project "Norms and Values in the European Migration and Refugee Crisis". His research focuses on cosmopolitan political philosophy and philosophy of international law, in particular on issues of transnational democracy, human rights, migration and climate justice. His most recent publications include: "Can There Be Special Rights for Some Citizens?", Critical Review of International Social and Political Philosophy, 2025; Cosmopolitan Norms and European Values: Ethical Perspectives on Europe's Refugee Policy, co-edited with Marie Göbel, New York, London: Routledge 2024; Migration. Herausforderungen im Spannungsfeld von Theorie und Praxis, co-edited with Oliviero Angeli and Hans Vorländer, Baden-Baden: Nomos 2024.

Uchenna Okeja is Professor in the Faculty of Humanities at Nelson Mandela University and Visiting Professor and Presidential Fellow at Yale University. He was previously professor and chair of the philosophy department at Rhodes University and philosophy lecturer at Goethe University Frankfurt, Germany. He has held fellowships and visiting professorships at Harvard University, Utrecht University, University of Chicago, Institute for Advanced Studies in the Humanities at Bad Homburg, Peter Wall Institute for Advanced Studies at UBC Vancouver, University of St. Gallen and Stellenbosch Institute for Advanced Study. His background is in political philosophy, ethics and critical theory. He is editor of Routledge Handbook of African Political Philosophy (London: Routledge 2023) and author of Deliberative Agency (Bloomington: Indiana University Press 2022).

Johanna Gördemann was a research associate at the Department of Philosophy at the University of Duisburg-Essen (Germany) and researcher in the Horizon 2020 project "Norms and Values in the European Migration and Refugee Crisis" (2018–2021). Her research focuses on political philosophy and philosophy of law with a particular emphasis on migration and immigration. Among her publications are "Human Rights and Corporate Reinsurance: From Ensuring Rights to Insuring Risks" (with Ch. Scheper), New Political Economy, 2022; "The European Union and the Global Compacts on Refugees and Migration: A Philosophical Critique" (with F. Boucher), Interventions, 2021 and "Ein Plädoyer für eine gerechte Welt?", Zeitschrift für philosophische Literatur, 2017.

Contributors

Afe Adogame Princeton Theological Seminary, Princeton, NJ, USA

Eilidh Beaton School of Divinity, History, Philosophy, and Art History, University of Aberdeen, Aberdeen, UK

Chris Boom Equal Freedom Institute, New Orleans, LA, USA

François Boucher Institute of Philosophy, KU Leuven, Leuven, Belgium

Benjamin Boudou Department of Public Law and Political Sciences, University of Rennes, Rennes, France

Andreas Cassee Department of Philosophy, University of Mannheim, Mannheim, Germany

Max Gabriel Cherem Department of Philosophy, Kalamazoo College, Kalamazoo, MI, USA

Phillip Cole School of Social Sciences, University of the West of England, Bristol, UK

Julian Culp Department of History and Politics, The American University of Paris, Paris, France

Oliver Eberl Department of Political Science, Philipps University Marburg, Marburg, Germany

Sarah-Lea Effert Department of Philosophy, University of Duisburg-Essen, Duisburg, Germany

Mareike Gebhardt Department of Political Science, Münster University, Münster, Germany

Mollie Gerver Department of Political Economy, King's College London, London, UK

Sophie Guérard de Latour École Normale Supérieure de Lyon, Triangle (UMR5206), Lyon, France

Therese Herrmann University of Duisburg-Essen, Essen, Germany

Peter W. Higgins Department of History and Philosophy, Eastern Michigan University, Ypsilanti, MI, USA

Naoto Higuchi School of Human Sciences, Waseda University, Tokyo, Japan

Iseult Honohan School of Politics and International Relations, University College Dublin, Dublin, Ireland

Daniel Kersting Department of Philosophy, Friedrich Schiller University Jena, Jena, Germany

Silja Klepp Department of Geography, Kiel University, Kiel, Germany

Patti Tamara Lenard Graduate School of Public and International Affairs, University of Ottawa, Ottawa, Canada

Desiree Lim Department of Philosophy, Pennsylvania State University, University Park, PA, USA

Matthew Lister Faculty of Law, Bond University, Robina, QLD, Australia

Bernard Matolino Department of Philosophy, University of Pretoria, Pretoria, South Africa

Julian F. Müller Department of Philosophy, University of Graz, Graz, Austria

David Owen School of Economic, Social and Political Sciences, University of Southampton, Southampton, UK

Amy Reed-Sandoval Department of Philosophy, University of Nevada, Las Vegas, NV, USA

Moritz Riemann Department of Philosophy, Kiel University, Kiel, Germany

Johan Rochel University of Fribourg, Fribourg, Switzerland

Gilberto M. A. Rodrigues Department of International Relations, Federal University of ABC, São Bernardo do Campo, Brazil

Alex Sager Department of Philosophy, Portland State University, Portland, OR, USA

Jose Blanes Sala Department of International Relations, Federal University of ABC, São Bernardo do Campo, Brazil

Nausikaa Schirilla Catholic University of Applied Sciences Freiburg, Freiburg, Germany

Gottfried Schweiger Centre for Ethics and Poverty Research, University of Salzburg, Salzburg, Austria

James Souter School of Politics and International Studies, University of Leeds, Leeds, UK

Brigitte Suter Department of Global Political Studies, Malmö University, Malmö, Sweden

Georgiana Turculet Department of Law, University Pompeu Fabra, Barcelona, Spain

Anja Weiß Department of Sociology, University of Duisburg-Essen, Duisburg, Germany

Part I
Foundations of Migration Ethics

Normative Foundations of Migration Ethics

Phillip Cole

Abstract In this chapter, I identify and examine what I take to be the normative foundations of Migration Ethics. In Sect. 1, I argue that those foundations are derived from liberal democratic theory, with the Moral Equality Principle as key. In Sect. 2, I explore the implications of the Moral Equality Principle for Immigration Ethics. I also identify the Rationality Principle, that ethical disputes should be settled by rational argument and evidence. I argue that the force of these two principles places work in Migration Ethics in four possible boxes: (1) immigration controls cannot be morally justified—an open borders position; (2) they can be morally justified, but not by appeal to ideas of nation and nationality as these cannot be rationally grounded; (3) the ideas of nation and nationality can be rationally grounded and so can provide a moral justification for immigration controls; (4) the Rationality Principle must be set aside or compromised in some way in order to provide a moral justification for immigration controls. All work in Immigration Ethics, I suggest, fits into one of these boxes. In Sect. 3, I explore the implications of the fourth approach, examining the extent to which political theorists can endorse the suspension of the rationality requirement in putting forward ethical arguments. Finally, in Sect. 4, I examine future challenges for Immigration Ethics, identifying three in particular: the first is deeper engagement with empirical scholarship; the second is to look at issues of migration outside of the context of the Global North; and the third is to give greater attention to forcible displacement and the range of forms it can take.

The ethics of migration is a comparatively new field of study. It was only in the late 1980s that articles and collections of essays first began to appear, and it was Joseph Carens' essay "Aliens and Citizens. The Case for Open Borders", published in 1987, that marks the birth of migration being put forward as a serious ethical issue (Carens 1987). Even then, it was difficult to have the question of migration taken seriously in academic journals, and in the late 1990s a journal editor refused to send a prospective article I had written to referees, as it was so obviously, he said, a

P. Cole (✉)
School of Social Sciences, University of the West of England, Bristol, UK
e-mail: Phil.Cole@uwe.ac.uk

A. Niederberger et al. (eds.), *Handbook of Migration Ethics*,
https://doi.org/10.1007/978-3-031-89877-8_1

non-issue. As far as I am aware, my book *Philosophies of Exclusion. Liberal Political Theory and Immigration* (Cole 2000) was the first monograph of political philosophy devoted entirely to the topic. Since then, however, it has expanded into one of the most vital areas of applied ethics, with the most recent full-length treatments—Carens's *The Ethics of Immigration* (Carens 2013) and David Miller's *Strangers in Our Midst. The Political Philosophy of Immigration* (Miller 2016)—gaining a great deal of attention throughout the field of ethical and political philosophy, and a new generation of scholars taking the Immigration Ethics debate forward in new directions.

1 The Democratic Presumption

Are there specific normative foundations underlying this wide body of work? I believe that there are, and my purpose in this chapter is to identify them and explore them and their implications for the shape of the migration ethics debate. A question to consider is whether those normative foundations are adequate, and whether we need to revise them and move beyond them. The first key foundation is the context of liberal democratic political practice—the migration ethics debate is based on the assumption that the states whose immigration policies are being morally assessed, or for which political philosophers are proposing new immigration policies, are liberal democracies. This is crucially important, because this means that the normative foundations of Migration Ethics are going to be partly constituted by the normative foundations of liberal democratic theory, at least in its ideal conception. Joseph Carens makes this clear, positioning his work within a theoretical framework made up of democratic ideas and principles found in the major liberal democratic states today. What emerges is a set of democratic principles and broad moral commitments, such as that all human beings are of equal moral worth; disagreements should normally be resolved through majority rule; all citizens are equal under the law; and people should not be discriminated against on the grounds of race, religion or gender (Carens 2013: 10).

There is, therefore, in Carens's work, and I believe in all Immigration Ethics emerging from the liberal democratic tradition, a Democratic Principles Presumption as Carens terms it. This, however, raises a challenge for Immigration Ethics, a challenge which Carens for one acknowledges, as he pursues a form of political contextualism, the view that these principles emerge from a specific political tradition, and so we cannot assume that they would hold for other political contexts. This means that the discussion is more or less limited to liberal democratic states in the Global North—the United States, Canada, Australia, New Zealand, and the democracies of Western Europe. This not only excludes states which are clearly not liberal democracies, but also recognises, says Carens, that there are liberal democracies which lie outside of this "normal" range, such as Japan and India. These liberal democracies work with very different political traditions, with which Carens admits he is not familiar and is not confident to suggest policies when it comes to migration. This means, he argues, that an Immigration Ethics that engages with these other contexts would need to find other starting assumptions. Carens says:

> For such a conversation to get off the ground we would need to search for other shared assumptions, but what form those would take, how we might proceed, and where we end up is something that could be determined only in the course of an exchange. (Carens 2013: 307)

This is a serious challenge for an Immigration Ethics which positions itself as part of a debate about global justice or as a contribution to international political theory. If Carens is correct, then we may have to accept that the Immigration Ethics we are working with has a limited scope and a limited audience. However, Alex Sager challenges this liberal democratic focus by offering a "critical cosmopolitan" framework for migration ethics, which, if it going to be genuinely critical *and* cosmopolitan, cannot be restricted in its scope in the way Carens suggests (Sager 2018).

2 The Normative Foundations

2.1 The Moral Equality Principle

A central assumption of this chapter is that any recognisable liberal political theory has at its centre a commitment to the moral equality of persons. The Moral Equality Principle holds that all persons are to be treated as equals when it comes to matters of justice, unless there is some difference that is morally relevant. What makes liberal political theory appealing to many is precisely this egalitarian principle, and it is hard to see it as anything other than a definitive moral commitment without which any political ideology fails to qualify as liberal. Amy Gutmann claims that what distinguishes modern political theory from ancient or classical theories is the commitment to individualism, and that the belief in human equality has its source in this individualism. An assumption of human equality is the "unifying theme of liberalism". However, she says, "no single postulate of human equality is to be found in classical liberal theory" (Gutmann 1980: 18). Will Kymlicka takes the "duty to treat people with equal consideration" as the basic principle upon which political theories can be assessed (Kymlicka 1990: 34). He takes seriously Ronald Dworkin's view that every plausible political theory is based upon equality as the ultimate political value. "A theory is egalitarian in this sense if it accepts that the interests of each member of the community matter, and matter equally" (Kymlicka 1990: 4). He recognises that the principle of moral equality is extremely abstract and therefore cannot act as the basis of any theory of justice; but: "Each theory of justice is not *deduced from* the ideal of equality, but rather *aspires* to it, and each theory can be judged by how well it succeeds in that aspiration" (Kymlicka 1990: 44). Therefore, while in one sense it is difficult to ground the Moral Equality Principle, it seems intuitively indispensable to a recognisable liberal theory.

One thing to notice is that Kymlicka believes that the Moral Equality Principle means a theory must accept that the interests of each member of the *community* matter equally. In fact, what the principle requires is that the interests *of all human beings* matter equally, and this restriction of principles of justice to a bounded community without argument is precisely the issue Immigration Ethics has focussed upon. It could be argued that what Kymlicka means is that *every* political community must accept that the interests of its members matter equally, and therefore universal egalitarianism is preserved—but this is to overlook the reality that any political community will have under its jurisdiction both members and variety of non-members, and what the Moral Equality Principle seems to require is that the interests of both members and those non-members ought to matter equally. The problem has been that liberal political theory has the Moral Equality Principle at its centre, but has traditionally concealed an assumption of a bounded community as the subject of justice. More recently theorists have put forward explicit arguments for the need for such bounded membership to ground practices of justice, but what now arises is a tension between this perceived need for limited membership and the Moral Equality Principle. Immigration Ethics seeks to address this tension in the context of a recognisable liberal moral framework.

Michael Walzer and others who could be described as "liberal communitarians" are amongst those who have argued that the existence of liberal democratic communities depends upon exclusive membership practices: without coercive borders—practices through which the state controls membership and movement—political communities could not exist. A world without such boundaries would either be one of economic anarchism or a global state. Neither option is attractive: the first would be the absence of anything that could be considered to be a political community with collective practices of solidarity between members; the second would require a state bureaucracy on a scale that defies imagination. And so coercive borders are needed in order to sustain political units on a scale which encourages the existence of liberal communal practices without excessive bureaucracy. Walzer famously argues: "The community itself is a good—conceivably the most important good—that gets distributed" (Walzer 1983: 29); and therefore membership of the community is itself a distributive good.

This means that while the liberal democratic project is driven by the Moral Equality Principle, the success of that project depends upon organised political communities with a level of cohesiveness and commitment among their members. Coercive borders are essential for the required cohesiveness and commitment. And so the Moral Equality Principle leads us to the need for coercive membership practices—immigration controls. The question is whether this is a deeply paradoxical problem for liberal theory, such that the liberal democratic project has a fundamental contradiction at its centre. This is why Migration Ethics, far from being a subfield within liberal theory, goes to its fundamental core.

2.2 *The Rationality Principle*

A less obvious normative foundation is what I call the Rationality Principle. This holds that any ethical dispute must be settled by rational argument or evidence. The assumption is that all human beings, as moral equals, are equally capable of rational thought, and so all political disputes are in principle capable of rational solution. What this rules out is appeal to a prior set of values that have authority over argument and evidence and limit their use—for example, appeal to a divine text taken to set out absolute and unquestionable moral truths. More broadly, appeal to non-rational or arbitrary criteria for "solving" political problems is, in some sense, illiberal. Of course, the democratic element of liberal theory shows that argument and evidence seldom leads to 100% agreement on a question, hence the need for democracy as a decision-making procedure. But still there is the assumption that voters will base their choices on the arguments and evidence they encounter; or, perhaps more strongly, acceptable democratic decisions are themselves limited by the Rationality Principle. This would mean that arbitrary or contingent factors are not acceptable as part of a policy, even if it would gain a majority in a democratic vote—in a truly liberal democracy there will be side-constraints on democratic decision-making so that, for example, explicitly racist decisions are ruled out as illegitimate. While decisions in the private sphere can, to some extent, be shaped by convictions that are on the face of it unreasonable or irrational (one can associate with whom one pleases on whatever grounds), in the political sphere a higher standard applies.

The Rationality Principle enters Immigration Ethics in relation to the issue we examined in the previous section, the relationship between the Moral Equality Principle and practices of exclusive membership. The argument there was that the achievement of what the Moral Equality Principle demands depends upon political communities where people work in solidarity, but such cohesive membership can only be achieved through exclusive membership practices—hence the need for immigration controls. The problem is that those practices must be such that they embody the Moral Equality Principle, or at least do not contradict it. If we find that we cannot find a set of exclusive membership practices which themselves respect the Moral Equality Principle, then the liberal project emerges as fundamentally self-contradictory. Liberal theory recognises the moral equality of all persons as fundamental, and yet the practices of citizenship and democracy it deduces from that principle may contradict it. And so the challenge for liberal political theory—communitarian or not—is clear: it must outline principles of limited membership that do not contradict the Moral Equality Principle. The attempt to arrive at such principles lies at the heart of Immigration Ethics.

This is especially challenging for those who argue that ideas such as "nationality" or "national identity" are essential for the cohesiveness of political communities, and so need to play a role in immigration controls. Walzer, for example, appeals to the notion of shared identity if political communities are going to achieve their purposes (see also Miller 1995, who appeals explicitly to the idea of "nationality").

That identity, for Walzer, is based on a shared understanding of who "we" are, based upon common meanings, language, history and culture. For Walzer, political communities depend upon a "sense of relatedness and mutuality" (Walzer 1983: 50). They are "communities of character", consisting of people "with some special commitment to one another and some special sense of their common life" (Walzer 1983: 62). If we take the Rationality Principle seriously, appeals to non-rational or arbitrary criteria are ruled out as illiberal, and the suspicion is that appeals to shared "character" or some kind of national identity are precisely appeals to the arbitrary and non-rational.

John Rawls is concerned to eliminate moral arbitrariness from liberal social and political practices. He condemns as unfair any distribution of resources which rests upon what he sees as the arbitrary properties of "social starting positions, natural advantages and historical contingencies" (Rawls 1978: 56). These, for Rawls, are the "fundamental" inequalities that a liberal theory of justice seeks to overcome. The arbitrary nature of resource distribution under capitalism therefore becomes the central problem for liberal theories of justice (Kymlicka 1990: 85–86). If, as Walzer has claimed, we see membership of the political community as a resource that needs to be distributed, then the same concerns about moral arbitrariness arise—who one's parents are and which side of a border one is born on are natural and historical processes which lie at the heart of liberal democratic membership practices, and yet both surely qualify as primary examples of what Rawls is talking about when it comes to morally arbitrary factors giving rise to fundamental inequalities. The only way to avoid this dilemma is to adopt a radically communitarian position, such that issues of distribution are only relevant within a given membership—no questions arise concerning the justice of distributions across the membership boundary. However, I am not aware of any liberal communitarian commentator on Migration Ethics or global justice who would take such a radically closed position—in effect, it rules out not only Migration Ethics, but any form of global or international ethics.

Migration Ethics therefore focusses on the issue of whether the membership question (whether states have the moral right to control membership) can be answered positively in a way that meets the requirements of the Moral Equality Principle, but also, crucially, the Rationality Principle—that is, without appealing to criteria that are arbitrary from a liberal moral point of view. There are four possible positions the liberal theorist could adopt here:

1. There are no non-arbitrary criteria for settling the membership question, and there is therefore no rational solution to it. The only liberal position is therefore to have no membership restrictions—open borders and open membership. The challenge for theorists taking this position is to explain how political communities can be sustained without membership restrictions: in other words, how the liberal project is possible in a world without immigration controls.
2. There are non-arbitrary criteria for settling the membership question, and therefore there is a rational solution to it: coercive membership can be morally justified. However, the notion of "national identity" is arbitrary, and so can have no role in this solution. The challenge for theorists taking this position is to identify

the non-arbitrary criteria that can act as the basis for membership without appealing to any notion of "national identity".

3. The membership question cannot be settled without some appeal to a notion of "national identity". However, a conception of national identity can be arrived at which is non-arbitrary and so meets the requirements of the Rationality Principle, and without violating the Moral Equality Principle. The challenge for political theorists taking this position is to explain how a conception of national identity can be rationally grounded.
4. There are no non-arbitrary criteria for settling the membership question, but it must, nevertheless, be settled—the open borders position is not acceptable. Also, it can only be settled by appeal to a notion of national identity. The need for political community in order to achieve the moral ideals of liberal egalitarianism overrides the Rationality Principle, and allows us to settle the membership question by appeal to the non-rational. The challenge for political theorists taking this position is to show how they can appeal to such non-rational notions and at the same time keep a moral and political distance from ideologies of membership they would themselves consider to be abhorrent—e.g. racist ideologies.

It is my contention that any liberal political theorist working on Migration Ethics is working within one of these boxes and so faces the specific challenges I have identified in order to find their way out.

3 Beyond Reason

Although theorists in boxes 2 and 3 face specific challenges, they share the same problem of resolving the tension between immigration controls and the Moral Equality Principle in a way that is consistent with the Rationality Principle. Elsewhere, I have argued that this cannot be done, and so we end up in box 1, that the only rationally coherent liberal position on the membership question is complete freedom of international movement (see Cole 2000). I will not go over those arguments here, but will draw attention to another strategy some theorists have employed, which is to set aside, to some degree, the Rationality Principle, and move into box 4. In order to remain within the scope of liberal theory, these theorists need to argue that the Rationality Principle has limited scope, such that appeal to arbitrary factors is morally permissible in Migration Ethics. This takes the form of arguing that some level of contingency is inevitable in political theory. This contingency enters at different levels and different places in different approaches, but the point being made is that there are some dimensions of political life that cannot be rationally justified. There are aspects of human existence that lie beyond the scope of reason—certain of our core beliefs and practices are not irrational, but are rather non-rational: we can give no reason for them but they are indispensable aspects of our identity and existence. The question is whether any such factors have a legitimate role to play in political theory when it comes to the moral justification of immigration controls, and also, as I pointed out earlier, whether we can appeal to such factors and at the same time rule out appeal to factors that would be completely unacceptable from a liberal point of view, e.g. those that are racist.

One starting place might be the national border: we have to take national borders as given in order to do political theory at all. To imagine otherwise is to move to a radically different political world where there are no nation-states, and would need a completely new way of doing political theory which would have no practical grasp on the critique of political arrangements as we find them, and so would make political theory completely irrelevant. And so the moral arbitrariness of where one finds oneself when it comes to a national border, inside or outside, has to be taken as given. However, as a matter of historical fact where national borders lie is *deeply* contingent—they have not been fixed by any practice that could be described as ethical or rational from a liberal point of view. Therefore we are not discussing an idealised, abstract liberal nation-state that fixed its territorial boundaries through a process that meets standards of fairness and justice; we are talking about national borders fixed by nature, war, political intrigue and other forms of violence and coercion. This is relevant for Migration Ethics because it is not merely on which side of the border one finds oneself that is arbitrary from a moral point of view, but also where that border falls and how it got there. Political theorists are, of course, interested in boundaries of membership as well as territory, but these are connected, as it is always one's relationship with a territorial space, and so with a set of borders, that plays some role in determining membership. This raises a worrying concern because now the ethical and rational arbitrariness of territorial boundaries is connected with national membership. This challenge can get very radical because the same concern holds for other questions of justice—why should the key obligations of justice stop at the national border? The answer has to be because we only owe these obligations to co-nationals, but if we cannot settle on a rational and ethical way to distinguish between co-nationals and non-nationals, then the whole project of liberal justice rests on a fundamental incoherence (a growing number of theorists are addressing the role of territory in issues of global justice—see Moore 2015 and Song 2018b).

This is perhaps why John Rawls in *The Law of Peoples* (Rawls 2001) is so adamant about the irrelevance of the question. There he says that national boundaries are "historically arbitrary" but that their role can be justified—"to fix on their arbitrariness is to fix on the wrong thing." He concludes: "There *must* be boundaries of some kind, which when viewed in isolation will seem arbitrary, and depend to some degree on historical circumstances" (Rawls 2001: 39). We have to start somewhere, and that somewhere has to be national borders or boundaries of national membership. The problem with this answer is that if we start here than we start with national boundaries that have been fixed not merely in arbitrary ways, but in deeply immoral ones, so how can a theory of justice rest on *these* boundaries. Also, Rawls says there must be boundaries of *some* kind, but why of *this* kind?

Michael Blake defends Rawls's position here, arguing that for Rawls, "membership is not itself a good whose distribution is to be evaluated by distributive justice; membership in a territorial state is required before his theory of distributive justice has applicability" (Blake 2012). And *The Law of Peoples* "allows for inequalities between states, even if the borders between them are arbitrary and the distribution of people between them is morally arbitrary; this reflects not a refusal to apply liberalism to international justice, but a particular vision of how the notion of justice ought to be understood" (Blake 2012). Certainly, Rawls must remove the question of membership from the question of distributive justice in order for his theory to get started, but some might think this means that it starts in the wrong place. It does not so much constitute a refusal to apply liberal theory to questions of international justice, but a fundamental failure to grasp the very question of international justice. Again, Immigration Ethics is asking core questions of the liberal project.

Blake himself makes an interesting contribution that brings arbitrariness into the centre of the debate. He argues that arbitrary factors can give rise to differences that are morally relevant (Blake 2005: 227). He concedes that my membership of a nation is arbitrary from the moral point of view, in that it is produced by factors beyond my control. However, "from the fact that the circumstances giving rise to a social or political difference are arbitrary, we cannot conclude that that difference is morally irrelevant" (Blake 2005: 227). The fact is that no matter how arbitrarily constructed, the national border marks out something of moral significance. "It marks out, among other things, the boundaries of shared liability to a political state" (Blake 2005: 227). I not only have rights and entitlements against the state, but it also has rights and entitlements against me. "If I am subject to the coercive political actions of a state, and you merely seek to become subject, it does not seem that you and I will necessarily be entitled to the same sort of justification from the state in question. The contingency in the world creates a difference with some moral relevance" (Blake 2005: 228). What follows from this is that: "While one's status as a citizen might ultimately be causally determined by arbitrary facts in the world, the status of citizen does create some distinct demands for justification" (Blake 2005: 228). A political system, in order to be legitimate, owes certain guarantees to those who fall under its authority. Those guarantees do not apply to those who are not subject to that authority. And so arbitrary differences in circumstances create a difference in political rights. "This difference, moreover, does not offend moral equality in the slightest" (Blake 2005: 229). Differences in the right to mobility also, therefore, do not offend moral equality, if they rest on this difference of shared liability to the state. To deny the right of a citizen of the USA to move from Buffalo to Boston, for example, would be fundamentally illiberal, given the political relationship in question. But it is not fundamentally illiberal to prevent a Canadian citizen from emigrating from Toronto to live in Boston. This will have negative consequences for the would-be migrant, but is not an affront to his moral equality. "Given that he does not live under the authority of the United States, but asks admittance to that status, it seems a mistake to consider him as equivalent to those who already stand in such a relationship of political authority" (Blake 2005: 229).

However, Blake's argument here only works if he makes the same move he attributes to Rawls, which is to remove the question of membership from the issue of distributive justice. I have argued in this chapter that the ethical challenge to immigration controls not only is based on concerns of distributive justice across borders but also goes much deeper. The question of distribution is not primarily to do with welfare, property or other resources, but with the membership that carries rights to those resources. It is the ethical coherence of the distinction between citizens and outsiders which is at stake, not merely that citizens of a particular state may enjoy better life prospects than an outsider. In Blake's approach, membership of a state is the theoretical starting point, but this is a particular kind of membership and of a particular kind of state—citizenship of a liberal democratic nation-state. What we can see, then, is that the starting point of the argument, the "given", is not made up of some core, fundamental basic concepts or ideas, but is packed with complex assumptions about the nature of the political/legal order of things, some of which seem to be highly localised to particular places and particular times. We return here to the challenge of contextualism we encountered in Sect. 2—that our version of Migration Ethics is only relevant to a limited range of liberal democratic states, and even then is too abstract to have any practical application to the circumstances in which they find themselves. We also return to the problem of territory, because what is assumed is legitimate occupation of territory such that there is legitimate authority over it. The question of how this can be reconciled with the historical reality of conquest and colonialism is not one that has a straightforward answer, but some kind of answer is required (See Song 2018a, b). Too often Migration Ethics seems to be about a world in which colonialism and enslavement did not take place, but they did, and that makes a difference in a world still fundamentally shaped by those practices.

4 Future Questions

In her paper "Political Theories of Migration", Sarah Song argues that there are three directions for future research for political theorists working on migration (Song 2018a). The first is deeper engagement with empirical scholarship; the second is to look at issues of migration outside of the context of the Global North; and the third is to give greater attention to forcible displacement and the range of forms it can take. I am in complete agreement that these are the issues Migration Ethics needs to attend to, and so in conclusion I will expand on Song's reasons for these priorities and add some of my own. First, on the need to engage more deeply with empirical scholarship, Song observes that: "Normative debates about migration rest in part on empirical claims, and if political theorists want their work to offer practical guidance in real-world debates, they must engage with the best empirical scholarship on the subject" (Song 2018a: 399). However, for me this engagement goes beyond the need to offer practical guidance to policy-makers. Even at its most "ideal", Migration Ethics must be based on a firm understanding of what migration is: what forms it takes, what drives it, what impact it has, its historical roots and connections with colonialism and enslavement, and the nature of the global political and economic system within which it now takes place. Even some of the most "Realist" approaches in Migration Ethics have sometimes been based upon extremely abstract understandings of migration, attributing motivations to migrants and suggesting impacts on sending and receiving nations with very little engagement with empirical evidence, and setting aside the history which still shapes global movements today. Migration Ethics must see itself as a part of a broader field of Migration Studies, working alongside and with researchers from other disciplines, rather than as a standalone field capable of reaching its own independent conclusions on these questions.

This relates to the second issue, of overcoming the contextualism which some theorists are working within and expanding our focus beyond the Global North. Song points out that migration is a global phenomenon, with large migration flows taking place in the Global South. Too often Migration Ethics has as its focus an idealised liberal democratic state towards which migration is flowing, rather than seeing migration in its broader global context. It is in danger of being what I describe as "insider theory"—theorising from inside a liberal democratic state in the Global North, and so shaping the migration question around the concerns and priorities of citizens of such states. It is the migrant, especially the migrant from the Global South, who is constructed as the problem to be solved in the interests of the citizen of the Global North liberal democracy. This can only give us a distorted and misleading version of Migration Ethics, irredeemably shaped by the historic power relations of a narrow set of nations.

The third issue Song draws attention to is that of forced displacement. The key distinction that needs to be overcome is that between economic migrants and refugees. According to this distinction, people who cross borders are either economic migrants, in which case they move voluntarily, have original membership of another state, and can return to that state if they are refused entry; or they are refugees who move involuntarily, but who still have original membership of another state, and could return there if the issues that forced them to leave are resolved. This distinction cannot withstand any level of critical scrutiny. In the first place, we have to recognise that those who migrate for economic reasons may not do so on any basis we would recognise as voluntarily. Song refers to "necessitous migrants", and Alexander Betts has referred to "survival migrants" (Betts 2013). People can be caught up in migration flows where they have little control over where they migrate to, or how, or why. In the second place, when it comes to refugees the expectation of return is stretched beyond credibility, once we realise that many are caught up in "protracted situations" which no end in sight, most of those confined to refugee camps (see Parekh 2017). According to the UNHCR, one third of the global refugee population were in protracted refugee situations by the end of 2017, and the average length of a PRS is now 26 years (https://www.unhcr.org/5d08d7ee7.pdf). In the third place, we must also question the assumption that all migrants have an original membership. This is to overlook the 10–15 million stateless persons in the world—no state recognises them as their responsibility (see Bloom et al. 2017). Song draws attention to the internally displaced, who make up the vast majority of the UNHCR's estimate of globally displaced persons but who, because they have not crossed a national border, cannot be regarded as refugees (of the 70.8 million people the UNHCR estimates were forcibly displaced in 2018, 41.3 million were internally displaced—https://www.unhcr.org/uk/figures-at-a-glance.html). And finally, the issue of environmental disaster and climate change threatens to cause displacement on a scale that overshadows the "traditional" conception of who counts as forcibly displaced. Migration Ethics, then, needs to expand its scope to take in issues concerning not only voluntarily migration and the situation of refugees, but also the internally displaced, the stateless, those displaced by environmental disaster and climate change, and the economically displaced, taking in issues of human trafficking and smuggling.

The fundamental challenge that emerges is that this body of work has, up to now, been based upon a narrow and overly abstract understanding of what migration is, and so has arrived at a narrow and overly abstract set of arguments and moral principles. This includes those thinkers who would describe themselves as "realists" rather than "idealists". There is a fundamental failure to understand the processes of migration, its causes, its impacts, the motivations behind it, and how it is experienced. Migration Ethics must arrive at an understanding of migration that captures all of these dimensions, and so provides a firmer basis upon which to ground theoretical and ethical work. This may lead to dramatic and radical transformations in Migration Ethics itself, a step-change in this body of work, such that everything we have done up to this point is recognised as merely preliminary and sketchy groundwork. The real work is yet to come.

Literature

Betts, A. 2013. State Fragility, Refugee Status and 'Survival Migration'. *Forced Migration Review* 43:4–6.

Blake, M. 2005. Immigration. In *A Companion to Applied Ethics*, ed. R. G. Frey and C. H. Wellman. Oxford: Blackwell.

Blake, M. 2012. Review of C. H. Wellman and P. Cole. Debating the Ethics of Immigration. Is There a Right to Exclude. *Notre Dame Philosophical Reviews*. http://ndpr.nd.edu/news/32280-debating-the-ethics-of-immigration-is-there-a-right-to-exclude. Last checked 27 June 2013.

Bloom, T., P. Cole, and K. Tonkiss, eds. 2017. *Understanding Statelessness*. London: Taylor & Francis.

Carens, J. 1987. Aliens and Citizens. The Case for Open Borders. In *The Rights of Minority Cultures*, ed. W. Kymlicka. Oxford: Oxford University Press.

Carens, J. 2013. *The Ethics of Immigration*. Oxford: Oxford University Press.

Cole, P. 2000. *Philosophies of Exclusion. The Ethics of Immigration Controls*. Edinburgh: Edinburgh University Press.

Gutmann, A. 1980. *Liberal Equality*. Cambridge: Cambridge University Press.

Kymlicka, W. 1990. *Contemporary Political Philosophy. An Introduction*. Oxford: Oxford University Press.

Miller, D. 1995. *On Nationality*. Oxford: Oxford University Press.

Miller, D. 2016. *Strangers in Our Midst. The Political Philosophy of Immigration*. Harvard: Harvard University Press.

Moore, M. 2015. *A Political Theory of Territory*. Oxford: Oxford University Press.

Parekh, S. 2017. *Refugees and the Ethics of Forced Displacement*. New York: Routledge.

Rawls, J. 1978. The Basic Structure as Subject. In *Values and Morals*, ed. A. Goldman and J. Kim. Boston, MA: D. Reidel.

Rawls, J. 2001. *The Law of Peoples*. Cambridge, MA: Harvard University Press.
Sager, A. 2018. *Towards a Cosmopolitan Ethics of Mobility. The Migrant's-Eye View of the World*. Cham: Palgrave Macmillan.
Song, S. 2018a. Political Theories of Migration. *Annual Review of Political Science* 21:385–402.
Song, S. 2018b. *Immigration and Democracy*. Oxford: Oxford University Press.
Walzer, M. 1983. *Spheres of Justice. A Defence of Pluralism and Equality*. New York: Basic.

Further Reading

Barry, B., and R. E. Goodin, eds. 1992. *Free Movement. Ethical Issues in the Transnational Migration of People and of Money*. University Park, PA: Pennsylvania State University Press.
Cole, P. 2011. Open Borders. An Ethical Defense. In *Debating the Ethics of Immigration. Is There a Right to Exclude?* ed. C. W. Wellman and P. Cole. Oxford: Oxford University Press.
Cole, P. 2014. Beyond Reason. The Philosophy and Politics of Immigration. In *New Challenges in Immigration Theory*, ed. C. E. G. Akakpo and P. T. Lenard. Abingdon: Routledge.
Fine, S. 2010. Freedom of Association Is Not the Answer. *Ethics* 120:338–356.
Fine, S., and L. Ypi, eds. 2016. *Migration in Political Theory. The Ethics of Movement and Membership*. Oxford: Oxford University Press.
Goodin, R. E. 1998. What Is So Special About Our Fellow Countrymen? *Ethics* 98:663–686.
Oberman, K. 2016. Immigration as a Human Right. In *Migration in Political Theory. The Ethics of Movement and Membership*, ed. S. Fine and L. Ypi. Oxford: Oxford University Press.
Pevnick, R. 2011. *Immigration and the Constraints of Justice. Between Open Borders and Absolute Sovereignty*. Cambridge: Cambridge University Press.
Wellman, C. H. 2011. Freedom of Association and the Right to Exclude. In *Debating the Ethics of Immigration. Is There a Right to Exclude?* ed. C. W. Wellman and P. Cole. Oxford: Oxford University Press.

Rights of Strangers in Europe Until 1800

Oliver Eberl

Abstract The concept of the stranger is central to the question of migration in European social and political history. In antiquity, state-building societies had strong social boundaries, and cities were very careful about who was allowed to cross the borderlines of their territory. In early modernity, after the end of the Middle Ages, the territorial state and the idea of territorial sovereignty became key factors in the management of migration of individuals—both citizens and foreigners—across borders. This paper will elucidate the concept of migration and explain its evolution in European social and political history from antiquity through the Enlightenment. The cosmopolitan concept of world citizenship is introduced as an attempt to overcome the dichotomic distinction between citizens and strangers which emerges with the territorial order of states.

The sociologist Georg Simmel (1950) distinguishes between three types of strangers in this time period: the wanderer, the trader and the enemy. Unlike the wanderer, who is here today and gone tomorrow, the trader is someone "who comes today and stays tomorrow". The trader is not an "owner of soil". Traders have a fixed function in the community, and as recurring acquaintances, they are eventually recognised as "equals". This kind of stranger is close to us and far from us at the same time. Simmel distinguishes a third form of strangeness where no common ground with the stranger is recognised: "[T]here is a kind of 'strangeness' that rejects the very commonness based on something more general which embraces the parties" (Simmel 1950: 407).

Simmel thus ruptures the range in which the stranger has been conceptionally captured in European social and political thought up to modern times: either as a temporary traveller, as a guest with special status, and finally as the Other or an enemy. To illustrate this latter category, Simmel cites the relationship of the Greeks to the "barbarians" (Simmel 1950: 407). As an unequal, the "barbarian" becomes a stranger,

O. Eberl (✉)
Department of Political Science, Philipps University Marburg,
Marburg, Germany
e-mail: oliver.eberl@uni-marburg.de

A. Niederberger et al. (eds.), *Handbook of Migration Ethics*,
https://doi.org/10.1007/978-3-031-89877-8_2

who is regarded as enemy and slave. In European thought up to the Enlightenment, the question of how strangers ought to be treated and what rights they are given by the local community has oscillated between the notion of guest rights and war against "barbarians". Strangers were seen as wanderers, guests (traders) or enemies and slaves ("barbarians"). This distinction was embraced in Europe's philosophical and legal doctrines during the epoch of European colonialism in early modernity and up until 1800 when some began to question its validity. The present article will show how during this period the concept of "guest" (a foreign person with a special status within the local state) and "barbarian" (as a concept used in a state's foreign relations) emerged. A particular emphasis will be placed on the ideas of Immanuel Kant, an Enlightenment philosopher who broke away from this pattern of conceiving of strangers and developed a novel conception of cosmopolitan law for world citizens, which was intended to improve the status of strangers and ultimately to end colonialism.

The dual relationship to the notion of a stranger, as an unknown wanderer and as an invited guest, is expressed in the Greek word *xenos*, which means both "stranger" and "guest". Hospitality, *xenía*, is a personal relationship based on reciprocity. It is connected with an obligation to care for and accommodate the guest, to protect and to assist them (Hiltbrunner 2005: 157). Analogously, the ambiguous relationship between stranger and enemy is expressed in the Greek word *xenos*. In etymological terms, *xenos* is connected with the Latin *hostis* and the Germanic *gast*. While the Latin *hostis* has a negative meaning, indicating hostility towards the enemy, the Germanic *gast* has a positive or neutral connotation (Spahn 1995: 40). Therefore, the guest is a friendly stranger, and the enemy a hostile stranger. The term "barbarian" clearly has a negative connotation. Initially used by Herodotus with some ethnographic neutrality, this term subsequently acquired an increasingly pejorative meaning. This culminates in Aristotle's description of the "barbarian" as a "natural slave". In Roman law, the term's negative connotation can be neutralised if the stranger is no longer construed in substantial terms as an Other, but rather as a Roman vassal or even Roman citizen. Finally, in the neo-Aristotelian discourses of early modernity, the term "barbarian" acquired a new meaning, especially in the context of European colonialism when it became the guiding concept for the subjugation of the indigenous peoples of the Americas (Eberl 2021).

Finally, during the Enlightenment, attempts were made at reformulating and positively reconceptualising the connection between colonialism and the concepts of "stranger" and "enemy" in cosmopolitan law. As Kant argues in *Perpetual Peace*, a cosmopolitan right of visiting a foreign land should be granted to all as equal world citizens. In this way, Kant attempts to overcome the elitist narrow-mindedness hitherto attached to the right to visit a foreign country as a guest but also to offset the colonialist consequences stemming from the concept of a stranger. Kant introduced a major innovation in Enlightenment discourse of world citizenship and cosmopolitan law which overcame the dichotomy between good strangers and bad strangers ("barbarians"). This created a conceptual space for a third category of persons—wanderers and traders—who enjoy a limited right to visit foreign countries. As a result, the Enlightenment put an end to a 2000-year-old dualism between hospitality and war against "barbarians" and made it accountable to the normative standard of cosmopolitan law and democracy. In Kant's cosmopolitan conception, world citizens have the same right of visiting a country other than their own, but no right of permanent residence there. This limitation of cosmopolitan law is connected to the local citizens' right of democratic self-government. This right of self-government generates tensions between citizens and strangers that are still palpable in debates over the status of the migrants in many contemporary democratic societies.

1 Strangers in the Old Testament: Alien Law and Love of Strangers

The rules of the first five books of the Old Testament, the Pentateuch, relate closely to the question of how to deal with strangers who reside in one's community. The third book of Moses prohibits discrimination against strangers: "And if a man from another country is living in your land with you, do not make life hard for him; Let him be to you as one of your countrymen and have love for him as for yourself; for you were living in a strange land, in the land of Egypt: I am the Lord your God" (Leviticus 19:33, 34). In Egypt, the Jews were of a different faith and thus technically foreigners. Leviticus curiously concludes that strangers of other faiths should be treated as natives. The Israelite free cities of antiquity included the stranger in their legal system: "You are to have the same law for a man of another nation living among you as for an Israelite; for I am the Lord your God" (Leviticus 24:22).

Hermann Cohen concludes that the principle of legal equality between strangers and Jews testifies to the separation of faith and "civic community" (Cohen 1992: 96–97). Above all, however, he appeals to the provision of the fifth book of Moses: "So be kind to the man from a strange country who is living among you, for you yourselves were living in a strange country in the land of Egypt" (Deuteronomy 10:18,19). This is taken to reveal a real "love of strangers" (Cohen 1992, 86). Some interpreters think that Cohen has shown "how the religion of ancient Judaism makes the most sophisticated ethical demand that any religion could ever develop: to consider and love the stranger as one's neighbour" (Loycke 1992: 120, my translation).

Other interpreters perceive only the Jews' fellow believers as their neighbours. In Leviticus 19:17,18, the alternate word for *rea*, neighbour, is *amit*, tribal member. God's love is to be shown to the tribal member. The stranger to whom the commandment of loving one's neighbour applies is thus more likely to be a member of the same co-religion at the time of the covenant legislation (Schreiner and Kampling 2000: 26). After the fall of the Northern Kingdom of Israel against the Assyrians in 722 BC, many Israelites fled to Jerusalem and the surrounding area and henceforth lived as strangers among the Israelites. They were called by the name *ger* derived from *gur*. The fifth book of Moses sets itself the task of integrating these Israelites into the local polity. In it, the word *rea* means "neighbour" or "fellow citizen", thus referring to someone who is "near". In contrast, a stranger from a foreign people is called *nokri* and is separated from foreign but integrated Israelites. Nonetheless, the rules of hospitality also apply to *nokri*. Travellers, merchants and strangers are granted temporary admission to the local community, but they remain strangers. *Nokri* is therefore a stranger who does not belong to Israel, in contrast to the Israelite who belongs. The stranger is not considered an enemy because enemies are only aggressors coming from outside and who have the intention to enslave the people of Israel, to drag them away, to seize their land or their crops, or who disrespects the God of Israel (Schreiner and Kampling 2000: 41). Overall, the Old and the New Testament depict the experience of being a stranger and therefore for the most part respect and protect the status of strangers and migrants (Picker 2020: 24). Moreover, in the Old Testament the question of the treatment of the stranger oscillates between the so-called near stranger, who is considered equal to the locals, and the stranger who is their unequal. This tension between two categories of strangers will intensify with the advent of the ancient Greek city-states.

2 Strangers in Greek Antiquity: "Barbarians" and Metics

The different social and political context in which ancient Greek thought developed, when compared to its biblical predecessors, accounts for the difference in treating strangers and enemies inside the Greek city (*polis*). The extensive network of Greek cities after the great colonisation (eighth to sixth centuries BC) created a certain understanding of common ground among the Greeks. Herodotus (1920) studied foreign peoples with interest, but on the whole they were met with rejection as unequals and were seen as the Other (Nippel 2002). Strangers who remained inside Greek cities were given the status of *metics*. They were personally free, but without political rights. Craftsmen and merchants formed the largest group among the metics. Metics had to pay an annual tax, and if this tax remained unpaid, they were sold into slavery. They were not allowed to buy land. Their legal status is thus a clear example of what we today consider discrimination: In Greek law, the killing of a slave, a passing stranger or a metic is always treated as unintentional, even if it was evidently intentional. This discrimination expresses the will to keep the resident strangers outside the citizens' community and not to allow any upward transformation in status (Bleicken 1995: 102–103). The metic always remains a stranger within the Greek *polis*, they cannot become a citizen. The status of the metics can be understood from a broader perspective, as a characteristic of how ancient Greek democracy dealt with strangers. The guest status was typical of aristocratic Greece, which required a personal, individual relationship with rich local families who were temporarily visited (Spahn 1995: 46). The status of the metic emerged in the tense phase of a process of internal democratisation accompanied by a stronger demarcation of the community of citizens against its "outside", comprising local tradesmen and craftsmen as well as political refugees. The Pericles' Citizenship Law of 451–50 BC, which required both parents of a child to be Athenians, represents a tightening of the rules. This narrowly defined rule for acquiring citizenship permanently prevented the naturalisation of the metics.

In terms of status, the metics stood above the stranger who was only a temporary visitor in Athens. A general legal protection applied to them. If a city had a treaty with Athens, this protection was even guaranteed under external agreements. An even better status protection was afforded to state guests who were under the institution of *proxeny*, which had developed from private hospitality (Spahn 1995: 105). If strangers wanted to reside longer in the host city, they had to apply for metic status. For this they needed the support of a citizen. The slaves had the lowest social status. A slave is the property of their master and enjoys no "independent legal capacities, marriage rights or freedom of movement" (Bleicken 1995: 108, my translation).

In the fourth century BC, Aristotle describes the slaves as "barbarians" who represent "natural slaves". Aristotle characterises these barbarians by an insufficient sense of freedom, a lack of reasoning ability and undisciplined emotions. Peoples who fall in this category cannot form a political order and are therefore better ruled despotically (as slaves) by the Greeks. In contrast, the Greeks command a sense of freedom, discipline, intelligence, political order, capacity for reason, philosophical and military skills. Aristotle associated the capacity for reason with knowledge of the Greek language and deduced from this that "barbarians" "have no perception of reason, but are entirely guided by appetite" (Aristotle, Politics, Book 1, 1254, b 23). Without any practical common sense and thus without an ability to "plan ahead", they were unfit for political existence. Hence, they had to be subjected to rule by the Greeks for their own good (Aristotle, Politics, Book 1, 1260a). In other words, the "barbarians" were seen as a class of persons governed by unruly emotions instead of practical sense (*phronesis*), which made it impossible for them to rule themselves politically. As they lacked rational judgement, it would be to their own advantage to be kept as slaves and to be ruled by a more advanced people, such as the Greeks were.

> Those men therefore who are as much inferior to others as the body is to the soul, are to be thus disposed of, as the proper use of them is their bodies, in which their excellence consists; and if what I have said be true, they are slaves by nature, and it is advantageous to them to be always under government. He then is by nature formed a slave who is qualified to become the chattel of another person, and on that account is so, and who has just reason enough to know that there is such a faculty, without being indued with the use of it. (Aristotle, Politics, Book I, Ch. 5, 1254 b)

Aristotle's idea that those led by reason (Greeks) ought to dominate those led by emotion ("barbarians") arises from the distinction between reason and affect as distinct parts of the soul. The reasonable person compensates for the missing or underdeveloped part of the soul of the person ruled by emotion; such rule is therefore despotic. From this it follows "that a herile and a political government are not the same, or that all governments are alike to each other, as some affirm; for one is adapted to the nature of freemen, the other to that of slaves" (Aristotle, Politics, Book I, Ch. 7, 1255 b).

This definition amounts to a degradation of the status of strangers outside Greece. Since all people outside Greece are "barbarians", strangers become enemies. This can be seen in the description of the war against "barbarians" in ancient Greek political thought. Plato distinguishes a war of the Greeks with strangers from a war between the Greeks.

> And therefore when Hellenes fight with barbarians and barbarians with Hellenes, they will be described by us as being at war when they fight, and by nature enemies, and this kind of antagonism should be called war; Hellenic warfare is only a kind of discord not intended to be lasting; but when Hellenes fight with one another we shall say that Hellas is then in a state of disorder and discord, they being by nature friends; and such enmity is to be called discord. (Plato, Republic, 5th Book, 470D)

The war against "barbarians" is waged in such a way that it ends in annihilation and enslavement; Greeks must not be enslaved by Greeks. Plato and Aristotle thus declare the inequality of "barbarians" and Greeks to be the basis of their subjugation. "Barbarians" have no rights as strangers, they are enemies. The guiding distinction for Greek political thought is therefore the one between local community and the outside world, not between citizens and strangers. "The distinction between Greek and barbarian overshadowed that between citizen and xenos" (Browning 2002: 261).

Plato and Aristotle compared strangers to enemies, to whom no rights are granted. Yet critical voices gradually arose among the Sophists and Stoics, who developed a tradition favourable to strangers: cosmopolitanism. Diogenes of Sinope coined the term "cosmopolites" (citizen of the cosmos), and subsequently Marcus Aurelius regarded individuals around the world as fellow citizens and spoke of universal citizenship (Cavallar 2002: 61).

3 Strangers in Roman Law: Integration and Exclusion

Roman law treats as basic not the demarcation between ourselves and our enemies, but the distinction between strangers and citizens. While Roman civil law (*ius civile*) formulates the ancient customary laws of the Romans and thus designates the privileges of Roman citizens, the law of nations (*ius gentium*) regulates the social interaction of strangers in the Roman Empire and that of Romans with strangers outside the Empire. Originally, the *ius gentium* included elements from other countries such as Babylon, Phoenicia and Greece, but over time it developed into a uniform legal system from which the later law of nations emerged.

During the Roman Republic (509–27 BC) citizenship was an elitist status for Romans, whereas all strangers were treated according to the law of their respective provinces. This meant that the privileges of Roman citizenship (acquiring and bequeathing property, a right to marry) remained closed to the latter group. During the reign of Augustus (27 BC–AD 14), the opportunities for non-citizens living in the provinces were improved. The climax of this development was the introduction of universal Roman citizenship in AD 212 with the Antonine Constitution. All strangers (*peregrini*) living in Rome now became Roman citizens, to whom the *ius civile* applied. Excluded from citizenship were only the *peregrini dedicitii*, stigmatised freedmen, "who had been branded or tortured or had fought in the arena" (Mathisen 2006: 1020). After this act of political and legal inclusion, non-stigmatised strangers had no more prospects of obtaining civic status, but only gradually possibilities of integration (such as marriage or land ownership) opened up for them. Rome was in principle pro-integration and allowed especially former soldiers to settle in Roman territory. After AD 247, the office of *praetor peregrinus* was established, who was responsible for disputes between strangers and Roman citizens and issued politically influential edicts.

Besides the inclusion of strangers into the *ius civile*, the status of a guest developed further. Hospitality was closely linked to the figure of the stranger who entered Rome as a temporary traveller: as a merchant, as a participant in holy festivals, as a state ambassador and finally as a scholar, who was always received by certain hosts. From this developed the relationship between guest and host, whom one always visited first and whom one could not simply neglect if one did not want to cause offence. The host had to protect the guest's life and property. Betraying or killing the guest was considered a violation of a most sacred duty.

Outwardly, the demarcation line between insiders (Romans) and outsiders (non-Romans) became more important again. In the late fifth century AD, the largely equal status of all Roman permanent residents within Rome gave rise to the equation of strangers and "barbarians" that was already familiar from Greek thought. Sidonius Apollinaris said about Rome: "In this state (*civitas*) alone of the entire world only barbarians and slaves are *peregrini*" (Epistulae, 1.6.3, quoted after Mathisen 2006: 1021). This means, that "after 212, those who were *peregrini* with respect to Roman citizenship were strangers from beyond the frontiers, the peoples known collectively as 'barbarians'" (Mathisen 2006: 1021). The stranger was thus almost consistently referred to as *peregrinus* and finally equated to "barbarian". Of all names for the stranger—*alienigenus* ("born elsewhere"), *externus* ("foreign" or "external"), *peregrinus* (stranger, foreigner), *provincialis* (inhabitant of a province) or *transmarinus* (overseas)—"only *barbarus* always had negative connotations" (Noy 2002: 2).

Another development was the importance of *hostis*. The term initially referred to non-Romans who lived peacefully in Rome, but the meaning changed later to designate strangers who threatened Rome from the outside. Cicero tried to explain this change in *De officiis* (Cicero 1913: 1:37). Since in his view the enemy is equal to the citizen, the former notion has both a positive and a negative meaning: This explains the second development towards *gast*, from gothic *ghostis* (stranger, guest). Thus, there is the friendly stranger (= guest) and the hostile stranger (= enemy), whereby in both cases no asymmetrical relationship arises between "stranger" and "Roman". On the contrary, it was precisely this relationship of equality that played the decisive role in Cicero's teaching of just war. The *iustus hostis* is the "just", i.e. morally authentic enemy, which is different from the villain, who cannot be a just enemy in war (Keller 2012: 113). The villain, on the other hand, is the "common foe of all the world", the *hostis humani generis*, the pirate (Cicero 1913, 3:107). This distinction has great significance for the later development of *ius gentium*. The doctrine of just war was influenced by it as well as the doctrine of crimes that affect everyone (piracy, slavery, genocide) and which therefore can be regarded as crimes against humanity.

4 Strangers in Early Modernity: The Rise of Colonialism and International Law

In the Middle Ages, without an overlapping imperial order, strangers in other countries stayed under their own laws. This means they did not have special rights as strangers. The judicial position of strangers was determined only by the monarch's right to grant privileges (*Königsschutz*). This position was primarily dependent on the economic value of the strangers, who had to pay taxes for their rights to the monarch. This is especially true for the Jews who gained emancipation only in the nineteenth century (Frisch 1910: 22–67). The transition to the territorial order of the early modern period is accompanied by the theory that acts fall under the law of the place where the act was committed.

The early modern period is marked not only by the emergence of territorial states, sectarian division, humanism and the Renaissance but also by European territorial expansion, which is accompanied by an internal aggression against strangers. In 1492, the so-called *Reconquista* of the Iberian Peninsula was completed. In 1496, all Muslims were expelled from Portugal, and in 1502, all Moors (Muslims) and Jews from Spain. The horizon of experience of early modern Europe was marked by expansion and contraction, as Jürgen Osterhammel calls it (Osterhammel 1997). The fear of strangers was combined with the need to triumph over them. European policy and statecraft were able to determine the rules for dealing with strangers. Where this did not succeed at once (as in China and Japan), it led to uncertainty and aggression. Europe itself had little experience with peaceful strangers who remained inside its territories (Osterhammel 1997: 387–388).

In the "New World" of the Americas, one met people whose social and political order differed markedly from the European order. This order was described with the idiom of deficiency, as had already been the case with Aristotle's description of the "barbarians". Aristotle's *topos* of the "barbarian" as a "slave of nature" was of fundamental importance in the dispute about how to deal with the Natives in the conquered lands of South America. Neo-Aristotelian scholars disputed the question of whether the so-called "barbarians" are protected by *ius gentium* and what kind of rights the Spaniards have towards them (Brunstetter and Zartner 2011; Hanke 1994; Pagden 1990). This was still dealt with in terms of *ius gentium*. The *ius gentium* regulates the rights and obligations of strangers residing in countries other than their own. Immigration and settlement law are its special subject. Further topics are seizure of property, slavery, martial law, law of envoys, marriage law, contracts of sale and exchange. It thus includes parts of modern private international law, but also those of public international law.

The *Conquista* of America can be explained by the definition of immigration rights in the *ius gentium* interpretation of Francisco de Vitoria. Vitoria broke with the former discourse, which had hitherto denied the Native Americans property rights over their territories because they were not Christians but "barbarians". If the "barbarians" lacked legal ownership of their land, this land represented a "masterless" area (*terra nullius*) which the Spaniards had a right to acquire. Vitoria rejected the Aristotelian or Christian justifications for the dispossession of the local inhabitants. In his view, the Spaniards had to accept the property of the Native Americans and could only have the status of strangers. According to Cavallar, Vitoria even went so far as to recognise the equality of all people. Vitoria "tries to assume an impartial viewpoint that considers the rights of foreigners such as the Native American 'barbarians' as if they were French fellow Christians or even Spaniards" (Cavallar 2002: 93). From Vitoria's point of view, everyone is entitled to enter a foreign country, to settle, to trade, to use its ports, to use its resources, etc. The only condition is that the exercise of this right must not cause harm to the inhabitants of the foreign country, or, in the words of Vitoria, "the Spaniards have the right to travel and dwell in those countries, so long as they do no harm to the barbarians" (Vitoria 1991: 278).

Vitoria establishes the right of the stranger to visit another country and attempt trade and exchange there with the *ius communicationis*, a right to natural partnership and communication. Usually, the right to travel and the right to trade were considered, but to these Vitoria added the right to dwell in the countries visited (Cavallar 2002: 108). In this way is the issue approached from the Spanish side: "if children born in the Indies of a Spanish father wish to become citizens (*cives*) of that community, they cannot be barred from citizenship or from the advantages enjoyed by the native citizens born of parents domiciled in that community" (Vitoria 1991: 281). It is crucial that the right of entry can be enforced. The denial of this right entails a right to just war since anyone whose rights are violated is entitled to go to war to restore them. Thus, although Vitoria's theory has a "xenophile" side, historically it served to promote the discourse of colonialism: "His *ius communicationis* was both the founding principle of a universal society composed by equal nations and the main legal ground for justifying the colonial conquest of the New World" (Chetail 2016: 905). It is very important to see that colonialism and the origins of *ius gentium*, which gave rise to the modern discourse of international law, are intertwined in this approach (Anghie 1996).

Vincent Chetail describes the subsequent development of international law as marked by the tension between sovereignty and hospitality. In his work *The Free Sea* (1609), Hugo Grotius adopts the insights of Vitoria's *ius communicationis* and calls for a universal right to travel by sea and visit other nations. Grotius supplements this general right with the right of free passage (Grotius 2009: 93). As with Vitoria, this right can be enforced if it is denied even though there is no abuse (Grotius 2005: 441). Chetail considers this to be a thesis that dismantles Vitoria's argument for the permissibility of colonialism because for Grotius the right "to remain in another country" (Chetail 2016: 908; cf. Grotius 2005: 446) refers above all to refugees. But if one rather thinks of the Dutch seafarers in Japan, his reasoning reveals colonialist connotations. At least, this is the reason for Kant to later challenge the right of entry into a foreign country and to limit the cosmopolitan right of foreigners to a right of visit.

The early modern era was marked by the rise of the state, princely sovereignty and territorial closure. For religious reasons, refugees became a political factor in Europe, and the Peace of Augsburg (1555) established the right of emigration (*ius emigrandi*), which was confirmed by the Peace of Westphalia (1648). As Samuel von Pufendorf explains in *The Law of Nature and Nations* (1672): "[I]t is left in the power of all states, to take such measures about the admission of strangers, as they think convenient: those being ever excepted, who are driven on the coasts by necessity, or by any cause that deserves pity and compassion. Not but that it is barbarous to treat, in the same cruel manner, those who visit us as friends, and those who assault us as enemies" (Pufendorf 1749: 873, quoted from Chetail 2016: 910). Pufendorf's aim was to restrict the right of free movement, as Vitoria had justified it, and to place its control in the hands of the sovereign (Chetail, 2016: 911). The decision on admission of strangers became a free discretion of the state, which was often determined to its own advantage. Christian von Wolff confirms and intensifies this line of argumentation in his work *Jus Gentium Methodo Scientifica Pertractatum* (1749). Not only could strangers who attempted to enter the territory of the state be punished (Wolff 1934: 149, §293), but Wolff also saw the territory of the state as its property, from which the stranger could be excluded (Wolff 1934: 150, §295; cf. Chetail 2016: 912). Princely sovereignty was territorial sovereignty. The right to be admitted for reasons of health, science or trading was dependent on the host state and could not be enforced (Wolff 1934: 157–177, §§310–346). Princely sovereignty for Pufendorf and Wolff applies only to questions concerning the migration of strangers. In conceptual terms, they represent the state-bound alternative to Vitoria's global *ius communicationis*.

According to Chetail (2016), in *The Law of Nations* (1758) Emer de Vattel reconciled both positions. In this work on international law, Vattel devoted a separate section to the "norms for the treatment of foreigners" (Vattel 2008: Book II, Ch. VIII, §§99–115). Like Vitoria, Vattel assumes the existence of an "original community" (Vattel 2008: Book II, Ch. VIII, §117). From this original community Vattel derives a right to enter the land of another property owner in case of emergency, even against the owner's will (Vattel 2008, Book II, Ch. VIII, §123). This entails a further right to live in a foreign country (§125). And although Vattel takes the territorial sovereignty of the princes for granted, like Grotius he recognises a right of innocent passage (Chetail 2016: 918). From this, the question follows whether strangers have independent rights against states. However, the democratisation of the state after the democratic revolutions of the eighteenth century changes this question into what rights of self-government democratic societies hold against strangers. This question has been central in discussions of democracy until the present day.

5 The Enlightenment: Social Contract and Hospitality Law

During the Enlightenment, the question of the rights of individuals moving across state borders—or migrants in contemporary terms—was characterised by two theoretical developments. The first was the attempt to theoretically justify the popular basis of state authority by means of a social contract, and the second was an attempt to reconceptualise the relations with strangers as relations between world citizens.

From this arose a tension between the rights of citizens to decide on the status of strangers in their own republic and the cosmopolitan right of world citizens to move freely on the Earth's surface "to present oneself for society", as Kant says (Kant 1996a: 328 (8:358)).

The rights of world citizens stand in contrast to attempts to strengthen the rights of strangers by reviving the right of hospitality as an alternative. Shouldn't the right to hospitality be a guarantee for a peaceful contact with strangers? Tacitus in his *Germania* had highly praised the hospitality of the Germanic peoples and presented it as an example (Tacitus 2013: Ch. 21). What is usually read as Tacitus's social criticism of his decadent society in the first century AD is at the same time the starting point for a tradition of veneration of the customs and ways of life of "backward" non-state peoples, which Europe has lost during the process of civilisation. Montaigne's essay on the Native Americans, Rousseau's discourses, Lahontan's travelogue from North America or Bougainville's Tahiti report exemplify highlights of the veneration of the strangers and their friendly customs. These counternarratives to the still prevailing barbarian discourse represent a criticism of their domestic societies (Graeber and Wengrow 2021). But not everyone shared the praise of the noble customs of the non-state indigenous peoples. Around 1760, theorists of progress and civil society, such as Adam Smith and Adam Ferguson, began to claim that "hospitality should be associated with 'barbaric' phases of social development" (Osterhammel 1997: 417); for example, with the nomadic phase of herding. Like William Falconer and William Robertson, or the Marquis de Condorcet, they were of the opinion that hospitality belongs to a pre-bourgeois—i.e. pre-civilised—stage and is therefore "economically irrational in more complex societies" (Osterhammel 1997: 418). If, in this view, hospitality is thus socially outdated because it has been depersonalised and economised, Rousseau and others saw in this tendency a decay of original values. But there is no going back to the old epoch of noble hospitality, not even for Rousseau. Instead, he looks forward with his theory of popular sovereignty.

The French *Encyclopaedia* still sees among strangers simply "one who was born under another dominion and in a country other than the country in which he is located" (Jacourt 2017: 6:71). In a combination of mercantilist state interests and humanist sentiments, it advocates the settlement of strangers to increase local prosperity and craftsmanship. This does not pose a problem for citizenship; it can be assumed that these foreign craftsmen will be treated as strangers who intend to become residents. But with the historical and social developments leading to the democratisation of the state, the problem changed. Rousseau's theory of the social contract, which sees the state as the product of popular will, created a special status of obedience even for those individuals who did not take part in the founding of the state: "If then there are opponents when the social compact is made, their opposition does not invalidate the contract, but merely prevents them from being included in it. They are strangers among citizens. When the State is instituted, residence constitutes consent; to dwell within its territory is to submit to the Sovereign" (Rousseau 1923: Book 4, Ch. 2). For Rousseau, the state, i.e. the political body of the republic, now confronts the strangers as a unit (Rousseau 1923: Book 1, Ch. 7). The old right to hospitality has become obsolete. It is no longer the individual guest relationship that determines admission to the state, but the will of the popularly elected sovereign, as laid down in the constitution.

As in the democratic constitutions of antiquity and the Roman Republic, the Enlightenment's own version of democratic theory, or popular sovereignty, raises the question of belonging as a question of citizenship and not one of descent. Is a person who has a residence in a foreign country allowed to become its citizen? The French Republic of 1793 will largely eliminate the tension between citizens and strangers by means of naturalisation regulations that set very low criteria for the acquisition of citizenship. Article 4 of the French Constitution of 1793 reads: "Every man born and living in France, of twenty-one years of age, and every alien, who has attained the age of twenty-one, and has been domiciled in France one year, and lives from his labour; or has acquired property; or has married a French woman; or has adopted a child; or supports an aged man" is permitted to exercise civil rights (French Constitution 1883, Art. 4) . However, this very integration-friendly provision has remained the exception in the history of democracy (Fahrmeir 2007).

Social contract theory effectively changed the meaning of the concept of "enemy": The stranger as an individual could no longer become an enemy. Wars are only fought between states and only these can be enemies, as Rousseau claims against Hobbes (Rousseau 1923: Book I, Ch. 4). Thus, war is the subject of international law, and the treatment of strangers is a matter for state law. This means that the relationship of individuals to states is left to the arbitrariness of the states.

6 Immanuel Kant: Anti-Colonialism and Cosmopolitan Law

Kant attempts to fill the gap between the constitutional law of the state and international law that regulates the relations of states through the concept of cosmopolitan law. Cosmopolitan law is supposed to secure the rights of individuals in a world of states and, at the same time, to protect peoples from colonialism. The Enlightenment is—to a certain degree—largely critical of colonialism and slavery (Muthu 2003). Kant condemns colonialism on the grounds that in the newly "discovered" countries the protagonists of colonialism "counted the inhabitants as nothing" (Kant 1996a: 328 (8:358)). He not only opposes colonial interests (gaining land and wealth), methods (conquest, theft or fraud) and consequences (wars, famine, riots) but also any moral justification of violent Christianisation and civilisation (Kant 1996a: 328 (8:358–359); 1996b: 489 (6:352–353)). Even the introduction of a universal legal status for the benefit of mankind in distant parts of the world cannot justify the violence used in such cases. Kant asks: "in newly discovered lands, may a nation undertake to settle (*accolatus*) and take possession in the neighborhood of a people that has already settled in the region, even without its consent?" (Kant 1996b: 489 (6:353)). He demands a contract for all settlements which cannot avoid "encroachment" with the already residing people and explicitly mentions "American Indians" and "shepherds and hunters" (Kant 1996b: 490 (6:353)). Such a contract protects stateless peoples against foreign usurpation of their land.

Kant frames the argument for cosmopolitan right in the context of a broader discussion of the self-understanding of enlightened European society, which sees itself as representing world citizenship (Kleingeld 2011). The abuses of the right to hospitality by European states, who used it as a pretext for colonial conquest, led Kant to formulate cosmopolitan law as a restriction of the right to hospitality (Niesen 2007). Within the framework of cosmopolitan right, transnational freedom of movement is "limited to conditions of universal hospitality" (Kant 1996a: 328 (8:358)). This also means "the right of a foreigner not to be treated with hostility because he has arrived on the land of another" (Kant 1996a: 328 (8:358)). With this, Kant reverses the principle of defence against damage according to Vitoria's view of *ius gentium* that the expulsion of strangers is permitted if they are not thereby harmed. For Vitoria, the Spaniards were allowed to stay and move freely as long as they did not harm the local inhabitants. For Kant, Japan and China are allowed to expel all merchants and sailors if this does not threaten the sailors' lives. For Kant, strangers no longer have the principle of universal hospitality on their side. "What he [i.e. the stranger] can claim is not the right to be a guest, (for this a special beneficent pact would be required, making him a member of the household for a certain time), but the right to visit; this right, to present oneself for society, belongs to all human beings by virtue of the right of possession in common of the earth's surface" (Kant 1996a: 328 (8:358)). Kant's cosmopolitan law and his ideas for a league of nations inspired activists of abolition, international arbitration and disarmament in the eighteenth and nineteenth centuries (Eberl 2023).

Today, the question of how cosmopolitan law relates to refugees and asylum seekers as well as to strangers seeking residence is pertinent to current debates. Poststructuralist Jacques Derrida accuses Kant of violating the "absolute right of hospitality" by demanding a special treaty imposing conditions of the visit and giving this absolute right up in favour of hospitality that is merely conditional (Derrida 2000). The ethical law of absolute hospitality, for Derrida, stands above the laws of the territorial state theorised by Kant. Derrida thus articulates an unease with the territorial principle of sovereignty that once again allows European states to arbitrarily shape their external relations. This principle prevents any attempt by strangers to visit a state without regard to emergency situations and suits attempts to decline the admittance of refugees.

It is possible, however, to understand cosmopolitan world citizenship as the right of a stranger who intends to stay permanently in a foreign community (Benhabib 2004). If this represents a right to be accepted and, at some point, to be naturalised in another state, then the boundaries in a global community of world citizens become blurred between the categories of strangers who are allowed to stay and those who come to visit temporarily or who claim a refugee status in extremis. Seyla Benhabib justifies the basic right of world citizenship not only as a right of transnational mobility but also as a right to remain in a foreign country of residence. Benhabib's approach reintroduces the centrality of individual freedom and derives from it the demand that permanent residency (denizenship) must also lead to a political voice (citizenship). In effect, this argument leads from cosmopolitan law back to state law, whereby the shaping of political order in relation to foreigners and strangers becomes a question of establishing inclusive democratic processes and political participation.

This discussion about cosmopolitan rights of world citizens has been criticised as part of an imperial project upheld by colonial powers (Mendieta 2009). In this view, the epistemic and geographical grounding of cosmopolitan rights is Eurocentric and colonialist. In this context, Kant's critique of colonialism is seen from a postcolonial view in contrast to his concept of "race", which is taken as his contribution to the imperial project. From a cosmopolitan point of view, Kant's critique of colonialism and his idea of a cosmopolitan right still offer an important defence of the rights of strangers, migrants and refugees and paves the way for a self-critique of Enlightenment philosophy as Eurocentric philosophy.

7 Conclusion

Until 1800, the right of strangers was a right granted to people from other countries or cultures by the city, the monarch or the state. With the development of the territorial order of states in the early modern period, strangers became the issue of international contracts. But only cosmopolitan law granted strangers rights as citizens of the world, irrespective of their home country or status. The right of a world citizen to seek contact to other world citizens is restricted by the will of the community to accept this as an offer of communication and interchange. With the democratisation of states, this tension has become even more urgent because the general will of a democratic community is as justified as the rights of world citizens. Today, under the condition of increasing transnational migration, it is obvious that the concept of strangers seems outdated and should be transformed into that of world citizens. To this end, we need to elaborate on how such a status of world citizen can be implemented in international and national law, which rights such a status would entail, and how it can avoid reproducing global asymmetries.

Literature

Anghie, A. 1996. Francisco de Vitoria and the Colonial Origins of International Law. *Social and Legal Studies* 5:321–336.

Aristotle. 1912. *Politics. A Treatise on Government*, Translated by William Ellis. London: JM Dent. https://www.gutenberg.org/files/6762/6762-h/6762-h.htm. Last checked 9 Aug 2022.

Benhabib, S. 2004. *The Rights of Others. Aliens, Residents, and Citizens*. Cambridge: Cambridge University Press.

Bleicken, J. 1995. *Die athenische Demokratie*. Paderborn: UTB.

Browning, R. 2002. Greeks and Others. From Antiquity to the Renaissance. In *Greeks and Barbarians*, ed. T. Harrison, 257–277. Edinburgh: Edinburgh University Press.

Brunstetter, D. R., and D. Zartner. 2011. Just War against Barbarians. Revisiting the Valladolid Debates Between Sepúlveda and Las Casas. *Political Studies* 59:1–20.
Cavallar, G. 2002. *The Rights of Strangers. Theories of International Hospitality, the Global Community and Political Justice Since Vitoria*. London: Ashgate.
Chetail, V. 2016. Sovereignty and Migration in the Law of Nations. And Intellectual History of Hospitality from Vitoria to Vattel. *European Journal of International Law* 27:901–922.
Cicero, M. T. 1913. *De Officiis*. Cambridge: Harvard University Press. http://www.perseus.tufts.edu/hopper/text?doc=Perseus%3Atext%3A2007.01.0048. Last checked 9 Aug 2022.
Cohen, H. 1992. Die Nächstenliebe im Talmud. In *Der Gast, der bleibt. Dimensionen von Georg Simmels Analyse des Fremdseins*, ed. A. Loycke, 80–102. Frankfurt: Campus.
Derrida, J. 2000. *Of Hospitality. Anne Dufourmantelle Invites Jacques Derrida to Respond*. Stanford: Stanford University Pres.
Eberl, O. 2021. *Naturzustand und Barbarei. Begründung und Kritik des Staates im Zeichen des Kolonialismus*. Hamburg: Hamburger Edition.
Eberl, O. 2023. The Cosmopolitan Challenge. Cosmopolitan Ideas in the Eighteenth and Nineteenth Century. In *Palgrave Handbook in International Political Theory*, ed. H. Williams et al., 185–203. London: Palgrave Macmillan.
Fahrmeir, A. 2007. *Citizenship. The Rise and Fall of a Modern Concept*. New Haven: Yale University Press.
French Constitution. 1883. *Appendix to Francis Lieber. On Civil Liberty and Self-Government*. Ed. T. Woolsey. Philadelphia. https://oll.libertyfund.org/pages/1793-french-republic-constitution-of-1793. Last checked 9 Aug 2022.
Frisch, H. V. 1910. *Das Fremdenrecht. Die staatsrechtliche Stellung der Fremden*. Berlin.
Graeber, D., and D. Wengrow. 2021. *The Dawn of Everything. A New History of Humanity*. New York: Farrar, Straus and Giroux.
Grotius, H. 2005. *The Rights of War and Peace*. Ed. R. Tuck. Indianapolis.
Grotius, H. 2009. *Mare Liberum 1609–2009*. Ed. R. F. Leiden. Boston.
Hanke, L. 1994. *All Mankind Is One. A Study of the Disputation Between Bartolome De Las Casas and Juan Gines De Sepulveda in 1550 on the Intellectual and Religious Capacity of the American Indians*. DeKalb: Northern Illinois University Pres.
Herodotus. 1920. *Histories. With an English translation by A. D. Godley*. Cambridge: Harvard University Press.
Hiltbrunner, O. 2005. *Gastfreundschaft in der Antike und im frühen Christentum*. Wissenschaftliche Buchgesellschaft Darmstadt.
Jacourt, C. 2017. Étranger, Droit Politique. In *Encyclopédie, ou dictionnaire raisonné des sciences, des arts et des métiers, etc.*, ed. D. Diderot and J. d'Alembert, 6:71. Chicago. https://artflsrv03.uchicago.edu/philologic4/encyclopedie0521/navigate/6/270/. Last checked 9 Aug 2022.
Kant, I. 1996a. Toward Perpetual Peace. In *Practical Philosophy, The Cambridge Edition of the Works of Immanuel Kant*, ed. M. Gregor. Cambridge: Cambridge University Press.
Kant, I. 1996b. The Metaphysics of Morals. In *Practical Philosophy, The Cambridge Edition of the Works of Immanuel Kant*, ed. M. Gregor. Cambridge: Cambridge University Press.
Keller, A. 2012. *Cicero und der gerechte Krieg. Eine ethisch-staatsphilosophische Untersuchung*. Stuttgart: Kohlhammer.
Kleingeld, P. 2011. *Kant and Cosmopolitanism. The Philosophical Ideal of World Citizenship*. Cambridge: Cambridge University Press.
Loycke, A. 1992. *Der Gast, der bleibt. Dimensionen von Georg Simmels Analyse des Fremdseins*. In *Der Gast, der bleibt. Dimensionen von Georg Simmels Analyse des Fremdseins*, ed. A. Loycke, 103–123. Frankfurt: Campus.
Mathisen, R. W. 2006. Peregrini, Barbari, and Cives Romani. Concepts of Citizenship and the Legal Identity of Barbarians in the Later Roman Empire. *The American Historical Review* 111:1011–1040.
Mendieta, E. 2009. From Imperial to Dialogical Cosmopolitanism. *Ethics & Global Politics* 2:241–258.

Muthu, S. 2003. *Enlightenment Against Empire*. Princeton: Princeton University Press.
Niesen, P. 2007. Colonialism and Hospitality. *Politics and Ethics Review* 3:90–108.
Nippel, W. 2002. The Construction of the 'Other'. In *Greeks and Barbarians*, ed. T. Harrison, 278–310. Edinburgh: Edinburgh University Press.
Noy, D. 2002. *Foreigners at Rome. Citizens and Strangers*. London: Duckworth.
Osterhammel, J. 1997. Gastfreiheit und Fremdenabwehr. Interkulturelle Ambivalenzen in der Frühen Neuzeit. In *Furcht und Faszination. Facetten der Fremdheit*, ed. H. Münkler and B. Ladwig, 379–436. Berlin: Akademie.
Pagden, A. 1990. Dispossessing the Barbarian. The Language of Spanish Thomism and the Debate over the Property Rights of the American Indians. In *The Languages of Political Theory in Early Modern Europe*, ed. A. Pagden, 79–98. Cambridge: Cambridge University Press.
Picker, C. 2020. *Flüchtlingsethik*. Leiden: Brill.
Plato. 1888. *The Republic*. London: Macmillan. http://eremita.di.uminho.pt/gutenberg/5/5/2/0/55201/55201-h/55201-h.htm#BookV. Last checked 9 Aug 2022.
Pufendorf, S. V. 1749. *The Law of Nature and Nations or a General System of the Most Important Principles of Morality, Jurisprudence and Politics*. London.
Rousseau, J. J. 1923. *The Social Contract and Discourses by Jean-Jacques Rousseau*. London: J.M. Dent. https://oll.libertyfund.org/titles/638#Rousseau_0132_472. Last checked 9 Aug 2022.
Schreiner, J., and R. Kampling. 2000. Der Nächste, der Fremde, der Feind. Perspektiven des Alten und Neuen Testaments. In *Die Neue Echter Bibel*. Würzburg: Echter.
Simmel, G. 1950. The Stranger. In *The Sociology of Georg Simmel*, ed. K. Wolff, 402–408. New York: Free.
Spahn, P. 1995. Fremde und Metöken in der Athenischen Demokratie. In *Mit Fremden leben. Eine Kulturgeschichte von der Antike bis zur Gegenwart*, ed. A. Demandt, 37–56. München.
Tacitus. 2013. The Germany and the Agricola of Tacitus. In *The Oxford Translation Revised*. https://www.gutenberg.org/files/7524/7524-h/7524-h.htm. Last checked 9 Aug 2022.
Vattel, E. d. 2008. *The Law of Nations, Or, Principles of the Law of Nature, Applied to the Conduct and Affairs of Nations and Sovereigns*. Ed. B. Kapossy and R. Whatmore. Indianapolis. https://oll.libertyfund.org/titles/vattel-the-law-of-nations-lf-ed. Last checked 9 Aug 2022.
Vitoria, F. d. 1991. On the American Indians. In *Political Writings*, ed. A. Pagden and J. Lawrance, 231–292. Cambridge: Cambridge University Press.
Wolff, C. 1934. *Jus Gentium Methodo Scientifica Pertractatum*. Translated by J. H. Drake. Oxford: Clarendon. https://archive.org/details/in.ernet.dli.2015.89798/page/n9/mode/2up. Last checked 9 Aug 2022.

Migration Ethics in East Asia

Naoto Higuchi

Abstract This chapter examines the historical path that has led to the status quo of migration in East Asia, focusing on Japan, South Korea, Taiwan and Hong Kong as major migrant destinations. I illustrate the three-layered structure of migration in this region in the following way. (1) What divides this region is the legacy of colonialism and the ongoing effects of the Cold War: these have greatly influenced the status of intra-regional migrants. (2) The developmental states, in which rights may be sacrificed for economic growth and order, have sponsored guest worker programmes that recruit temporary labour migrants, who are denied rights such as family reunification and access to permanent residence. (3) Feminisation of migration in this region is characterised by the importance of marriage migrants as well as domestic and care workers because marriage has become one of the very few legal channels for people seeking to settle in these societies. It has been said that strict immigration controls in East Asia are based on the ethno-nationalist ethics that justify restrictions on the rights of migrants, but a careful look at the three-layered structure reveals that ethnic nationalism is not the basic principle underlying them. What determines migration ethics in this region are the logics of economically motivated developmental states and geopolitical confrontations surrounding postcolonialism and the Cold War.

East Asia had been a region of net emigration since the nineteenth century: China sent a great number of migrants to Southeast Asia and North America; the Japanese government promoted emigration to North and South America; and many migrated from the Korean Peninsula to North America, the former Soviet Union, China and Japan. This was a result of incorporating this region into the periphery of the modern world system. Although the population flow from China and North Korea is still ongoing, rapid economic growth in this region transformed Japan, South Korea and Taiwan into net migrant-receiving economies (Hong Kong has been a city of

N. Higuchi (✉)
School of Human Sciences, Waseda University, Tokyo, Japan
e-mail: higuchinaoto@waseda.jp

A. Niederberger et al. (eds.), *Handbook of Migration Ethics*,
https://doi.org/10.1007/978-3-031-89877-8_3

immigrants for a century, Wong 2007). These countries have been described as new destinations for migrants.

While it has been said that restrictive migration policies in East Asia are based on strong ethnonational identities exclusive to non-member ethnics, I argue the importance of two alternative principles that determine political measures: economic growth and geopolitical conditions. The former is exemplified by the notion of "developmental states", in which rights may be sacrificed for economic growth and order (Seol and Skrentny 2009a). The results are restrictive immigration and refugee policies (Chung 2014) that reluctantly introduced state-sponsored temporary guest worker programmes (Surak 2017).

Developmental states brought about another feature of migration in East Asia. Restrictive policies towards family integration led to highly gendered patterns of migration in accordance with the structure of migrant labour markets. These trends were accelerated by East Asian compressed modernisation, which refers to the development of first modernity followed by a transition to second modernity almost at once, within the space of just one or two decades (Han and Shim 2010). This has brought about rapidly changing family structures, such as later marriage, decreasing birth rates, and ageing, which have resulted in increased numbers of transnational marriages and chronic shortages of care workers.

The notion of developmental states, however, is not sufficient to examine the background of migration ethics in this region. East Asia has two faces: one the centre of global manufacturing with new demands for migrant workers and the other a conflict-laden area still suffering from adverse legacies of prewar colonialism and the Cold War. In terms of the latter, migration was not unaffected by this. While industrialised economies of East Asia have become destinations for migrants from Southeast Asia, South Asia and South America since the 1980s, most migrants until the 1970s were from neighbouring countries such as China and Korea. The status of these intra-regional migrants has been strongly influenced by colonialism and the ongoing effects of the Cold War. In the following sections, I outline the three-layered structure of migration in East Asia.

1 The Status of Migrants Under the Postcolonial Cold War Regime

East Asia is a unique region for two reasons. On the one hand, a significant division has existed between the former colonial empire (Japan) and its colonies (Korea and China), which has caused complex patterns of ethnic mixing and unmixing in the region. On the other hand, even after the 1990s, a Cold War structure remained in East Asia, although China, Japan and Korea are now leading world economies and have become equal economic players. In this section, I argue how these geopolitical conditions have brought about the patterns of migration we see in East Asia.

1.1 Historical Conditions

The origin of the migration issues within East Asia can be found in the colonisation of Korea and China by the Empire of Japan. After Japan narrowly escaped colonisation by the great powers of the West, it emerged in the region as a new imperialist state: Japan won the First Sino-Japanese War (1894–95) and ceded Taiwan; the Russo-Japanese War (1904–05) resulted in the cession of southern Sakhalin by Japan; Korea was annexed to Japan (1910); and it further founded Manchukuo as its de facto colony (1932). The expansion of the Japanese empire produced two different flows of migrants: one was rural-to-urban migration from Korea and Taiwan to Japan, and the other was rural-to-rural migration from Korea and Japan to Manchuria and southern Sakhalin (Brubaker and Kim 2011; Kim 2016). Ethnic mixing in Japan's expanding territory accompanied initiatives by the country to integrate itself as a multiethnic empire (Oguma 2002).

After Japan's defeat in World War II, followed by the dismantling of the Japanese empire, many migrants found themselves divided by newly emerging borders across these colonies. While the expansion of the empire brought about ethnic mixing in its territory, its demise resulted in ethnic unmixing (Brubaker and Kim 2011). In fact, the majority of these migrants repatriated within a few years after the end of World War II, but unmixing remained incomplete in the sense that a considerable number of Koreans continued to live in their new lands, which complicated the decolonisation process. In this context, especially important were Korean minorities in Japan, Northeast China (former Manchuria) and Sakhalin, who followed different tracks in accordance with the policies of their host countries and the relations between these countries and the Korean Peninsula. For example, while Koreans in China held citizenship of the People's Republic of China (PRC) and political autonomy that enabled formal education in Korean, their counterparts in Japan were deprived of Japanese nationality and denied recognition as a minority group (Min 1992).

Furthermore, the Cold War turned China and the Korean Peninsula into a front line of East-West confrontation. In 1948, China was divided into the PRC and Taiwan, and the Korean Peninsula into the North and the South. Borders drawn under the Cold War intertwined with those brought about by the collapse of the Japanese empire. To make matters more complicated, lack of diplomatic relations between the two camps prohibited intra-regional migration. Japanese migrants in Manchuria, for example, were repatriated under official repatriation programmes between 1946 and 1948, and after the PRC was founded this repatriation programme was run by the Red Cross between 1953 and 1958. However, those left behind could not return to Japan until the normalisation of diplomatic relations between Japan and China in 1972 (Ward 2006).

1.2 Unfinished Colonial Settlement: The Case of Japan

Legacies of colonialism have influenced migration flows in East Asia. The most clearly apparent among these is Hong Kong. While it is now known for its relatively liberal migration policies under a *laissez-faire* economy, Hong Kong had been regulating population flow from the PRC during the colonial era. The border is still controlled even since its reversion to China in 1997: mainland Chinese were forced to wait years for family reunification in Hong Kong when they married Hong Kong citizens (Newendorp 2008).

But Japan is the most typical case of a country still suffering from the postcolonial predicament: this century, the country has seen the rise of hate groups targeting Korean residents despite them being socioeconomically well integrated (Higuchi 2016). The root of the problem extends back to the migration policies of the postwar Japanese government. These were characterised by an underlying motivation to expel Koreans and the resultant lack of initiatives to integrate them. Behind this fundamental principle was the initial decolonisation process during the American occupation (1945–52). Although it is often said that the occupation reform brought about comprehensive democratisation of Japan, migration policy (mostly targeted at Koreans) was an exception.

The Japanese government made use of the US anti-communism doctrine to gain approval for a policy of "unmixing" Koreans as a politically dangerous group inclined to communism. The Japanese government, with the acquiescence of SCAP (Supreme Commander for the Allied Powers), oriented itself towards the repatriation of Koreans and the removal of their rights (Morris-Suzuki 2010). When the Treaty of Peace with Japan came into effect in 1952, people from Japan's former colonies could neither keep Japanese citizenship nor were they granted permanent resident status. Instead, they were controlled by a tentative legal status under Law No. 126 (Kim 2006).

Under the Cold War regime, the Korean Peninsula was divided into two successor states, South Korea and North Korea, which greatly influenced the legal status of Korean residents. Their status was one of the topics prioritised at the negotiations to normalise diplomatic relations between Japan and South Korea. However, as the goal of the Japanese government at this time was unmixing, its initial proposals at the Japan-South Korea negotiations were restrictive: fresh applications would have to be required for Korean residents to qualify for permanent residence and welfare recipients were to be forcibly repatriated. In reality, the Japanese government regarded Korean residents as a burden and thus provided substantial assistance to efforts to send people to North Korea (Morris-Suzuki 2007).

After the start of Park Chung-hee's administration in South Korea, the Association of Koreans in Japan came under its control and had its support, which enabled joint negotiations to resist Japan's efforts at unmixing. As a result of the negotiations, the Treaty Permanent Residency status was created for those with South Korean nationality in 1965, and their status thus became relatively stable. However, those without South Korean citizenship (mostly affiliated to the pro-North Korea General

Association of Koreans in Japan) retained tentative status, which produced a division in legal status among Koreans in Japan.

It was the ratification of the International Covenant on Economic, Social and Cultural Rights in 1979 and the Convention Relating to the Status of Refugees in 1981 that promoted partial change in these conditions. This led to legal changes in the form of the Immigration Control and Refugee Recognition Act, which resulted in Koreans without South Korean nationality gaining permanent resident status. The final change came about as a result of negotiations with the South Korean government on the legal status of third-generation South Koreans, designed to review the 1965 normalisation of diplomatic relations 25 years on. The final result was the Special Act on Immigration Control, enacted in 1991, which gave permanent resident status to both South and North Koreans for the first time.

The Japanese government failed to see the status of Koreans as its own issue to resolve. In fact, there were three major points of progress in the legal status of Koreans, but none were initiated by the Japanese government: bilateral negotiations between South Korea and Japan resulted in permanent resident status in 1965 and 1991, and ratification of UN conventions led to permanent resident status for North Koreans in 1982. It took 46 years from the end of World War II to settle the legal status of Koreans as former colonial citizens. The issue of legal status ought to have been decided then, but the Japanese government had been unwilling to grant stable legal status to Koreans. Furthermore, the Japanese government has been dismissing campaigns by South Korean nationals to demand voting rights. The reluctant decolonisation process allowed governmental censorship of pro-North Korea associations and thereby hate speech against Korean residents (Higuchi 2016).

1.3 Membership of Co-nationals Under East Asian Geopolitical Conditions

The abovementioned divergent status of North and South Koreans in Japan reflected Cold War allegiances. However, China and Korea have been more deeply influenced by the East Asian geopolitical structure. Attitudes of divided nations towards their counterparts are ambivalent: while they are bitter enemies at the forefront of opposing camps, they are past co-nationals to be potentially integrated in the future.

Such ambivalence is clearly seen in the case of Taiwan: according to Taiwan's constitution, although mainland Chinese are not regarded as foreigners, cross-Strait migration was banned. It was after Taiwan's martial law was lifted in 1987 that the number of Chinese smugglers increased. Although employers overwhelmingly preferred to hire mainland Chinese over workers from other countries, the Taiwanese government did not allow mainland Chinese to work as unskilled workers (Friedman 2015). The policy is at least partly based on geopolitical concerns of the Taiwanese government regarding population pressure, national security and social stability (Skrentny et al. 2007). This is why the government permitted very few mainland

Chinese to visit Taiwan before 2008. The exception was marriage migrants and other categories of family reunification. Yet the Taiwanese government reluctantly accepted them, and thus mainland Chinese were put at a disadvantage compared with other migrants in terms of entry permits and naturalisation (Friedman 2015; Wang 2011).

The case of South Korea is more complex: there are two sources of co-national population flows that have been differently handled. The first is refugees from North Korea, who are granted South Korean citizenship upon entry (Lee 2015). This is because the South Korean government does not define North Koreans as foreigners. In fact, this membership principle is the same as that of mainland Chinese in Taiwan, but the different policies stem from differences in the nature and quantity of population flows. In the context of the Cold War, North Korean refugees showcased South Korea's victory over its northern opponent. In addition, the number of escapees from North Korea never exceeded 3000 per year (Lee 2015).

In contrast, policies towards Korean Chinese migrants (and a smaller number of Koreans in the former Soviet Union) have been much more restrictive. As a result of emigration to other areas of the Japanese empire and North America, the Korean diaspora, with a population of six million, is one of the largest in the world (Park and Chang 2005). Korean Chinese are the biggest group among them, but they have been less favourably treated than their counterparts in Japan and North America. It was in the late 1980s that South Korea began to accept Korean Chinese. This was caused by a combination of attenuation of the Cold War and South Korea's rapid economic growth (Kim 2016).

At first, the Korean state welcomed this group and offered them permanent residence and the possibility of citizenship for those descended from anti-Japanese activists. But Chinese leaders opposed what they saw as an affront to Chinese sovereignty (this has also been a factor that prohibited privileged access to citizenship for Korean Chinese). In 1990, the policy was revised, offering instead only 3-month tourist visas (Seol and Skrentny 2009b). However, unlike the Taiwanese guest worker programme, the South Korean government launched a guest worker programme called "Industrial Trainees", in which Korean Chinese were more privileged: they were afforded a separate large quota and paid higher wages (Seol and Skrentny 2009a). Nevertheless, the government neither granted citizenship nor paved the way for permanent residence to Korean Chinese except for marriage migrants. Thus the anti-familialist stance towards Korean Chinese workers coexists with a familialist stance towards Korean Chinese brides (Freeman 2011).

The inhospitality of the South Korean government towards Korean Chinese can be attributed to at least three reasons. First, Korean Chinese lost Korean nationality when they acquired Chinese citizenship with the establishment of the PRC. In this sense, they are co-ethnics but clearly foreign nationals. Second, South Korean policymakers thought it necessary to control the massive influx of Korean Chinese (unlike other overseas Koreans living in wealthier states such as the United States and Japan) as it considered them poor and unskilled migrants. This is why the Overseas Korean Act enacted in 1999 excluded Korean Chinese. Third, there was a national security concern that North Korea might use the allowance as a route for infiltration, thereby causing immediate security threats (Sktentny et al. 2007).

These examples show that while the loosening of the Cold War structure opened the door for co-ethnic migration across the Iron Curtain, the structure still affects East Asia: Taiwan and South Korea remain reluctant hosts which manage to discourage long-term settlement of their co-ethnics.

2 Labour Migration to the Developmental States

The postwar East Asian "economic miracle" transformed countries in the region from sending to receiving countries of migration within a short period. This means economic growth has been an overarching imperative that has greatly influenced the ethics of migration policies in these economies. At first, lack of immigration policies in these newly receiving economies (with the exception of Hong Kong) resulted in an initial increase in irregular migrants (Athukorala and Manning 1999). The countries later accommodated the demand by combining the introduction of co-ethnic migrants and short-term guest workers, and although the conditions of importation and the balance between the two differed to a considerable degree, they had much in common. Firstly, the acceptance of co-ethnic migrants was followed by the change of the Cold War structure and determined by the logic of nationalism, while it is said that East Asian governments use co-ethnic preferences instrumentally for economic goals (Skrentny et al. 2007). Secondly, they have managed to discourage long-term settlement of unskilled workers, which caused human rights issues in this region. Thirdly, they also saw the rise of policy debates on multiculturalism in response to the increasing diversity of traditionally homogeneous societies.

2.1 From Blood Ties to Neoliberal Exclusion: The Case of Return Migration to Japan

Japan saw debates on "foreign workers" in the late 1980s when it suffered most seriously from labour shortages. The result was the revision of the Immigration and Refugee Recognition Act in 1990, which rejected the importation of low-skilled migrants but opened the door for third-generation descendants of Japanese emigrants, who lived in South American countries such as Brazil and Peru. It seems natural that the Japanese government depended on "blood ties" as the new source of unskilled labour. In fact, the overwhelming majority of Japanese descendants had been working in factories as precarious agency workers.

However, the privileged visa status of Japanese descendants was a by-product of the Special Act on Immigration Control mentioned in the previous section. The Ministry of Justice had the idea of granting special visas to Japanese descendants when Japan and South Korea started negotiations regarding the status of Korean residents in Japan in 1985 (Kajita et al. 2005). This was before the foreign worker debate in 1988, and policymakers had never thought of importing Japanese descendants to solve the labour shortage. Instead, they had in mind balancing the status of third-generation Koreans and Japanese descendants based on jus sanguinis, which is the basic principle of nationality law in East Asia.

Japanese descendants enjoyed privileged status in the sense that they could work anywhere in Japan and were eligible for permanent residence and family reunification. However, the governmental measures towards them were little more than a laissez-faire approach: the Japanese government welcomed Japanese descendants but lacked integration policies. As a result, they could not achieve upward mobility despite living in the country for an extended period. Children suffered from low levels of educational achievement (Kajita et al. 2005). The "failure" of the experiment led policymakers to adopt a more restrictive approach to Japanese descendants, although Japan was facing a structural labour shortage due to its declining birth rate.

In 2018, the Ministry of Justice granted a new visa for fourth-generation Japanese descendants, who were allowed to enter solely as children of the third generation. At first glance, this may look like expansion of jus sanguinis to the fourth generation, but the newly instituted status is blatantly based on meritocracy and is becoming similar to guest worker programmes (Takaya 2019). The visa allows only a 5-year stay and cannot be extended. Visa applicants are requested to pass the N4 level of the Japanese Language Proficiency Test and a higher level after a 3-year stay in Japan. After three decades of experience as a host country for migrants, Japan's reliance on "blood ties" was replaced by neoliberal meritocracy.

2.2 *Guest Worker Programmes of Developmental States*

Japan and South Korea granted preferential status to co-ethnic migrants: Japanese descendants (up to the third generation) enjoy highly secure visas and are allowed family reunification. In addition, all East Asian economies referred to in this chapter have policies to attract the highly skilled, which permit renewable visas and family reunification as well as granting citizenship within a short period. Nevertheless, what characterises migration policies in this region is the recruitment of *temporary* migrant workers, who are denied rights such as family reunification and freedom of choice in employment (Seol and Skrentny 2009a). All these economies introduced state-led guest worker programmes, but with certain variations among them (Surak 2017).

Taiwan opened the "front door" to temporary migrant workers: the government passed the Employment Services Act in 1992. At first, the maximum duration of a migrant worker's contract was 3 years, and each worker could benefit from the visa only once (Lan 2006). Because this rotational system was too strict to be sustainable, the duration has been gradually extended and the 3-year contract can now be renewed up to four times (Surak 2017).

It is Japan's trainee and Technical Intern Training Programme, however, that better exemplifies the paradigm of guest worker programmes in East Asia. The Japanese government has still never approved the introduction of unskilled migrant workers: instead, it opened the "side door" to respond to labour demands. Officially, Japanese descendants mentioned in the previous section come to visit their families in Japan and then use this opportunity to work. Likewise, trainees and technical interns are not defined as unskilled labour. Launched in 1993, the formal purpose of the programme is international cooperation with developing countries: trainees are given 1-year on-the-job training and then serve their apprenticeship as technical interns for 2 years. Because the programme's official aim is to transfer advanced skills from Japan to sending countries, technical interns are expected to put their skills to practical use in their countries of origin.

The programme is designed to satisfy several contradictory purposes. First, the strict rotation system (a one-time 3-year contract that is non-renewable) is legitimised in the name of technical skills transfer. Second, the official purpose of the training justifies the extremely low wage. Third, the closed-door policy is firmly maintained because neither Japanese descendants nor technical interns are officially regarded as unskilled migrant workers. This is why the Japanese government insisted on the programme, even though it has been harshly criticised.

Nevertheless, Japan saw a major policy change in 2018: the government established new visa categories known as Specified Skilled Workers 1 and 2 (Takaya 2019). The most striking feature of the new policy is opening the "front door" to unskilled workers: although the technical intern programme is still maintained, interns can extend their stay to obtain a working visa (Specified Skilled Worker 1) after their 3-year internship. In addition, the Specified Skilled Worker 2 visa is renewable, and workers are eligible for family reunion and can apply for a permanent residence visa. This does not mean, however, that Japan has transcended the legacy of the developmental state. The government changed its migration policy simply because it needed to fulfil an increasing demand for migrant workers: it neither considered the human rights of migrants nor responded to growing criticism. In fact, the government still insists on the Technical Intern Training Programme, which is consistent with the goal of the developmental state. Furthermore, migrant workers are expected to climb up the ladder from technical intern to Specified Skilled Worker 2 on their own: the state provides neither adequate language courses nor vocational training.

South Korea is unique in the sense that it partly followed the path taken by Japan and then looked for its own route (Surak 2017). The primary policy at the initial stage was the Industrial Trainee System adopted in 1991, which took its cues from Japan. However, a great number of absconders and harsh criticisms from civil society diminished the system's legitimacy. The South Korean government finally abolished it in 2004 to open the "front door" by adopting the Employment Permit System. However, it is no more open than the Taiwanese guest worker programme in the sense that it limits the maximum length of stay to 10 years. Although employment permit workers can theoretically apply for permanent residence, it is virtually impossible to meet the conditions regarding income, occupation and assets. In terms of access to permanent residence, Japan's new guest worker programme is slightly more inclusive than those of other East Asian migrant-receiving economies. Likewise, Japan's policy towards ethnic Japanese migrants is more open because Korean Chinese migrants are required to earn at least the average national income if they apply for permanent residence visas in South Korea.

2.3 *Limits of Multiculturalism in Developmental States*

East Asian advanced economies have regarded themselves as homogeneous nation-states, but since the 1990s they came to promote some sort of multicultural policies, with the exception of Hong Kong. However, the reality is exactly what Sookyung Kim (2015) described as a "reversed" gap in migration policies. The gap hypothesis in migration research usually refers to a situation whereby tough rhetoric to regulate migration is preferred, but this does not stop migration flows. In the case of East Asian multiculturalism, policies proclaim benevolent and generous goals, but the actual implementation is restrictive (Kim 2015: 74).

In 1997, Taiwan was the first to recognise itself as a multicultural nation, but this move was not associated with the rights of migrants but those of aborigines (Kim and Oh 2012). While policies of the Taiwanese government towards migrants were previously assimilative, they have shifted to multiculturalism. However, promotion of multiculturalism was based on "neoliberal" motives: cultural plurality is desirable as far as it is supportive of economic growth. In fact, children of marital migrants have been expected to serve as cultural and linguistic bridges to strengthen economic relations with Southeast Asian countries because most of their mothers are from Southeast Asian countries such as Vietnam, the Philippines and Indonesia (Lan 2019).

In contrast, Japan's idea of multicultural coexistence is rooted in grassroots activities, not an initiative imposed from above. It was the Great Hanshin Earthquake in 1995 that spread the notion of multicultural coexistence in the country. Unlike the Great Kanto Earthquake in 1923, in which panicked Japanese massacred thousands of Koreans, in 1995 migrants and Japanese are said to have helped each other. Since then, the notion became widely used by NGOs and then diffused to local governments. In Japan, migrant organisations are generally weak, except those of Koreans, but pro-migrant organisations have been actively engaged in providing services such as consultation, language courses, lobbying and protest actions (Shipper 2008). The national government finally published a report on the promotion of multicultural coexistence at the local level in 2006. When the Japanese government formally decided to import unskilled migrant workers in 2018, the cabinet released guidelines for acceptance of and *coexistence* with foreign human resources.

Nevertheless, a series of policies does not mean any change in governmental stance towards promoting the integration of migrants. Critiques of multicultural coexistence have pointed out the following issues (Kajita et al. 2005; Takaya 2019). First, the notion takes for granted that the government is not responsible for taking the initiative to promote equal and harmonious relationships between culturally different groups. It is not the government but peoples that are responsible for achieving multicultural coexistence. Indeed, the rights of ethnic minorities are not acknowledged in the context of multicultural coexistence: neither of the abovementioned policy documents refer to human rights of foreign residents. Second, the idea of multicultural coexistence emphasises that migrants are local residents rather than foreign labourers, which has had two negative effects. (1) Local governments tend to exclude temporary contract labourers (mostly technical interns) from policies for cultural diversity because they do not regard it necessary to care for those only in the country for work. (2) The idea emphasises the cultural difference between the Japanese and migrants alone, which leads to ignorance of socioeconomic inequalities between the two groups. Third, policies for multicultural coexistence are reluctant to deal with the rights of Koreans surrounding colonial settlement. Apart from a few exceptions which have dealt with social movements of Koreans, local governments avoid making reference to the history of former colonial citizens.

Multiculturalism in South Korea followed a different track. Similar to Japan, South Korea is a country with weak migrant organisations (except for Korean Chinese, Kim 2016) and strong pro-migrant organisations. In the 1990s, NGOs supporting migrants were keen to learn from their Japanese counterparts and imported the idea of multiculturalism from Japan (Kim 2015). They then gained leverage to become one of the driving forces in abolishing the Industrial Trainee System in 2004 during the progressive Roh Moo Hyun administration. In this period, South Korean migration policy made remarkable progress: the Diet passed the Act on Treatment of Foreigners in Korea in 2007 and the Act on Support for Multicultural Families in 2008. This is substantial progress compared with Japan and Taiwan, which still lack basic laws to promote inclusive policies towards migrants.

However, Korean policies around multiculturalism are either segregationist (vis-à-vis unskilled migrant workers) or assimilationist (vis-à-vis foreign spouses of Koreans) (Kim 2015). The policies are solely targeted at migrants who have familial or blood ties with Koreans. On the one hand, foreign spouses can access citizenship after they learn Korean culture and become assimilated in the process. On the other hand, Korean Chinese are favoured because they are thought to be more easily integrated than non-Korean unskilled workers.

These examples clearly show the limits of multiculturalism in East Asia, which can survive in so far as it is beneficial to the goals of developmental states. Taiwanese multiculturalism is used to promote economic connections with Southeast Asian countries; the Japanese government promotes multicultural coexistence because it evades rights-based arguments on migration; and in South Korea, multiculturalism is used to strengthen the reproduction of the Korean nation. Overall, soft talk on multiculturalism is used as a cover story for hard policies.

3 Incorporating Migrants into the Reproduction Process

Restrictive migration policies in East Asia have also prevented migrant domestic and care workers from accessing long-term settlement. While foreign brides are, in contrast, relatively welcome to immigrate in order to engage in reproductive labour and assimilate into familialist regimes, developmental states in this region further mobilise them in economic activities so that they can contribute to economic growth.

3.1 Marriage Migrants and Settlement

The guest worker programmes examined in the previous section have resulted in the salience of marriage migration in East Asia as marriage has become one of the very few legal channels for people seeking to settle in these societies. Although the number is smaller than that of labour migrants, marital migrants are important because they are situated at the intersection of gender, ethnicity, nationality and class (Yang and Lu 2010). On the one hand, most of them are brides married to lower-class male citizens, and thus they are incorporated into the lower tier of ethnic-class hierarchies in host economies. On the other hand, they are expected to contribute reproduction of the nation as spouses and/or parents of nationals.

With the exception of the Chinese spouses of Taiwanese and Hong Kong citizens (Friedman 2015; Newendorp 2008), East Asian governments have thus favourably (or at least neutrally) accepted marriage migrants. Although they are cautious about "fake marriages", they regard cross-border marriage as a solution to low fertility rates and shortage of wives. In fact, cross-border marriages account for more than one third (Hong Kong in 2005), 32% (Taiwan in 2003), 14% (South Korea in 2005) and 6% (Japan in 2004) of the total marriage rates in these countries (Yang and Lu 2010). Some Japanese local governments in rural areas went further to sponsor arranged marriages with Filipinas. Nevertheless, Japan has been rather indifferent to the increase in foreign brides. Instead, it is South Korea's policy reaction that is worth mentioning.

The aforementioned Act on Support for Multicultural Families, which mainly targets intermarriage between Korean men and foreign women, exemplifies the policy orientation of the South Korean government. While it denies settlement of unskilled migrant workers, the government shows a clear intention to integrate marriage migrants. One aspect of this policy is related to the intersection between nationalism and gender. Familial affiliation with Korean nationals is a prerequisite for enjoying cultural diversity because the law defines a multicultural family as an international couple between a Korean national and a marriage immigrant and their families (Kim 2015). Marriage migrants are respected in so far as they bear the burden of reproductive labour (domestic work, having and raising children, and caring for the elderly) and thus are incorporated into the reproduction process of the Korean nation. In the same vein, intermarriage between Koreans and co-ethnics is still relatively welcomed: the South Korean government promoted marriage between Korean Chinese and Koreans as a symbol of national reunification across borders (Freeman 2011; Kim 2016).

The intersection of class and ethnicity is another context surrounding integration of marriage migrants. Part of the marginalisation that migrant wives have experienced in South Korea results from social class. Most of them are married to working-class men facing precarious economic conditions (Choo 2016). One solution to this is the participation of marriage migrants in the labour market: various integration programmes are primarily aimed at assimilating foreign brides into Korean family life, but also at strengthening their position in the labour market. The result is that the labour market participation rate is even higher among marriage migrants than among women in the general population (Yang and Lu 2010). These women thereby shoulder the triple responsibility of being good wives/mothers, diligent workers and bona fide citizens in the future.

3.2 *Workers in the Reproduction Sphere*

Another side of the feminisation of migration in East Asia is the increasing number of migrant workers engaged in jobs in the reproduction sphere. Although policies greatly vary between receiving economies, almost all temporary female migrants are assigned precarious jobs with no access to permanent settlement. An irregular example is migrant "entertainers" in Japan, a visa category through which Filipinas began to immigrate in the early 1980s (Parreñas 2011). These women officially came to Japan as professional dancers or singers but in fact worked as hostesses. However, following the report of the US Department of State in 2004 that identified Filipina entertainers as sexually trafficked persons, the Japanese government stopped accepting such entertainers. Although they were only allowed to stay in Japan for up to 6 months, their work as hostesses brought about many marriages with Japanese men. This is the primary reason why Filipinos are the fourth largest migrant group in Japan.

It is Hong Kong that has the longest history of importing migrant workers in the reproduction sphere. Wealthy families in Hong Kong have employed Chinese domestic workers since the nineteenth century. Although there has been population flow from mainland China to Hong Kong (Wong 2007), Chinese migrants preferred to work in the manufacturing sector because it offered higher wages. This is why the Hong Kong government introduced migrant domestic workers from the Philippines (followed by Thailand and Indonesia) in the mid-1970s (Athukorala and Manning 1999; Constable 2007). The number of migrant domestic workers increased to five percent of the total population because of the increased wealth of middle-class families working in the service sector.

Though not to the extent of Hong Kong, Taiwan also introduced reproductive migrant workers engaged in domestic and care labour largely from Indonesia (Surak 2017). This is because of Taiwan's familialist care regime, which relies heavily on individual families and migrant care workers instead of governmental care services. Although the Taiwanese government stopped accepting domestic workers due to opposition by native childcare workers, the number of migrant care workers became the biggest among the four economies (Chien 2018).

Although care work has been an ethnic niche for foreign brides (both in Japan and South Korea) and Korean Chinese irregular migrants (in South Korea), the two governments had generally closed the door to domestic and care workers. Now, the South Korean government has opened the domestic and care labour market to migrant workers, but this is permitted only for co-ethnic migrants. While Japan has accepted qualified migrant care workers since 2016, the requirements for this are too strict, leading to the acceptance of only a negligible number of migrants. In terms of the formal acceptance of migrant domestic and care workers, there is no sign of policy convergence in the four economies. The variation largely comes from the difference in welfare regimes (Chien 2018). While Hong Kong and Taiwan, which provide poorer governmental welfare services (laissez-faire or familialist regimes), import live-in domestic or care workers, Japan and South Korea are more reluctant to introduce migrant domestic or care workers.

4 Nationalist Ethics of Migration and Beyond

In this chapter, I have examined how three different principles matter for migration ethics in East Asian economies. Unlike conventional views on the region, economic growth and the geopolitical conditions, not ethnonational identities, are the most important factors that underlie the status of migrants there.

Strict immigration controls in East Asia seem to be based on the ethno-nationalist ethics that justify restrictions on rights of migrants. Common-sense knowledge that migrant destinations in East Asia have persisted in the ideal of a homogeneous society has also reinforced such an understanding. However, a careful look at migration policies reveals that ethnic nationalism is not the basic principle underpinning them. Co-ethnic migrants are privileged, provided that they do not pose security threats: Chinese migrants are far from welcomed in Taiwan, while Japan opened the door for descendants of the Japanese in South America.

"Blood ties" are, therefore, of secondary importance in this region. What matters more than nationalist ethics are the logics of economically motivated developmental states and geopolitical confrontations that are legacies of colonialism and the Cold War. The former has resulted in the granting of restrictive rights to migrants so that receiving governments can maximise the economic benefits of introducing guest worker programmes. The latter arouses discrimination against migrants from hostile states even when they are co-ethnics: this exemplifies the different contexts between intra-regional and inter-regional migration (mostly from Southeast Asia).

In contrast to restrictive policies at the state level, civil societies in this region have been more open than governments. Comparative surveys show that citizens of Japan, South Korea and Taiwan are relatively positive towards acceptance and integration of migrants. Pro-migrant groups have compensated for the weakness of migrant organisations to promote the rights of migrants. In addition, there are no significant anti-immigrant parties that assert more exclusive measures to migration. This means that most challenge the current order of migration policies to defend migrant rights. Civil society in East Asia is not powerful enough to change the historical path of restrictive policies rooted in the developmental nature of states and their specific geopolitical conditions, but it is more mature than states in terms of embracing the pro-migration cause.

Literature

Athukorala, P.-C., and C. Manning. 1999. *Structural Change and International Migration in East Asia. Adjusting to Labour Scarcity*. Oxford: Oxford University Press.

Brubaker, R., and J. Kim. 2011. Transborder Membership Politics in Germany and Korea. *European Journal of Sociology* 52:21–75.

Chien, Y.-C. 2018. The Struggle for Recognition. The Politics of Migrant Care Worker Policies in Taiwan. *Critical Sociology* 44:1147–1161.

Choo, H. Y. 2016. *Decentering Citizenship. Gender, Labor, and Migrant Rights in South Korea*. Stanford: Stanford University Press.

Chung, E. A. 2014. Japan and South Korea. Immigration Control and Immigrant Incorporation. In *Controlling Immigration. A Global Perspective*, ed. J. F. Hollifield, P. L. Martin, and P. Orrenius. Stanford: Stanford University Press.

Constable, N. 2007. *Maid to Order in Hong Kong. Stories of Migrant Workers*. Ithaca: Cornell University Press.

Freeman, C. 2011. *Making and Faking Kinship. Marriage and Labor Migration between China and South Korea*. Ithaca: Cornell University Press.

Friedman, S. L. 2015. *Exceptional States. Chinese Immigrants and Taiwanese Sovereignty*. Berkeley: University of California Press.

Han, S.-J., and Y.-H. Shim. 2010. Redefining Second Modernity for East Asia. A Critical Assessment. *British Journal of Sociology* 61:455–488.

Higuchi, N. 2016. *Japan's Ultra-right*. Balwyn North: Trans Pacific Press.

Kajita, T., K. Tanno, and N. Higuchi. 2005. *Kao no Mienai Teijuka. Nikkei Brazil jin to Kokka, Shijo, Imin Network (Invisible Residents. Japanese Brazilians vis-à-vis State, Market and Migrant Network)*. Nagoya: Nagoya University Press.

Kim, B. 2006. From Exclusion to Inclusion? The Legal Treatment of 'Foreigners' in Contemporary Japan. *Immigrants & Minorities* 24:51–73.

Kim, S. 2015. Soft Talk, Hard Realities. Multiculturalism as the South Korean Government's Decoupled Response to International Migration. *Asian and Pacific Migration Journal* 24:51–78.

Kim, J. 2016. *Contested Embrace. Transborder Membership Politics in Twentieth-Century Korea*. Stanford: Stanford University Press.

Kim, H.-R., and I. Oh. 2012. Foreigners Cometh! Paths to Multiculturalism in Japan, Korea and Taiwan. *Asian and Pacific Migration Journal* 21:105–133.

Lan, P.-C. 2006. *Global Cinderellas. Migrant Domestics and Newly Rich Employers in Taiwan*. Durham: Duke University Press.

Lan, P.-C. 2019. From Reproductive Assimilation to Neoliberal Multiculturalism. Framing and Regulating Immigrant Mothers and Children in Taiwan. *Journal of Intercultural Studies* 40:318–333.

Lee, J. 2015. Disciplinary Citizenship in South Korean NGOs' Narratives of Resettlement for North Korean Refugees. *Ethnic and Racial Studies* 38:2688–2704.

Min, P. 1992. Comparison of the Korean Minorities in China and Japan. *International Migration Review* 26:4–21.
Morris-Suzuki, T. 2007. *Exodus to North Korea. Shadows from Japan's Cold War*. Lanham: Rowman & Littlefield.
Morris-Suzuki, T. 2010. *Borderline Japan. Foreigners and Frontier Controls in the Postwar Era*. Cambridge: Cambridge University Press.
Newendorp, N. D. 2008. *Uneasy Reunions. Immigration, Citizenship, and Family Life in Post-1997 Hong Kong*. Stanford: Stanford University Press.
Oguma, E. 2002. *A Genealogy of 'Japanese' Self-Images*. Melbourne: Trans Pacific Press.
Park, J.-S., and P. Y. Chang. 2005. Contention in the Construction of a Global Korean Community. The Case of the Overseas Korean Act. *Journal of Korean Studies* 10:1–27.
Parreñas, R. 2011. *Illicit Flirtations. Labor, Migration, and Sex Trafficking in Tokyo*. Stanford: Stanford University Press.
Seol, D.-H., and J. D. Skrentny. 2009a. Why Is There So Little Migrant Settlement in East Asia? *International Migration Review* 43:578–620.
Seol, D.-H., and J. D. Skrentny. 2009b. Ethnic Return Migration and Hierarchical Nationhood. Korean Chinese Foreign Workers in South Korea. *Ethnicities* 9:147–174.
Shipper, A. 2008. *Fighting for Foreigners. Immigration and Its Impact on Japanese Democracy*. Ithaca: Cornell University Press.
Skrentny, J. D., S. Chan, J. Fox, and D. Kim. 2007. Defining Nations in Asia and Europe. A Comparative Analysis of Ethnic Return Migration Policy. *International Migration Review* 41:793–825.
Surak, K. 2017. Migration Industries and the State. Guestwork Programs in East Asia. *International Migration Review* 52:487–523.
Takaya, S., ed. 2019. *Imin Seisaku Towa Nanika. Nihon no Genjitsu kara Kangaeru (What Is Migration Policy? An Inquiry into States of the Art of Japan)*. Kyōto: Jinbun Shoin.
Wang, H. 2011. Immigration Trends and Policy Changes in Taiwan. *Asian and Pacific Migration Journal* 20:169–194.
Ward, R. 2006. Japaneseness, Multiple Exile and the Japanese Citizens Abandoned in China. *Japanese Studies* 26:139–151.
Wong, S. 2007. *Exploring 'Unseen' Social Capital in Community Participation. Everyday Lives of Poor Mainland Chinese Migrants in Hong Kong*. Amsterdam: Amsterdam University Press.
Yang, W.-S., and M. C.-W. Lu, eds. 2010. *Asian Cross-Border Marriage Migration. Demographic Patterns and Social Issues*. Amsterdam: Amsterdam University Press.

Migration Ethics in South America

Gilberto M. A. Rodrigues and Jose Blanes Sala

Abstract Migrants make an important contribution to the historical formation and identity of Latin America. Over time, a culture of hospitality has been established and cultivated in the region. This fact does not exclude contradictions and ambiguities regarding migration ethics there. This chapter presents the main facts that have marked the history of migration in Latin America, particularly in South America including both voluntary and forced migration. There have been different developments in the perceptions and treatments of both types of migration because international law and the international organisations mandated to deal with the issue (UNHCR, IOM, OAS) have developed different regimes and levels of commitments and obligations towards them. Although international, regional and human rights organisations have claimed that a human rights approach is applied to migrants in general, a broad concept of forced migration, including the right to asylum, protection policies and refugee integration, has seen a peculiar development in Latin America, with a more profound impact on migration ethics. In this regard, the notion of "solidarity" was expressed and adopted in the Cartagena Regime applied for refugees in the region. This paper is divided into six sections: (1) From Independence to World War II; (2) The OAS and the 1951 Geneva Convention; (3) The Cold War (1947–1989), the Emergence of Authoritarian Regimes and their Impacts on Migration; (4) From the Post-Cold War Era to the Era of the War on Terror (1990/2001–); (5) Contemporary Issues Concerning Globalisation, Regionalism, Migration and Its Contradictions; (6) Future Research, Relevance for Current Migration Ethics.

It is normal in Latin America—including in academia—to say that its countries were forged by migrants. Also part of the region's culture, and its self-image, is the idea that its society is open to receiving migrants with an ethic of hospitality. This article aims to explain and discuss through a historical lens the assumption that

G. M. A. Rodrigues (✉) · J. B. Sala
Department of International Relations, Federal University of ABC,
São Bernardo do Campo, Brazil
e-mail: gilberto.rodrigues@ufabc.edu.br

A. Niederberger et al. (eds.), *Handbook of Migration Ethics*,
https://doi.org/10.1007/978-3-031-89877-8_4

Latin America has a tradition of hospitality towards migrants. Has hospitality been shown equally to both voluntary and forced migrants? What are the trends and recent developments that promote "a tradition of protection" of forced migrants (FMR 2017)? And what are the facts that challenge this perceived identity of regional hospitality?

The article also points out how the ethic of solidarity (mainly related to the principle of solidarity, i.e. to protect forced migrants by keeping borders open and allowing them to enter and be safe in the territory) has evolved into an ethic of hospitality (which implies the notion of equal rights and the procedures and instruments needed to guarantee and implement this equality).

In colonial times, Spaniards and Portuguese began to occupy by force a vast region already populated by originary peoples. This was the beginning of an extensive European exploitation of human beings and natural resources in Latin America. Portugal was the main actor that ran a lucrative business trafficking enslaved people from Africa, with the aim of providing labour for extractivism and then agriculture on the new continent. After the abolition of slavery, induced migration flow was fostered by Latin American countries during the nineteenth century, immediately after the independence of Spain and Portugal and the emergence and consolidation of new sovereign nations. In most cases, migration from Europe and Asia was stimulated by eugenic policies aimed at replacing the socio-economic presence and labour force of Afro-descendant and autochthonous peoples. However, collective and private memories of ancestral roots on other continents are still alive in many families and support a broad sense of openness towards an ethic of hospitality.

Moreover, the region developed an ethic of solidarity together with a tradition of protection of forced migrants in contemporary times, which had taken place since the Second World War—with some ambiguities—but took on particular characteristics in the 1970s when the region suffered military dictatorships during the Cold War. Authoritarian regimes of the East-West conflict labelled migrants as potential threats, including as terrorists, a regressive move in human rights that was reflected in national laws centred on securitarian conceptions. In this period, geopolitics transformed borders into security spaces instead of borders as territories of integration.

With the process of re-democratisation in the 1980s, new laws on migration and asylum were adopted and defined a new approach to the human rights of migrants. Alongside this, in regional organisations, such as the Organisation of American States (OAS), the role of the Inter-American System has proved very important in promoting, protecting and receiving—favouring an ethic of hospitality—migrants and forced migrants, through the decisions of its monitoring and judicial mechanisms—the Inter-American Commission on Human Rights and the Inter-American Court of Human Rights.

UNHCR has played an important role in the region and supported the emergence of new mechanisms for the protection of refugees and integration policies. In fact, the Cartagena Regime and its declarations and action plans (1984, 1994, 2004, 2014), which contain principles and governance mechanisms that link states, international organisations and civil society, has led the region to adopt a new protagonism in addressing regional and global challenges regarding forced migration.

States in the region have engaged significantly in the Global Compact for Migration and Global Compact for Refugees processes. However, new social and political developments have shown some contradictions and tensions between a democratic, human rights approach and a conservative, far-right approach, including the xenophobic rhetoric and actions of some governments and part of their societies towards migrants and refugees. Today, there are national migration laws much more in line with a human rights approach and a regional tradition of protection, which favour the integration of migrants. Those laws are recognized by the UNHCR, the IOM and experts in migration and refugee policy as important advances. However, all this legacy has been challenged at the same time by regressive movements by conservative and far-right governments and part of their societies.

A research agenda and new developments in this field include issues concerning both migrants and forced migrants, among them the role of local governments in the reception and integration of migrants, political participation and political rights of migrants.

We will then present a brief history in order to demonstrate the change in the ethical standard of migrations on the American continent over time: from the first claims of Bartolomeu de las Casas to the most recent state legislations.

Before that, it is worth noting a conceptual issue of utmost importance. The initial ethical standard, based on simple solidarity, originating in the eighteenth century and reaching its peak in the two decades following the Second World War—during the period of the last decolonisation process—, derived mainly from the idea of cooperation as an instrument for solving major world problems and had reached a point of exhaustion. It needed a new and deeper foundation for ethical conduct to achieve reasonable efficacy, especially with regard to the global problems arising from forced migration.

Some philosophical reflections played a fundamental role in shaping what we now call the migratory ethic of hospitality. In this sense, it is worth mentioning some European authors with a profound influence on Latin American thought.

The work of Hannah Arendt and her ideas developed in the United States, the country of reception, for example, shed light on what she considers a "second birth", demanding the possibility of political participation for people without nationality. For Jürgen Habermas, following similar reasoning and also using the idea of "public space", it is impossible not to include "the other" in the search for a cosmopolitanism, based on the Kantian ideal, on the basis of communication and dialogue.

Equally relevant were the contributions of authors from phenomenology and existentialism, such as Emmanuel Levinas (1980), who made a crucial contribution to the understanding of hospitality by discussing the idea of alterity, seeking to lay the epistemological foundations for the true discovery of the "other", as well as the Franco-Algerian Jacques Derrida Anne Dufourmantelle (1997), who very often used the concept of hospitality as a true itinerary of the abandonment of oneself in his reflections on the "metaphysics of presence".

All these authors, and their works, significantly influenced Latin American authors on a multidisciplinary scale, either in literature (one example is the Uruguayan Eduardo Galeano) or in academia (examples are the Brazilian Paulo

Freire (1974) and, more recently, the Argentinean Ana Paula Penchaszadeh (2014)), to name but a few, who discuss the ethic of hospitality and contribute to the understanding of an ethical attitude towards the migration phenomenon.

1 From Independence to the Second World War

In colonial times (1492–1800s), the Spanish and Portuguese kingdoms began to occupy by force the vast region of the Americas, from present-day Mexico to Argentina, which were already populated by native peoples, some with very complex and well developed societies. Since then, the native peoples have been exploited and, in many cases, exterminated (Galeano 1997). Because they had no biological defences against them, many of these originary peoples died from opportunistic diseases introduced by Europeans.

At that time, the Catholic Church played a central role in defining the ethics of the Spanish conquistadors towards the native peoples of the Americas. There were two opposing Spanish theologians in the theological debate: Bartolomeu de las Casas (1484–1566), a Dominican theologian who preached in Mexico, argued that the natives had a soul, which in practice meant that the conquistadors should respect their dignity and rights. His intellectual opponent, the Dominican theologian Ginés de Sepúlveda (1489–1573), argued that indigenous people were inferior and had no soul, and could therefore be subjugated like animals. Mutatis mutandis, this germinal debate in Latin America has not lost its relevance in relation to contemporary migration ethics, in some situations.

A small and sparsely populated nation, Portugal managed a huge traffic of enslaved people from Africa as labour for extractivism and agriculture in Brazil and other regions of Latin America. The newly independent Spanish-American states quickly abolished slavery and adopted liberal constitutional and legal parameters, influenced and under pressure from England. Brazil was the last country to do so, in 1888.

1.1 Figures on Migration in Latin America

According to J. Moya (2018), "The Europeans who arrived between the mid nineteenth century and the world depression of 1930, represent the largest population inflow in Latin America history." In fact, postcolonial European immigration represents 70% of total immigration to Latin America in the period 1492–2015 (Table 1).

European migration was greatly encouraged by the political and economic elites of the newly independent Latin American states. There were two ideas behind this movement: a racist one, which aimed to "whiten" the nations, and an economic one, which sought a well-prepared labour force to work in agriculture.

Table 1 Migrations to Latin America, 1492–2015

	Number	% of total
From Africa to Brazil	3,527,000	15
From Africa to Spanish America	1,235,000	5
From Spain to Colonial Sp. Ame.	900,000	4
From Portugal to Colonial Brazil	700,000	3
Postcolonial Asian immigration	855,000	4
Postcolonial European immigration	16,820,000	70
Total	24,037,000	100

Source: Moya (2018)

Another particular wave of immigration not included in Moya's abovementioned work occurred after the War of Secession in the US, in the 1860s. There was also a migratory flow of population from the southern states of the US—former Confederates—to Latin America (Oliveira and Rodrigues 2020). They left the US in search of new opportunities abroad or even in an attempt to escape discrimination or some kind of perceived persecution. Emperor Pedro II of Brazil offered incentives to these migrants because he was interested in developing cotton planting and the associated industry in his country; he also benefited from their knowledge in the use of weapons and war logistics—Brazil was at war with Paraguay at that time (1860–1865).

There was also an important migratory flow from the Middle East (the region occupied by the Ottoman Empire). Because of this, migrants from Lebanon and Syria, the main countries of origin of these Arab migrants to Latin America, were called "Turks" due to their passports issued by the Ottoman Empire (Amado 2012). From Asia, especially Japan, another important migratory flow took place at the beginning of the nineteenth century to Brazil and Peru.

In Brazil, with the proclamation of the Republic (1889), the federative states were authorised to manage immigration contracts with European nations in order to bring people to work on the coffee plantations.

The main country of origin for most migrants from Europe to Latin America was Italy, with well over a third of all immigrants arriving in the region during the period 1820–1960. Spain came a close second, followed by Portugal. The "Westernisation" of Latin America (reflected in the very name of the region) happened firstly as a result of the colonisation process mainly by European Mediterranean countries and secondly due to the will of Latin American governments that encouraged European immigration—in a very selective migration policy to fulfil a nation-building project (Table 2).

The above figures show that immigrants—and particularly European immigrants—have constituted a significant portion of the Latin American population over time and imprinted their influence on the cultural identity of most Latin American countries. Jürgen Habermas (2000), for example, has highlighted how Europe has benefited from the flow of these immigrants to the New World throughout modern and contemporary history.

Table 2 Origins of European immigrants to Latin America, 1820–1960

Origin	Number	% of total
Italy	6,710,000	39.9
Spain	5,380,000	32
Portugal	1,850,000	11
Germany	470,000	2.8
Eastern European Jews	420,000	2.5
Levant	410,000	2.4
France	360,000	2.1
Other Europe	1,220,000	7.3
Total	16,820,000	

Source: Moya (2018)

1.2 Pan-American Conferences and Forced Migration

The debate on the beginning of Latin American regional multilateralism in the 1800s includes historical and legendary personalities such as Simon Bolivar, "El Libertador", who advocated and worked to establish a great Latin American homeland ("La Patria Grande"), and James Monroe, President of the United States, who in 1823 declared "America for Americans", known since then as the Monroe Doctrine. Although with different purposes, both ideas contributed to developing different aspects of regionalism. However, in a pragmatic movement, the Pan-American Conferences began at the initiative of the US government, with the aim of increasing trade relations with Latin America. The first Pan-American Conference was held in Washington, DC, in 1889–1890. The conferences then evolved to include other issues such as health and harmonisation of law. During the fourth conference, in Buenos Aires in 1910, the Pan-American Union was approved, with headquarters in Washington, DC. This regional multilateral culture developed since the early twentieth century in the region was an important asset in forging a regional identity that would contribute to a political and legal culture of asylum and refugee protection.

With regard to asylum, the region pioneered the approval of a set of regional agreements in the framework of the Pan-American Conferences, such as the Treaty on International Penal Law (Montevideo, 1889), the Convention on Asylum (Havana, 1928), the Convention on Political Asylum (Montevideo, 1933) and the Treaty on Political Asylum and Refuge (Montevideo, 1939). These documents revealed a regional concern to recognise asylum as an important legal institute, which would later be recognised as a regional customary law (Arredondo 2012).

1.3 Refugees During World War II: Antisemitism vs. Humanitarianism

Between the end of the First World War and the beginning and development of the Second World War, some Latin American governments with sympathies for fascism, such as Argentina, Chile and Brazil, adopted antisemitic laws and regulations for issuing visas to Jews. Extradition permits were also issued by Supreme Courts: the case of Olga Benario, a communist activist extradited by the Brazilian government to Berlin, was one of the most controversial. She was killed in the Bernburg camp in 1942 (Morais 2007).

The double game played by President Getúlio Vargas in Brazil and President Lazaro Cardenas in Mexico was part of a similar strategy by nationalist governments to take advantage of both sides during the ideological rifts between the Axis powers and the US and the other liberal nations in World War II, before both countries decided to declare war on Germany. But even with these regulations restricting or prohibiting receiving Jews from Europe, there were movements and acts of resistance that made a difference by supporting an ethic of hospitality against these restrictions.

In this regard, the case of Brazilian Ambassador Luiz de Souza Dantas is emblematic. Already a senior diplomat and a former foreign minister, Souza Dantas was head of Brazil's mission in Paris during the Second World War. When the French capital was transferred to Vichy, Souza Dantas issued hundreds of humanitarian visas to Jews who were saved by his personal interpretation of the international context, based on an ethic of humanitarianism. He violated his country's restrictive and discriminatory immigration regulations in the name of a humanitarian principle. Souza Dantas made the first speech to the United Nations Assembly in 1946 as head of Brazil's delegation, but soon retired and received no posthumous recognition in his country until many years later (Koifman 2002). Because of his courage and humanitarian work to save Jews from the Holocaust, he was awarded the "Righteous Among the Nations" medal by Yad Vashem, the World Holocaust Remembrance Center (Yad Vashem 2003).

2 The OAS and the Geneva Convention of 1951

2.1 *The OAS: Democracy and Human Rights*

Building on the Pan-American Conferences and the existing Pan-American Union (as mentioned in Sect. 1.2), the Organization of American States (OAS) was created in 1948 with the aim of promoting cooperation among the countries of the American Hemisphere on a wide range of issues, including democracy, human rights and trade. Democracy was not part of the debate in the UN system that emerged after the war, due to the perception that it was a Western, rather than universal, concept. But in the Americas, by contrast, democracy was part of the rhetoric of most Latin American governments, but it was also part of the US strategy to avoid and limit the left and to foster anti-communist influence in the region.

The OAS Charter inaugurated a new human rights monitoring mechanism through the Inter-American Commission on Human Rights (IACHR), an independent body with a 4-year term and seven commissioners elected by the OAS General Assembly to carry out its work. The IACHR would become the main regional body for the protection and promotion of human rights. Its role in migration and forced migration principles and standards would prove crucial in the future.

2.2 *The 1951 Geneva Convention—Refugees: Geographical and Temporal Limits*

In 1943, the philosopher Hannah Arendt (2009) wrote “We Refugees”, an important essay on the plight of Jews in the Second World War, originally published in the Menorah Journal. Awareness regarding refugees reached a very high level at the end of the war when the Allies learned in detail about the concentration camps and the genocide that took place there. Soon after, the 1951 Geneva Convention relating to the Status of Refugees was one of the first conventions adopted under the aegis of the UN and linked to its core principles. However, its conception was limited by two restrictive clauses: one geographical—refugees would only be recognised if they were victims of events occurring on European territory; the other limited in time—refugees would only be recognised if related to events occurring before 1 January 1951—which included the First and Second World Wars. Only two Latin American countries signed the Convention after its approval in 1951 and before its entry into force in 1954: Brazil and Colombia. However, of the current 34 members of the OAS, 26 Latin American and Caribbean states have adopted it.

In fact, the region had already developed a legal culture of political asylum that favoured a commitment to hospitality towards forced migrants and the emerging International Refugee Law.

3 The Cold War (1947–1989), the Emergence of Authoritarian Regimes and Their Impacts on Migration

One of the main outcomes in Latin America of the Cold War (1947–1989) was the emergence of authoritarian regimes through military and civil-military coups d’état intended to combat progressive leftist political regimes and their leaders. These coups were directly supported by the US government in an attempt to prevent “new Cubas” in the region. Cuba became an important Cold War ally of the Soviet Union in the 1960s and a major influencer of left-wing parties and movements in the region. The result of American hegemony in the region was at the expense of an ethic in migration policies. As the US developed its hegemonic policy over the region, the National Security Doctrine, promoted by Washington among Latin American countries, especially among their armed forces elites, introduced the idea of “foreign as enemy”. Military dictatorships obtained political power through coups that began in Guatemala in 1954 and continued through to Chile in 1973 (Allende 1985). These regimes adopted securitisation as the main idea of migration policies and laws. As a result of the influence of US securitisation policies in Latin

America, their countries limited migration to a security issue, with little room for hospitality and solidarity at the government level.

3.1 Migration Laws Based on the National Security Doctrine

Based on the National Security Doctrine spread by Pentagon policies and embraced by Latin American anti-communist military dictatorships, restrictive and discriminatory laws on migration were passed, most of them without democratic debates in Congress. The laws were designed to strictly control migrants, their entry, as well as their freedom, facilitating arbitrary detentions, deportations and expulsions.

With the exception of Costa Rica, Panama and Venezuela, the only countries in Latin America that were democratic regimes uninterruptedly throughout the Cold War (and became important host countries for intraregional asylum in that period), other countries in the region adopted these restrictive migration laws linked to the National Security Doctrine.

3.2 The Asylum Case (1950) Before the International Court of Justice (ICJ)

In the context of the Cold War, Peruvian leftist and nationalist politician Victor Raúl Haya de la Torre, founder of the APRA (American Popular Revolutionary Alliance), was persecuted by the military after his movement lost a 1-day civil war in 1949. He then sought diplomatic asylum at the Embassy of Colombia in Lima. The Colombian government unilaterally offered asylum and asked the Peruvian government to grant him safe passage. Peru refused to accept this decision and did not agree to allow him safe passage to Colombia. The International Court of Justice (ICJ) ruled that Colombia was entitled to qualify the asylum seeker as a political persecutor (rather than a criminal, as the Peruvian authorities accused him). The decision acknowledged that Colombia was not obliged to hand him over, nor was Peru obliged to offer safe passage. The rights of both countries were, in a sense, recognised.

But the ICJ decision led the United Nations to discuss and approve in the same year (1954) two conventions on this matter, one on diplomatic asylum and one on territorial asylum. The right of asylum in these conventions was not the same right covered by the 1951 Geneva Convention under the UN refugee law regime.

3.3 The 1967 Protocol to the 1951 Geneva Convention

The 1967 Protocol eliminated the geographical and temporal clauses that had hitherto limited the scope of application of the refugee definition. It was adopted on the margins of the decolonisation process that increased the number of new countries in the United Nations and demanded the recognition of a new world not reflected in a then Eurocentric multilateralism. The Protocol received a significant number of 26 adoptions from Latin American countries, according to current UN data.

3.4 The American Convention on Human Rights: 1969

Another important development that began to strengthen a multidimensional system of migrant rights and asylum was the adoption in 1969 of the American Convention on Human Rights (ACHR), known as the Pact of San José de Costa Rica. The Convention regulates civil and political rights in the region (social, economic and cultural rights are mentioned, but regulated in another document, the Protocol of San Salvador 1988).

On asylum and migration, the ACHR, in Section 22, regulates "Freedom of Movement and Residence". Section 22.7 regulates the right to asylum in the following terms: "Everyone has the right to seek and obtain asylum in a foreign territory, in accordance with the laws of the State and international conventions, in case of persecution for political offences or related ordinary crimes." On migration, Section 22.2 states: "Everyone has the right to leave freely any country, including his own." In addition, the ACHR created the Inter-American Court of Human Rights, a judicial tribunal that would be installed 10 years later in 1979.

3.5 The Cartagena Declaration (1984): A Turning Point for Regional Refugee Law and Policy

With the continuing civil war in Colombia and the rise of civil wars in Central America (El Salvador, Guatemala and Nicaragua), the latter due to US President Ronald Reagan's counterinsurgency policy based on the Domino Theory, millions of people—internally displaced persons and refugees—had to leave their homelands and flee from mass human rights violations by armed state security forces and guerrillas in these countries. As legal bases, the 1951 Geneva Convention and the 1967 Protocol were not sufficient to apply to these situations almost never considered under the classical definition.

With a huge humanitarian crisis underway, the UNHCR Regional Office in the Americas and a group of Latin American academics and human rights activists met in Cartagena de las Indias, Colombia, in a colloquium to discuss the regional humanitarian crisis and possible legal solutions. From that non-governmental meeting, a document was issued—the 1984 Cartagena Declaration—which called on Latin American countries to consider broadening the classic refugee definition to include "widespread violation of human rights". The solution was not particularly innovative as the then Organisation of African Unity (now the African Union) had already adopted a similar decision years earlier. But the transcendent influence of the 1984 Cartagena Declaration made it a landmark document for the development of a new regime for refugee policy in the region.

Even though the 1984 Cartagena Declaration was not binding for states and did not include their direct participation in its drafting and adoption, it quickly became an alternative for the region to accommodate a new window of legal interpretation and thus be able to receive more refugees. Many Latin American countries have incorporated the notion of widespread human rights violations into their legislation as an expanded definition (in relation to the 1951 Geneva Convention definition) of refugee status in their national laws. Over 40 years, the Cartagena Regime, without having binding instruments and mechanisms (considered by some jurists as soft law), has consolidated a regional ethic of solidarity for the reception of refugees from the region and beyond. In 2004, the Declaration of Mexico that integrates the Cartagena Regime established three framework projects based on the principle of regional solidarity: Solidarity Borders, Solidarity Cities and Solidarity Settlements (Rodrigues 2021).

4 From the Post-Cold War Era to the Era of the War on Terror (1989–2001+)

The détente triggered by the end of the Cold War generated significant gain for many regions and territories on the continent. On some borders, for example, the national security approach that reproduced similar tensions of confrontational East-West borders made way for borders of integration, connecting the real and vivid life taking place on many borders. In the case of South America, a huge shift has occurred in this regard, with Argentina and Brazil leaving behind mutual mistrust and beginning to promote integration.

In fact, new initiatives of regional integration processes, such as Mercosur (Treaty of Asunción, 1991), linking Argentina, Brazil, Paraguay and Uruguay, and other already existing ones such as the Andean Pact, with Bolivia, Colombia, Ecuador, Peru and Venezuela, found momentum in a new wave of regionalism that promised not only more trade but also more democracy and more mobility of people.

4.1 New Migration Laws with a Human Rights Approach

The end of the Cold War in 1989, with the fall of the Berlin Wall, has opened up space for transcendent changes in the approach to human rights by governments and civil society in Latin America. This is happening not only in international relations but also in national policies, including those on immigration. In Latin America, new democratic governments have begun to implement constitutional and legal reforms

Table 3 Mercosur countries' migration laws—a comparison

Mercosur countries	Argentina	Brazil	Paraguay	Uruguay	Venezuela
Previous migration laws and their focus	22.439/1981 National security	6815/1980 National security	470/1974 National security	9.604/1936 National security and selective migration	19.329/1937 National security
Post-Cold War migration laws and their focus	25.871/2004 Human rights and non-discrimination	13.445/2017 Human rights and non-discrimination	978/1996 National development	18.250/2008 Human rights and non-discrimination	37.944/2004 Human rights

Source: Adapted by the authors from Rodrigues and Silva (2018)

that have placed human and fundamental rights at the centre of the legal system. The right to asylum has been incorporated into most of the newly reformed constitutions, and new refugee and migration laws have been passed.

With regard to migration laws, there have been significant changes in their focus: from a national security approach to a human rights approach.

In Table 3, the change in Mercosur's migration legislation from a security paradigm to one focused on human rights can be seen. Rodrigues and Silva (2018) discussed whether the new migration laws of Mercosur countries could be considered a new human rights paradigm , with a shift from a security approach to a human rights focus. The authors' analysis shows that there are no doubts about the democratic and human rights approach that these laws represent, since they were widely debated with civil society and received important contributions from the UN and its specialised organisations (IOM, UNHCR, ILO). But with the exception of the Uruguayan law, considered the best migration law in Latin America by the ILO (EFE 2018), the other laws have human rights approaches and important progressive developments, but also some security legacies—and new perceptions of security—that govern migration policies in the region.

4.2 Narcotraffic and New Migration Controls

International drug trafficking was declared a threat to US National Security by President Ronald Reagan's administration, which initiated the "war on drugs" in the late 1980s. Since then, this US security policy has affected migration laws and migration control in Latin America. The three largest drug producers in the region—Colombia, Peru and Bolivia—have suffered varying levels of intervention in their sovereignties by aggressive and invasive US policies to eradicate drug crops and prosecute drug traffickers.

In Colombia, the scenario was much more complex, involving not only drug capos such as the notorious Pablo Escobar, head of the Medellin Cartel, but also guerrillas such as the FARC (Revolutionary Armed Forces of Colombia) and the ELN (National Liberation Army), who at one point allied with or benefited from the illegal drug business. During the 1990s, Colombia was even called a "narco-democracy" due to the involvement of many politicians in the drug business network, which acted by financing political electoral campaigns. Because of this, the level of pressure on the Colombian authorities was so high—both from outside and inside the country, as underlined by the kidnapping of important Colombian personalities—that Congress amended the Constitution to allow the extradition of Colombian citizens to the US to be legally prosecuted and incarcerated in the country. This was the first time that a Latin American country allowed its nationals to be extradited to another country—a cornerstone of regional sovereignty (Garcia Marquez 1998).

As a result of this complex situation, Colombian citizens, as well as Peruvian and Bolivian citizens, were labelled and stereotyped as potential drug traffickers or couriers. This situation brought severe restrictions on the nationals of these countries in migration controls, although these controls were restricted to border controls and not expressed in visa obligations in Latin America, unlike in other European and North American countries.

4.3 The War on Terrorism and Regressive Migration Policies

After the terrorist attacks of 9/11, and due to the reaction of the US government, a new regressive wave of migration controls threatened the region, although Latin America was not in the middle of the conflict. The impact on new transport regulations affected the whole region in its flux to the US and Canada, but the country most affected was Mexico, because of its borders with the US. But an indirect outcome was a new securitisation process regarding the flow of various goods and of people. Nevertheless, most Latin American countries were against the invasion of Iraq and the war on terrorism and favoured a less discriminatory approach to migrants for this reason in comparison to other Western countries.

5 Contemporary Issues Concerning Globalisation, Regionalism, Migration and Their Contradictions

5.1 Regional Human Rights Norms and Standards and the Role of Inter-American Institutions

The role of the Inter-American System for the protection of migrants and refugees has become a cornerstone for human rights standards in the region. The Inter-American Commission on Human Rights (IACHR) issued a document (OAS 2015) containing a set of norms and standards that constitute an important asset for migration ethics.

The document lists a number of norms and standards: (1) general obligations towards persons in the context of migration that include (a) obligations to respect and ensure the exercise of human rights, (b) the duty to adopt domestic legal provisions, (c) the right to equal protection and non-discrimination; (2) the prohibition of slavery, servitude and trafficking in human beings; (3) freedom of movement and residence; (4) the right to a fair trial in deportation or extradition proceedings; (5) the right to family life in immigration proceedings; (6) the right not to be subjected to cruel, inhuman or degrading treatment or punishment as a result of deportation; (7) the right to personal liberty and procedural guarantees in immigration detention; (8) the right to seek and receive asylum; (9) the principle of non-refoulement; (10) the right to nationality; (11) the right to property.

5.2 Integration Processes and Migration Issues

After the Cold War, there was an emergence of new integration arrangements that evolved throughout the region. Building on existing organisations in South and Central America—such as Mercosur, the Andean Pact, SICA and CARICOM, the development of integration processes reached a wide range of issues. Citizens of those countries have in many cases benefited from acts that have simplified and/or abolished bureaucracy for travel or residence. But migration policies remain a sovereign policy within the framework of national laws.

In Mercosur, there was an attempt to create a common citizenship, based on the aforementioned mobility facilities. However, as Blanes Sala and Rodrigues (2015) have pointed out, Mercosur citizenship cannot prevail over national rule without a supranational legal system and regional bodies that can manage and decide on conflicts involving national authorities.

5.3 Cartagena Regime and Its Agenda

With the support of the UNHCR Regional Office, the Cartagena Regime has become an intergovernmental process between Latin American countries, including civil society and humanitarian NGOs. A decennial evaluation process has been implemented since 1994, when Cartagena + 10 took place in San José, Costa Rica; in 2004, there was Cartagena + 20 in Mexico, and the last one, Cartagena + 30, took place in Brasilia, Brazil in 2014. At each of these assessment conferences, new declarations were issued, along with new 10-year action plans. In this sense, the Cartagena Regime not only set new parameters for regional refugee law practices but also created a regional policy to be implemented in Latin America with the political commitment of all actors involved. The Cartagena Regime has evolved on the basis of the principle of solidarity which, in turn, has propelled refugee protection and migration ethics in the direction of hospitality.

New policy instruments were designed to deal with the new demands of forced migration. The Mexico Declaration and Plan of Action of 2004 underlined the principle of regional solidarity within the region and with other regions and continents. Three main areas were chosen to implement the solidarity plans: Solidarity Borders; Solidarity Cities; and Solidarity Resettlement.

The Brazil Declaration and the 2014 Action Plan stressed the need to support refugees and displaced persons who are victims of organised crime, with special attention to the Northern Triangle in Central America (Honduras, Guatemala and El Salvador), which is affected by gangs (*marras*) that threaten the human rights of hundreds of thousands of people in these countries.

5.4 Peace Accords in Colombia: 2016

The armed conflict in Colombia has been one of the most complex and difficult to resolve in the world. It has produced millions of Colombian refugees and internally displaced persons for over 60 years. In this regard, the Colombian Peace Accords negotiated and signed between the Colombian government of President Juan Manuel Santos and the FARC (Revolutionary Armed Forces of Colombia) in 2016 were concluded as a first step towards ending a significant part of the conflict (other guerrilla and paramilitary groups remain active). President Santos was awarded the Nobel Peace Prize in 2017. However, due to powerful internal critics of the agreements and political changes in the country, since 2018 the peace agreements have become vulnerable, and violence has increased again.

5.5 Global Compacts: Migration and Refugees

During the diplomatic process for the Global Compact on Refugees, Latin American countries made a significant contribution to the final document through the regional caucus. In the document entitled "The 100 Points of Brasilia", approved in February 2018, the region offered a very comprehensive set of recommendations that mirrors its own experience. In the preamble, the document says: "We highlight the importance to ensure the unconditional respect, protection and promotion of the human rights of asylum-seekers, refugees, displaced persons, and stateless people, with particular attention to groups in a situation of vulnerability" (UNHCR 2018). It is important to underline the word "unconditional", which is fully in line with migration ethics regarding refugees.

5.6 *Migration Awareness vs. Regressive Migration Policies in the Wake of New Right-Wing and Far-Right Governments*

With new right-wing and far-right governments in Latin America, migration policies have been progressively affected by new restrictive regulations. Despite the fact that most countries in the region have shifted the focus of their laws from a security approach to a human rights one, which means the recognition of a broad set of migrant rights and protection against discrimination, law regulations made by executive decrees have changed migration procedures in Argentina, Chile and Brazil, among others, reintroducing new security interpretations that call for a kind of permanent state of exception.

The rhetoric against immigrants is part of an ideological concept in which immigrants are a burden and potentially a threat to security. In Brazil, President Bolsonaro (2019–2022) labelled social movements and protesters as terrorists. In many cases, immigrants are also seen as criminals.

5.7 *Venezuela Crisis as a Turning Point in Intraregional Forced Migration*

The crisis in Venezuela has significantly changed the profile of forced migration in Latin America (Rodrigues 2019). Even the conflict in Colombia, which has lasted for more than 60 years, did not have such a level of forced migration flux in a short period of time like the Venezuelan one has produced since 2015. In South America, the main countries affected are Colombia and Brazil due to their borders with Venezuela. For many Venezuelans, both countries are transit territories which facilitate travel to others as second host countries, such as Peru, Ecuador, Chile and Argentina. The pressure on Brazil and Colombia has been incredibly high in comparison to other neighbouring countries. Nevertheless, they did not impose legal restrictions on receiving Venezuelans, instead adhering to the principle of open borders in humanitarian situations. Moreover, Brazil decided in late 2019 to recognise thousands of Venezuelans applying for the broad definition of widespread human rights violations through the prima facie recognition, a kind of fast-track, collective procedure. By contrast, Chile, Ecuador and Peru began to impose legal restrictions for entry and residence.

6 Future Research, Importance for Current Migration Ethics

With its recognised tradition in protection, Latin America is historically a region of hospitality for migrants and refugees. The principle of solidarity, reflected in the opening of borders, the recognition of equal rights for nationals and immigrants by the new migration laws and the specific programmes existing under the Cartagena Regime are important evidence of this. However, a different emphasis has been placed on the protection and integration of migrants and refugees, with the latter receiving much more legal support through legislation and policies.

There are many important questions for a research agenda on migration ethics in Latin America. Despite the fact that the region has a long tradition in the protection and integration of migrants and refugees, which includes the right to asylum, some of the current problems are due to regressive policies that challenge the migration and refugee laws already in place. Many of these problems relate to executive regulations of the laws, which in many cases are made by national or federal authorities and are much narrower than the laws themselves.

Another important issue concerning the protection of forced migrants is the humanitarian visa, a kind of complementary protection for those who could not be recognised as refugees (such as the Haitians in Brazil); or a means of applying a fast-track procedure, as Brazil used for Syrians (FMR 2017); Afghans and Ukrainians.

Among the present and future research topics in the field of Migration Ethics in South America, or with impact in this region, it is worth mentioning:

- The threat of criminalization of migrants by state authorities, despite existing human rights-focused laws.
- Discrimination and xenophobia against migrants enhanced by the growth of ultra-right-wing groups and movements.
- The need for protection and durable solutions, including reception and local integration for displaced and environmental/climatic migrants.
- Local protection and integration solutions for Venezuelan migrants and refugees in light of the serious humanitarian, political and economic crisis in Venezuela and its serious impact on the region.
- The role of regional and local governments in migration policy is another important research topic, due to the centrality that states and provinces, and municipalities have been assuming, either by international policies (Global Compacts, SDGs, etc.), or by their realities to directly receive migrants in their territory.

Therefore, it is worth concluding in the present analysis of migration ethics observed as currently practised in Latin America, despite all the obstacles described above, that the prevailing ethic of hospitality goes far beyond solidarity and simple protection. The mentioned ethics aims to overcome even the idea of inclusion. The word that best describes this effort would be "reception" (*acogida* in Spanish, *acolhimento* in Portuguese). While solidarity can only be achieved through a set of efforts aimed at preserving one's own interests, reception—inherent in hospitality—presupposes true generosity, where simple interest is not enough; on the contrary, it is necessary to renounce one's own interests in search of the dignity of the other, always accepting difference and diversity.

Literature

Allende, I. 1985. *The House of Spirits*. New York: Alfred A. Knopf.
Amado, J. 2012. *La découverte de l'Amérique par les Turcs*. Paris: Stock.
Arendt, H. 2009. We, refugees. In *The Jewish Writings*, ed. H. Arendt. New York: Schocken Books.
Arredondo, R., ed. 2012. *Manual de Derecho Internacional Público*. Buenos Aires: La Ley.
Blanes Sala, J. B., and G. M. A. Rodrigues. 2015. O conceito de cidadania no Mercosul. In *Paradiplomacia em movimento. perspectivas em homenagem aos 20 anos da atuação da Rede Mercocidades*, ed. J. F. Gomes and R. S. Corradi. Porto Alegre: Editora Uniritter.
Derrida, J., and A. Dufourmantelle. 1997. *L'hospitalité*. Paris: Calmann-Lévy.
EFE. 2018. *La OIT dice que la ley uruguaya es la mejor de Latinoamerica*. Montevideo, July 26, 2018. https://www.efe.com/efe/cono-sur/sociedad/la-oit-dice-que-ley-migratoria-uruguaya-es-mejor-de-latinoamerica/50000760-3702227.
Forced Migration Review (FMR). 2017. *Latin America and the Caribbean. Building on a Tradition of Protection*, 56. https://www.fmreview.org/latinamerica-caribbean.
Freire, P. 1974. *Pedagogia do Oprimido*. São Paulo.
Galeano, E. 1997. *Open Veins of Latin America. Five Centuries of the Pillage of a Continent*. New York: Monthly Review Press.
Garcia Marquez, G. 1998. *News of a Kidnaping*. London: Penguin.
Habermas, J. 2000. *The Inclusion of the Other. Studies in Political Theory*. Cambridge, MA: MIT.
Koifman, F. 2002. *Quixote nas trevas. O Embaixador Souza Dantas e os refugiados do nazismo*. Rio de Janeiro: Editora Record.
Levinas, E. 1980. *Totalité et Infini*. The Hague: Martinus Nijhoff.
Morais, F. 2007. *Olga. Revolutionary and Martyr*. New York: Grove Weidenfeld.
Moya, J. 2018. Migration and the Historical Formation of Latin America in a Global Perspective. *Sociologias* 49. http://www.scielo.br/scielo.php?script=sci_arttext&pid=S1517-45222018000300024.
OAS. Inter-American Commission on Human Rights. 2015. *Human Mobility Inter-American Standards. Human Rights of Migrants, Refugees, Stateless Persons, Victims of Human Trafficking and Internally Displaced Persons: Norms and Standards of the Inter-American Human Rights System*. OEA/Ser.L/V/IL, Doc.46/15. Washington, DC. http://www.oas.org/en/iachr/reports/pdfs/HumanMobility.pdf.
Oliveira, A. C., and G. M. A. Rodrigues. 2020. Proteção internacional de migrantes forçados e a agenda brasileira. histórico, temas e atores. In *Migrações Internacionais. Experiências e desafios para a proteção e a promoção de direitos humanos no Brasil*, ed. G. Redin, 144–171. Santa Maria: Editora UFSM. https://editoraufsm.com.br/migracoes-internacionais-529.html.
Penchaszadeh, A. P. 2014. *Política y hospitalidade. disquisiciones urgentes sobre la figura del extranjero*. Buenos Aires: EUDEBA.
Rodrigues, G. M. A. 2019. *Refugiados—O grande desafio humanitário*. São Paulo: Moderna.
Rodrigues, G. M. A. 2021. South America and the Cartagena Regime. A Comprehensive Approach to Forced Migration Responses. In *The EU Pact on Migration and Asylum in light of the United Nations Global Compact on Refugees. International Experiences on Containment and Mobility and their Impacts on Trust and Rights*, ed. S. Carrera and A. Geddes, 157–169. S. Domenico di Fiesoli: EUI.
Rodrigues, G. M. A., and L. F. Silva. 2018. Legislação Migratória nos países do Mercosul: um novo paradigma com enfoque em Direitos Humanos? *Refúgio, Migrações e Cidadania—Caderno de Debates* 13:19–35. https://www.migrante.org.br/publicacoes/caderno-de-debates/caderno-de-debates-13-refugio-migracoes-e-cidadania/.
UNHCR. 2018. *The 100 Points of Brasilia Inputs from Latin America and the Caribbean to the Global Compact on Refugees*. Global Compact for Refugees Platform. http://www.globalcrrf.org/crrf_document/the-100-points-of-brasilia-inputs-from-latin-america-and-the-caribbean-to-the-global-compact-on-refugees-20-february-2018/.
Yad Vashem. 2003. Luis Martins de Souza Dantas. https://www.yadvashem.org/righteous/stories/dantas.html.

Nationalism and Migration Ethics

Uchenna Okeja

Abstract Imaginations of nationalism constitute an important aspect of migration ethics. Although nationalism is frequently invoked in the discourse on the ethics of immigration through reference to concepts like national identity, it is unclear whether national identity and immigration are incompatible. In this chapter, I discuss the relationship between nationalism and immigration. Specifically, I analyse two approaches to migration where nationalism plays a critical role. The first approach is the framing of immigration ethics in contemporary political theory and the second is populist sensational discourses. The critical role of nationalism manifests in the first approach through methodological nationalism. In the second case, nationalism plays a crucial role due to the influence it exerts on public opinion and policy. I show that nationalism inspired arguments about migration are unsatisfactory because they focus on a very narrow aspect of migration. The desideratum for further research I propose is therefore that theorists of migration should be more critical about the concept of nationalism. They should theorise migration from a standpoint that considers nations and borders as factors that have histories that are interlinked with broader social and political developments.

Migration debates often entail deep disagreements about national belonging. This happens because the notion of an immigrant is intertwined with an imagination of an outsider living among insiders that supposedly share a national identity. The relationship of immigrants to their host-societies is defined therefore by the degree they can claim to belong in the sense of integration within specific territories. To understand the influence of nationalism on migration discourses, a viable starting point would be to consider how the complexity of nationalism lends support to different arguments about migration ethics. I consider in this chapter how nationalism is mobilised in framing arguments about the ethics of immigration and in populist sensational discourses that aim to shape public opinion and policies about migration.

U. Okeja (✉)
Nelson Mandela University, Gqeberha, South Africa
e-mail: uchenna.okeja@mandela.ac.za

A. Niederberger et al. (eds.), *Handbook of Migration Ethics*,
https://doi.org/10.1007/978-3-031-89877-8_5

To articulate a perspective on nationalism and the ethics of immigration, I will first explore three ideas of nationalism and stipulate the ways they are mobilised in framing ethical positions about immigration. I will then discuss two accounts of the ethics of immigration to show how they assume methodological nationalism as foundation. To explain the role nationalism plays in populist sensationalism, I discuss recent contestations about public opinion and immigration policy. Although I will not aim to argue for a compelling notion of nationalism that ought to inform arguments about migration, I will attempt to show that nationalism inspired arguments offer unsatisfactory grounds to reject or embrace immigration.

The discussion of nationalism in this chapter is primarily descriptive. As such, I focus on explicating how nationalism influences our thinking about migration. This means that I do not try to conceive an original interpretation of nationalism or argue for an understanding of the concept that should guide our thinking about migration in contemporary political theorising. Lastly, it is important to note that although I refer to experiences of people in different parts of the world, I do not attempt to construct a history of nationalism that should apply globally or in every tradition of political thought.

1 Three Ideas of Nationalism

Nationalism is defined in many ways and the definitions assume specific imaginations of the nation. A renowned theorist of nationalism, Benedict Anderson, articulated three ideas at the core of how we imagine nations. The first is that nations are thought to be limited. The reason is that "even the largest of them, encompassing perhaps a billion living human beings, has finite, if elastic, boundaries, beyond which lie other nations" (Anderson 2006: 7). Going by this understanding, nations are bounded, and essentially linked to a territory. Beyond having this feature, nations are also non proselytising. This explains why there are no nationalists who dream to recruit into their nation "all the members of the human race" (Anderson 2006: 7).

The second imagination is that a nation is sovereign. According to Anderson, the idea of a nation "was born in an age in which Enlightenment and Revolution were destroying the legitimacy of the divine-ordained, hierarchical dynastic realm" (Anderson 2006: 7). As an idea of political thinking, the concept of the nation emerged at a point in "history when even the most devout adherents of any universal religion were inescapably confronted with the living pluralism of such religions" (Anderson 2006: 7). The concept of the nation, therefore, expresses a desire for a freedom secured by the "sovereign state [as its] gage and emblem" (Anderson 2006: 7). Conceived this way, freedom acquires a meaning that is not subject to any religious creed or faith.

The third core idea associated with the concept of a nation is that it is a community. Thus, "regardless of the actual inequality and exploitation that may prevail in each, the nation is always conceived as a deep, horizontal comradeship" (Anderson 2006: 7). Anderson argues that this imagination explains why millions of people are "willing to die for such limited imaginings". To belong to a nation is to be a part of a community on this view. From this, so many inferences are drawn, including that one who is a national of a given territory has duties of patriotism. And being patriotic, means to be so devoted to one's territorial belongingness as to be willing to defend it with one's life.

Although these ideas about nations have informed migration policies in different epochs of human history, they feature most prominently in debates about immigration and integration, which could be seen as the two major issues in the ethics of migration. We can therefore understand the claims of theorists of the ethics of immigration by exploring how they rely on imaginations of nationalism to construct their arguments. We can do this by considering how ideas about the nation shape the perceptions of theorists of migration regarding who is an immigrant, the action immigration demands from the state and how to justify it. To answer these questions, it is useful to look more closely at the three ideas about the nation recounted above. We need to understand what each idea amounts to when considered at from a perspective that is decidedly political. Said in another way, the question to answer is what it means for political thinking to say that the concept of a nation is at core related to limited territory, sovereignty, and community. This will reveal why the nation as a concept has been mobilised in different ways in thinking about migration.

It would not be an exaggeration to say that more than any other concept, nationalism played a crucial role in modern political history in shaping the political destinies of millions of people due to the complexities traceable to the interaction of European nations with people in other parts of the world. This assertion holds to the extent that the actions of the imperialist European nations and successor postcolonial states can be said to have exerted enormous influence on the political destinies of majority of the people within and outside Europe. Against this background, it is plausible to argue that nationalism is an essential part of the political imagination of migration in modern political thought. As Piet Emmer noted, "one of the most important and visible remnants of the expansion of Europe is the change in the ethnic composition of the world" (Emmer 1992: 2). The change Emmer referred to would not have come about without the large-scale movement of people, especially after 1880 which is generally considered the point in history when there was a reduction in the hinderances to transportation of both goods and people (Emmer 1992: 3).

Christian Joppke argues that "under mercantilism, labour scarcity had motivated absolutist states to prohibit exit, limiting migration to the forced incorporation of slaves or indented labour. With exit prohibited, entry (and thus, immigration) could not even be an issue" (Joppke 1999: 2). This orientation ultimately gave way to a wave of liberalism "in which the removal of barriers to exit coincided with non-existing restrictions to entry." This wave of "liberalism was the classic hour of immigration, with unprecedented population movements within and outside Europe, the outside movement giving rise to the transoceanic new settler nations of Canada, Australia, and the United States" (Joppke 1999: 2). After 1880, Joppke notes, Western nations moved back to immigration restriction, establishing as a result "today's fundamental contradiction of how states deal with international migration: a generalized commitment to freedom of exit is combined with severe restriction on entry" (Joppke 1999: 2). From this, it is plausible to infer that nation-building efforts account for the prevailing attitude of states to immigration. Although efforts at nation-building unfolded in different ways around the world, new and old states approached this task through mobilisation of nationalist sentiments.

The concepts of nation and nationalism are certainly related, but they are not equivalents. The nation as an idea, Calhoun argues, is an old term. In fact, "before the modern era, it meant only people linked by place of birth and culture. It signalled nothing about the relationship of such identity to larger or smaller groupings; neither did it carry clear political connotations" (Calhoun 1997: 9). Calhoun underscores that his point should not be interpreted as a claim that the best way to define nationalism is by tracing it to a proto-origin to which we can trace the spread of the idea. In his view, there is no such proto-origin of the idea of nationalism. To understand nationalism, whether in relation to the ethics of immigration or any other phenomenon, Calhoun argues that we should approach it by looking at the discursive formation of the concept (Calhoun 1997: 7). In other words, his suggestion is that we should attempt to explain the concept of nationalism by untangling the families of rhetorics that shape the constitution of its meaning. This means to distil an explanation of the concept from analysis of the ways it is used to explain different experiences.

On this view, attempts to explain the significance of nationalism in debates about the ethics of immigration do not necessarily require a construction of a theory of nationalism because "nationalism is a rhetoric for speaking about too many different things" (Calhoun 1997: 8). Even without a unified theory of nationalism, it is possible to understand the role nationalism plays in theorizing the ethics of immigration by analysing how theorists of immigration construct their arguments. For many theorists of immigration, nationalism plays the kind of role that the concept of homo economicus played in classical economics. In other words, it is axiomatic to assume that nationalism is necessary for theorizing immigration. With this, we return to the point I made earlier about what it means for political thinking about immigration to say that the concept of a nation is at core related to limited territory, sovereignty, and community.

Considering the discussion of the idea of the nation and its relationship to the phenomenon of immigration, it would be reasonable to say the following: for political thinking, the imagination that a nation at core implies limited territory, sovereignty and community means that its mobilisation in framing the ethics of immigration reveals an orientation to thinking. This orientation takes the nation as a fact, imagines immigration as a challenge introduced from the outside, and assumes that states have a duty to take action on immigration because migration is a voluntary action of individuals. Thinking about migration through the lens of the nation as a sovereign political community that exists within bounded territories therefore produces a discourse in which prominence is given to certain issues at the cost of excluding others. By considering two approaches to theorizing the ethics of immigration, I show below the relationship between nationalism and the ethics of immigration. The two approaches I consider are methodological nationalism and populist sensationalism.

2 Methodological Nationalism and Two Perspectives on the Ethics of Immigration

There are two dominant ways theorists argue about the ethics of immigration. The first approach refers to a family of discourses that is conventionally regarded as arguments for open borders. The second approach relates to the opposite set of arguments conceived to justify restrictions to the movement of people. These two approaches have established themselves as the dominant ways we think about the ethics of immigration for several reasons. A common assumption of the two approaches is traceable to the imagination that there is such a thing as the right of a political community to decide on issues related to membership. Michael Walzer and Joseph Carens are generally considered the leading theorists of the ethics of immigration because of the divergent views they hold about the rights of a political community to regulate membership. Whereas Walzer argues that states have a right to impose restrictions on immigration, Carens makes a case for a limited right to exclude and a duty to have a much more open border.

For Walzer, "the primary good that we distribute to one another is membership in some human community. And what we do with regard to membership structures all other distributive choices: it determines with whom we make those choices, from whom we require obedience and collect taxes, to whom we allocate goods and services" (Walzer 1983: 31). If immigration is taken here as one of the ways that membership in human community is constituted or reconstituted, then, what Walzer is suggesting is that immigration poses a fundamental challenge. We see the extent this is true in the sustained arguments about the action that states are permitted or prohibited from taking in relation to immigration.

As indicated already, varieties of open-borders argument constitute one of the two broad perspectives on the ethics of immigration. A careful consideration of the background of the debate about open borders should give us a picture of the issues at stake. Lea Ypi provides an apt description of what is at stake in these debates by noting that "justice in immigration indicates when restrictions on incoming freedom of movement are unjustified and provides a principled way of assessing the distribution of benefits and responsibilities between migrants and citizens of host societies". Relatedly, justice in emigration pertains to "when restrictions on outgoing freedom of movement are unjustified and provides a principled way of assessing the distribution of benefits and responsibilities between migrants and citizens of sending societies" (Ypi 2008: 391). Against the background of this description, the open borders debate is a discourse on whether it is justified to impose restrictions on freedom of movement. Beyond the permissibility of restrictions to freedom of movement, there is also the question of the distribution of benefits and responsibilities among all the parties affected by immigration.

In *Spheres of Justice* (1983), Walzer defends the right of states to close their borders for reasons related to self-determination and communal independence. He considers "the right to choose an admissions policy" a basic freedom of states which is "not merely a matter of acting in the world, exercising sovereignty, and pursuing national interests" (Walzer 1983: 61). Choosing an admissions policy, Walzer argues, reflects "the deepest meaning of self-determination" (Walzer 1983: 62). The states that are unable to decide who to admit or exclude "could not be communities of character, historically stable, ongoing associations of men and women with some special commitment to one another and some special sense of their common life" (Walzer 1983: 62). For Walzer, although the right to self-determination is fundamental, it is limited by two factors, namely, the internal decision of citizens of a state and the external principle of mutual aid (Walzer 1983: 62).

Walzer's ideas provide the foundation for the development of many recent works on restricted immigration as most of them propose arguments grounded on such ideas as freedom of association and the duty to protect the economy, culture, religion, and political institutions of the community. Theorists of ethics of immigration articulate their arguments in different ways, however, for the proponents of restricted immigration, a basic assumption is that the state is a sovereign entity that has limited territory. The members of the state are also assumed to engage in mutual cooperation and evince a form of social solidarity that makes them a community.

On the side of the proponents of open borders, Joseph Carens stands out for many reasons. Carens argues that borders should be either completely open or much more open than they currently are. This conclusion is premised on several arguments Carens developed to challenge the prevailing policies and attitudes to immigration in liberal democratic societies. The background of the argument for open borders is the imagination that "citizenship in Western liberal democracies is the modern equivalent of feudal privilege—an inherited status that greatly enhances one's life chances" (Carens 1987: 252). This way of thinking is often taken to mean that the nation is a kind of property that is collectively owned by citizens. Against this view, Carens argues that "if the notion of collective ownership is used to justify keeping aliens out, it opens the possibility of using the same notion to justify redistributing income or whatever else the majority decides" (Carens 1987: 254). This argument can be regarded as the inconsistency argument because the main issue is that immigration restriction is incompatible with some of the fundamental principles of liberal democracy.

To make a positive case for open borders, Carens adapted the original position proposed by John Rawls. In doing this, he assumeed "a global view of the original position" to advance the claim that if people were to choose behind a veil of ignorance, not knowing the citizenship they would have, they "would insist that the right to migrate be included in the system of basic liberties". They would do this because of the same reasons they "would insist that the right to religious freedom be included: it might prove essential to one's plan of life" (Carens 1987: 258). Migration on this view will be unrestricted. The only exception would be that restriction on migration is permissible for the sake of protecting liberty, in the sense preventing a threat to public order. Looking beyond the argument drawn from ideal theory, Carens proposed that there is little ground for a strict restriction of migration. If at all, the reasonable restrictions on migration that could be derived from so-called realistic concerns about immigration can only "imply a much less restrictive policy than the one currently in force which is shaped by so many other considerations besides the need to maintain public order" (Carens 1987: 260). The restrictions on migration that can be justified in a nonideal theory are thus "severely limited" (Carens 1987: 262).

For Carens, utilitarian principles support open borders in liberal democratic societies. This is because "under current conditions when so many millions of poor and oppressed people feel they have so much to gain from migration to the advanced industrial states, it seems hard to believe that a utilitarian calculus which took the interests of aliens seriously would justify significantly greater limits on immigration than the one entailed by the public order restriction implied by the Rawlsian approach" (Carens 1987: 264). The main point Carens defends is that restrictions to freedom of movement are largely unjustified. Besides Carens, there are many recent arguments that attempt to show why borders should be open. These arguments rely on an imagination of freedom of movement as a fundamental human right. A common feature is that the different arguments for open borders assume that the state is a sovereign entity that, even though is limited territorially, should consider all persons as potential members of the community it embodies.

To understand the role nationalism plays in debates about the ethics of immigration, it is pertinent to closely consider the importance of thinking about migration from an ethical perspective. In other words, why is it important to theorise immigration as a moral problem? There are at least three reasons that explain the interest of theorists in the ethics of migration. The first is the idea that thinking about ethical questions raised by migration is necessary if we want to develop just policies. The second is the understanding that fairness towards strangers requires an understanding of the ethical problems posed by migration. The third reason is intellectual curiosity which is necessary for us to establish whether our cherished political ideology represents a coherent and consistent framework that can help us to find answers to new political challenges.

Although the first and second reasons that explain the importance of theorising immigration from an ethical standpoint are crucial, it is the third reason that accounts for the outsized role that nationalism plays in the debates about the ethics of migration. The approach many scholars take in formulating ethical theories of immigration is influenced by nationalism because they are invested in showing whether or to what extent a political ideology provides a consistent answer to the challenge of immigration. Given that the imagination of the state as the basic form of political organisation is cherished, in the sense of constituting a widely shared ideology in contemporary political theory, many scholars theorise the ethics of migration from a perspective that puts the state at the center. Methodological nationalism is for this reason a problem for the dominant perspectives on the ethics of migration.

Alex Sager highlighted the challenge of methodological nationalism in his observation that "despite their theoretical differences, political theorists of migration are largely committed to a social ontology that presupposes the nation state as the basic agent and unit of analysis. They assume an international system of sovereign states with limited consideration of other structures and actors" (Sager 2016: 46). He argues that this approach to thinking about migration "affects the political theory of migration by framing distributive justice in an international framework that treats each state as an autonomous mona" (Sager 2016: 49). Furthermore, the approach misleads political theorists by conditioning them to see "international migration as an anomaly" (Sager 2016: 50). What is salient in Sager's contention about methodological nationalism is his argument that political theories of immigration, even if not nationalist in the derogatory sense, are fundamentally beholden to nationalism at the level of conceptual framework.

3 Nationalism, Public Opinion and Migration Policies

The second approach to migration where nationalism plays a role is sensational discourses. In most cases, the aim of these discourses is to influence public opinion and shape public policy. To explain the role of nationalism in this approach to shaping people's orientation to immigration, a good starting point would be to consider a few examples. Let us begin with experiences in some African countries. On 19 November 1969, the Ministry of Information in Ghana informed the public that it had come "to the notice of the Government that several aliens, both Africans and non-Africans in Ghana, do not possess the requisite residence permits in conformity with the laws of Ghana". The Statement noted further that "there are others, too, who are engaging in business of all kinds contrary to the term of their visiting permits". Responding to the 'problem' posed by aliens, the government issued a directive "that all aliens in the first category, that is those without residence permits, should leave Ghana within fourteen days, that is no later than December 2, 1969" (Sudarkasa 1979: 141).

This Directive issued by the Prime Minister of Ghana, Kofi Abrefa Busia, was surely a strong reaction to immigration. It was however not unique as many African countries reacted to the question of immigration through implementation of similar policies. In postcolonial Africa, "the expulsion of non-European resident strangers" was a problem after "the dismantling of colonial empires in Africa and Asia" (Challenor 1979: 67). Apart from Ghana, there were other cases of expulsion of African immigrants. For instance, Ivory Coast expelled immigrants from Dahomey, present day Benin Republic, in 1958. Congo Brazzaville expelled African immigrants in 1962, and Niger implemented similar policies in 1964. Sierra Leone and Guinea sent away Ghanaian immigrants from their territory in the late 1960s, just as Nigerians were driven out of Cameroon in 1967. While the Directive issued by Kofi Abrefa Busia's government in 1969 led to the expulsion of over a million West Africans, countries like Kenya, Malawi, and Uganda compelled Asians to leave their territory around the same period.

Certainly, the expulsion of immigrants was not limited to Africa. The point I want to highlight is the role that nationalism played in these expulsions of immigrants. The deployment of populist rhetoric was a way to guarantee support for the implementation of restrictive immigration policies in most of the countries mentioned above. Populist rhetoric about immigration appeals fundamentally to sentiments embedded in conceptions of national identity. In these rhetorics, nationalist sentiments are usually expressed through reference to "national character", "national culture", and "our way of life". Notwithstanding, national identity and the sentiments it embodies is always undergoing important conceptual transformation due to external and internal factors. Due to constant transformation of the imagination of nationalism, new ideas emerge regarding what counts as migration or what constitutes acceptable forms of immigration.

This explains why at some point in history, nationalist sentiments were mobilized to frame the moral challenge of migration as the duty of states to prevent emigration, and at other times, as the right of states to control the "deluge" of immigrants that are members of a specific race, religion, and language community. In South Africa, for instance, the changing understanding of national identity accounts in part for the reason xenophobia is increasingly linked to immigration. Overall, although nationalism is a localised phenomenon, the sentiment it embodies can have a decisive impact on public opinion and immigration policies. This point is important for the discourse on the ethics of migration because it underscores the relevance of taking seriously the intricate connections between public opinion and individual attitudes towards immigration.

The infusion of sensationalist rhetoric about nationalism into public opinion with the goal of influencing individual moral attitudes towards immigration is evident in Donald Trump's first presidency and his successful campaign for re-election in 2024. Trump's sensationalist rhetoric about immigrants often included the ascription of labels like "rapists", "thugs", "animals", and people from "shithole countries". These labels were mobilised to translate the nationalist imagination of the desirable immigrant. In other words, "the [first] Trump Administration's immigration policies reflect[ed] its white nationalistic rhetoric" (Srikantiah and Sinnar 2018–2019: 200). The perception that white Christian nationalism was an important part of immigration policy under Donald Trump's first presidency arises from his use of disparaging language to describe non-white immigrants and the bans imposed on the admission of such immigrants in the Executive Orders he issued. Besides Donald Trump, there are many other examples of political leaders that appeal to nationalist sentiments in the way they approach immigration.

One of the concerns that theorists of the ethics of immigration raise about populist rhetorics that infuse nationalist sentiments into public opinion about migration is that it is dangerous. The history of nationalism in Europe shows that there is almost no limit to the destruction that could arise when nationalist sentiments are mobilised to target specific immigrants. Notwithstanding the need to take seriously the destructive potential of nationalism, it must be acknowledged that national identity is an important component of the way people think about reality and their role in it. This explains why the concept of citizenship of a country, which could be taken as an expression of national identity, is more than mere political status. Given this, it is crucial to think about the influence of nationalism on public opinion and immigration policies beyond the banal sensationalism involved. By this I mean that it is necessary for theorists of the ethics of immigration to consider the implications of nationalism in everyday life. To take this aspect of nationalism into account, one must ask questions like the following: what does it mean that in a liberal democracy, people's attitude in their dealings with others is shaped by nationalist rhetoric about specific groups of immigrants? Although theorists of ethics of immigration consider the problem of nationalism through analysis of multiculturalism, authoritarianism and justice, sufficient effort has not been devoted to the exploration of the implications of the experience of nationalism in daily life in the framing of the ethics of immigration. Gillian Brock (2021) seems to be the only exception in this regard.

4 Desiderata for Future Research

Earlier, I noted that nationalism inspired arguments about immigration are unsatisfactory because the sense of immigration they assume is too narrow. Since my recommendation is that theorists of immigration ethics should be more critical about the concept of nationalism, it is important to consider why it is plausible to suggest that nationalism inspired arguments about migration ethics rely on a narrow view of immigration. To contextualize my proposal, I will explain below what it means to have a more critical outlook on the concept of nationalism. I have shown above that nationalism manifests in debates about the ethics of immigration in the discussion of what is now generally regarded as methodological nationalism. The role of methodological nationalism in immigration ethics is criticized because its focus on the role of the state leads to a significant neglect of other important factors and actors. I think that beyond the shortcoming of excessive focus on the state, nationalism inspired arguments about immigration ethics are problematic because they rely on a very narrow conception of immigration.

Firstly, in the quest to enunciate the implications of sovereignty for immigration, theorists reduce the phenomenon of migration to a kind of movement that unfolds according to a specific, predictable pattern. They assume that the phenomenon is a movement across international borders. This explains why the theories informed by methodological nationalism consider the focus of the ethics of immigration to be the right action to take to regulate a future event, that is, the act of crossing international borders that will occur in the future. The issue, however, is that the ethics of immigration is as much about regulating potential crossing of borders that has not yet occurred as it is about dealing with immigration that has already occurred. An important limitation of thinking about immigration ethics through the perspective of methodological nationalism is that such an approach compels theorists to take a disaggregated approach instead of a wholistic perspective. This is an important shortcoming because a robust conception of the ethics of immigration must do more than offer putative discussion of what constitutes right action by the state in relation to certain features or problems of immigration. A theory of the ethics of immigration should offer at minimum an orientation to thinking about migration as a human experience—and not merely a problem to be solved by specific kinds of states.

Secondly, nationalism inspired theories of the ethics of immigration often assume that the main issue to consider in thinking about immigration is the "agency" of states. Theorists that adopt this perspective tend to consider sovereignty, policing of borders and the rightness or otherwise of immigration restrictions as the main questions of immigration ethics. The problem with this assumption is that we cannot conceive a just policy of immigration by focusing squarely on challenges that arise due to the action that states take in the interest of protecting national borders or ensuring sovereign integrity. Taking a perspective that is mainly focused on the agency of states could lead to a situation where migration is reduced to an aberration from a supposedly universal 'norm' of nativism. Furthermore, thinking about the ethics of immigration from this perspective is inadequate because it leads theorists to ignore relevant historical events that contribute significantly to the current movement of people across international borders. The narrow view of immigration that informs nationalism inspired theories of the ethics of immigration seem to me to account for the tendency to exclude the migration of expatriates from the range of movements that matter in theorizing the ethics of immigration (Okeja 2017: 141).

The adoption of a limited conception of immigration is a problem for populist discourses on immigration because a major issue in such discourses is the assumption that immigrants are radically different from citizens due to their lack of certain features, character, and dispositions. Besides the tyranny that this inaccurate supposition of a homogeneity about members of "the nation" could lead to, this framing of immigration severely limits the scope of meaning that is attributed to the concept of immigration. It reduces immigration to a concept that refers to the radical difference between people who supposedly share a national identity and those who do not share or are incapable of participating in such identity. The consequences of this reduction of scope of meaning are numerous, but the most concerning is that it distorts the basis of discourse, namely, a shared conceptual understanding of the phenomenon of immigration.

Having established that nationalism inspired arguments hinge on a limited understanding of immigration, I will now conclude with a note on what it means to suggest that theorists of the ethics of immigration should be more critical about the concept of immigration. At least two things are crucial for a critical orientation to the concept. The first is to take seriously the fact that liberal political thinking is one among many ways of thinking about political questions. Katrin Flikschuh is correct to argue that "liberal life may simply be one among several equally available ways of life, no less contingently valid, in principle, than any of the others" (Flikschuh 2017: 114). The implication of this view is that theorists of the ethics of immigration need to consider the implications of the experience of the state as a failed universal in non-Western contexts for the ideas they propose.

The second thing a critical orientation to the concept of immigration entails is to be attentive to the fact that the concept has a history. It is important, to put it differently, to show awareness of the fact that immigration refers to experiences of real people. Against this background, a critical orientation to the concept of immigration demands moving away from the imagination that immigration is the opposite of sedentariness, which should be the norm for all human beings. Giving up on this assumption about immigration would make it possible to begin to think about immigration as more than just movement across the borders of some powerful countries that glamorize themselves as "liberal democracies". Taking this critical approach to the concept of immigration enables theorists of the ethics of immigration to spell out the implications of the personhood of migrants for the question of justice in immigration.

To overcome the shortcomings of nationalist inspired arguments, future research on the ethics of immigration should aim at developing ideas that are grounded in a broader understanding of the phenomenon. This is essential because the goal of the ethics of immigration is to help us to determine the right orientation we should cultivate due to the recognition that immigration is a human experience. Operating with a limited understanding of immigration as I have shown would continue to blunt the potential of ethics of immigration to serve as a guide to practical action. Since immigration raises a practical concern, the ethics of immigration must aim to do more than generate theories that are of little to no practical use. The limitation of nationalism inspired arguments shows that there are many important factors a robust theory of ethics of immigration must consider.

Literature

Anderson, B. 2006. *Imagined Communities. Reflections on the Origin and Spread of Nationalism*. London: Verso.

Brock, G. 2021. *Migration and Political Theory*. Cambridge: Cambridge University Press.

Calhoun, C. 1997. *Nationalism*. Minneapolis: University of Minnesota Press.

Carens, J. 1987. Aliens and Citizens. The Case for Open Borders. *The Review of Politics* 49:251–273.

Challenor, H. S. 1979. Strangers as Colonial Intermediaries: The Dahomeyans in Francophone Africa. In Strangers in African Societies, ed. W. A. Shack and E. P. Skinner, 67 - 83. Berkeley: University of California Press.

Emmer, P. C. 1992. European Expansion and Migration. The European Colonial Past and Intercontinental Migration. An Overview. In *European Expansion and Migration. Essays on the Intercontinental Migration from Africa, Asia, and Europe*, ed. P. C. Emmer and M. Mörner, 1–12. New York: Berg.

Flikschuh, K. 2017. *What Is Orientation in Global Thinking? A Kantian Inquiry*. Cambridge: Cambridge University Press.

Joppke, C. 1999. *Immigration and the Nation State. The United States, Germany and Great Britain*. Oxford: Oxford University Press.

Okeja, U. 2017. Reverse Migration, Brain Drain and Global Justice. *South African Journal of Philosophy* 36:133–143.

Sager, A. 2016. Methodological Nationalism, Migration and Political Theory. *Political Studies* 64:42–59.

Srikantiah, J., and S. Sinnar. 2018–2019. White Nationalism as Immigration Policy. *Stanford Law Review Online* 71:197–209.

Sudarkasa, N. 1979. From Stranger to Alien: The Socio-Political History of the Nigerian Yoruba in Ghana, 1900 - 1970. In *Strangers in African Societies*, ed. W. A. Shack and E. P. Skinner, 141–167. Berkeley: University of California Press.

Walzer, M. 1983. *Spheres of Justice. A Defense of Pluralism and Equality*. New York: Basic.

Ypi, L. 2008. Justice in Migration. A Closed Border Utopia? *The Journal of Political Philosophy* 16:391–418.

Part II
Migration Ethics in Philosophy

Migration Ethics and/in Applied Ethics

Julian F. Müller

Abstract This article discusses migration ethics in the context of applied ethics. The first part offers a characterisation of applied ethics. In the second part, the article summarises the three standard methods of applied ethics and discusses whether and to what extent these methods are employed in migration ethics. The third part draws attention to the fact that migration ethics as a field has not yet begun to reflect on its own methods. The final part argues that to develop its methodological foundations, applied ethics might benefit from consciously incorporating the methods of political philosophy for conceptualising and resolving moral disagreements.

Do nation-states have a right to close their borders for economic immigrants? And what about refugees (Carens 2013)? Indeed, are there any principled moral reasons to distinguish between the claims of refugees and economic migrants? Given that there is a human right to claiming asylum, are states justified in combatting human smugglers transporting refugees (Müller 2021)? Assuming that a particular nation-state cannot fulfil its duty to provide shelter for refugees, are there any other ways that nation-states can discharge their duty to provide asylum (Müller 2016; Betts and Collier 2018)?

The events of the European migrant crisis between 2014 and 2016 forcefully put these important moral questions back on the table. When millions—dislodged by the Syrian Civil War and other simmering conflicts—crossed the Mediterranean Sea in hope of safety and dignified life prospects, Europe was no longer able to ignore the humanitarian catastrophes playing out in its backyard. Even though ethical issues pertaining to immigration and asylum have been on the radar of philosophers ever since Joseph Carens's (1987) seminal article "Aliens and Citizens. The Case for Open Borders", the migration crisis added fuel to these philosophical debates. Since climate change will produce waves of climate-induced migration from poorer to more affluent regions of the world (Dietrich and Wündisch 2015), migration ethics, moreover, will likely continue to attract the attention of philosophers for the coming decades.

J. F. Müller (✉)
Department of Philosophy, University of Graz, Graz, Austria
e-mail: julian.mueller@uni-graz.at

A. Niederberger et al. (eds.), *Handbook of Migration Ethics*,
https://doi.org/10.1007/978-3-031-89877-8_6

The ethical questions raised by mass migration fall squarely under the purview of applied ethics. To understand what distinguishes applied ethics from normative ethics, it is useful to first introduce the distinction between metaethics, descriptive and normative ethics. Metaethics is not concerned with the justification or description of moral rules and principles but with the presuppositions of moral judgements, speech acts and practices. It asks questions such as: Are there moral facts? What is the nature of these moral facts? Or: How can we acquire moral knowledge? In contrast, descriptive ethics is concerned with describing, cataloguing and explaining moral attitudes and beliefs of groups. Various academic disciplines contribute to the research agenda of descriptive ethics, such as anthropology, sociology, evolutionary biology and various sub-disciplines of psychology. Descriptive ethics, therefore, is interested in moral attitudes, beliefs and commitments of actual people past and present. This brings us to normative ethics. In most general terms, normative ethics is concerned with the question of what the right moral attitudes, beliefs and commitments are. An important part of normative ethics consists in justifying moral principles, rules and ideals. How, then, might we characterise the difference between applied and normative ethics? One common way is to focus on their respective object of inquiry. The goal of normative ethics is to clarify, analyse, explicate and justify the core evaluative criteria, principles and maxims of our shared moral practice. Philosophers working in normative ethics are interested, for instance, in working out different coherent conceptions of evaluative criteria, such as equality, fairness, freedom, justice or welfare. In developing these evaluative criteria for judging the choice-worthiness of actions, rules and institutional sets, theorists regularly need to draw and elaborate on basic moral concepts such as agency, autonomy, authenticity, care, coercion, consent, human flourishing, personhood, rights, virtue, and so on. A further goal of normative ethics is therefore to provide a deeper understanding of our shared moral concepts. What drives philosophical research in normative ethics are mostly theoretical as opposed to practical questions. That is to say: the problems that drive normative ethics are theory-driven. Whereas the questions in normative ethics are theory-driven, the questions that are tackled by research in applied ethics are often of immediate public concern. As Borchers (2012: 26) notes, public and semipublic fora produce the problems to which applied ethics attends. The moral problems that are passed on to applied ethics mainly come in two forms: the problem consists either in a public disagreement on a certain topic, meaning that there are already a number of positions on the topic, or there is a perceived need to clarify a moral issue, meaning that there are no clear positions on the issue yet. The public role of applied ethics in both cases is to clarify the issue, present different ways to conceive the problem or to develop candidate solutions for the problem at stake. Applied ethics and normative ethics can thus be differentiated according to the source and nature of the problem. A second way to differentiate normative and applied ethics is closely related to the respective nature of the problems that both disciplines attempt to settle. Theoretical problems call for conceptual solutions, while public disagreements usually concern practical problems such as: How many refugees should the Europe Union accept? Is it morally permissible to accept or reject refugees on the basis of ethnicity or religion? How should refugees be distributed within the European Union? Should refugees be permitted to work in their host country? Should migrant workers have a right to vote in the host country? In general, practical questions demand practical as opposed to theoretical answers. The public disagreements to which applied ethics attends are usually about determining the appropriate response to a certain moral issue. These responses might take the form of policies, regulations, codes of conducts, or informal social rules. This said, applied ethicists usually aim to spell out the moral criteria that apply to a case in question and in that aim to identify the set of permissible solutions rather than suggesting any particular one. Given that public disagreements are about practical solutions to problems of public concern, engaging with applied ethics forces philosophers to get up from their metaphorical armchairs. To contribute to policy questions, philosophers working in applied ethics must thus become acquainted with the empirical literature that pertains to a specific question. Philosophers working on the question of whether states have the authority to restrict immigration, for instance, need to consider the economic, sociological, cultural and political implications of various border regimes. Sometimes it is argued that philosophers should abstain from intervening in policy discussions because by trade they do not have the required expertise in the various disciplines that bear on the respective policy issue. However, the argument seems to cut both ways. Policy recommendations are necessarily normative statements such as: The European Union needs to fight commercial human smugglers! Germany needs to patrol its borders and close loopholes in the European asylum system! There should be an upper limit to claims to asylum! However, given that social scientists usually do not possess any formal background in ethics, it seems questionable why they should be in a better position to provide policy recommendations. Since policy recommendations necessarily involve normative and empirical questions, interdisciplinary efforts seem to be called for. There are at least two objections to this line of argument. First, social scientists might counter that they can provide policy recommendations in the form of if-then statements and thus circumvent taking any stance on normative questions. For instance, an economist might say "*if* Germany wants to maximise economic growth over the long term, *then* it should accept more economic migrants of a certain type". However, the policy question that needs answering is typically that of what a certain political agent should do, all things considered. Thus, social scientists can only hope to give a partial answer to the question at stake. The second objection is deeper, it asks whether there is something like moral expertise in the first place. However, even though there is considerable disagreement about the metaethical question of whether there are moral facts—and hence disagreement about whether ethicists possess moral expertise in a strong sense—it cannot be denied that ethicists are experts in clarifying, sorting and reconstructing moral arguments. Ethicists thus at least have moral expertise in a weak sense. To summarise: normative ethics and applied ethics can be distinguished according to the nature of the problems that they set out to solve (theoretical questions vs public disagreements), the kind of solutions aimed for (conceptual refinement vs binding rules) and the knowledge required for the solution (conceptual knowledge vs conceptual and empirical knowledge).

1 General Methodological Particularities of Applied Ethics

Giving an overview of the methods of applied ethics, the methodological particularities of applied ethics and even more so the methodological particularities of applied ethics in the context of migration ethics is a challenging task. There are almost no sustained theoretical treatments on the nature and methods of applied ethics to begin with. Most treatments of applied ethics are either primers that feature a short overview chapter about the methods and then go on to discuss various topics (such as bioethics, business ethics, and so on) or book-length treatments of particular applied ethics questions. However, the latter treatments usually do not feature chapters on applied ethics or the methods of applied ethics, but rather engage with the respective topic within the confines of a particular given debate. Some fields of applied ethics, such as business ethics and bioethics, feature profound and long-standing scholarly debates about methods and methodology. Others, especially less well-established fields, do not. Migration ethics belongs to the latter category. Surveying the field of migration ethics, it is hard to find any dedicated treatments of method. What makes the task seem even more difficult is that there is not even a settled understanding of the term "method" in applied ethics (Paulo 2016: 5). For the purposes of this article, the term "method" should be understood simply as a "procedure to attain a certain end". The definition has the virtue that it stays close to its original meaning and captures our pre-theoretical understanding of the term sufficiently well. Applied ethics mainly aims to solve practical problems. The methods of applied ethics, then, are procedures that aim to produce justified proposals (such as rules, laws and practices) to solve practical problems. Within the current body of applied ethics, three main methods can be identified: the top-down method, the bottom-up method and the method of reflective equilibrium. In the following, I will present each method in turn and provide examples for the use of the respective method in migration ethics.

Top-Down Method

The top-down method, sometimes called the deductive method, to applied ethics seeks to resolve practical problems by reference to a certain established moral rule or principle. Moral rules and principles can be expressed as conditional statements. The antecedent of a moral principle, then, spells out the condition(s) under which the consequent is triggered. The top-down model simply asks whether the specific case in question falls under the conditions named in the antecedent. The following schematic exemplifies how to derive a specific policy from moral criteria (spelled out in a moral rule) in a deductive way:

1. If policy *p* meets the moral criteria *c*, then implementing policy *p* is morally permissible.
2. Policy *p* meets the moral criteria *c*. Thus:
3. Implementing policy *p* is morally permissible.

The virtue of the top-down method is its simplicity. To answer the question whether rich countries should introduce an open borders regime, the top-down method, given that the required moral principle is at hand, produces a straightforward answer. Take, for instance, utilitarianism. Given that the economist Michael Clemens (2011) is right and the gains "from reducing emigration barriers are likely to be [...] measured in tens of trillions of dollars", a utilitarian might simply argue that an open borders regime should be established on a global level because an open border regime will likely maximise overall utility (on the premise that monetary gains are a good proxy for utility). However, not only utilitarianism allows for such straightforward deductive reasoning. A stylised deduction from human rights to questions of legitimate border regimes might be as follows: governments are only permitted to implement border regimes that do not violate human rights. There is a human right to free immigration. Thus, closed border regimes are impermissible (because they violate the human right to free immigration). For a nuanced discussion of whether there is a human right to immigrate, see Oberman (2016). One of the problems with the deductive method is that at best it seems to clarify and ground particular lines of argument in a given public debate. However, if the goal of applied ethics is to find convincing policy proposals to solve public disagreements, it is unlikely that this can be done by restating—although in a more rigorous form—one of the positions in the public debate. This might be one of the reasons why in practice very few theorists, if any, advocate the deductive model of policy or rule justification (Düwell 2002: 245).

Bottom-Up Method

Whereas most debates in migration ethics deal with the obligations of certain institutions—such as nation-states or international organisations—many subdisciplines in applied ethics are concerned with the (particular) obligations of professionals, i.e. individual agents. Medical ethics, for instance, deals with the specific moral conundrums that medical professionals face, and business ethics often deals with the particular moral obligations of managers. The bottom-up method has developed mainly in the fields of applied ethics that reflect on the special obligations of individual professionals. In a first step, bottom-up models take as inputs the informal and formal moral rules that a specific profession has developed in response to certain recurring moral conundrums and dilemmas. In a second step, the bottom-up model aims to articulate, clarify and develop the moral rules that govern the practice of a certain profession. Rules that proved morally valuable in governing concrete situations are "generalised in the sense that their application is extended to further cases" (Knoepffler 2010: 52, my translation). The upshot of this effort might be a code of conduct or a set of moral criteria that in turn helps professionals to adjudicate moral controversies and clear up prior moral ambiguities. Because the emphasis is on reconstructing and articulating existing moral codes, the bottom-up method bears close resemblance to the method of narrow reflective equilibrium (Daniels 2020). One of the central moral objections to the bottom-up method is that it fails to critically engage with established moral practices. By relying heavily on problem description and professional judgements within a certain profession, the bottom-up model runs the risk of regurgitating the biases and ethical blind spots of the respective target practice.

Since the bottom-up model seems to be tailored to subfields of applied ethics that focus on the obligations of particular groups of professionals, it does not seem to readily lend itself to questions of migration ethics since the latter tends to be concerned with the obligations of institutional agents. Even though the bottom-up method is not regularly employed, it is a part of the toolbox of philosophers working in migration ethics. Take, for example, David Miller's (2014) seminal paper "Immigration. The Case for Limits". The paper starts by pointing out that "[m]ost liberal, conservative, and social democratic politicians support quite strict immigration controls in practice, but they generally refrain from spelling out the justification for such controls." Hence, one of the goals of Miller's paper seems to be to reconstruct and articulate the argument in favour of migration controls that animates those politicians and their constituencies.

Coherentist Method

Many philosophers have argued that all ethics is based on a coherentist model (Brink 2014; Scanlon 2003). Indeed, some philosophers have even gone so far as to claim that the coherentist model underlies all philosophical reasoning (DePaul 2011). The coherentist model, nevertheless, denotes a distinct approach to applied ethics and can be best accentuated by contrasting it with the bottom-up approach. The bottom-up model can be understood as consisting of at least two branches that can be distinguished according to the inputs. Whereas the reconstructive branch takes as inputs mainly the problem description, individual judgements and moral codes of practitioners, the intuitionist branch takes as inputs the philosopher's trained judgements about particular cases in order to work up to more general principles. Both branches thus aim to reach what Daniels calls a narrow reflective equilibrium. In juxtaposition to the bottom-up model, the coherentist model aims to reach a wide rather than a narrow reflective equilibrium. Forming a wide reflective equilibrium about certain moral questions means that the theorist considers not only the particular judgements from a certain community (reconstructive branch) or her own intuitions about certain cases (intuitionist branch) but also takes into consideration a much broader set of inputs. An important part of forming a wide reflective equilibrium is to take into consideration competing moral intuitions and principles as well as religious doctrines that pertain to the question that the theorist aims to solve. The goal of the theorist is then to reconstruct from these various inputs a coherent position vis-à-vis a certain moral problem. In his seminal paper "Aliens and Citizens. The Case for Open Borders", Carens, for instance, takes as inputs contemporary approaches to political philosophy, such as the Rawlsian, Nozickean and utilitarian approaches, and aims to show that they all converge in a laissez-faire position to immigration.

2 Relevance of Applied Ethics Methodology in Migration Ethics

Despite the fact that some early contributions to immigration ethics date back to the 1980s, immigration ethics only received sustained attention from the profession after the onset of the European migrant crisis. Perhaps this explains why philosophers working on migration ethics have so far paid little to no attention to methodological issues. The fields in applied ethics that feature long-standing and quite intricate debates about methods and methodology, for instance bioethics and business ethics, tend to have a much longer history and thus tend to be institutionally much more entrenched than migration ethics. Many universities have dedicated chairs in business and bioethics and a number of dedicated journals provide space for the particular methodological discussions that pertain to these fields. However, these specialised debates do not seem to have produced much in terms of general frameworks that can be readily adapted and deployed for the purposes of other subdisciplines. One reason why there is little reflection on the question of which methods bind together the various subdisciplines under the label of applied ethics might be that philosophers working in applied ethics, to put it in the words of Tom L. Beauchamp (2005: 12), "do what philosophers have always done." The founding father of bioethics argues that applied philosophers in the end "analyze concepts, examine the hidden presuppositions of moral opinions and theories, offer criticism and constructive accounts of the moral phenomena in question, and criticize strategies that are used to justify beliefs, policies, and actions. They seek a reasoned defense of a moral viewpoint, and they use proposed moral frameworks to distinguish justified moral claims from unjustified ones." If Beauchamp is right, then what sets apart applied ethics from normative ethics is neither the engagement with conceptual work nor—at least so far—any distinct method.

3 Desiderata for Future Methodological Development

Migration ethics is different from various traditional fields in applied ethics in that it does not focus on the ethical quandaries of professionals but mainly enquires about the legitimacy of various statutes and laws that relate to immigration. One might wonder, then, whether migration ethics should not best be conceived as belonging to political philosophy. Having opened the drawer of political philosophy, however, one might start to wonder how the relation between political philosophy and applied ethics is best conceived. Even though there is no real debate, many theorists who have commented on that question have arrived at the conclusion that political philosophy is a branch of applied ethics. For instance, Shelly Kagan (1998: 3) writes: "I think the entire field of political philosophy can legitimately be viewed [...] as one (vitally important) branch of applied ethics [...]." However, there are also dissenting voices (Fenner 2010). What sets moral philosophy and political philosophy apart is arguably that political philosophy starts from the fact that there is reasonable disagreement about what the correct moral principles are. As Charles Larmore (2013: 306) puts it: "You have your moral views, I have mine, and each of us is convinced that he is right, standing ready to show the other the error of his ways. But once we confront the problem of how people like us are to live together, we enter the terrain of political philosophy." In this view, political philosophy is quintessentially about how to deal with the fact of moral and political disagreement. Over the last decade, political philosophy has cultivated an impressive toolset for thinking about and dealing with moral disagreement. To build the methodological foundations of applied ethics, it would then only seem natural to develop the foundations of applied ethics by incorporating the concepts and methods established by political philosophers. However, upon closer examination, introductions into applied ethics do not usually cite any of the tools developed in political philosophy with the exception perhaps of deliberative approaches to disagreement. Thus, whether or not political philosophy is construed as a subfield of applied ethics, theorising on the methods of applied ethics has yet to tap into the methods and tools of political philosophy.

In the wake of the European migrant crisis, Thomas Schramme (2015) argued forcefully that philosophy is not in a position to productively intervene in the public debate about immigration because it lacks the appropriate methods. Philosophers engaging with the migration crisis, Schramme argued, essentially "shoot from the hip." As this article shows, Schramme is not wrong, there are few if any methods that are particular to applied ethics. Indeed, if the top-down model boils down to deductive reasoning, whereas the bottom-up model and the coherentist model are two versions of reflective equilibrium, Beauchamp might be right in his assessment that philosophers in applied ethics are essentially using the same methods as philosophers in general. At least for migration ethics, a field that is held together by shared topics, Beauchamp's diagnosis seems to be correct. If Schramme is right and applied ethics first needs to elaborate a specific set of methods to be in a position to craft proposals to pressing moral problems, then elaborating a method of applied ethics that gives centre stage to political disagreement and how to approach political disagreement in structured ways seems to be a promising line for further research. To develop applied ethics into a field that can respond to its publicly assigned task more readily, forging such a method, in turn, might be a promising first step.

Literature

Beauchamp, T. L. 2005. The Nature of Applied Ethics. In *A Companion to Applied Ethics*, ed. R. G. Frey and C. H. Wellman, 1–16. Malden: Wiley-Blackwell.

Betts, A., and P. Collier. 2018. *Refuge. Transforming a Broken Refugee System*. London: Penguin.

Borchers, D. 2012. Angewandte Ethik als Entreé in die Philosophie. In *Angewandte Philosophie*, ed. C. Runtenberg and J. Rohbeck, 21–38. Dresden: Thelem.

Brink, D. O. 2014. Principles and Intuitions in Ethics. Historical and Contemporary Perspectives. *Ethics* 124:665–694.

Carens, J. H. 1987. Aliens and Citizens. The Case for Open Borders. *The Review of Politics* 49:251–273.
Carens, J. H. 2013. *The Ethics of Immigration*. Oxford: Oxford University Press.
Clemens, M. A. 2011. Economics and Emigration. Trillion-Dollar Bills on the Sidewalk? *Journal of Economic Perspectives* 25:83–106.
Daniels, N. 2020. Reflective Equilibrium. In *The Stanford Encyclopedia of Philosophy*, ed. E. N. Zalta. Stanford: Stanford University.
DePaul, M. R. 2011. *Balance and Refinement. Beyond Coherence Methods of Moral Inquiry*. Abingdon: Routledge.
Dietrich, F., and J. Wündisch. 2015. Territory Lost—Climate Change and the Violation of Self-Determination Rights. *Moral Philosophy and Politics* 2:83–105.
Düwell, M. 2002. Angewandte oder Bereichsspezifische Ethik. In *Handbuch Ethik*, ed. M. H. Werner, M. Düwell, and C. Hübenthal, 243–247. Stuttgart: Metzler.
Fenner, D. 2010. *Einführung in die Angewandte Ethik*. Stuttgart: Verlag Eugen Ulmer.
Kagan, S. 1998. *Normative Ethics*, Dimensions of Philosophy Series. Boulder: Westview Press.
Knoepffler, N. 2010. *Angewandte Ethik. Ein systematischer Leitfaden*. Stuttgart: UTB.
Larmore, Charles. 2013. What Is Political Philosophy? *Journal of Moral Philosophy* 10:276–306.
Miller, D. 2014. Immigration. The Case for Limits. In *Contemporary Debates in Applied Ethics*, ed. A. I. Cohen and C. H. Wellman, 363–375. Malden: Wiley-Blackwell.
Müller, J. F. 2016. Advancing Justice by Appealing to Self-Interest. The Case for Charter Cities. *Moral Philosophy and Politics* 3:223–245.
Müller, J. F. 2021. The Ethics of Commercial Human Smuggling. *European Journal of Political Theory* 20:138–156.
Oberman, K. 2016. Immigration as a Human Right. In *Migration in Political Theory. The Ethics of Movement and Membership*, ed. S. Fine and L. Ypi, 32–56. Oxford: Oxford University Press.
Paulo, N. 2016. *The Confluence of Philosophy and Law in Applied Ethics*. London: Palgrave Macmillan.
Scanlon, T. M. 2003. Rawls on Justification. In *The Cambridge Companion to Rawls. The Cambridge Companions to Philosophy, Religion and Culture*, ed. S. R. Freeman, 139–167. Cambridge: Cambridge University Press.
Schramme, T. 2015. Wenn Philosophen aus der Hüfte schießen. *Zeitschrift für Praktische Philosophie* 2:377–384.

Migration Ethics and/in Political Philosophy

Iseult Honohan

Abstract Until recently political philosophy addressed the normative issues of justice, freedom, democracy and citizenship mainly within the frame of reference of a bounded state. Growing awareness that these issues transcend this frame, and the increased political salience of international migration have brought discussions of migration to the fore. Fundamental issues include the justification for and constraints on states' control over entry to and residence within their borders, the grounds for awarding or withholding citizenship, and the relative weight of the claims of states and citizens vis-a-vis outsiders in general and special cases such as refugees, family members and stateless persons. There is a tension between liberal approaches based on universal individual freedom and equality, and those based on communal commitments or democratic self-determination, sometimes framed as a tension between justice or human rights, and community or democracy. It is also contested whether political philosophers should aim to provide foundational ideals, critical norms or concrete policy guidance, to what extent appropriate answers depend on particular contexts, and what count as salient contextual considerations. Areas requiring further research include (a) articulating more precisely what may be distinctive about *territorial jurisdiction* and *self-determination* for the justification and limits of states' rights to control boundaries and to determine the citizenship status of individuals, (b) transcending *methodological nationalism* to consider the treatment of migration, participation and membership issues at sub-national, transnational and supranational levels and (c) addressing the particular implications for justice of the increasing and diverse migration likely to accompany *climate change*.

Migration has recently come to be widely discussed in political philosophy, as its focus has extended beyond the frame of the individual state to global issues. Central questions here include: Do states have an intrinsic right to control who enters their territory and who are their members, on the grounds of political authority or

I. Honohan (✉)
School of Politics and International Relations, University College Dublin, Dublin, Ireland
e-mail: iseult.honohan@ucd.ie

A. Niederberger et al. (eds.), *Handbook of Migration Ethics*,
https://doi.org/10.1007/978-3-031-89877-8_7

self-determination? Or is there a human right of freedom of movement between states? On what basis should citizenship, membership of a state, be allocated?

1 Methodology in Political Philosophy

Despite a long-standing variety of approaches to political philosophy, explicit reflection on methodology in the field has been limited. One of the most prominent strands, *analytical political philosophy*, is concerned with clarifying key concepts, evaluating ideals and analysing arguments relevant to normative issues arising in politics, including justice, legitimacy, freedom, equality, democracy, rights, and citizenship. Many see it as a branch of moral philosophy applied to the relationship between citizens and the state (McDermott 2008). Until recently, this analysis was carried out almost exclusively within the frame of the state. Liberal thought in principle applied universally, but assumed individuals in a bounded society, and mainly addressed the relations between the state and individuals within that society, rather than across state boundaries.

Political philosophy has also mainly focused on ideals and the permissible actions of individuals and states, rather than on the structure and working of institutions (Waldron 2016). Thus, no hard and fast line divides contemporary political from moral philosophy. Political philosophy may be more likely than moral philosophy to assume that we need to live with a plurality of reasonable perspectives, and seek areas of shared understanding as the basis of authoritative decisions (List and Valentini 2016). Increasingly, however, self-described *realists* have insisted on a sharper differentiation of politics from morality, seeing politics as a realm of inherent disagreement and conflict, in which power, order and legitimacy are the central considerations, rather than principles derived from morality, understood as concerned with right and wrong at the individual level, and justice at the collective level (Geuss 2008; Sleat 2018). These divisions are neither purely methodological nor purely normative, but often represent an amalgam of the two.

Approaches to philosophy other than the analytical approach should be noted: In *historically oriented* political theory, arguments in ancient and modern texts are understood either as contributions to a conversation on enduring issues across time, or, more recently, as specific discourses in their local historical context. Post-Marxist *critical theory* examines norms as they emerge from conflict and inequality within social structures and relationships. A more *interpretative* approach seeks to articulate the implicit meanings or values embedded in practices rather than applying predetermined principles. *Feminists*, *multi-culturalists* and *critical race theorists* draw attention to unjust background conditions that have rendered gender, cultural or racial oppression invisible and have not been thematised in mainstream political philosophy. Several of these approaches have had influence both on recent developments in analytical political philosophy and its treatment of migration, and on wider migration debates. This article focuses, however, in particular on analytical political philosophy, where greater change in the approach has been necessitated by the movement beyond ideal and state-bounded approaches.

Since the mid-twentieth century John Rawls's liberal theory of justice has shaped the development of analytical political philosophy (Rawls 1971). He adopted an avowedly ideal approach, one that aims to outline the character of a just society in which freedom and equality alike are honoured, and this was explicitly within the confines of a bounded state. Extensive debates have been devoted to how equality (of opportunities, of outcomes or of relations), freedom (as non-interference, non-domination, or autonomy) and other values, should be conceptualised.

Communitarian and multicultural critics rejected the individualist tenor of Rawlsian justice, as overlooking the constitutive significance of groups and culture for individual flourishing. This reflected a significant turn towards examining justice in relationships as well as in distribution. Indeed, it can be argued, with great significance for migration ethics, that the *scope of distribution—what determines who are the members among whom goods are distributed—is more fundamental than how they are distributed (Walzer* 1983).

Along with substantive critique, the aims and methods of political philosophy were reconsidered—partly in a specific debate on the Rawlsian distinction between "ideal" and "non-ideal" theory (focusing on how one should act in a society when it is broadly just, and when it is not, respectively). Sometimes, if enacted in non-ideal contexts, pursuing ideal solutions will be ineffective, or even lead to worse results than a second-best action. Increasingly a broader contrast has been drawn between "ideal" and more "practical" approaches. It has been asked: should political philosophy focus on ideal social solutions, critically examine current politics and practice, or provide specific guidance for policy in existing conditions? Thus, some emphasised remedying particularly stark *in*justices, and others addressed more detailed issues of justice (in areas from bioethics to humanitarian intervention).

The influence of Rawls led others to explore the implications of his theory for questions of global justice, applied to all humans. This puts the frame of the nation-state in question, and gives rise to normative issues that transcend borders, including migration, in a new field of international political theory.

2 Approaches to Political Philosophy in Migration Ethics

Growing awareness that issues of justice transcend the frame of a single society, and the increased political salience of migration have brought it centre stage in political philosophy. Among the most fundamental questions at stake are: whether, and on what grounds, states may exclude, or select among, would-be immigrants, and award or withhold citizenship; and what weight should be given to claims of outsiders in general, and of special cases such as refugees, family members, and stateless persons. More specific issues have also received increasing attention, including the rights of long-term residents, temporary workers and irregular immigrants, and the integration of immigrants. Thus discussions of migration ethics in political philosophy have reflected both its close overlap with moral philosophy and recent trends towards more applied approaches. They have also featured the distinction between ideal and non-ideal theory, especially concerning how states should or may act with respect to migrants when other states do, or do not, observe their obligations.

On the most widely accepted public view, states have a sovereign right over entry and residence within their borders. But while many theorists accept the premise of a world of separate states as the circumstances under which migration issues acquire their contemporary significance, the right of states to control borders is one of the most contested issues in political philosophy of migration.

Important differences are found in the perspectives and interests that different approaches take as their starting point: those of the state, of citizens, of would be-migrants, or of humanity in general, and the relative weights attached to those interests. In addition, the question whether theorists should develop ideals or provide more concrete and feasible policy guidance is a particularly live issue in the political philosophy of migration (Carens 2013; Miller 2016).

2.1 Universalist vs. Particularist Approaches

The political philosophy of migration displays a tension between two broad approaches: liberal universalist (or cosmopolitan) approaches, based on principles of individual freedom and equality applied to all humans (see Cole 2000; and chapter "International Freedom of Movement" in this Handbook), and particularist approaches giving independent weight to communal commitments (see Walzer 1983). This is sometimes framed as a tension between justice or human rights, on the one hand, and community or democratic self-determination, on the other. Yet the wide range of substantive positions on the relation of freedom and border controls demonstrates how answers to questions about migration are not fully determined by the adoption of universalist or particularist perspectives.

From the first, universalist perspective, Rawlsian principles of freedom and equality seem to imply that the state must give equal consideration to all (Carens 1987). Yet, while concerns for global justice and equality have been the basis of arguments for open borders, more open migration may be regarded as only one, and not perhaps the most effective, way to pursue these (see chapter "Migration, Global Justice and Development" in this Handbook). For some, freedom may be seen as intrinsically limited by border controls. Thus, in what has been called a "cantilever" argument, the right of freedom of movement recognised within states may be claimed to apply globally and even to be a human right (Carens 2013; Oberman 2016). Yet this may be challenged, as there are limits to all freedom of movement, and moving between states may not be seen as a fundamental human interest (Miller 2016).

The principle of freedom of association has been used both to support and to oppose free movement across borders. In the first case this is based on the right of members to invite outsiders into the territory (Steiner 1992); in the second case freedom to associate (like choosing a life-partner) implies society's freedom *not* to associate with any particular person (Wellman 2008). Working out the implications of freedom of association is complicated by the fact that the state is *un*like an association or club in several important ways. Membership is not strictly voluntary, but usually acquired at birth; controls on migration raise the question of how someone may become part of the institutional structure of a state. The impact of exclusion—from the territory as well as from association—is even more significant than exclusion from other groups, as evident particularly in the impact on those whom no state will admit (Fine 2010).

Among the analogies extensively invoked in migration debates, another is the right to exclude from private property. The objection that this elides territorial jurisdiction with property ownership has been addressed by approaches that attempt to identify more specifically political grounds for the state's right to exclude, including the costs of jurisdictional protection that states incur (Blake 2001) and the associative ownership of political institutions contributed to by citizens over time (Pevnick 2011).

The implications of freedom for border controls are also complicated by different understandings of freedom. Libertarian arguments, based on freedom from interference, are often invoked to support open borders. On another account borders directly coerce only those who actually seek entry (Miller 2016). If, however, they threaten to coerce all outsiders, they may, on principles of democratic freedom, at least require justification to those subject to them (Abizadeh 2008).

Even if such a hypothetical threat of coercion cannot be seen as a substantial invasion of freedom as non-interference, it may constitute an arbitrary *capacity* to interfere, or a "dominating" power in the neo-republican sense. But republicans too are divided on whether border controls are dominating. Some see it as a political necessity that non-domination can be promoted only within a state with border controls (Pettit 2012). For others, these constitute domination beyond the state and call for institutional and contestatory powers to limit the state's arbitrary powers not only to exclude, but also to impose the status of alien on outsiders (Costa 2016; Owen 2014).

The second, particularist, perspective starts from the growing interest, noted earlier, in relationships and membership in political philosophy. Here the political community carries an intrinsic normative weight that must be taken into consideration in migration ethics. The most prominent particularist arguments for migration controls have been based on community and commitments to fellow citizens, rather than state sovereignty. State sovereignty is historically and conceptually more a matter of supreme authority within the state, territorial integrity and autonomy versus other states than of border controls over individuals. From this particularist perspective, political communities are distinctive communities of character, constituted by shared understandings and mutual commitment, including their conception of what justice or equality entails. Only if the state is bounded can this character be sustained. Self-determination provides the grounds for the freedom to admit or exclude outsiders (Walzer 1983).

This argument reflects the important insight that some moral obligations arise in relationships (rather than being universally owed to humanity), as with the commitments to family and friends. The recurring question is whether membership of a political community in general or a nation-state in particular constitutes such a relationship, what kinds of mutual obligations it may justify, and how these might be reconciled with universal commitments.

Political philosophers have increasingly dismissed nationality, based on ascribed characteristics such as race and ethnicity, as a justified claim for closure. But for some liberals the importance to individuals of a cultural context provides grounds to limit migration that could threaten the sustainability of a society's culture. Others understand migration controls as instrumentally necessary for democratic self-government. For liberal nationalists, democratic institutions and practices, or redistributive welfare policies depend on a common *political* culture, rather than a thick cultural community (Miller 2016). Yet here too it is argued that high levels of migration, especially if bringing greater diversity of culture or religion, may undermine the social cohesion necessary for democratic institutions. This implies that immigrants should be admitted only to the extent that they can be integrated into the political culture of the host society. Critics of these positions point to the natural evolution of cultures over time, and query the connections of common culture, social cohesion, and democracy.

Among defenders of border controls, it is debated to what extent exceptions should be recognised. Many recognise constraints on the right to exclude: humanitarian duties of mutual aid to others include providing aid to those in urgent need, and admitting as refugees those who cannot be assisted externally. Whether defined as those persecuted by their state, or more broadly as those who lack the protection of their human rights, it is widely argued that states have an obligation to admit refugees (Walzer 1983; Miller 2016). Discussion of refugees in particular may be seen as located in non-ideal theory, since they would not exist if all states were broadly internally just. Some, however assert that there is no duty to admit refugees—what is required instead is to ensure protection in their own country, even through external intervention (Wellman 2008). Others debate whether family members of citizens and residents have relational or human rights claims to entry, and what weight to give these relative to claims of refugees (Gibney 2004).

Thus, it remains in question whether theorists should be concerned with working out the application of universal principles, or reconciling universal principles with norms that arise within special relationships, and how.

2.2 *Ideal vs. Realistic Approaches*

There is a second tension in political philosophy of migration between ideal and more realistic approaches. Should it aim to provide ideal solutions or more realistic policy recommendations, and to what extent should it take account of empirical realities in order to give concrete guidance in actual contexts (see chapter "Migration

Ethics Between Ideal and Non-Ideal Theory" in this Handbook)? In current discussions of migration, it may be noted, the emphasis is more on issues of feasibility than on power, order and legitimacy central to the broader movement of political realism.

In migration debates, the contrast could be framed as one between two approaches: "top down"—starting from abstract or general moral principles and then applying them to practical contexts—and "bottom up"—starting from practical contexts or experience and generating approaches informed by those contexts. This may mean thinking critically about current practices and policies to determine which policies, if not ideal, may be acceptable or permissible (Carens 2013), drawing on empirical social sciences, or engaging with the experiences of migrants (Fine 2019). In the context of migration those characterised more specifically as "realists" argue that, rather than proposing infeasible ideals of global freedom of movement or open borders, theorists should address what is currently feasible, while also giving consideration to the alternative claims and urgent needs of outsiders (Miller 2016). Issues that arise are: which constraints on the capacity to accept immigrants should be recognised: for example, the size and wealth of states, their capacity to integrate, and the impact of immigration on the economic, social or political life of host countries? And should we distinguish "hard"—e.g. geographic—constraints that have to be taken as given from "soft"—e.g. institutional—constraints that may be subject to political change (Lawford-Smith 2013). This has led to debates on what count as the empirical and political facts that constrain feasibility. But the divisions here may be seen as representing less a divergence of methodological approach than one of normative commitments.

If states should accept immigrants up to the point where the numbers undermine possibilities of integration, public order, democratic practices or welfare services, empirical research is required to establish what this level might be. In particular, empirical evidence has yet to establish firmly the links between social cohesion and democratic practices proposed as a constraint on the extent of feasible immigration. Particularly contentious is the extent to which current public opinion should be regarded as a hard constraint relative, for example, to the urgency of needs of refugees, and even if based on misinformation about the volume of migration, its economic impact and its relationship to crime, for example. More broadly, does respecting democracy mean accepting a particular expressed will of the people, or being committed to including all in decisions to which they are subject?

Since migration is highly political salient, and raises new transnational issues of justice with which political philosophy has only recently engaged, it seems important that theorists should work on all fronts—to develop ideals and critical norms, to attend to the evidence provided by empirical social and political science, to engage with policy practice and issues, to consider the voices and experience of all affected, thus expanding the methodological repertoire of political philosophy beyond ideal theorising.

2.3 *Citizenship*

In the contemporary world the legal status of citizenship, or membership, awarded by states, is the most secure guarantee of entry to a state. While citizenship acquisition is thus highly relevant to migration, it gives rise to some distinct issues, and debates about legal citizenship have generally focused more on concrete legal and policy considerations than have philosophical debates on migration, though the implications of this gap have been little discussed. This may be related to the fact that normative liberal and republican accounts of citizenship have tended to focus primarily on the content of citizenship rather than its boundaries, and (even if potentially universally inclusive) have not provided an account of the grounds on which a specific formal citizenship should be acquired. It could be argued also that political philosophy has focused more on the broader issues of migration than on citizenship because of their visible urgency, or that, when citizenship has been discussed, it has taken second place to the related but distinct question of the composition of the demos.

Citizenship is not only a formal legal status, but also a bundle of rights and duties that centrally includes membership of a self-governing community. Arguments on access to legal citizenship partially overlap with, but are different from, arguments about the rights of political participation in decision-making, or who should be members of the demos: all affected by political decisions, all subjected to the authority of the state, or all long term stakeholders in the future of the country (Bauböck 2018).

For some theorists today the availability of formal citizenship is less relevant than the increasing recognition of human rights. Others see access to citizenship as crucial to the protection of individuals and thus needing to be more readily accessible; yet for others, citizenship represents the full value of membership, and is devalued if too easily acquired (see chapter "Migration and Borders" in this Handbook). The question of the appropriate grounds for holding a specific citizenship arises in a number of contexts including naturalisation, dual citizenship, citizenship withdrawal, and birthright citizenship. While theorists generally dismiss ascription on the basis of ethnic or cultural affinities, alternative normative grounds proposed include: social attachment or membership, contract or voluntary consent, identity, loyalty, contribution, or political interdependence in subjection to the authority of the state.

While borders control entry and residence, membership boundaries control the guarantee of that entry and residence. States have sometimes been seen as more restrictive on initial entry and residence than on membership—described as being "hard on the outside and soft on the inside" (Bosniak 2006). But the conditions that states may require for the naturalisation of migrants have perhaps been the most widely discussed aspect of citizenship. Whether the state should be sovereign in awarding citizenship, or it should follow from immigration is disputed. While Walzer argues that a state can choose who to admit, he insists that excluding long term residents from full membership is tantamount to slavery. Some see voting rights for migrants as an alternative to citizenship, others argue for mandatory citizenship on the grounds of political interdependence (de Schutter and Ypi 2015).

More applied political philosophy has addressed issues of required residence periods, multiple citizenship, and increasingly stringent conditions of integration, in particular integration tests, all reflecting political issues of identity, loyalty, social cohesion, and democratic public culture. Such conditions are contrasted to fast tracking procedures for high-skills candidates, and the marketisation of citizenship.

The fact that most people in the world gain their citizenship through birth, either by descent or birth in a territory, has been criticised as leading to radically unequal opportunities, and, with migration, meaning that globally many are not citizens of the country in which they live. Accordingly, the case has been made that citizenship should be awarded in the country with which a person is most closely connected, rather than on the dominant birthright regimes (Shachar 2009).

Citizenship policies also render a significant number of people *stateless* from birth, not only limiting their security of residence and re-entry, but leaving them without the protection of a state (Bloom et al. 2017). Also at issue are the increasingly common state provisions for withdrawing citizenship, often on security grounds, that can result in deportation and even statelessness.

3 Topics for Future Research in Migration Ethics

In discussion of migration, the ideal focus of analytical political philosophy, and its overlap with applied ethics have been criticised and it has been modified by exponents of non-ideal theory and, to a lesser extent, by political realists. It has increasingly engaged with more concrete contexts of migration and concerns of feasibility, but it is another question whether there is a distinctively "political realist" methodology (focused on order and legitimacy) rather than different normative perspectives at work here.

I conclude by indicating three areas for future research.

(a) There is a need for more precise articulation of what may be distinctive about *territorial jurisdiction* as the basis of states' rights to control boundaries and to determine the status of individuals, compared with other associations (clubs, families, marriage) and institutions (property), analogies which have hitherto featured prominently in political philosophy of migration. This requires more examination of the basis on which states have such territorial control, including the idea of "occupation" (Moore 2015; Stilz 2019).

Though arguments for association, nationality, culture, property, and collective ownership of institutions have been shown to be problematic, there remains considerable power in the claim of a political community to *self-determination*, detached from these bases. But questions remain about the locus of self-determination, and what it entails, and, in particular whether, why (and when) the self-determination of one set of people overrides the interests of others. And, in any case, does justified partiality to fellow citizens and self determination extend to controlling entrance and residence in the state (Song 2018)?

(b) Migration research needs to recognise more explicitly the historical and contingent nature of the international order of nation-states and border regimes, and to transcend what has been aptly termed *methodological nationalism* predominant in political science, for example, to consider the treatment of migration, participation and citizenship issues at sub-national, transnational and supra national levels (Sager 2016; Bauböck 2018). There has been an increasing tendency to highlight the idealisation involved in the view of a world of discrete and equal nation-states with clear borders and clear distinctions between insiders and outsiders in which each person has a place as a citizen that some, ostensibly realistic, approaches have adopted.

(c) If trends in migration continue, it may be that the future world will be one of hypermobile, rather than the present relatively sedentary, populations. An increasing dimension of migration relates to *climate change*: this will change the game for many issues of migration in the numbers and diversity of migrants, as well as the nature of their claims. Already there is a refugee crisis in which international agreement on procedures for distributing responsibilities to refugees have broken down. With climate change, the issue of who counts as a refugee (contrasted to an economic migrant), already a controversial issue, will become more complicated (Owen 2020).

All three issues suggest a need to reorientate the focus of the political philosophy of migration to address the kind of obligations and collective action needed among states and citizens globally.

Literature

Abizadeh, A. 2008. Democratic Theory and Border Coercion. No Right to Unilaterally Control Your Own Borders. *Political Theory* 36:37–65.

Bauböck, R. 2018. *Democratic Inclusion*. Manchester: Manchester University Press.

Blake, M. 2001. Distributive Justice, State Coercion and Autonomy. *Philosophy and Public Affairs* 30:257–296.

Bloom, T., K. Tonkiss, and P. Cole, eds. 2017. *Understanding Statelessness*. Abingdon: Routledge.

Bosniak, L. 2006. *The Citizen and the Alien. Dilemmas of Contemporary Membership*. Princeton, NJ: Princeton University Press.

Carens, J. 1987. Aliens and Citizens. The Case for Open Borders. *Review of Politics* 49:251–273.

Carens, J. 2013. *The Ethics of Immigration*. Oxford: Oxford University Press.

Cole, P. 2000. *Philosophies of Exclusion. Liberal Political Theory and Immigration*. Edinburgh: Edinburgh University Press.

Costa, M. V. 2016. Republican Liberty and Border Controls. *Critical Review of International Social and Political Philosophy* 19:400–415.

De Schutter, H., and L. Ypi. 2015. Mandatory Citizenship for Immigrants. *British Journal of Political Science* 45:235–251.

Fine, S. 2010. Freedom of Association is not the Answer. *Ethics* 120:338–356.

Fine, S. 2019. Refugees, Safety and a Decent Human Life. *Proceedings of the Aristotelian Society* 119:25–52.
Geuss, R. 2008. *Philosophy and Real Politics*. Princeton, NJ: Princeton University Press.
Gibney, M. 2004. *The Ethics and Politics of Asylum. Liberal Democracy and the Response to Refugees*. Cambridge: Cambridge University Press.
Lawford-Smith, H. 2013. Understanding Political Feasibility. *Journal of Political Philosophy* 21:243–259.
List, C., and L. Valentini. 2016. The Methodology of Political Theory. In *Oxford Handbook of Philosophical Methodology*, ed. H. Cappelen, T. Szabó Gendler, and J. Hawthorne. Oxford: Oxford University Press.
McDermott, D. 2008. Analytical Political Philosophy. In *Political Theory. Methods and Approaches*, ed. D. Leopold and M. Stears. Oxford: Oxford University Press.
Miller, David. 2016. *Strangers in Our Midst. The Political Philosophy of Immigration*. Cambridge, MA: Harvard University Press.
Moore, M. 2015. *A Political Theory of Territory*. Oxford: Oxford University Press.
Oberman, K. 2016. Immigration as a Human Right. In *Migration in Political Theory. The Ethics of Movement and Membership*, ed. S. Fine and L. Ypi, 32–56. Oxford: Oxford University Press.
Owen, D. 2014. Republicanism and the Constitution of Migrant Statuses. *Critical Review of International Social and Political Philosophy* 17:90–110.
Owen, D. 2020. *What Do We Owe to Refugees?* Cambridge: Polity Press.
Pettit, P. 2012. *On the People's Terms*. Cambridge: Cambridge University Press.
Pevnick, R. 2011. *Immigration and the Constraints of Justice. Between Open Borders and Absolute Sovereignty*. Cambridge: Cambridge University Press.
Rawls, J. 1971. *A Theory of Justice*. Cambridge, MA: Harvard University Press.
Sager, A. 2016. Methodological Nationalism, Migration and Political Theory. *Political Studies* 64:42–59.
Shachar, A. 2009. *The Birthright Lottery. Citizenship and Global Inequality*. Cambridge, MA: Harvard University Press.
Sleat, M., ed. 2018. *Politics Recovered. Realist Thought in Theory and Practice*. New York: Columbia University Press.
Song, S. 2018. *Immigration and Democracy*. Oxford: Oxford University Press.
Steiner, H. 1992. Libertarianism and the Transnational Migration of People. In *Free Movement. Ethical Issues in the Transnational Migration of People and of Money*, ed. B. Barry and R. E. Goodin, 87–94. University Park, PA: Pennsylvania State University Press.
Stilz, A. 2019. *Territorial Sovereignty. A Philosophical Exploration*. Oxford: Oxford University Press.
Waldron, J. 2016. *Political Political Theory*. Cambridge, MA: Harvard University Press.
Walzer, M. 1983. *Spheres of Justice*. Oxford: Oxford University. Press.
Wellman, C. H. 2008. Immigration and Freedom of Association. *Ethics* 119:109–141.

Migration Ethics and/in Philosophy of Law

François Boucher

Abstract In this chapter, I discuss the relevance of theories of rights for the ethics of migration by focusing on issues of entry (mostly leaving aside questions of membership, for the sake of concision). In the first section, I introduce the main elements of the analysis of the concept of rights. I discuss the special weight of rights, the connection between rights and obligations, the basic constitutive elements of rights, and the competing will theory and interest theory of rights. In the second section, I show how disagreements between the interest and will theory of rights shape debates about the right to immigrate and the right to exclude. This section also draws attention to the elements constituting the right to immigrate and the right to exclude, as well as to their corresponding duties. In the final section, I suggest that legal philosophy can shed further light on debates about migration regarding rights restrictions and rights differentiation.

1 Theories of Rights. Methodological Particularities

Discussions about rights pervade our political morality. The topic of migration is no exception to this, and debates within the ethics of immigration are often framed in the language of rights: is there a right to international freedom of movement? Do states have the right to exclude would-be migrants? Should immigrants be granted the same rights as citizens? Such use of the language of rights is not without consequences since the concept of rights has very distinctive features that make rights-based moral theories differ substantially from other forms of morality, such as goal-based and duty-based theories (Dworkin 1978, 169–173; Feinberg 1970; Raz 1986).

Duty-based moralities stipulate what people are required to do as a matter of obligation, emphasising conformity with rules such as "thou shalt not kill". From the point of view of duty-based theories, nonconformity with deontic rules is the essence of wrongdoing (Waldron 1984: 13). By contrast, rights-based moralities

F. Boucher (✉)
Institute of Philosophy, KU Leuven, Leuven, Belgium

A. Niederberger et al. (eds.), *Handbook of Migration Ethics*,
https://doi.org/10.1007/978-3-031-89877-8_8

locate the wrongdoing in the harm done to the individual victim. When a right has been violated, someone has been wronged. Rights, indeed, have a special importance for the right holder herself. Goal-based theories are concerned with the maximisation of utility or with promotion of desirable states of affairs. By contrast, rights-based moralities are grounded in individualistic considerations and are concerned with protecting individuals from the pursuit of the good by others or by society as a whole. Rawls, for instance, clearly posits his theory of justice as fairness as a rights-based theory when he states that "each person possesses an inviolability founded on justice that even the welfare of society as a whole cannot override" and that "justice denies that the loss of freedom for some is made right by a greater good shared by others" (Rawls 1971: 3–4). Rights-based moralities do not focus on what is desirable to do (the goals to be pursued), but rather on what is permissible to do.

A rights-based approach to migration is therefore mostly independent from discussions about the benefits and costs of migration. From this perspective, that more open borders would make both host and sending societies more prosperous does nothing to support the claim that there is a right to immigrate. Similarly, there is no contradiction in claiming that states have a right to exclude immigrants and that would benefit from immigration. The right to exclude means that it is *permissible* to deny entry to would-be migrants (or to deny them access to full citizenship), not that it is *advantageous* to do so or that there is a duty to do so. Similarly, the right to immigrate means that it is permissible for one to immigrate to another country.

Rights have special weight or a special force. They protect individuals in the sense that they are "trumps" that override utility considerations related to the desirability of certain societal goals (Dworkin 1978: 90–94; 1984: 152). Rights are resistant to trade-offs; they deny that one's freedom or interests can be sacrificed for the greater good of others. However, rights are not absolute, they only resist the balancing of interests and desirable outcomes up to a certain point (Dworkin 1978: 191–192; Thompson 1990: 149–175). That is, someone's right can be limited in order to protect someone else's rights, or it may be that the pursuit of some goal is so urgent and so important that it becomes justified to restrict people's rights. Still in those cases of legitimate rights restrictions, some proportionality criteria must be observed: the restriction of the right must be the least restrictive way to indeed achieve the urgent goal and the fulfillment of the goal must still be of a value great enough to compensate for the loss of freedom (Letsas 2015; see also Thompson 1990: 149–175).

Rights also have a special force in the sense that they are grounds for imposing duties onto other persons. One's right is another's liability to justified coercion. As Hart says: "a very important feature of a moral right is that the possessor of it is conceived as having a moral justification for limiting the freedom of another" (Hart 1955: 178). Rights have correlative duties. Yet, having a right is more than the mere fact that another has a correlative duty, it also includes the right-holder's capacity to influence whether and when the duty bearer discharges his duty (Feinberg 1992: 155). At the most general level, rights have two forms: active and passive. One has an active right to do X when others have the negative duty not to interfere with one's

Xing. One has a passive right to X when another party has the duty to provide one with X (Wenar 2005: 233). Analytical jurists also distinguish four main elements constituting rights (the simple "atoms" from which more complex "molecular" rights are made). Hohfeld (1964 [1919]) thus distinguished (1) claims (claim-rights) by virtue of which someone can demand a particular thing from another, where that other person has a correlative duty to comply with that demand; (2) liberties (or "privileges") to do or to refrain from certain things; (3) powers by which people can change their existing claim and privileges; and (4) immunities by which someone is protected from the powers of others.

There is a tendency to assimilate all rights to claim-rights. In this view, X's claim-right to P means that Y has the duty to provide X with P, or that Y has the duty not to prevent X from enjoying P. Yet, many legal philosophers hold that rights (moral or legal) consist of a variety of combinations of basic elements constituting rights (called "Hohfeldian incidents") (Lyons 1970; Scanlon 2003; Wenar 2005; cf. Hohfeld 1964 [1919]: 36; Thompson 1990: 37–60). The concept of claim-rights seems to capture the normative relations at stake in rights against specific harms (like the right to life), rights to certain goods or services (the right to education imposes a duty on the state, to offer publicly financed education), or rights to the performance of agreed-upon actions, such as the creditor's right to be paid back. Yet, some important rights are best described as pairs of privileges: my right to religious freedom means that I have no duty to practice a religion, and no duty not to practice a religion; the rights to free speech and freedom of association are similar (Wenar 2005: 227–230). For some, such freedom rights also have an immunity element that protects their holders against the liability to having one's normative situation changed by other agents (Lyons 1970: 49–54).

One of the most salient aspects of rights analysis for normative debates about migration relates more to the function of rights than to their forms. There are two rival accounts of the function of rights, of what rights are meant to do to their bearers: the interest theory and the will theory (for overviews of this long-lasting debate, see Waldron 1984; Edmundson 2004: 119–132; Frydrych 2018). The interest theory maintains that the function of rights is to protect the interests of the right holder (Raz 1986; MacCormick 1977; Kramer 2017). For instance, Raz asserts that "'X has a right' if and only if X can have rights, and other things being equal, an aspect of X's well-being (his interest) is a sufficient reason for holding some other person(s) to be under a duty" (1986: 166). Supporters of the interest theory must start from an account of human well-being or flourishing to establish a list of basic needs or interests that are central enough to a decent human life so that their fulfillment justifies holding another agent under the duty to secure the right holder's interest. Thus, for interest theories, rights must pass a threshold of sufficiency: to assert that there is a right, it must be shown that the interest is sufficiently strong or important to justify holding another agent under a duty (Wenar 2005: 242).

The will theory, by contrast, asserts that the function of rights is to protect individual autonomy (the main proponents of the will theory include Hart 1955, Feinberg 1970 and Wellman 1985). Respecting individual rights is necessary to treat individuals as free and equal persons. Hart, for instance, conceives that a right holder is a "small-scale sovereign" to whom duties are owed (1982: 183). According to him, what a right does to the right holder is to provide her with "a moral justification for limiting the freedom of another", but this justification is not grounded in the (sufficiently strong) interests of the right holder but in the view that "a certain distribution of freedom will be maintained" if the right holder has the choice to set certain limits on how others act (Hart 1955: 178). Here, rights define a space within which the individual has the prerogative to act (or not) without others' interference or to demand (or not) that others act in a certain way. For will theorists, non-interference within this space of freedom and choice for the individual is the basis of respect for persons qua persons, or of human dignity. Feinberg, for instance, asserts that having rights enables us to stand up like free and equal humans, to "look others in the eye, and to feel in some fundamental way the equal of anyone" (1970: 252).

From a normative point of view, the two theories mostly differ in that the will theory starts with a presumption in favor of liberty (one's freedom can only be restricted when one is under the duty to respect someone else's rights), and thus places the burden of proof on those who wish to limit liberty. By contrast, in the interest theory the burden of proof is on those asserting the existence of rights, as they need to explain why an interest is strong enough to ground another's duty.

2 Theories of Rights and Immigration

Proponents of the right to immigrate have developed arguments relying either on the will theory, or the interest theory of rights. In denying that there is such a right, critics have drawn on the weaknesses of each theory. Proponents and critics have also drawn on different features of the form of rights. The same can be said of the proponents of the right to self-determination and to exclude immigrants (and their critics).

Many philosophical discussions of the right to immigrate rest on the interest theory (Oberman 2016; Miller 2016b; Song 2018). They assert that people have a right to immigrate if they have a fundamental interest in moving across borders and in settling in a different society, provided that the interest is strong enough to impose a duty upon receiving societies to admit immigrants. Such arguments often adopt the so-called "Cantilever strategy" (Miller 2016b: 16–17). They proceed by establishing that the right to internal (domestic) freedom of movement protects significant interests, and then claim that international freedom of movement protects the same interests. Oberman, for instance, claims that internal freedom of movement protects personal and political fundamental interests (2016: 35–36). People have a personal interest in accessing a wide range of life options in the domains of friendship, love, family life, work, civic associative life, culture, religion, and so on. This in turn requires ability to move across space, to visit other persons, attend meetings and gatherings, and to leave one's city and relocate to a different part, in order, for instance, to take a job or join a community that is more culturally vibrant. People also have a political interest in enjoying "free and effective political processes" (Oberman 2016: 36), which entails being able to attend demonstrations and speeches, to gather information directly, to discuss with other people and join associations. But the interest in having life options and access to free political processes also requires freedom to cross boundaries and to immigrate. As Carens explains, in addition to opening important economic opportunities abroad, international freedom of movement is necessary for the pursuit of many other personal interests, as "one might fall in love with a citizen from another land, one might belong to a religion which has few followers in one's native land and many in another, one might seek cultural opportunities that are only available in another society" (1987: 258).

Critics of this account of the right to immigrate draw on the weakness of the interest theory of rights. Recall that interest-based arguments for rights must show that the interest meets a sufficiency threshold to justify the imposition of a correlative duty. However, the capacity of one's interests to generate a duty does not only depend on the importance of the interest, it also depends "on the absence of conflicting considerations" (Song 2018: 97), such as undue burdens on the potential duty bearer or conflicting rights held by the potential duty bearer (see also Miller 2016b: 16–17). Therefore, critics of interest-based defenses of the right to immigrate make one of the two following claims: (1) the interest in international freedom of movement is not strong enough; (2) there are countervailing considerations indicating that the duty imposed upon receiving societies is too burdensome, or conflicts with their rights. Song takes the second road by claiming that the states' right to self-determination overrides the interest in international mobility (2018: 96–98). I discuss accounts of the right to self-determination later in this section.

Miller takes the first road, arguing that the truly vital personal interest at stake with freedom of movement is the interest in having access to an *adequate* range of life options (2016a: 52; b: 19–21). Although people have a vital interest in being able to move from one region or city to another in order to have adequate economic, social, cultural and religious options, the ability to move to a different country provides more than an adequate range of options. This extra range of options made possible by international mobility is a luxury, a preference, not a vital interest. As such, it is not strong enough to justify imposing a duty to a receiving society. However, critics of the adequacy view claim that it would permit severe restrictions of internal freedom of movement. If the residents of a small country, both in size of territory and population, have access to adequate life options, then it seems to entail that a larger state such as the United States would not be violating the right to (internal) freedom of movement by restricting movement between its federated states (Oberman 2016: 39; Song 2018: 96). Critics also argue that people do not merely have an interest in having an adequate range of life options, they have an interest in "accessing the *full* range of existing life options" (Oberman 2016: 38) and the extra amount of life options made possible by immigration is not a matter of mere preference.

Another way to do away with the adequacy view of freedom of movement simply rejects the very concept of rights underpinning the argument. Several supporters of the right to immigrate do not presuppose that the function of rights is to protect interests. Instead, they embrace a will theory of rights that asserts that the function of rights is to protect persons as autonomous beings. This is the route taken by Kukathas who grounds the right to immigrate in "a presumption in favor of free movement" and on the view that "it is intervention or interference in the activities of others that requires justification" (2012: 655). In this view, there is no need to establish that international mobility is required to protect vital interests and that accessing the full range of life options is not a matter of simple preference. Respect for persons requires that we refrain from restricting their freedom, unless we have very strong reasons (e.g. Hart's preservation of an equal distribution of freedom). Even if accessing the full range of life options made possible by international freedom of movement is not a matter of basic, vital interests, immigration controls fail to respect migrants as persons because the mere fact that doing something is a preference and not a vital interest is not in itself a valid reason for preventing someone from doing it.

Cole also develops a defense of the right to immigrate based on the will theory. He asserts that the capacity to immigrate is "an essential component of human agency" (2011: 297). It enables people to exercise control over their life and to be the authors of their own life. However, for him, the right to immigrate is not simply about a liberal concern for the intrinsic value of respecting people as choosers. He also grounds his analysis on a critical perspective which recognises that legal rights and positive law are a double-edged sword: they can promote liberal and humanistic values, but they also reflect existing relations of power and contribute to maintaining hierarchies already in place (see Ivison 2008). In this view, we should not only conceive of rights as protecting pre-political freedoms, but also as protections against typical threats linked to abuses of power. As Cole remarks, border controls function mainly to exclude the global poor from rich states (2011: 298). As a result, would-be migrants from poorer countries enjoy lesser freedom of movement and fewer economic opportunities. The right to immigrate, in this context, can be defended as a protection against the tendency of wealthier states to exclude the global poor. In other words, it can be defended as a tool to redistribute power and enable the most vulnerable to better control their lives (see also Kukathas 2012: 664–665).

One important difficulty with a will-based argument for the right to immigrate is the difficulty of identifying who has the corresponding duty to admit a specific immigrant. In international law, the right to emigrate imposes a duty on a clearly identifiable agent (a negative duty not to prevent exit imposed on one's native state), but it is hard to identify which state has the duty to admit an immigrant (Benhabib 2006: 30). The will theory rests on the capacity of individuals to make claims against other agents and to choose when to force the bearer of a corresponding duty to perform her duty. This makes it hard for will theorists to claim that there is a right to immigrate, as would-be immigrants lack the capacity to (choose to) hold any state under the corresponding duty to admit them.

However, this objection misunderstands the distinction between legal and moral rights. If it is true that there is no legal right to immigrate in the current international system, it does not follow that there is no moral right to immigrate. Moral rights demand to be institutionally implemented. Therefore, to say that there is a moral right to immigrate means that there should be an international legal system assigning corresponding duties to states. This can for instance take the form of an agreement between states on principles of responsibility assignment or of the creation of an international agency with the authority to determine which state should admit which immigrant (Cole 2011: 201). Moreover, one can question the view that someone's right to immigrate is conditional upon the identification of a specific agent bearing the corresponding duty of admission. This view seems to reduce the right to immigrate to a single Hohfeldian incident: a claim-right, which has a passive form: X's right to Z means that X can demand that Y provides Z to X. This form adequately describes rights to certain goods and services. However, the right to immigrate contains a component of liberty (privilege) that is not captured by this. The freedom to move across boundaries is indeed akin to fundamental freedoms such as freedom of conscience, of expression, and of association. Freedom of movement is a pair of liberties (no duty not to move, no duty to move) (see Wenar 2005). If that is correct, then what is needed to secure immigrants' ability to choose when to activate corresponding duties is not so much an international agency assigning each immigrant to a specific state, but rather an international legal system granting a constitutional immunity to immigrants against the liability to have their pair of movement-related liberties taken away by states.

Proponents and critics of the states' right to exclude have also framed their arguments in the terms of the interest and will theories. Those debates focus on whether collective self-determination includes the effective capacity to control borders and exclude would-be immigrants. One way of justifying the right to exclude starts from the claim that people have an interest in belonging to self-determining political communities, capable of making various policy choices (Miller 2016a: 62–72). The right to exclude is central to self-determination because immigration can adversely affect people's interest in collective self-determination. For instance, immigration might increase cultural diversity, which in turn allegedly weakens the social trust needed to support many social policies and institutions of the welfare state, thereby undermining the community's capacity to pursue a given political agenda (Miller 2016a: 64–65). Critics argue both that (a) the empirical evidence supporting the view that diversity undermines the social trust upon which the welfare state rests is at best very thin (Pevnick 2009), and (b) that the boundaries of culturally homogenous nations do not match those of states, namely the purported bearers of the right to exclude (Fine 2013: 264).

Other accounts directly tie collective self-determination with human agency and individual autonomy. Wellman, for instance, grounds self-determination in the respect for persons as autonomous beings and as bearers of special responsibilities (2011: 22–23). Self-determination enables the members of a political community to be the authors of their own lives. It also enables political communities to carry-on the important special responsibilities to set up political and legal institutions designed to protect and promote the rights of their members. Disabling states in the area of migration control fails to express respect to people. It amounts to forcing people to associate with others, which fails to respect them as autonomous persons just as forcing one person to marry another does. Those objecting to this line of reasoning cast doubt on the analogy between the state and associations that undoubtedly have a right to exclude (Fine 2010). The state is not a voluntary association and it is unlikely that accepting migrants would force individuals to associate with them in the usual sense of forming clubs, churches, and so on.

Instead of appealing to freedom of association, Blake draws on the idea of a legal system (2013). For him, one of the morally salient aspect of states, in relation to immigration, is that they have jurisdiction over a territory. The state is the final authority over a piece of land; it determines which laws apply and it is responsible for enforcing and interpreting them. States are complex legal systems which include various institutions that play an essential rule in the protection and fulfillment of the rights of their members (and of all those on their territory). This means that when migrants cross boundaries, they acquire new legal rights, rights that they can demand to be enforced by the receiving society. At the same time, receiving states acquire new legal obligations to respect the human rights of migrants on their territory. Admission also sets them on a course to eventually acquire the obligation to protect full-citizenship rights of naturalised migrants (if there is an obligation to offer naturalisation after some time, see Carens 2013: 158–162). It also generates obligations to respect the special rights of temporary migrants who do not plan to stay indefinitely and acquire citizenship (Ottonelli and Torresi 2012). Seen in this light, the right to immigrate is not simply a claim-right or a pair of liberties backed by immunity. It is also a power-right, that is, an effective capacity to modify existing rights and duties (what has also been noticed by some proponents of open borders, see Kukathas 2012: 653). Thus, one can say, drawing on Blake jurisdictional account, that self-determination is an immunity-right that protects members of a political community against being liable to acquiring new and unwanted obligations as a result of migration.

Song also links self-determination to concerns for human agency and the capacity of people to direct their lives as they see fit (2018). Yet, she argues that Blake and Wellman fail to appreciate the democratic and irreducibly collective nature of the right to self-determination. For her, uncontrolled immigration threatens self-determination not so much because it imposes unwanted legal obligation and forms of association upon the individual members of a political community. Nor does it undermine self-determination because of its alleged effects on cultural diversity and social trust. Rather, self-determination is a democratic ideal: it enables a demos to rule itself. In this perspective, immigration control is therefore essential to it because without it, immigrants "sidestep the political process by which members of the political community can define who the collective self is" (Song 2018: 66).

Wellman, Blake and Song do not ground their justification of the right to exclude upon a conception of humans' vital interests or basic needs or on the expected consequences of open borders on people's well-being. They rather frame their arguments in terms aligned with the will theory and emphasise the importance of the right to exclude for respecting people as the authors of their lives. However, critics of immigration controls have opposed the state's right to exclude on the same ground. They argue that immigration policy must also consider the agency and equal freedom of people who are denied entry in country. Starting from the claim that to be legitimate, the exercise of coercive political power must be justified to those subjected to it, they argue that since borders coercively exclude would be immigrants, immigration policies cannot be the result of unilateral decisions by individual states, they can only acquire legitimacy through "cosmopolitan democratic institutions in which borders received actual justifications addressed to both citizens and foreigners" (Abizadeh 2008: 48).

3 Desiderata for Future Developments

Proponents of border controls generally do not believe that self-determination entails an absolute right to exclude. Typically, they claim that this right admits certain restrictions so that, for instance, states cannot legitimately exclude those fleeing persecution or extreme deprivation (Blake 2013: 125–126; Miller 2016a: 76–93; Song 2018: 69). Similarly, supporters of the right to immigrate often call for a "qualified right" to immigrate that can be subjected to certain limitations (Cole 2011: 247; Carens 2013: 250–252). But if rights are especially weighty considerations designed to trump the pursuit of collective goods, how is it possible to limit them without betraying their main purpose? What principles should guide practices of rights restrictions? Legal philosophers have claimed that rights restrictions can be legitimate if they are proportional. Their views can shed light on legitimate limitations to the right to immigrate (or to exclude immigrants). For example, with the proportionality framework, although some restriction of this right can be justified, restricting the right to immigrate to fulfill the goal of protecting public order and preventing chaos is not a proper justification when the threat of chaos is purely hypothetical. Restricting immigration to protect national identity can be seen as disproportional because it is not the least restricting measure likely to fulfill the objective, as integration can achieve the same result with fewer rights infringements. Carens considers such a framework when explaining why the right to immigrate need not be absolute (2013: 251).

I have tried to highlight how theories of rights can shed light on core problems in the ethics of immigration by focusing on issues of entry, leaving aside issues of membership. Once immigrants are admitted, which rights should they have? Should they quickly be granted full citizenship rights? Should citizenship rights be disaggregated so that some rights, but not all, can be granted to immigrants while they are on their way to acquire citizenship? Should temporary migrants be granted a specific bundle of rights adapted to their specific situation? A full account of the relevance of theories of rights for the ethics of migration should also explore how the interest and will theories of rights offer different grounds for rights differentiation. For example, Ottonelli and Torresi suggest that respect for human agency requires us to grant temporary migrants a specific bundle of rights that contains less than full citizenship rights, but includes unique elements tailored to the specific situation of temporary migrants (2012).

Literature

Abizadeh, A. 2008. Democratic Theory and Border Control. No Right to Unilaterally Control Your Own Border. *Political Theory* 36:37–65.

Benhabib, S. 2006. *Another Cosmopolitanism*. Oxford: Oxford University Press.

Blake, M. 2013. Immigration, Jurisdiction and Exclusion. *Philosophy and Public Affairs* 41:103–130.

Carens, J. 1987. Aliens and Citizens. The Case for Open Borders. *The Review of Politics* 49:251–273.

Carens, J. 2013. *The Ethics of Immigration*. Oxford: Oxford University Press.

Cole, P. 2011. Open Borders. An Ethical Defense. In *Debating the Ethics of Immigration. Is There a Right to Exclude?* ed. P. Cole and C. H. Wellman, 159–313. Oxford: Oxford University Press.

Dworkin, R. 1978. *Taking Rights Seriously*. Cambridge, MA: Harvard University Press.

Dworkin, R. 1984. Rights as Trumps. In *Theories of Rights*, ed. J. Waldron, 153–167. Oxford: Oxford University Press.

Edmundson, W. A. 2004. *Introduction to Rights*. Cambridge: Cambridge University Press.

Feinberg, J. 1970. The Nature and Value of Rights. *Journal of Value Inquiry* 4:245–257.

Feinberg, J. 1992. In Defence of Moral Rights. *Oxford Journal of Legal Studies* 12:149–169.

Fine, S. 2010. Freedom of Association Is Not the Answer. *Ethics* 120:338–356.

Fine, S. 2013. The Ethics of Immigration. Self-Determination and the Right to Exclude. *Philosophy Compass* 8:254–268.

Frydrych, D. 2018. The Theories of Rights Debate. *Jurisprudence. An International Journal of legal and Political Thought* 9:566–588.

Hart, H. L. A. 1955. Are There Any Natural Rights? *The Philosophical Review* 64:175–191.

Hart, H. L. A. 1982. Legal Rights. In *Essays on Bentham. Jurisprudence and Political Philosophy*, ed. H. L. A. Hart, 162–193. Oxford: Oxford University Press.

Hohfeld, W. N. 1964 [1919]. *Fundamental Legal Conceptions as Applied in Judicial Reasoning*. Westport, CT: Greenwood Press.

Ivison, D. 2008. *Rights*. Stocksfield: Acumen.

Kramer, M. 2017. In Defence of the Interest Theory of Right-Holding. Rejoinders to Leif Wenar on Rights. In *New Essays on the Nature of Rights*, ed. M. McBride, 49–84. Oxford: Bloomsbury.

Kukathas, C. 2012. Why Open Borders? *Ethical Perspectives* 19:650–675.

Letsas, G. 2015. Rescuing Proportionality. In *Philosophical Foundations of Human Rights*, ed. R. Cruft, S. M. Liao, and M. Renzo, 316–340. Oxford: Oxford University Press.

Lyons, D. 1970. The Correlativity of Rights and Duties. *Noûs* 4:45–55.

MacCormick, N. 1977. Rights in Legislation. In *Law, Morality and Society. Essays in Honour of H. L. A. Hart*, ed. P. Hacker and J. Raz, 189–209. Oxford: Clarendon.

Miller, D. 2016a. *Strangers in Our Midst. The Political Philosophy of Immigration*. Cambridge, MA: Harvard University Press.

Miller, D. 2016b. Is There a Human Right to Immigrate? In *Migration in Political Theory. The Ethics of Movement and Membership*, ed. S. Fine and L. Ypi, 11–31. Oxford: Oxford University Press.

Oberman, K. 2016. Immigration as a Human Right. In *Migration in Political Theory. The Ethics of Movement and Membership*, ed. S. Fine and L. Ypi, 32–56. Oxford: Oxford University Press.

Ottonelli, V., and T. Torresi. 2012. Inclusivist Egalitarian Liberalism and Temporary Migration. A Dilemma. *The Journal of Political Philosophy* 20:202–224.

Pevnick, R. 2009. Social Trust and the Ethics of Immigration Policy. *The Journal of Political Philosophy* 17:146–167.

Rawls, J. 1971. *A Theory of Justice*. Cambridge, MA: Harvard University Press.

Raz, J. 1986. *The Morality of Freedom*. Oxford: Clarendon.

Scanlon, T. 2003. Rights, Goals and Fairness. In *The Difficulty of Tolerance. Essays in Political Philosophy*, ed. T. Scanlon, 26–41. Cambridge: Cambridge University Press.

Song, S. 2018. *Immigration and Democracy*. Oxford: Oxford University Press.

Thompson, J. J. 1990. *The Realm of Rights*. Cambridge, MA: Harvard University Press.
Waldron, J. 1984. Introduction. In *Theories of Rights*, ed. J. Waldron, 1–20. Oxford: Oxford University Press.
Wellman, C. 1985. *A Theory of Rights*. Totowa, NJ: Rowman & Allanheld.
Wellman, C. H. 2011. Freedom of Association and the Right to Exclude. In *Debating the Ethics of Immigration. Is There a Right to Exclude?* ed. P. Cole and C. H. Wellman, 13–155. Oxford: Oxford University Press.
Wenar, L. 2005. The Nature of Rights. *Philosophy and Public Affairs* 33:223–252.

Migration Ethics and/in Democratic Theory

Daniel Kersting

Abstract Migration and democracy are in a tense relationship. On the one hand, migrants have, in terms of human rights, the same legitimate claim to equal moral consideration as every citizen of a state they wish to enter. On the other hand, citizens of democratic communities have the *de facto* right to prevent migrants from entering their territory and to exclude them from political membership. In the following article, this tension is described against the background of four different normative models of democracy—liberal, communitarian, deliberative and agonistic—which are introduced and also discussed from a transnational or cosmopolitan perspective in the first section (I). Taking into account these different approaches in democratic theory, the second section (II) discusses three systematic questions central to the research field, namely (a) the question of the legitimacy of territorial exclusions, (b) the question of the possibilities of political membership and (c) the question of the connection between democracy and refugee protection. Finally, in a brief outlook (III), the paper argues for further research to reflect the relationship between migration and democracy more strongly from a cosmopolitan perspective, combining insights from theories that attempt to establish a legitimate global democratic order with radical democratic and deliberative elements. This could open up a perspective on migration in which migration both calls the democratic "we" and its boundaries permanently into question and thereby pushes forward processes of democratisation of borders and boundaries towards a more inclusive and more legitimate democratic order.

Worldwide flight and migration movements expose democratic states to the question of the legitimacy of their external (territorial) as well as internal (civil) borders and the resulting inclusions and exclusions: Do states have the right to exclude migrants at their own discretion from access to territory or to deprive parts of their population of citizenship for years and decades? To answer this question, both advocates and critics of state sovereignty repeatedly refer to the idea of democracy. Some

D. Kersting (✉)
Department of Philosophy, Friedrich Schiller University Jena, Jena, Germany
e-mail: daniel.kersting@uni-jena.de

A. Niederberger et al. (eds.), *Handbook of Migration Ethics*,
https://doi.org/10.1007/978-3-031-89877-8_9

see the "right to exclusion" as a direct element of democratic self-government; others see the idea of democracy as having the potential to politically include the excluded and to democratise border regimes. Obviously, these different positions are based on different understandings of democracy, and the plausibility of these understandings ultimately determines the persuasive power of the respective position. In order to understand and critically evaluate the different positions, I will first give a brief overview of the most important approaches to democratic theory and their implications for migration policy.

1 Democratic Theory in a Globalised World

The term "democracy" comes from the Greek and literally means popular rule (*demos*: people and *kratein*: to rule). The central idea that is still associated with this term today is that of self-government of free and equal individuals. However, what is meant by "self-government", especially of a "people", what practices or procedures characterise it and in what institutional form it best develops are both theoretically controversial and politically contested.

1.1 Four Normative Models of Democracy

When we speak of modern democracy, we usually refer to the idea of *liberal democracy*, which is historically rooted in two different political traditions, liberalism and republicanism. Liberalism emphasises the protection of individual freedoms and the private sphere against the political power of the majority; republicanism, conversely, defends the idea of self-legislation of political communities and considers the freedom of the individual to be realised only in her active role as a citizen. Liberal democratic states have historically attempted to unite the two traditions by delegating the democratic power of the people to representatives and by constraining the executive power of the state by binding it to basic and human rights under the rule of law. This allowed a complex system of checks and balances to develop, from which state power derives its legitimacy. This model of democracy continues to shape the self-image of most Western states to this day. At the same time, liberal democracy has become the object of a wide variety of philosophical, political and sociological criticisms, from which independent and influential approaches to democracy have increasingly developed since the 1980s.

One of the most powerful social and political countermovements to liberal democracy is *communitarianism.* Similar to republicanism, according to the communitarian understanding of democracy, political decisions achieve their legitimacy through their orientation toward the common good as an idea of the good life. This process is enabled by an ethos of shared values and cultural practices, which is understood as the social glue that holds societies together (Walzer 1983). Instead of ceding political power to representatives, many communitarians advocate a more participatory or "strong democracy" (Barber 1984) that envisions politics "as a way of living" (Barber 1984: 118). Not the anonymous mass society, but the civil society of strong communities—encompassing families, villages and cities up to the nation as a whole—is the site of democratic processes and experience. Because of this strong community orientation, communitarianism considers borders to be constitutive for democracy; and because too much diversity could endanger the cohesion of a society, immigration tends to be perceived as a threat rather than an enrichment for democracy.

A third *deliberative* model of democracy sees it essentially as a communicative process of collective problem-solving. The genuine place of democracy is the public sphere, which mediates between civil society and the government. According to Habermas (1994), the deliberative model of democracy represents a middle course between liberalism and communitarianism or rather republicanism. The individual and the community, private and political autonomy are no longer seen as opposites, but prove to be "*co-originally* constituted" (Habermas 1996: 122) within the democratic process. Correspondingly, the legitimacy and rationality of democratic decisions do not result from an orientation toward aggregated individual interests or an assumed common will, but solely from the "processes of collective deliberation conducted rationally and fairly among free and equal individuals" (Benhabib 1996: 69). Because of this procedural character of democracy, boundaries have a different status in this model than in the other ones: Since political decisions must in principle be kept open for further deliberation, even ruling border regimes are conceived as provisional and reversible. In principle, they can be questioned and renegotiated by all participants in the discourse community, which is considered to be unlimited.

Such a processual and open understanding of democracy also characterises *radical democratic* and *agonistic* approaches. In contrast to authors such as Habermas, however, agonists doubt that democratic participation can be achieved discursively alone or could be satisfactorily guaranteed by any institutional political order. According to them, the "heart of democracy" lies rather "in the existence of counterpowers that limit the power of the state" (Balibar 2014: 283) or that of supranational units, such as the European Union. According to this approach, democracy is thus primarily a practice of "disagreement" (Rancière 1999), understood in terms of exclusions and resulting conflicts that the ruling democratic regimes engender throughout their violent and oppressive territorial and civil boundaries. How inclusive a democracy is thus depends to a large extent on the success of the struggles of migrants and refugees themselves.

1.2 Transnational and Cosmopolitan Theories of Democracy

Since the early 1990s, there have been increasing attempts in political philosophy to free the concept of democracy from the corset of "methodical nationalism" and to determine how democracy could be shaped in a globalised world. The starting point is the observation that, in the course of globalisation, cross-border effects of state action are increasing (e.g. climate change, financial crises, environmental disasters and, of course, flight and migration), which can no longer be adequately controlled by nation-states alone, so that the solution of global problems requires the formation of corresponding supranational institutions. The interest in democratic theory is mostly focused on how these institutions and their decisions can be endowed with legitimacy: Advocates of cosmopolitan federalism (Benhabib 2004; Habermas 2006) argue that the political power of states should be controlled and possibly limited by constitutionalising international law and strengthening supranational institutions such as the United Nations. For others, this vision of a federal and subsidiary world republic does not go far enough. According to them, what is needed is the formation of an at least rudimentary world state, "a transnational, common structure of political action which alone, ultimately, can support the politics of self-determination" (Held 1995: 232). What both approaches have in common is that they proceed from a globally uniform legal space. According to the authors, this does not conflict with the diversity, heterogeneity and federalism of the different levels at which democratic life takes place, but is rather its prerequisite.

Anyhow, it is precisely this legal and political monism that ignites much of the criticism of democratic cosmopolitanism. According to the standard objection, cosmopolitanism merely projects the model of the state onto the global level, thereby taking over all the problems of the liberal-democratic state without solving them: for example, the inability of states to deal with conflicts and a plurality of perspectives and values. For this reason, some argue for abandoning the idea of cosmopolitanism altogether and replacing it with the vision of a multipolar world order of different power centres (Mouffe 2009). Others try to take the criticism into account by conceiving the idea of cosmopolitan democracy as a pluralistic order of different *demoi* (Bohmann 2010), in which conflicts can be fought out in a variety of perspectives and political contexts.

Although all these designs of a cosmopolitan or multipolar world order have undergone considerable differentiation in some cases, they contain only a few clues as to how their ideas could be applied to the topic of flight and migration. The situation is different with radical democratic approaches, which do not link the idea of cosmopolitanism to the establishment of a global order, but rather to the permanent questioning of this order and its supposed legitimacy. Such a "subaltern cosmopolitanism" (Zeng 2014) is rooted in specific contexts and practices and takes its starting point from the real experiences of exclusion and oppression of the actors. Contrary to a "top-down" model of politics, which is only concerned with the implementation of objectives designed in theory, this understanding of politics aims at a "cosmopolitanism from below", consisting "of a politics by which particular forms of exclusion, domination, exploitation and marginalisation are challenged by those who suffer them" (Ingram 2016: 73). Accordingly, migration and the numerous conflicts it generates in liberal democracies are no longer seen as an "external" problem of democracy, which could only be solved within the framework of a cosmopolitan order. Political problems rising from migration are rather seen as a result of liberal-democratic institutions in a globalised world itself, which in turn allows for questioning the current world order.

2 Democratic Theory and Migration

Many scholars hold the view that migration and democracy find themselves in a tense relationship (e.g. Benhabib 2004). This tension is often interpreted as the expression of an unresolved conflict between the two promises of freedom that liberal democracies attempt to reconcile: the liberal promise of extensive individual freedom, which results in the claim to global freedom of movement, and the democratic promise of collective or political autonomy, which *prima facie* seems to imply the right to exclude others from the territory of a political community and from its membership. Beyond this "standard view", however, there have been numerous attempts to problematise the manifold exclusions that modern states carry out and justify in the name of democracy. These different approaches will be presented in the following with reference to three relevant systematic questions of this research field: to the question of (1) the legitimacy of territorial exclusions, (2) the possibilities of membership and (3) the connection between democracy and refugee protection.

2.1 Is There a Democratic Right to Territorial Exclusion?

Authors who advocate a "right to exclusion" often justify it with reference to democratic theory. The central argument is that in order for citizens to exercise political control over central cultural, social, legal and political conditions of their common life within a territory, they also need the jurisdictional authority to regulate immigration according to their own needs (Moore 2015: 196). Collective self-determination thus implies a "right to exclusion". According to Michael Walzer, "admission and exclusion" are even "at the core of communal independence. They suggest the deepest meaning of self-determination" and "without them, there could not be *communities of character*" at all (Walzer 1983: 62).

Differences between these approaches result, among others, from how the "self" of collective self-determination is interpreted, which in turn influences the justifications for the particular right in question: While Margaret Moore, for example, defines the collective subject "people" exclusively in political terms (Moore 2015: 54ff), Walzer's communitarian argumentation is based on the assumption that the cultural character of a community is constitutive for the clarification of normative questions of justice, and that the members of this community therefore have the right to decide for themselves what kind of community they want to be (Walzer 1983). David Miller argues in a similar way, even though he does not anchor the right to exclusion in a concept of culture, but rather in a morally loaded concept of the nation, out of which a "reasonable partiality towards compatriots" is founded (Miller 2005). The respective positions are supported by analogies in which certain aspects of the political, cultural or national community are compared, for example, with the active political living of settlers on an isolated island (Moore 2015: 194ff), with the social ties in families and friendships (Miller 2005: 67ff) or with neighbourhoods or clubs (Walzer 1983: 35ff).

However, these analogies also show the weaknesses of the respective approaches: The concept of cultural, national or political community that serves as the basis for the respective approach may be appropriate to the idyllic community life of a group of English settlers on a previously uninhabited island. But this image has little to do with the coexistence of people in modern mass societies mediated by abstract law. And even if certain aspects of the analogies might also apply to modern or postmodern societies, it remains to be discussed whether these aspects are also desirable in normative terms: In view of the resurgence of nationalism and racism in Western societies, one could also argue the other way round that it would be desirable to overcome nationalist partiality and to question homogenising and essentialist notions of cultural identity. Moreover, it does not follow from the observation that collective self-determination is intrinsically valuable that it can be enforced even against the will of individuals. In order for the right to collective self-determination to become a legitimate right of exclusion, an additional premise is needed: that the right to collective self-determination is not only one good among others, but a claim of overwhelming importance. Whether such a primacy of political over individual autonomy can be justified, however, is disputed.

Joseph Carens developed a prominent counterargument in his highly regarded essay *Aliens and Citizens* (1987). Carens brings the Rawlsian thought experiment of the "original position" to the global level and asks what basic freedoms the parties would agree on if they were to be prevented by the "veil of ignorance" from knowing in which country they were born and what citizenship they had. Carens is convinced that they would include the freedom of movement among states in a global system of equal liberties, so that borders should in principle be open (Carens 1987). With this liberal argumentation—which now advocates the primacy of individual freedom—Carens nevertheless follows the same pattern as the democratic-theoretical approaches outlined above, with the consequence that in his argumentation, too, the relationship between migration and democracy must be understood primarily as a competitive relationship between different interests or freedoms. While such a view may be appropriate to the real world of our time, it is not conceptually compelling, as deliberative and radical democratic approaches argue.

Arash Abizadeh (2008) believes that the right to unilateral border controls is incompatible not only with liberalism but also with a deeper sense of democratic legitimacy. As a universal principle of democratic legitimation, the idea of popular sovereignty requires that state coercion has to be "justified to all those over whom it is exercised" (Abizadeh 2008: 45), which is also referred to as the *all coerced* or *subjected principle*. Since immigration controls are a case of state coercion, "justification for a particular regime of border control is owed not just to those whom the boundary marks as members, but to nonmembers as well" (Abizadeh 2008: 45). The prevailing practice of unilateral border controls would therefore simply be undemocratic. Abizadeh sees a possible correction in the formation of cosmopolitan democratic institutions "that have jurisdiction either to determine entry policy or legitimately to delegate jurisdiction over entry policy to particular states (or other institutions)" (Abizadeh 2008: 48).

This position is often objected to as being unrealistic in terms of its implementation and regarding its normative consequences both "over-inclusive" and too unspecific: If any subjection to state coercion were to give rise to political co-determination, tourists would also have to be enfranchised in their holiday country because they too are subject to many of the country's laws that have been coerced during their stay. On the other hand, Abizadeh's definition of coercion is so unspecific that, depending on the specification of the term, one may well doubt whether the refusal of entry really does have to be interpreted in a normatively relevant sense as subjection to coercion (Miller 2010).

In order to avoid these problems, many authors interpret the principle of democratic legitimacy not as the principle of being subjected to coercion, but rather to law (Habermas 1996). Whether the application of this principle could actually establish a right of migrants to co-determination over the border regime of another country remains to be discussed. However, it should be undisputed that, according to this principle, the current practice of unilateral border controls has at least a democratic deficit that states must remedy in order to act in accordance with their normative principles. This is, at the same time, a step towards mediating migration and democracy and opens the way for a "*democratization of borders*" (Balibar 2004: 49).

Radical democratic approaches ask how such democratisation could be promoted. The starting point for their considerations are the normative claims of the migrants themselves, their political self-organisation as well as their struggles for global freedom of movement and political participation. This focus on democratic practices gives "migration" a different status than in the theories discussed above. While in those theories migration is primarily seen as a problem for democracies, here it becomes the "motor" of democratic processes, in which migrants are no longer treated as mere "objects" of foreign rule, but appear as actors of a democratic practice. In this way, radical democratic approaches also enable a methodological change of perspective: Instead of treating migration one-sidedly from a democratic theory perspective, as is traditionally the case, these approaches develop a vocabulary that can be used to critically question democracies and theories of democracy from a migrant perspective.

It is questionable, however, whether these approaches, with their strong focus on migrant struggles, do not in turn narrow the political in a one-sided, activist way and therefore only allow those migrants to be seen as political actors who are activists. A radical democratic politics of migration would also have to ask self-critically about the exclusions of its own understanding of politics and to develop participation opportunities for all those who—for whatever reason—cannot (any longer) fight for their rights. In this context, questions of citizenship and political participation rights of migrants become virulent—questions that no longer only concern the "external" but the "internal" boundaries of political communities and the claim to membership.

2.2 Migration and Political Membership

"Migration" implies not only leaving one's own country and crossing international borders but also arriving in another country to settle temporarily or permanently. For most immigrants, this is the beginning of a long and exhausting struggle for cultural and legal recognition. During this time, many have a precarious residence status, are threatened with deportation or forced into illegality. But even those who have a longer-term perspective of staying in the countries of arrival or are allowed to settle permanently have hardly any political rights, above all no right to vote and thus no access to the procedures and institutions of political self-determination. This access is usually linked to citizenship, which in most countries is still determined either by the nationality or ethnicity of the parents (*ius sanguinis*) or by the place of birth (*ius soli*) and thus granted to migrants only as an exception (Shachar 2012). Thus, facing the massive socio-economic inequality between states, some authors have compared citizenship in rich liberal democracies with a "feudal privilege—an inherited status that greatly enhances one's life chances" (Carens 1987: 252). However, this practice of granting privileges is not only difficult to justify in terms of justice but also in terms of democracy because it denies a part of the population the exercise of political self-determination and thus subjects them to foreign rule.

There is a controversy about the principle according to which the demos can legitimately be composed or self-constituted. Some argue that the *all subjected principle* mentioned above should be understood as the principle of democratic inclusion. This would have three remarkable consequences: First, contrary to widespread intuition, members of a political community would then not be free to decide arbitrarily on its composition. In contrast, political exclusions would only be legitimate if they did not violate this normative principle. Secondly, the principle formulates a purely political criterion for membership; other criteria, such as the possession of certain competences (language skills, knowledge of the history and culture of a country, etc.), would be normatively irrelevant and corresponding "naturalisation tests" would be illegitimate. Thirdly, many migrants, refugees and also illegalised persons, who are now excluded from political participation in modern democratic nation states, are, in fact, part of the demos when applying this principle.

In principle, there are different ways of granting political membership: Some propose to decouple political rights and membership from their national corset, since today "rights, participation, and representation in a polity, are increasingly matters beyond the vocabulary of national citizenship" (Soysal 1994: 165). Turkish migrants in Berlin represent an example of this form of postnational membership. Others argue that political rights, such as the right to vote, should be granted through citizenship, whereby the act of the so-called "naturalisation" can be understood as "a unidirectional and graduated transformation or 're-birth': from alien to citizen" (Shachar 2012: 1013).

However, as Balibar pointedly notes, political rights and citizenship are never merely "granted or conceded from above, but are, in an essential respect, constructed from below" (Balibar 2004: 48). Although the inclusive potential of modern citizenship cannot be denied, historically this potential has only ever been able to unfold through concrete struggles e.g. of workers, women and black persons who asserted their democratic claim to equal freedom against the established political classes (Balibar 2014). Placing the struggles of migrants, refugees and illegalised persons in this tradition, their struggles would also have to be understood as to represent an "essential component of citizenship" (Balibar 2004: 49). They could be described in a deliberative perspective as "democratic iterations" through which "universalist rights claims and principles are contested and contextualised, invoked and revoked, posited and positioned throughout legal and political institution as well as in the associations of civil society" (Benhabib 2004: 179). Or they could be understood more agonistically as practices of civil disobedience in which noncitizens "constitute themselves as those with 'the right to claim rights'" (Isin 2009: 371)—and therefore *de facto as* citizens. Some even see in these struggles "the emergence of a new figure of citizenship" (2009: 381), the "activist citizen" that holds citizenship "constantly in flux" (377). However, one should be critical of too much idealism. Despite all the enthusiasm for the fluidification of legal statuses, one should not overlook the real dependencies in which illegalised persons and refugees actually find themselves. Moreover, fluidification of legal status could also lead to a weakening of legal rights, which are needed by the most vulnerable members of society.

2.3 Democracy and Refugee Protection

While current international law leaves the states largely free to determine their own immigration policy, state action in dealing with refugees is bound by a large number of international legal norms and international treaties. In contrast to "migrants", potential "refugees" may not be turned back at the border of a state *(non-refoulement)* (Art. 33 Geneva Refugee Convention) and, once they have been recognised as refugees, are granted encompassing civil and social rights in the state of refuge. This normative concept of the refugee is defined in Art. 1 of the Geneva Refugee Convention (GRC), which has now been signed by 146 states. In conjunction with the Universal Declaration of Human Rights (UDHR), it established an individual right to protection for refugees against a foreign state and is therefore seen by many as a paradigmatic example of an albeit rudimentary cosmopolitan legal system.

However, the reality of international refugee protection is far from these ideal legal provisions. For there are apparently many ways in which states can undermine, override or simply ignore their obligations under international law. Hence, also in the field of refugee protection, sovereignty and human rights are in tension, and from a democratic theory perspective the question arises again of how those who are unilaterally subject to the violence of ruling border regimes can exert political influence on them. Some have suggested establishing permanent representation of refugees in international committees. It is questionable, however, who should be able to represent such a heterogeneous group as "the refugees" and whether the representing body would also be equipped with sufficient power to actually bring about change. Others plead for a break with the liberal logic of representation and locate the political power for change precisely in the concrete actions of refugees and their self-organisation (Balibar 2004; Isin 2009).

In fact, however, the democratic power of refugees is more than limited: Refugees around the world still live in highly unsecure situations, e.g. in overcrowded camps, where they often remain dependent on humanitarian aid for years. Camp situations are not only psychosocially extremely stressful and degrading but also subject people to complete political domination. The camps on the Greek islands, are only the most visible expression of a policy of domination under which people are not even able to establish conditions for protecting themselves with physical distance and hygiene against a deadly virus that is threatening the world. But even established procedures of refugee reception, such as UNHCR's resettlement programmes, or the desperate search by the European Union for a mechanism to distribute refugees fairly among EU member states, can be questioned in terms of their democratic legitimacy, because they only seek for "justice between states", while ignoring the interests of the refugees concerned (Gibney 2015). However, a democratic solution would not only have to recognise "that refugees have a powerful and legitimate interest in where they receive asylum" (Gibney 2015: 461) but also to ensure that they have "the capacity to deliberate on and change the terms of democratic cooperation, and thus have normative power over the distribution of normative powers" (Bohman 2010: 9).

3 Towards a Democratic Cosmopolitanism from Below and from Above

Research on how migration and refugee protection can be democratically shaped in the global space is still in its beginnings. In many places, the preceding discussion suggests that above all cosmopolitan as well as deliberative and radical-democratic approaches seem promising for the further treatment of the topic. This is because they overcome the problematic "methodological nationalism" of the other approaches and try to combine the core idea of democracy—the idea of political self-determination of particular communities—with the universalist claim to equal liberty. Cosmopolitan approaches have already presented elaborate proposals in the institutional design of a more democratic world order, but have so far hardly applied them to the issues of flight and migration; since they think cosmopolitanism primarily "from above", they find it difficult to integrate migrant perspectives. Deliberative and especially radical democratic approaches develop their ideas of democracy the other way round, based on concrete experiences of migration and flight and the migrant struggles for participation, but have often neglected considerations of institutional theory. Their "cosmopolitanism from below" (Ingram 2016) allows for strong criticism of current migration regimes, but gives little positive response to the question of how to change the responsible institutions and shape migration policy. Future research in democratic theory and migration could start here and ask how the strengths of both approaches can be combined so that those who today are still subject to the violence of ruling migration and democracy regimes can in future participate in political life as free and equal.

Literature

Abizadeh, A. 2008. Democratic Theory and Border Coercion. No Right to Unilaterally Control Your Own Borders. *Political Theory* 36:37–65.

Balibar, É. 2004. *We, the People of Europe? Reflections on Transnational Citizenship*. Princeton.

Balibar, É. 2014. *Equaliberty. Political Essays*. Durham: Duke University Press.

Barber, B. R. 1984. *Strong Democracy. Participatory Politics for a New Age*. Berkeley: University of California Press.

Benhabib, S. 1996. Toward a Deliberative Model of Democratic Legitimacy. In *Democracy and Difference. Contesting the Boundaries of the Political*, ed. S. Benhabib, 67–94. Princeton: Princeton University Press.

Benhabib, S. 2004. *The Rights of Others. Aliens, Residents, and Citizens*. Cambridge: Cambridge University Press.

Bohman, J. 2010. *Democracy across Borders. From Dêmos to Dêmoi*. Cambridge, MA: MIT Press.

Carens, J. H. 1987. Aliens and Citizens. The Case for Open Borders. *The Review of Politics* 49:251–273.

Gibney, M. J. 2015. Refugees and Justice Between States. *European Journal of Political Theory* 14:448–463.

Habermas, J. 1994. Three Normative Models of Democracy. *Constellations* 1:1–10.

Habermas, J. 1996. *Between Facts and Norms. Contributions to a Discourse Theory of Law and Democracy*. Cambridge: Polity Press.

Habermas, J. 2006. *The Divided West*. Malden: Cambridge.

Held, D. 1995. *Democracy and the Global Order. From the Modern State to Cosmopolitan Governance*. Stanford: Stanford University Press.

Ingram, J. D. 2016. Cosmopolitanism from Below. Universalism as Contestation. *Critical Horizons* 17:66–78.

Isin, E. F. 2009. Citizenship in Flux. The Figure of the Activist Citizen. *Subjectivity* 29:367–388.

Miller, D. 2005. Reasonable Partiality Towards Compatriots. *Ethical Theory and Moral Practice* 8:63–81.

Moore, M. 2015. *A Political Theory of Territory*. Oxford: Oxford University Press.

Miller, D. 2010. Why Immigration Controls Are Not Coercive. A Reply to Arash Abizadeh. *Political Theory* 38:111–120.

Mouffe, C. 2009. Democracy in a Multipolar World. *Millennium* 37:549–561.

Rancière, J. 1999. *Disagreement. Politics and Philosophy.* Minneapolis: University of Minnesota Press.
Shachar, A. 2012. Citizenship. In *The Oxford Handbook of Comparative Constitutional Law*, ed. M. Rosenfeld and A. Sajó, 1002–1019. Oxford: Oxford University Press.
Soysal, Y. N. 1994. *Limits of Citizenship. Migrants and Postnational Membership in Europe.* Chicago: University of Chicago.
Walzer, M. 1983. *Spheres of Justice. A Defense of Pluralism and Equality.* New York: Basic Books.
Zeng, M. 2014. Subaltern Cosmopolitanism. Concept and Approaches. *The Sociological Review* 62:137–148.

Migration Ethics Between Ideal and Non-Ideal Theory

Julian Culp

Abstract Ideal and non-ideal theory refer to distinct, but not necessarily mutually exclusive, ways of engaging in ethical theory, political philosophy and political theory. Ideal theory concerns the justification of some ideal normative requirements—ideal principles of social justice, for example. Non-ideal theory, by contrast, refers to the grounding of normative requirements in situations in which the ideal normative requirements are not met or are violated. Some ethicists, philosophers and theorists defend a complementary understanding of the relationship between ideal and non-ideal theory. Others, by contrast, maintain exclusive understandings of ideal and non-ideal theory that view them as theoretical endeavours which are independent from and compete with each other.

This article shows the importance of the ideal/non-ideal theory distinction in the ethics of migration by focusing on the open v. closed borders debate. It illustrates that both ideal and non-ideal theoretic argumentations can be employed to either defend or reject states' moral right to close their borders. In addition, the article explains positions that combine an ideal theoretic argumentation in favour of open borders with a non-ideal theoretic argumentation in favour of closed borders and, vice versa, positions that combine an ideal theoretic argumentation in favour of closed borders with a non-ideal theoretic argumentation in favour of open borders. In this way, the article reveals that clarifying the precise use of ideal and non-ideal theory is necessary in order to properly understand and assess the various positions in the ethics of migration.

J. Culp (✉)
Department of History and Politics, The American University of Paris, Paris, France
e-mail: jculp@aup.edu

A. Niederberger et al. (eds.), *Handbook of Migration Ethics*,
https://doi.org/10.1007/978-3-031-89877-8_10

1 General Methodological Particularities of Philosophical Discipline

John Rawls (1971: 7–11, 142–150, 243–251, 350–355, 541–544) introduced the distinction between ideal theory and non-ideal theory to differentiate between two different kinds of subject matter and corresponding methodologies of theorising justice. The subject matter of an ideal theory of justice, according to Rawls (1971: 9), is the identification of a "perfectly just society". The methodology that ideal theory uses to identify such a society is to assume full compliance with the normative requirements of ideal justice as well as, by implication, favourable social or natural conditions that allow full compliance. As Rawls (1971: 245) puts it, the "ideal part assumes strict compliance and works out the principles that characterize a well-ordered society under favorable conditions". A non-ideal theory of justice, by contrast, has as its subject matter the determination of the demands of justice within a status quo that is unjust or not just. Accordingly, its methodology involves assuming partial compliance with the demands of ideal justice or that social or natural conditions prevent the realisation of ideal justice. Non-ideal theory "studies", so Rawls, "the principles that govern how we are to deal with injustice" (1971: 8).

Rawls does not, however, conceive of ideal and non-ideal theory as two mutually exclusive ways of theorising justice. Instead, Rawls (1971: 245) understands ideal and non-ideal theory as fulfilling two complementary roles within a single theory of justice. More specifically, Rawls conceptualises non-ideal theory as a morality of transition that asks which practical steps are necessary in order to eventually realise ideal justice in the long run. Thus, non-ideal theory, for Rawls, is not simply a matter of determining which changes would reduce injustice in the here and now. Rather, it is a matter of evaluating alternative courses of action relative to their contribution to achieving ideal justice. Hence, Rawls (1999: 89–90) views non-ideal theory as dependent on an ideal theory: "[Nonideal] theory presupposes that ideal theory is already on hand. For until the ideal is identified, at least in outline—and that is all we should expect—nonideal theory lacks an objective, an aim, by reference to which its queries can be answered."

In contrast to Rawls's complementary understanding of the relation between ideal theory and non-ideal theory, many normative theorists propose either an exclusively ideal theory or an exclusively non-ideal theory approach to theorising justice. G. A. Cohen (2008), for example, employs an ideal theory of justice on the grounds of his claim that the justification of principles of justice must not rely on any empirical facts. For example, the "luck egalitarian" principle of justice, which states that it is unjust if some are worse off than others through no fault of their own, is fact-free in the sense that its validity does not depend on any empirical facts about the world. Thus, if one were to justify certain normative statements regarding justice on the basis of empirical considerations such as facts about non-compliance with a certain principle of justice, or about the existence of unfavourable conditions, then one would not be justifying principles of justice proper, but instead, as Cohen (2008: 317) puts it, mere "rules of regulation." This is because, so Cohen (2008: 309), "facts cast normative light by reflecting the light that fact-free first principles shine on them." Another case in point is David Estlund (2011), who maintains that an ideal theory of justice is valuable even if it has no practical value. According to Estlund (2011: 415), an ideal theory of justice is valuable because it enables us to understand something "of great value." An ideal theory is thus valuable even if it does not specify which practical steps should be undertaken in the here and now and is hence not practically action-guiding.

By contrast, philosophers such as Colin Farrelly (2007) defend an exclusively non-ideal theory approach to theorising justice. They assume that a theory of justice must be action-guiding and believe that only non-ideal theory, but not ideal theory, is practically relevant. Farrelly (2007: 853), for example, criticises Rawls's ideal theory because "[b]y ignoring the realities of non-compliance and scarcity of resources, Rawls's theory [...] insulates itself from the most pressing concerns that face every real society." Along these lines, Amartya Sen (2006: 221) has maintained that ideal theory is *not sufficient* for identifying states of affairs that would be less unjust than the status quo. Knowing all the conditions of a perfectly just society does not allow us to comparatively rank two societies that do not meet these ideal conditions. This is because, as Richard Lipsey and Kevin Lancaster (1956/7) have shown in their economic analysis of the problem of the second-best, the second-best alternative need not be that alternative which satisfies more of the ideal conditions than all other alternatives. Instead, under conditions in which the best option is not available, the second-best option can be one which satisfies less rather than more of the ideal conditions of the first-best option. Consider, for example, that economic efficiency was optimally realised in a society in which all economic goods were produced in free markets. But assume that in none of the existing societies all economic goods are produced in markets. Lipsey and Lancaster's central insight is that in this situation the society in which more economic goods are produced in markets than in any other society is not necessarily the most efficient society. This is why thinkers such as Sen believe that the conditions of ideal justice by themselves do not inform us which of two less than fully just states of affairs is more or less just. "To consider an analogy," Sen further explains, "the fact that a person regards the *Mona Lisa* as the best picture in the world, does not reveal how she would rank a Gauguin against a Van Gogh" (2006: 221). In addition, Sen holds that an ideal theory of justice is also *not necessary* for engaging in the non-ideal theorising of justice. It is superfluous to know what ideal justice involves in order to rank two states of affairs relative to how unjust they are. As Sen (2006: 222) puts it, "it is not at all obvious why in making the judgment that some social arrangement x is better than an alternative y, we have to invoke the identification that some quite different alternative z is the 'best' or the 'right' social arrangement." Besides, Sen (ibid.) also draws an analogy to the different heights of mountains and posits that we do not need to know the exact height of the highest mountain, Mount Everest, in order to determine which of two smaller mountains is the higher one.

Nevertheless, many theorists defend (a version of) Rawls's view that ideal theory and non-ideal theory should be viewed as complementary. To begin, they point out that the justification of a perfectly just society yields normative insights, e.g. about how to trade-off equality vis-à-vis liberty, that are also relevant under non-ideal circumstances (cf. Swift 2008; Hinsch 2011: 373–374). What is more, they highlight that understanding the conditions of ideal justice is necessary in order to avoid sacrificing the realisation of ideal justice in the long term for the sake of achieving more rather than less justice in the short or medium term (Simmons 2010: 21–24; Hinsch 2011: 374). That is, we may need to know what ideal justice involves in order to be able to choose a course of action which may be less just in the short term, but which is more promising to achieve full justice in the long run. For example, if ideal international justice consisted of a world of independent sovereign states, then it would not be wise to adopt non-ideal redistributive measures that would reduce international economic inequality if such measures would also deepen international domination and thus inhibit the formation of an ideally just world of independent sovereign states. Without recognising the importance of the requirements of ideal justice, by contrast, an exclusively non-ideal theorist might suggest that reducing international economic inequality is the central requirement of justice because this diminishes injustice relative to the status quo. Finally, some theorists argue that ideal theory matters for motivational reasons (Culp 2019). They emphasise that ideal theory can demonstrate that an ideally just society is in principle realisable and that this recognition of the potential realisability of an ideally just society can mobilise hope and keep cynicism at bay.

Although the ideal and non-ideal theory distinction originally concerned different ways of approaching questions of (in-)justice, many use this distinction more generally to differentiate between different ways of doing political philosophy, ethics or political theory. Those who employ this distinction in this broader sense have also endorsed similar views regarding the adequacy of either exclusively relying on ideal theory, or exclusively relying on non-ideal theory, or relying on a complementary understanding of ideal and non-ideal theory. For example, Geoffrey Brennan and Philip Pettit (2005) defend an exclusively non-ideal theory approach to political philosophy on the grounds that the above-mentioned problem of the second-best would render ideal theorising of politics practically futile. Another case in point is Charles Mills (2005), who favours an exclusively non-ideal theory approach to ethics. Mills views an ideal theory approach to ethics as ideological because it idealises or assumes away morally significant features of social reality such as material inequality or historical injustices such as slavery. While this ideal theory approach serves the interests of the theorists that put forward ideal ethical theories, Mills argues, it does not offer adequate responses as to how to understand, assess and reduce actually existing ethical problems. Finally, Robert Goodin (1995) also recognises that a solid empirical understanding of the status quo is necessary in order to articulate contextual, non-ideal political theories that are practically action-guiding in the here and now. He insists, nevertheless, that an ideal political theory is needed to determine the political values on which practically action-guiding principles and policies can be grounded.

2 Relevance of/Controversies About the Methodology of This Discipline in Migration Ethics

Recognising the distinct ways in which ethicists, political philosophers and political theorists rely on the ideal/non-ideal theory distinction is crucial for understanding and assessing normative positions on migration. To illustrate, consider the open vs closed borders debate in political philosophy, which concerns the question whether states should have a moral right to exclude potential immigrants. The focus of this debate is whether a state should be allowed to prevent foreigners to immigrate into—that is, enter and live permanently in—its territory. The debate does not therefore necessarily address or relate to further normative questions about immigration, such as who should count as a refugee and which moral rights refugees have, which obligations host countries have vis-à-vis guest workers (and their children), which moral rights irregular migrants have, and which selection criteria states must employ when determining who should enter. This open v. closed borders debate consists of ideal theoretic and non-ideal theoretic argumentations, as well as argumentations that combine ideal and non-ideal theory.

2.1 Ideal and Non-Ideal Theoretic Argumentations for Open Borders

To begin, consider an ideal theoretic argumentation for open borders such as that of Joseph Carens (2013: 229). Carens (2013: 230) explains that he "set[s] aside worries about what to do if some people or some states are unwilling to do what they should" and that he is "not making a policy proposal that [...] might be adopted (in the immediate future) [...] by presidents or prime ministers or public officials charged with making immigration policy." Carens is therefore neither primarily concerned with how to respond to injustice nor with guiding political action (although he assumes that his ideal theory is practically relevant, albeit in indirect ways). Instead, Carens (2013: 229) aims to identify what "is right as a matter of principle." Echoing Estlund's claim that understanding something (such as justice) is what matters, Carens (2013: 229) points out that "it is important to gain a critical perspective on the way in which collective choices are constrained, even if we cannot do much to alter those constraints." Furthermore, Carens (2013: 230) maintains that even if his position does not instruct policymakers on what to do, it nevertheless puts into question the "complacency" of those European and North American citizens who "think that they are morally entitled collectively to what they have [...] and entitled to protect it by keeping others out." In this way, Carens's ideal theory informs the way in which we should morally judge the contemporary political world. Thus, a political philosophy of immigration is valuable that helps us to grasp which regime of state borders would be ideal, even if it fails to be politically action-guiding in the here and now. Those who question an (exclusively) ideal theoretic approach to ethics or politics, however, will find such an ideal theoretic argumentation for open borders wanting. They will insist that it is unsatisfactory because it fails to practically guide individual and political action. Hence, the way in which one evaluates the methodological reasons in favour of ideal theory and non-ideal theory, respectively, directly affects how one assesses certain positions in the ethics and politics of immigration.

If one endorses a complementary understanding of the relation between ideal and non-ideal theory, then it is also possible to combine an open borders position in ideal theory with a defence of either open or closed borders in non-ideal theory. Thomas Christiano (2008), for example, has argued that permitting wealthy liberal democratic states to close their borders might be the best way of achieving a world of open borders in the long run. If states were granted the right to exclude potential immigrants, at least temporarily, then they would be in a better position to develop accountable international institutions that would be effective in reducing the global incidence of extreme economic poverty. And the lowering of this incidence of extreme economic poverty, in turn, would enable liberal democratic states to open their borders because they would no longer expect a massive influx of immigrants. However, some philosophers might question Christiano's empirical claims to how closed borders will improve international institutions. *Pace* Christiano, they might even argue that international institutions will only develop in the way in which Christiano envisions if an open borders policy is adopted. For without the kind of intermingling of individuals and groups from various nationalities that open borders promote, it might not be possible to bring about the kind of cosmopolitan mindset that the support of international institutions requires. Hence, the complementary understanding of the relation between ideal and non-ideal theory allows being committed to an ideal theory that favours open borders while also endorsing a non-ideal theory that defends either closed or open borders.

In addition, it is also possible to defend an exclusively non-ideal theoretic argumentation for open borders. For example, a philosopher might claim that an open borders policy should be chosen because it (maximally) reduces injustices in the here and now, regardless of whether this policy is best for realising either open or closed borders in the long run. For example, open borders might be viewed as more just than closed borders simply because they reduce the level of global interpersonal economic inequality relative to a situation of closed borders in which states have attempted to reduce such inequality by means of bi- and multilateral aid. Such a position is especially powerful if one holds that governmental foreign aid is an ineffective or counterproductive instrument to reduce global interpersonal economic inequality, whereas an open borders policy is an adequate means to reduce such inequality.

2.2 *Ideal and Non-Ideal Theoretic Argumentations for Closed Borders*

Christopher Wellman (2008) pursues an ideal theoretic argumentation for closed borders. Based on Michael Walzer's (1983) defence of a general right to political self-determination, he maintains that freedom of association morally entitles legitimate states to decide whether to open or close their borders. In the same way that we appreciate that it is valuable if individuals can decide freely with whom they associate, e.g. when choosing whom to marry, we should also recognise the moral importance of legitimate states' liberty to decide whom to let in and whom to exclude. Wellman's position represents an ideal theoretic argumentation for closed borders because he specifies that it is *only legitimate states* which enjoy a *merely presumptive moral right* to close their borders. That is, states' moral right to close their borders is only valid under the idealising conditions that states are legitimate and that there are no overriding moral considerations in favour of opening borders. Thus, for example, if restricting the state's right to close its borders is necessary to ensure a child's reunification with her father, then it might be morally necessary to suppress the state's moral right to exclude foreigners.

Like Wellman, John Rawls (1999) also defends the moral right of states to control their borders as a matter of ideal theory. Rawls's ideal theory is a *political liberal* conception of international justice, according to which a just world consists of politically self-determining societies that respect human rights and cooperate peacefully with each other on fair terms. The right to exclude foreigners expresses societies' political self-determination and ensures that societies choose population and environmental policies that sustainably protect their territory. In such an ideal world, Rawls (1999: 8–9) argues, many of the domestic and international injustices that trigger transnational migration would not exist. For example, in a world of internally well-ordered societies, individuals would neither be subject to religious or political persecution nor would they suffer from starvation or experience the population pressure that can arise from women's subjection. Hence, those situations in which open borders appear as morally important would not be present. Rawls realises, however, that these ideal conditions do not exist, and thus he recognises the importance of complementing his ideal theory with a non-ideal theory of international justice. But in his non-ideal theory, Rawls does not explicitly address normative questions of immigration. Instead, Rawls (1999: 105–113) discusses in considerable detail the ways in which well-ordered societies should help non-well-ordered societies through international development policies. This may suggest that Rawls views a closed borders regime as preferable under non-ideal conditions.

Also conceivable, however, is complementing Rawls's ideal theory of international justice with the position that under non-ideal circumstances states do not have the right to close their borders because an open borders policy is best in order to realise the ideal theoretic requirements of closed borders in the long run. This non-ideal argumentation may state, for example, that open borders would increase the overall level of economic productivity and that this would eventually enable states to be politically self-determining and to respect human rights.

A critic of ideal theory such as Mills (2005), however, could argue that Rawls's failure to discuss normative questions of immigration in non-ideal theory is not accidental but the result of ideal theory's ideological effects. By focusing predominantly on the conditions of an ideally just international order that consists of politically self-determining societies, Rawls fails to recognise the urgency and ubiquity of transnational migration in the non-ideal status quo. In this way, ideal theory generates a biased conception of the morally relevant features of the actual social and political reality.

Finally, a purely comparative, non-transitional non-ideal theory may state that a closed borders policy should be chosen simply because it (maximally) reduces injustices in the here and now. In this case, the kind of comparative assessment that Sen favours is necessary, which determines how unjust one unjust state of affairs is relative to another unjust state of affairs. A case in point for such a non-ideal theoretic position is the work of David Miller (2016), who points out many normative problems that opening state borders would generate. For example, Miller (2016: 62–67) explains that immigration would increase cultural diversity and thereby undermine national self-determination, especially if it threatened the existing public culture through the creation of "parallel societies". In addition, Miller (2016: 64–65) highlights that immigration can lower the existing levels of interpersonal trust and trust in political institutions. He goes on to argue that this could transform public deliberation into self-interested bargaining and lower the provision of public goods as well as the level of domestic economic redistribution. When arguing in this manner, Miller (2016: 70) refrains from stating that "there is actually a *right* to self-determination" in ideal theory. He does not develop his non-ideal theoretic position on closed borders against the backdrop of an ideal theory that defends the rights of states to close their borders. Instead, Miller starts from the competing normative considerations in favour of and against closed borders and aims to balance these appropriately under the existing conditions. This is why his non-ideal theory is of a comparative rather than a transitional kind. Hence, in order to clearly state a position in favour of closed borders in non-ideal theory, it is necessary to specify the particular type of non-ideal theory that is used.

3 Desiderata for Future Methodological Development, (Possible) Importance for Future Research in Migration Ethics

One basic methodological desideratum in the development of ideal theory and non-ideal theory is that normative theorists state clearly not only whether they are engaging in ideal theory or in non-ideal theory but also to which particular conception of ideal theory, non-ideal theory and their relation to each other they are committed. Otherwise it will be difficult to understand and assess what the normative theorist is trying to achieve. For example, there is a difference between a comparative and a transitional understanding of non-ideal theory, which implies that a non-ideal position has a distinct meaning depending on whether it represents either a comparative or a transitional non-ideal theoretic argumentation.

An additional desideratum is tolerance for methodological diversity based on the recognition of the reasonable plurality of the aims of ethical, philosophical or political theorising. That is, there may not be the one correct way of building an ethical, philosophical or political theory, simply because there is a reasonable plurality of the aims that philosophers or theorists may pursue. While some may have a legitimate interest in identifying what would reduce injustice in the here and now, others may have a similarly legitimate interest in understanding what ideal justice would require. By analogy, consider that some artists may view their work as deeply political and aimed at contributing to the reduction of current injustices. Other artists, by contrast, may simply be motivated to produce art that is intrinsically beautiful. We may want to accept both forms of art as legitimate, even if we prefer one type of art over the other. Likewise, we may also recognise different types of normative theories as legitimate, even if we find one type more appealing than the other. By contrast, a significant part of the methodological debate about ideal theory and non-ideal theory seems to assume that there is only one correct way of engaging in normative theory. Those who make this assumption should at least justify why we should accept this assumption, given that it is controversial.

In the domain of migration ethics, like in many other domains of ethics and political philosophy, there seems to have been a certain tendency as of late in favour of non-ideal theory. The main reason for this appears to be that this methodology is able to provide normative insights that are practically (more) action-guiding. Non-ideal theory is considered as having more impact, and having an impact is increasingly viewed as an important feature of academic research. However, if it is true that there is a plurality of reasonable aims that a normative theory may pursue, then it is not clear why having an impact is a more important feature of a normative theory than, say, a normative theory's capacity to illuminate what morality or justice ideally require. What is more, it is questionable whether in the present context of highly specialised academic work ethicists and philosophers are particularly well-placed to engage in the empirically oriented type of research that non-ideal theory requires. In addition, the focus on practically action-guiding research can entail the neglect or bracketing of fundamental philosophical and normative disagreements. Especially in the context of transnational migration, however, in which individuals and groups from many different regions and cultures interact with each other, it is fundamentally important to carefully deliberate how to respond to such disagreements and, correspondingly, which normative principles, if any, avoid the problem of parochialism. Thus, for example, it is unsatisfactory to simply assume the validity of some liberal or Western ethical principle for the sake of generating a non-ideal theory of migration that is practically relevant. Instead, it may be better to explore what kind of normative principle is appropriate in a highly diverse context such as that of transnational migration—even if this exploration does not yield any immediately relevant political principles.

Literature

Brennan, G., and P. Pettit. 2005. The Feasibility Issue. In *The Oxford Handbook of Contemporary Philosophy*, ed. F. Jackson and M. Smith, 258–279. Oxford: Oxford University Press.

Carens, J. 2013. *The Ethics of Immigration*. Oxford: Oxford University Press.

Christiano, T. 2008. Immigration, Political Community and Cosmopolitanism. *San Diego Law Review* 45:933–961.

Cohen, G. A. 2008. *Rescuing Justice and Equality*. Cambridge, MA: Harvard University Press.

Culp, J. 2019. Zur praktischen Relevanz des politischen Liberalismus – eine Verteidigung am Beispiel gerechter Bildungspolitik. In *Liberalismus*, ed. K. Fischer and S. Huhnholz, 387–407. Baden-Baden: Nomos.

Estlund, D. 2011. What Good Is It? Unrealistic Political Theory and the Value of Intellectual Work. *Analyse & Kritik* 33:395–416.

Farrelly, C. 2007. Justice in Ideal Theory. A Refutation. *Political Studies* 55:844–864.

Goodin, R. 1995. Political Ideals and Political Practice. *British Journal of Political Science* 25:37–56.

Hinsch, W. 2011. Ideal Justice and Rational Dissent. A Critique of Amartya Sen's The Idea of Justice. *Analyse & Kritik* 33:371–386.

Lipsey, R., and K. Lancaster. 1956/7. The General Theory of Second Best. *The Review of Economic Studies* 24:11–32.

Miller, D. 2016. *Strangers in Our Midst – The Political Philosophy of Immigration*. Cambridge, MA: Harvard University Press.

Mills, C. 2005. Ideal Theory as Ideology. *Hypatia* 20:165–184.

Rawls, J. 1971. *A Theory of Justice*. Cambridge, MA: Harvard University Press.

Rawls, J. 1999. *The Law of Peoples*. Cambridge, MA: Harvard University Press.

Sen, A. 2006. What Do We Want From a Theory of Justice? *Journal of Philosophy* 103:215–238.

Simmons, J. 2010. Ideal and Nonideal Theory. *Philosophy and Public Affairs* 38:5–36.

Swift, A. 2008. The Value of Philosophy in Nonideal Circumstances. *Social Theory and Practice* 34:363–387.

Walzer, M. 1983. *Spheres of Justice*. New York: Basic Books.

Wellman, C. 2008. Immigration and Freedom of Association. *Ethics* 119:109–141.

Part III
Major Controversies in Migration Ethics

Humanity and Mobility

Alex Sager

Abstract Human mobility poses a major challenge for social and normative theories developed during the age of nation-building. Many of these theories suffer from the cognitive biases of sedentarism, presupposing that populations are largely immobile, and of methodological nationalism, assuming a state-centred view of culture, territory, and human relationships. This raises conceptual issues for social theories that attempt to understand migration (a transnational phenomenon) through categories that presuppose largely homogenous, sedentary populations and rigid state borders.

These cognitive biases create normative challenges by treating mobility as exceptional and pathological, in need of regulation and restriction. This perspective conflicts with widely recognised rights to mobility and with the fact that humans are a mobile species. People move in response to violence, persecution, and the inability to satisfy basic needs, as well as for reasons of family unification and economic interests. The use of force to prevent people from exercising their rights and from accessing important goods requires moral justification. This in turn suggests the need to situate mobility in a broader theory of global justice, as well as post-national social theory.

The history of humanity is a history of mobility, but the history of political thought and of the social sciences has been one of stasis. This contradiction creates a challenge for social scientists attempting to understand human mobility and for normative theorists seeking to determine rights and obligations toward mobile individuals and populations. Mobility is a basic right, necessary to access work, education, and family and social networks; nonetheless, states and private actors restrict it for morally dubious reasons. The regulation and restriction of human mobility raises theoretical and conceptual issues about how we should understand mobility and mobile

A. Sager (✉)
Department of Philosophy, Portland State University, Portland, OR, USA
e-mail: asager@pdx.edu

A. Niederberger et al. (eds.), *Handbook of Migration Ethics*,
https://doi.org/10.1007/978-3-031-89877-8_11

people. It also raises fundamental questions in ethics and in social and political philosophy. In both cases, coming to terms with mobility challenges us to rethink our social and normative theories.

1 General Issues

Humans are a mobile species. The International Organization for Migration estimates there are 244 million international migrants, as well as more than 740 million internal migrants (International Organization for Migration 2018). 1326 million international tourists crossed borders in 2017 (United Nations World Tourism Organization 2018). Moreover, mobility plays a core role in the lives of people not identified as migrants. Every day many of us travel miles for work, school, and leisure. Despite the centrality of mobility to people's lives, we have yet to fully come to terms with how to study it or how to create appropriate institutions for a mobile world.

Human mobility raises conceptual, epistemic, and normative issues for the social sciences and for social and political philosophy. How should we understand and represent mobility and, in turn, how is mobility used to categorise groups of people? Even if we restrict ourselves to physical mobility (and do not directly address social mobility) and to human mobility (putting aside the transportation of commodities, the movement of ideas, or the migrations of non-human animals), much mobility goes unnoticed or unanalyzed. Studying mobility is challenging because many of the tools we use to understand the world capture data at a fixed time and place. Mobility is usually reconstructed from static, often partial, data which informs us about departure and arrival, but tell us little about the actual journey. Moreover, we only become aware of mobility when it has social salience.

The neglect of mobility in the social sciences has led some researchers to call for a "mobility turn" or a "new mobilities paradigm" (Blunt 2007; Cresswell 2006; Sheller and Urry 2006). These scholars not only draw attention to types of mobility that have been traditionally neglected, but attempt to see topics like human migration differently. They are sensitive to how mobility is represented—or, in some cases, not represented—and how it is experienced (Cresswell 2010). Mythologies of the automobile and fantasies about international tourism (often with colonialist iconography) contrast with dehumanising depictions of impoverished migrants. This connects to how people experience mobility differently because of their social position and context. Citizenship and social status determine whether a border checkpoint is met with anxiety or indifference. Gender and race also shape mobility, determining where people are permitted to go and under what conditions. Ableism similarly conditions accessibility and opportunity.

Mobility cannot be studied without examining physical and social infrastructures and technologies that enable it. It depends on roads, airports, ports, and communication networks, as well as laws and policing. Studying mobility requires identifying points of departure and arrival, as well as points during transit in which travelers temporarily come to rest. We also need to understand how mobility is embodied and how people's bodies are regulated by social and political structures and norms, often along gendered, racial, and class lines.

Finally, we cannot understand mobility without understanding how states have shaped our understanding of it. One way of doing so is through an investigation of "methodological nationalism", a set of the epistemic biases that arise from approaching the world from a state-centred view (Beck 2000; Sager 2016; Wimmer and Glick Schiller 2002). Under methodological nationalism, the social sciences have adopted and reinforced nation-building projects by uncritically assuming that states (rather than subnational, transnational, or global agents or structures) are the unit of analysis and by conflating society with the nation-state's imaginary community. Nation-building systematically erased and distorted our understanding of migration by theorising migrants in relation to (allegedly) homogenous and (again, allegedly) sedentary national populations. Written history is a product of states. As a result, the representation and regulation of migration often serves nation-building projects. In many cases, states have represented mobile populations as savages or barbarians and settled nomads so they can be taxed, conscripted, and, as has often been the case, enslaved (Scott 2009). In other cases, mobility has a role in constituting states, sometimes through the journeys of mythical founders (Honig 2003) or through settler colony mythologies that facilitate the marginalisation or erasure of indigenous populations. Expulsion and ethnic cleansing have also played a central role in nation-state formation (Mann 2005). States choose whether they celebrate the elimination of indigenous populations or the expulsion of ethnic minorities or whether they disguise, bury, and ignore past atrocities. Forced mobility becomes a tool for removing populations designated as outsiders or foreign.

Furthermore, the technologies that we use to understand mobility are normally developed by states for their purposes. Most of the information we use to study migration is gathered by states. Migrants have played a central role in history, but have a peripheral or distorted role in historiography. The history of mobile groups must be partly reconstructed from often fragmented records, distorted through the perspective of state officials. One challenge is that migrants leave few traces in the archives and in the archaeological record (Hoerder 2011). Mobile people are often not captured in censuses and, even when their presence is noted, it usually means assigning them a fictional, fixed address.

Studying mobility depends on state-developed technologies such as passports and visas. This means that transnational or global phenomena disappear or must be reconstructed from state records. State interests determine the categories used to study migration and the data collected. These include categories such as tourist, student, temporary worker, and nomad in which the perceived transience of these populations defines their degree of membership. In contrast, permanent residents and (the controversially named) "undocumented", "illegal" or "unauthorised" immigrants are seen as immobile. States' practices of illegalising and criminalising segments of the immigrant population make it difficult to accurately determine the size of unauthorised populations that seek to remain undetected.

Methodological nationalism also leads us to presuppose distinctions between international and internal mobility and to study them independently. When Ravelstein formulated his laws of migration in the nineteenth century, he did not distinguish between internal and international migration (Ravelstein 1885, 1889). Today migration studies have been largely confined to international migration—migration is seen as a process in which people travel between states. Internal migration is mainly studied by disciplines such as urban studies. This is problematic given that there are many parallels between internal and international migration. Arguably, the practical and normative impacts of internal migration outweigh those of international migration, as the move from farms and villages to cities has ushered in major structural changes to economies and ways of life. Finally, international law and organisations distinguish between refugees and internally displaced persons based on whether they have crossed an international border, even if every other aspect of their plight is identical.

This final point demonstrates how epistemic and conceptual choices invariably have normative implications. Categories are not just ways of understanding the world, but of intervening through law, policy, and policing. Access to refugee status is not only symbolic, it can also mean access to forms of protection and rights that are not extended to displaced people without it. Mobility determines our capacity to access social goods. People with advanced degrees and significant incomes are able to acquire visas and to immigrate or travel. Furthermore, mobility is also gendered and racialised with formal mobility rights undermined by physical risk and social stigma (Bosworth et al. 2018). Finally, the choice not to move is another type of privilege with social class determining those who are able to reside close to their places of work or to work from home.

Methodological nationalism is at the core of many normative debates surrounding international migration. Under methodological nationalism, much migration across international borders is seen as abnormal and problematic, demanding restriction and regulation. Migration is often framed primarily as a security issue and migrants are often reduced to the impersonal and threatening hydraulic metaphors of flows, floods, and waves. Journalists, politicians, and academics describing migration quickly lapse into the language of crisis with little evidence or analysis (Sager 2018a).

2 Topics, Positions, and Arguments

The right to move freely within the borders of one's country and to emigrate are fundamental human rights, enshrined in the Universal Declaration of Human Rights and in many state constitutions. Mobility is a fundamental right for its own sake (it is not coincidental that prison's punishment is to strip people's freedom of movement). Moreover, without the ability to move, it is difficult to access food, shelter, and employment, to maintain intimate relationships, or to participate in democratic institutions. In fact, most basic rights depend on mobility. Nonetheless, mobile and transnational populations are often persecuted and mobility is restricted and regulated in many ways.

Mobility rights are exercised locally in particular places and depend on infrastructure. Even when people enjoy formally equal mobility rights, their ability to put them into practice is inequitably distributed by ability, race, ethnicity, gender, and class. This is perhaps starkest for people with disabilities who may be excluded from social, political, and economic opportunities because architecture and transportation networks fail to accommodate them. Security guards discourage or prevent racialised minorities from accessing some neighbourhoods or public spaces. Mobile communities such as the Roma and English Travelers are frequently persecuted and expelled or, alternatively, forcibly settled. Gender discrimination, which sometimes combines with laws regulating what women can and cannot wear in public spaces, leads women to restrict their movement. Limited or non-existent public transportation restricts the use of public spaces to wealthier citizens with cars. Furthermore, public works and development projects displace societies' most economically and socially vulnerable members. Houseless people experience forced movement as they're removed from street corners or evicted from public property and forced immobility as they're confined to certain areas of the city and pressured into shelters or convicted of vagrancy and imprisoned.

Though the right of states to control immigration is mostly taken for granted, restrictive border controls sit uneasily with a commitment to moral equality and to fundamental rights such as freedom of movement, opportunity, and association. A fundamental principle in political philosophy is that the use of force to restrict freedom needs justification. People with citizenship from the rich world and the means to travel are courted and celebrated as tourists while their states spend billions of dollars to enforce immigration policies against poorer people with the wrong (or no) passports seeking safety or opportunity. They routinely imprison and deport immigrants, often with limited legal protections or recourse to legal counsel (Nethery and Silverman 2015). They build walls that funnel migrants toward dangerous routes, risking their lives crossing seas in flimsy boats or trekking through deserts (De Genova and Peutz 2010). They also contract with private corporations and foreign states—often ones with abysmal human rights records such as Libya and South Sudan—to prevent immigrants from reaching their territories. These practices obligate migrants, including refugees fleeing persecution and violence, to rely on smugglers.

Beyond the issue of justifying the use of force, place of birth is arbitrary from a moral perspective. Though we are not responsible for where we are born, place of birth plays a central, sometimes decisive, role in determining one's life chances. This is difficult to reconcile with a commitment to the equal moral worth of all humans. Migration controls prevent people from pursuing educational and economic opportunities and contribute to annual losses of hundreds of billions, if not trillions, of dollars to the global economy; individuals forfeit thousands of dollars in annual lost income because they are not allowed to pursue work abroad (Clemens et al. 2019).

The difficulty of reconciling a commitment to freedom and equality with a system that assigns membership by birth and gives states the discretion to use force to prevent people from migrating have led some scholars to call for open borders (Carens 2013; Chapter "International Freedom of Movement" in this Handbook). Some critics meet calls for open borders with concerns about the potentially harmful effects of open borders on political institutions and security. Philosophical opponents of open borders have justified restrictions by appealing to the democratic right to self-determination, the right to privilege co-nationals' economic wellbeing, and the right to shape national culture through immigration policy (Miller 2016; Wellman and Cole 2011). These debates, pitting national sovereignty and self-determination against individual freedom and opportunity, are ongoing. Many of these debates have primarily involved philosophical analysis and thought experiments, rather than empirical research. Though the evidence is still limited, recent research on real-world cases where borders have been opened suggests that some concerns about open borders may be overstated (Fetzer 2016; Haas et al. 2019).

Emphasising how mobility constitutes many of our institutions and activities adds another dimension. While freedom of movement does play a key role in these debates, it is often deployed from a perspective that treats stasis as the norm. There is usually limited attention to social scientists who centre mobility in their empirical and ontological investigations. Rather than simply insisting on a right to move, this shift in emphasis and perspective seeks to normalise mobility so that it is not so much a right to be demanded or contested, but an essential aspect of everyone's life (and not just something undertaken by migrants). This has normative implications. Instead of seeing mobility as an exception that existing institutions should accommodate or resist, this approach takes mobility as an axiom that should guide institutional evaluation and design.

Forced migration further complicates conceptual and normative debates about immigration policy. The United Nations High Commissioner for Refugees estimates that there are currently 25.4 million refugees (including 5.4 million Palestinian refugees under the United Nations Relief and Works Agency for Palestine) and 3.1 million asylum seekers. Additionally, there are 40 million people who have been displaced within the borders of their own country. Philosophers and legal scholars who defend extensive state rights to restrict and regulate immigration usually acknowledge significant moral and legal responsibilities to refugees.

Nonetheless, the definition of who counts as a refugee (cf. Chapters "Refugees and International Law" and "Humanitarianism and Its Problems" in this Handbook) raises conceptual issues. The Convention and Protocol on the Status of Refugees defines refugees as people who flee persecution due "to the well-founded fear of being persecuted for reasons of race, religion, nationality, or membership of a particular social group or political opinion…" (Article 1). This definition faces a number of challenges. First, the Convention definition emerged in response to the mass displacement in Europe after the Second World War. It may be inadequate to address humanitarian disasters in other regions and contexts as the regional treatises like the 1969 OAU Refugee Convention in Africa and the 1984 Cartagena Declaration in Latin America suggest. Second, this definition of refugee, if interpreted narrowly, encompasses only a small segment of people who are forced to move. People moving because of war, natural disaster, gender-based violence, environmental degradation, and economic destitution often do not meet the legal criteria for refugee status despite facing violence or being unable to secure their basic needs. This has led to proposals to expand our understanding of who counts as a refugee to cover more people from states that fail to secure their human rights (Gibney 2004; Maley 2016).

Even if we retain a narrow understanding of who counts as a refugee, affluent states are not meeting their obligations toward them. These states have taken extreme measures to ensure that even people who would qualify as refugees under the Convention and Protocol and under their domestic laws cannot lodge a claim to asylum. Applying for asylum usually requires setting foot on national territory, so states have placed barriers to prevent refugees' arrival (FitzGerald 2019). These include airline sanctions so that people without a visa cannot travel by air, maritime interception and return of ships, offshore processing of refugee claims—often in horrific conditions—and deals with other states to prevent travel. Refugees are forced to turn to smugglers to travel illicitly, often at great risk.

Though the European Union and United States often claim that there is a refugee crisis that is straining their resources, most refugees flee to adjacent countries or are displaced within their own country. Millions of these refugees live at the margins of cities or in UNHCR-run camps. Forced migration is often followed by forced immobility in which many refugees lived in camps for years and even decades where they have few opportunities for education or employment. Unable to return to their country of origin, they are simultaneously prevented from integrating into the local society. Only a small number of these refugees—according to the UNHCR, 55,676 in 2018—are resettled in third countries. Millions of people remain immobilised in camps with no realistic hope to return to the communities they fled and little chance of restarting their lives through resettlement. This is a moral failure for the international community. Focus on mobility, rather than questions of asylum and international duties to displaced people, should draw our attention to how problematic our policies are. Among other things, they deprive people of a basic tool, movement, to seek their livelihoods and construct their lives.

Normative theorists have dealt at length with questions of border controls and the obligations toward refugees. The bias of methodological nationalism has given undue focus to international migration, neglecting how many of the forces that influence international migration also cause people to move internally. One of the most vivid and tragic examples of this is internally displaced people. The UNHCR estimates that there were 40 million people displaced within the borders of their own country because of armed conflict, generalised violence, human rights violations, or climate change. As mentioned above, the only significant difference between internally displaced people and refugees is often that refugees crossed an international border.

More generally, neoliberal globalisation has transformed economies, often with disastrous consequences for the most vulnerable people (Sassen 2014). Millions of people are moving from rural to urban environments unprepared for sudden population growth. The causality of these migrations is complex with local, national, and global factors. In particular, economic globalisation shaped by powerful states and international corporations plays an important role in development and, as a consequence, human displacement. This raises questions about obligations of global justice, for example to provide reparations to people whose livelihoods are destroyed through development. A theory of migration justice needs to be situated within a broader theory of global justice. It also has to place mobility—along with forced immobility—at the core of justice as we seek to construct institutions that treat movement as a normal part of human life. Mobility should be recognised as a necessary strategy that individuals, households, and communities use to avoid harms and seize opportunities. Treating human mobility as a normal part of life encourages us to incorporate it in planning so that receiving regions and districts can anticipate demographic changes.

At a more local level, gentrification displaces people in many cities. Because of methodological nationalism, this is often not viewed as migration, even though it involves people forced to leave their homes. Many of the same forces that lead people to emigrate also cause people to migrate internally as a response to economic, social, and environmental forces (cf. Chapter "Climate Mobility" in this Handbook) that compel people to move because they are unable to fulfill their basic needs (Sager 2018b). Despite these similarities, normative theorists treat internal and international mobility separately.

International migration and internal migration also intersect. International migrants are often among the most vulnerable groups within states, marginalised by state laws and policies that illegalise their presence (De Genova 2002). They find themselves in underserved, *de facto* segregated districts and neighbourhoods, often at the mercy of landlords. Structural injustice reproduces inequalities over generations as racialised immigrant groups interact with legal and educational systems that systematically disadvantage them (Hing 2009). Immigrant law has increasingly become merged with criminal law, criminalising unlawful entry and increasingly subjecting immigrants to deportation for minor offenses (Stumpf 2006). Though often not conceptualised as such, deportation is another form of forced migration. In some cases, such as the mass Mexican "repatriation" from the US in the 1930s, it resembles ethnic cleansing (Balderrama and Rodriguez 2006). Ethical theorists have begun to examine the justifiability of deportation, particularly for people who have settled in a territory and formed social connections (Carens 2013; Hosein 2016). Nonetheless, analysis is still needed.

3 Future Research

Mobility is constituted by the technologies that enable it and is accessible through the techniques we have for representing it. The study of international migration depends on ships and planes and highways, but also on the invention of the passport and the bureaucratic practices for monitoring border crossings (Torpey 2000). With the rise of GPS and surveillance, we may witness types of mobility that were until recently invisible taking on new social salience. The use of facial recognition technology in ports of entry, combined with algorithms that comb through troves to data to assign risk to individual travelers raise questions of privacy. The need for states to incorporate and to involve the private databases challenges Westphalian conceptions of sovereignty (Longo 2018). Cell phone data and tracking aps may play a role in ensuring public health, but create the possibility of police, government authorities, and corporations misusing private data. Reliance on technology to guarantee mobility rights also creates an issue of equity, in which people who do not have access to cell phones or do not participate in programs monitoring their movements may be excluded from important public goods.

Technological shifts raise conceptual and normative questions and are related to more fundamental questions about how to theorise mobility. We have not yet come to terms with how to integrate mobility within the social sciences and normative theory. An adequate normative theory of mobility depends on adequately descriptive and explanatory theories. One intriguing set of approaches attempts to reconceptualise political philosophy from the perspective of the migrant or the nomad (Deleuze and Guattari 1987; Nail 2015; Ringmar 2020). Nomads and migrants challenge the sedentary bias of states, as well as the alleged isomorphism between state and nation. Taking the claims of mobile peoples seriously would mean reconsidering some of the master categories of political thought: territory, sovereignty, citizenship, and perhaps the state itself (MacKay et al. 2014). They invite us to cease treating mobility as pathological and to ask how just institutions could accommodate mobile people and minimise inequities in freedom of movement. These efforts need to be pursued with care or they risk fetishising mobility and overlooking how both movement and stasis characterise our lives and societies. A politics of mobility should not be carried out independently of an analysis of the infrastructure that permits and structures movement and the need for people to come to rest to form and sustain communities. It must dispel the cognitive biases of methodological nationalism and sedentariness and posit new institutions that better facilitate mobility and support the lives of mobile people. This has yet to be achieved.

Literature

Balderrama, F. E., and R. Rodriguez. 2006. *Decade of Betrayal. Mexican Repatriation in the 1930s*. Albuquerque: University of New Mexico Press.

Beck, U. 2000. *What Is Globalization?* Cambridge: Polity Press.

Blunt, A. 2007. Cultural Geographies of Migration. Mobility, Transnationality and Diaspora. *Progress in Human Geography* 31:684–694.

Bosworth, M., A. Parmar, and Y. Vázquez, eds. 2018. *Race, Criminal Justice, and Migration Control. Enforcing the Boundaries of Belonging*. Oxford: Oxford University Press.

Carens, J. H. 2013. *The Ethics of Immigration*. New York: Oxford University Press.

Clemens, M. A., C. E. Montenegro, and L. Pritchett. 2019. The Place Premium. Bounding the Price Equivalent of Migration Barriers. *The Review of Economics and Statistics* 101:201–213.

Cresswell, T. 2006. *On the Move. Mobility in the Modern Western World*. London: Routledge.

Cresswell, T. 2010. Towards a Politics of Mobility. *Environment and Planning D. Society and Space* 28:17–31.
De Genova, N. P. 2002. Migration 'Illegality' and Deportability in Everyday Life. *Annual Review of Anthropology* 31:419–447.
De Genova, N. P., and N. Peutz, eds. 2010. *The Deportation Regime: Sovereignty, Space, and the Freedom of Movement*. Durham: Duke University Press.
Deleuze, G. and Guattari, F. 1987. *A Thousand Plateaus. Capitalism and Schizophrenia*. Minneapolis, London: Bloomsbury Academic.
Fetzer, J. S. 2016. *Open Borders and International Migration Policy. The Effects of Unrestricted Immigration in the United States, France, and Ireland*. Houndmills, Basingstoke, Hampshire: Palgrave Pivot.
FitzGerald, D. S. 2019. *Refuge Beyond Reach. How Rich Democracies Repel Asylum Seekers*. Oxford: Oxford University Press.
Gibney, M. J. 2004. *The Ethics and Politics of Asylum. Liberal Democracy and the Response to Refugees*. Cambridge: Cambridge University Press.
de Haas, H., Vezzoli, S., and Villares-Varel, M. 2019. "Opening the Floodgates? European Migration under Restrictive and Liberal Border Regimes 1950–2010". In *Working Paper Series 150*. International Migration Institute Network.
Hing, B. O. 2009. Institutional Racism, ICE Raids, and Immigration Reform. *University of San Francisco Law Review* 44:307–352.
Hoerder, D. 2011. *Cultures in Contact. World Migrations in the Second Millennium*. Durham: Duke University Press.
Honig, B. 2003. *Democracy and the Foreigner*. Princeton: Princeton University Press.
Hosein, A. 2016. Arguments for Regularization. In *The Ethics and Politics of Immigration. Core Issues and Emerging Trends*, ed. A. Sager, 159–179. London: Rowman & Littlefield International.
International Organization for Migration. 2018. *World Migration Report 2018*. International Organization for Migration.
Longo, M. 2018. *The Politics of Borders. Sovereignty, Security, and the Citizen after 9/11*. Cambridge: Cambridge University Press.
MacKay, J., J. Levin, G. de Carvalho, K. Cavoukian, and R. Cuthbert. 2014. Before and after Borders. The Nomadic Challenge to Sovereign Territoriality. *International Politics* 51:101–123.
Maley, W. 2016. *What Is a Refugee?* Oxford: Oxford University Press.
Mann, M. 2005. *The Dark Side of Democracy. Explaining Ethnic Cleansing*. New York: Cambridge University Press.
Miller, D. 2016. *Strangers in Our Midst. The Political Philosophy of Immigration*. Cambridge, MA: Harvard University Press.
Nail, T. 2015. *The Figure of the Migrant*. Stanford: Stanford University Press.
Nethery, A., and S. J. Silverman, eds. 2015. *Immigration Detention. The Migration of a Policy and Its Human Impact*. London: Routledge, Taylor & Francis Group.
Ravenstein, E. G. 1885. Laws of Migration. *Journal of the Royal Statistical Society* 48:167–227.
Ravenstein, E. G. 1889. Laws of Migration. *Journal of the Royal Statistical Society* 52:241–301.
Ringmar, E. 2020. The Anti-Nomadic Bias of Political Theory. In *Nomad-State Relationships in International Relations. Before and After Borders*, ed. J. Levin, 45–61. Cham: Palgrave Macmillan.
Sager, A. 2016. Methodological Nationalism, Migration and Political Theory. *Political Studies* 64:42–59.
Sager, A. 2018a. Ethics and Migration Crises. In *The Oxford Handbook of Migration Crises*, ed. C. Menjívar, I. Ness, and M. Ruiz, 589–602. Oxford: Oxford University Press.
Sager, A. 2018b. *Toward a Cosmopolitan Ethics of Mobility. The Migrant's-Eye View of the World*. Cham: Palgrave Macmillan.
Sassen, S. 2014. *Expulsions. Brutality and Complexity in the Global Economy*. Cambridge, MA: The Belknap Press of Harvard University Press.

Scott, J. C. 2009. *The Art of Not Being Governed. An Anarchist History of Upland Southeast Asia.* New Haven: Yale University Press.

Sheller, M., and J. Urry. 2006. The New Mobilities Paradigm. *Environment and Planning A* 38:207–226.

Stumpf, J. 2006. The Crimmigration Crisis. Immigrants, Crime, and Sovereign Power. *American University Law Review* 56:367–419.

Torpey, J. 2000. *The Invention of the Passport. Surveillance, Citizenship, and the State.* Cambridge: Cambridge University Press.

United Nationals World Tourism Organization. 2018. *UNWTO Tourism Highlights 2018 Edition.* Madrid: World Tourism Organization.

Wellman, C. H., and P. Cole. 2011. *Debating the Ethics of Immigration. Is There a Right to Exclude?* Oxford: Oxford University Press.

Wimmer, A., and N. Glick Schiller. 2002. Methodological Nationalism and Beyond. Nation-State Building, Migration and the Social Sciences. *Global Networks* 2:301–334.

Migration, Global Justice and Development

Andreas Niederberger

Abstract Many forms of contemporary migration appear to be reactions to global injustices. Theories of global justice, yet, have not paid much attention to migration for a long time and sometimes even see migration as a phenomenon that reinforces global injustice. This article first reviews some recent approaches that have been developed to justify claims to migration on the basis of principles of global justice or violations of such principles. These approaches, though, show difficulties in adequately responding to empirical migration, the impact of development aid or the factors that lead to current injustice. Against this background, the article asks whether the opposite position, namely, to prove that migration is undesirable due to demands for global justice, is more convincing. It turns out that here, too, migration is ultimately not rejected for reasons of justice, but because it does not help to alleviate the suffering caused by injustice. In a third step, the proposal to distribute the costs of forced migration in a globally fair manner is briefly examined. Given its difficulty to properly address global injustice as the main reason for forced migration, the article ends with a call to approach migration against the backdrop of a theory of global justice that focuses on the distribution of opportunities to exercise control over the global conditions.

Migration takes place in a world with great inequalities and interdependencies between states, regions and their populations. Incomes in the Global North are many times higher than the wages paid for similar work in the Global South. Poor countries' access to the global market is severely restricted or limited to raw materials or simple products and services. The resource privilege allows warlords or local elites to sell natural resources internationally, even if the local population has not authorized them to do so. Other states or international networks foment and promote supposedly local violence. Human rights violations are often only addressed in the international arena when it is politically and economically opportune, and democratization movements in many cases do not receive the necessary

A. Niederberger (✉)
Universität Duisburg-Essen, Institut für Philosophie, Essen, Nordrhein-Westfalen, Germany
e-mail: andreas.niederberger@uni-due.de

A. Niederberger et al. (eds.), *Handbook of Migration Ethics*,
https://doi.org/10.1007/978-3-031-89877-8_12

international support to succeed in their struggles for power. States from the Global North employ international organizations and law to assert their interests in other parts of the world, preventing the relevant institutions from providing everyone with the ability to control global interactions. In the context of anthropogenic climate change, the richer and more technologically advanced states refuse to take responsibility for their resource use and are only partially willing to make their technological capabilities available to poorer states and regions in a way that allows them to adapt to the impacts of climate change.

Migration is obviously related to these global conditions. Historically, it has contributed to the formation of colonialism and imperialism, enabling some parts of the world to achieve domination over others. Today, migrants respond to global inequalities and try to improve their situation by moving elsewhere. This, in turn, creates new difficulties and challenges, especially in the Global South, where most migrants remain, to ensure adequate legal, political, socioeconomic, and cultural participation for all in sometimes already precarious conditions. Typically, it is not the poorest and neediest who migrate, but local middle classes with better education and training. Despite their relatively good education, however, integration into the labour market and society in arrival contexts often demands an effort from migrants that makes it difficult for them to be politically active beyond that. Migrants are therefore not necessarily a major driving force in arrival contexts for the transformation of global relations. And although remittances provide transfers to contexts of origin, research suggests that this does not usually lead to broad local improvements. In many cases, they allow additional people to acquire education and skills that increase their chances of success in subsequent migration. This raises the question of whether migration helps improve global conditions or whether it tends to perpetuate their desolate state.

1 Should We Consider Migration in Light of the Global Justice Debate?

Reflecting on the inequalities and interdependencies mentioned above, the debate on global justice has been one of the focal points of philosophical research in recent decades (Brooks 2008). However, with the putative return of nationalism, populism, and national borders, and the unwillingness to engage in trans- and international solidarity, as seen since the global financial crisis of 2008 and the European "refugee crisis" of 2015, this debate has faded into the background. Few philosophical approaches to the refugee crisis have been linked to discussions of principles of global justice or transnational legitimacy. Most have concentrated either on helping people in need or on criteria for distributing refugees among European states. Yet, given the interconnections between global conditions and the reasons for migration, its practice, and its consequences, the question arises whether migration should not be considered more in light of theories of global justice (see also Weber and Tazreiter 2021; Niederberger 2024).

The debate on global justice began with a critique of John Rawls's limitation of his principles of justice to individual states. Brian Barry, Charles Beitz, and Thomas Pogge argued that in a decision-making situation such as Rawls models it in his original position, individuals would not be willing to accept a world with borders if it meant that they would find themselves in a situation where they would be permanently worse off than others in terms of opportunities (Niederberger 2013). If borders are to be allowed, then they must be complemented by redistributive principles, similar to those in individual states. Such principles would ensure that comparable living conditions prevail everywhere—at least in relevant respects. Global justice, then, does not necessarily imply a world state, but it would require substantial transfers between states, or at least a global order that prevents states from exploiting others or limiting their chances for development. If it turns out that the adverse circumstances described above are a major reason for the migration that is taking place today, it seems logical to understand it as a reaction to global injustice. Life chances are very unequally distributed and there are only weak global redistributive principles or international institutions that could overcome inequality. Consequently, migrants' claims could be understood as justified by broader claims for global justice.

However, even if responses to global injustice can explain migration, it is not immediately clear whether the possibility of migration is itself a demand of global justice. The fact that conditions are unjust and those affected have little or no ability to change them may make it rational for these agents to leave their societies. At the same time, one can ask whether such migration leads to globally more just conditions. It has already been briefly pointed out that migration can also mean the loss of labor, skills, and political activity in the contexts of origin, and that even remittances do not clearly improve the overall situation in these contexts—not to mention that it is unlikely that migration will lead to a change in the global political economy. Migration could therefore even exacerbate global injustice for the remaining populations and thus be an element of this injustice itself. Against this background, one must examine whether and how migration—possibly in different forms—is part of or contributes to achieving more just local and global conditions.

The questions of whether migration can be traced to global injustices and whether the claim to migration is itself part of a globally more just order assume that migrants are agents, and they focus on their duties and rights. In some cases, however, migration is also more or less directly forced by circumstances such as wars, natural disasters, or persecution. This migration can result in the imposition of costs on host contexts to accommodate and care for migrants that exceed their capacity and that other states, societies, or regions do not have to bear. Migration and its associated costs can therefore also be the subject of questions about what should be justly distributed on a global scale. The forced migration in question occurs primarily in the Global South and to a large extent between economically similarly weak and politically fragile contexts. The contingency of allocating migrants by local proximity therefore means not only that migrants' claims may not be met. It also means that the distribution of burdens may be unfair in the global context, first between South and North, and second between societies and states in the respective regions.

2 Important Controversies Regarding Migration and Global Justice

There are thus three main issues concerning the relationship between global justice and migration: first, whether there is a connection between global justice or injustice and migration; second, whether migration is an adequate means of responding to global injustice; and third, how the burdens that come with forced migration should be distributed in a globally just manner. These issues are at the heart of three controversies in migration ethics, each of which will be addressed below individually, but also in their interrelation.

2.1 Global Justice and the Right to Migrate

The main debate on the ethics of immigration, as led by Joseph Carens, David Miller, and others, focuses on what basic rights people have and whether they include the right to open borders or freedom of association. While this debate touches on the foundations of liberal theories of justice, it is not understood as a debate about the just or fair distribution of the relevant rights. Although Carens (2013: 233–236) emphasizes the importance he places on combating global inequalities and the goal of achieving global equality of opportunity, his argument for open borders is primarily tied to the general value of freedom, and freedom of movement across borders thus acquires the status of a human right or a liberal basic right. Carens is aware of the difference between the justifications. Others who argue for a right to open borders or global freedom of movement locate themselves in a theory of global justice, but conceive of it very broadly, implying that their approach rests on cosmopolitan or universalist foundations. Juss (2006: 1–32), for example, assumes in this vein that global freedom of movement is the most fundamental human right because without it all global aspirations remain “precarious.”

Eric Cavallero, by contrast, offers a reconstruction of migrants’ claims in a theory of global distributive justice. He assumes that global relations are just when international law does not systematically disadvantage anyone on the basis of national citizenship or origin. Such discrimination would exist if members of certain communities were less successful than non-members over the course of their lives and one could explain this lack of success by the effects of legal institutions triggered by group membership (Cavallero 2006: 98). Since the respective interests and the criteria for success vary greatly in different contexts, one cannot simply measure the success of life plans objectively. Cavallero therefore assumes that perceptions of adverse conditions that stand in the way of one’s life success translate into migration intentions. According to this reasoning, if the lack of success is primarily due to one’s nationality or birth context, people will seek to live their lives elsewhere. Immigration pressure, i.e., the relative strength of the propensity to migrate in a society, can thus be used to determine the justice or injustice of conditions. If more people—relative to the size of the population—want to immigrate than want to emigrate, one can conclude that the chances of succeeding in one's own life are higher in that state than in others. If, on the other hand, immigration and emigration interests are equally high or low, a state of equilibrium exists in which one can speak of a just distribution of opportunities (Cavallero 2006: 106–107).

Unlike Carens and other open-borders advocates, Cavallero does not presuppose that people have a fundamental right to migrate. But if the pressure for immigration is stronger than the willingness to emigrate, one must assume an unfair distribution of opportunities for a successful life, which in turn leads to claims by the disadvantaged against the country in question. However, these claims do not relate directly to migration, but to equal opportunities for leading a successful life. The addressee of the claims can therefore respond in at least two ways: either by opening the border until the equilibrium point is reached, or by providing development assistance in contexts with a higher propensity to emigrate (Cavallero 2006: 108–112).

The "immigration pressure model of global distributive justice" offers an interconnected assessment of migration and global justice: If migration pressure is one-sided, we can assume global injustice, and if there is global injustice, there may be claims to migration. This assessment presupposes, of course, that Cavallero's "cosmopolitan premise", i.e., the claim of all people to similar opportunities for success in their own lives, can be grounded in a general theory of global justice. Critics of this model, however, note that it operates with an idea of migration and its causes that has little to do with current empirical migration research: Alex Sager (2016: 49), for example, accuses Cavallero of "methodological nationalism" by conceiving of migration as a mere transition between nation-states. This hardly reflects the complicated relationships between migration and the level of development of societies. It does not take into account the aforementioned fact that it is usually not the poorest who migrate, but rather those who are relatively well off. Consequently, those who are supposed to generate the corresponding immigration pressure do not exert it. Development assistance as an alternative to migration also neglects the fact that in many cases it leads to greater interest in migration. Migration and migration interest are thus not simply correlated with a disadvantaged situation in the global context: Neither can we simply take them as indicators of global injustice, nor will greater justice necessarily lead to less migration.

More generally, one might say that comparative migration readiness is a poor measure of global inequality of opportunity: Many theories of global justice arrive at principles that tie the justice of the global order to a distribution of resources that ensures equal opportunities for all people to lead successful lives. At first glance, it may seem that this always already justifies the claim of all people to migrate if necessary and to be included in contexts where such opportunities exist. For many theorists, however, this is not the case. Especially when theories emphasize that we should not be looking only at globally equal ways of living, the need to migrate may mean that we are not successful in leading the life we are aiming for. Ideas of a successful life are often tied to specific contexts, and many theories of global justice therefore demand that all people globally have the opportunity to live a successful life, where the conditions for this success may be quite different locally and may include the right to remain in the very local context (Oberman 2011). Thus, people may not want to migrate because they do not see how they can live the lives they would like to live elsewhere (Brock 2009: 194).

If we want to account for migrants' claims through principles of global justice, we need to discuss these claims to some extent independently of the factors that concretely drive migration, and global equality of opportunity is not sufficient for the claims in question. Mathias Risse (2012: 153–156) makes such a proposal. According to him, all people can claim a proportionally similar use of the planet's resources, based on "common ownership of the Earth". The world is divided into different territories. Therefore, for each territory and for each generation, we should determine—at least ideally—the amount of resources available. If it turns out that this amount differs between territories, this over- or under-supply or use leads to the need for and justification of migration to the extent that it restores an equal distribution of proportionate use. Risse, too, emphasizes that this does not imply a direct individual right to migrate. Rather, migration policies must be negotiated between countries and take into account the functioning of the state. But development aid is not enough either, as it conflicts with self-determination regarding the use of resources.

Risse suggests how the necessity and permissibility of migration can be derived from globally equal claims on resource use. In doing so, he provides a reference point for international negotiations on migrant quotas that is not based on empirical assumptions about migration. Even under ideal conditions, migration may or even must occur to ensure similar claims to use for each new generation. Risse supposes that the normative reference point he develops is not necessarily at odds with empirical reasons for migration or willingness to receive migrants. This is because the duty to admit immigrants or to emigrate is underdetermined with respect to the specific individuals involved in each case, so that it can be fulfilled in different ways. Although the idea of separate territories for which we record population density and resources to be exploited might again lead to the charge of "methodological nationalism," this does not apply to Risse. In his case, states or a specific set of territories are merely the most obvious evaluative framework under present circumstances, which might well be revised under different conditions. Empirically complex migration patterns across multiple territories and with different implications for different contexts can inform assessments of the permissibility of particular migration practices in light of what are ultimately methodologically individualistic claims to the "proportionate use" of resources.

Nevertheless, this theory seems irritatingly "far-fetched". It provides an individualistic approach that locates normative claims at a level below the multiple relational and institutional factors that shape contemporary injustices. For Risse, one should concede differential productivity in resource use and therefore reject migrants' claims to the surplus value of resources or to goods and services made possible by such productive use. We should reject general claims of equal opportunity, and the theory should focus on available resources as such. In doing so, however, Risse loses sight of the fact that resource use, both locally and transnationally, is in many ways conditioned by institutions or by resource use elsewhere. Not every state, society, or market is equally well or fairly organized and constituted. In part, these differences result from local conditions, but often they also go back to historical developments and transnational structures. The ability of some states or regions of the world to make it impossible or very difficult for other states or regions to use the resources available to them as they see fit, is certainly a greater problem at present than the limited availability of resources. However, if forms of usability were also included in the definition of global justice, this would greatly reduce the difference from general equality of opportunity. And transnational institutions and regulations would certainly have to be added to ensure appropriate forms of resource usability.

2.2 *Migration as an Instrument for More Global Justice?*

In an early attempt to examine the relationship between global justice and migration, Robert Goodin (1992: 8) concluded that, from a global justice perspective, either all people globally are provided goods or money, or people have the right to move to the goods or money. The previous section has shown, however, that it is not so easy to elaborate in detail a theory of global justice that identifies principles or entitlements that explain whether and when migration is permissible—at least if freedom of movement is not simply understood as a general basic right. The difficulty, though, is not only in finding a convincing principle of global distributive justice according to which people might have a right to migrate. An even more serious problem is that it is difficult to explain how migration should contribute to greater global justice. Many theories of global justice can clearly identify what is unjust about the current situation. But given such an understanding of injustice, can we really say that emigration from unjust conditions will make them more just and better? Many proponents of theories of global justice answer this question in the negative, arguing that such an approach must reject migration at least as a remedy against injustice.

Thomas Pogge (1997: 17–18), for example, asks whether it is worth fighting for open or at least more open borders for migrants from the Global South in countries of the Global North. In his view, we should not reject open borders per se. But under current conditions, they could only be opened, if at all, with great effort against existing resistance in the Global North. From the perspective of a theory of global justice, a corresponding effort for greater border openness in the Global North would only have moral worth if it protected those currently exposed to unjust conditions from those conditions. Pogge reminds us that migration benefits only some, and possibly those who are already relatively well-off compared to the poorest. If, on the other hand, the resources that must be spent on integrating migrants could be used much more effectively in the Global South, as Pogge assumes, the funds should go directly there and not be used to support migrants in the Global North. In an almost paradoxical twist, Pogge (1997: 22–23) suspects that proponents of open borders may ultimately be driven by patriotism and care less about improving conditions in other countries than about the moral perfection of their own country.

Other advocates of theories of global justice support Pogge's argumentation. Gillian Brock (2009: 204–208), for instance, points out that remittances usually do not have any lasting effects in the societies of origin, but are spent mainly on consumer goods. Moreover, migrants are usually not willing to send money for very long. But this is not the only respect in which migrants do not benefit societies of origin. Brock (2009: 198–204; Brock 2015: 36–59) also emphasizes that migration often leads to a brain drain in poor societies, especially in areas that are particularly relevant for poverty reduction, such as health care. For the affected societies, migration often means not only no benefit but also a further loss of the ability to overcome existing poverty. Peter Higgins (2013: 65–67) adds that because of the often limited opportunities in the labour markets of contexts of arrival, ultimately those who migrate do not necessarily benefit either. Rather, they often become victims of exploitation, racism, and abuse. Patti Lenard and Christine Straehle (2011) thus argue that guest worker programs in particular, which block guest workers' path to citizenship in the country where they are temporarily working, ultimately serve the interests of the host country rather than those of migrants and run counter to their potential contribution to global justice. These programs leave little opportunity for guest workers to shape their employment and residence in a way that adequately responds to the economic pressures that originally led to their decision to migrate as guest workers.

Pogge and many others consider this reasoning obvious and deny that existing global injustices are a good reason for extensive migration rights. These observations contain important objections to the assumption that there is a direct link between migration and the improvement of living conditions in the Global South. However, with Kieran Oberman (2015: 240–243), one can ask whether the empirical evidence does not also point to forms of migration and effects of remittances that certainly have a significant impact on poverty reduction. In general, looking at the overall structure of the argument, one can criticize that in its rejection of migration rights as a response to global injustice, it suffers from being based on an overly simplistic idea of aid or development. Pogge and Brock emphasize that global injustice is (also) the result of institutions and domination, that is, the exercise of power by the Global North, sometimes together with local corrupt elites. Their arguments for efficiency and harm prevention, on the other hand, are based on the belief that there are aid measures that could meet the needs of the poorest with relatively little effort or by keeping well-educated people in the societies of the Global South. The latter may be true, but it would not make the world a more just place; it might reduce the suffering of some, but it would not eliminate the structural reasons for global injustice. If true, this is not an argument against the permissibility of migration in global justice theory. Rather, it shows that, given the suffering caused by global injustice, migration not only does not mean greater justice, but also does not necessarily promise to improve suffering.

2.3 Is There a Globally Just Distribution of the Costs and Burdens of Migration?

We see, thus, that the attempt to prove that migration runs counter to global justice is not convincing either. Neither can migration claims be simply justified by principles of global justice, nor should we reject such claims because they conflict with demands for a change in global conditions toward greater justice. Perhaps the connection between global justice and migration becomes clearer when we ask how to distribute the special burdens sometimes associated with forms of irregular migration.

Most forced and irregular migrants live in the Global South. In some cases, this does not pose an additional serious problem because the countries in question are weak states anyway, so that the internally displaced or new arrivals from neighbouring countries have to take care of themselves like the people already living there or they may even contribute to the local economy. Sometimes the migration in question has a long history, and its current irregularity is due primarily to political borders that are often an expression of the colonial past. In other cases, however, particularly in the case of larger and often poorly planned migrations resulting from wars, persecution, genocide, or disasters, the arrival of the migrant groups in question places a significant burden on the arrival contexts. Without international support, they must provide basic goods and services, and it is not uncommon, for example, for larger refugee camps to become political and security problems for host states and societies (Milner 2014: 155).

For a long time, these large groups of forced migrants were not considered an important issue in theories of global justice. However, at the latest with the crisis of European refugee policy in 2015 and the failure of the European system of allocating refugee responsibility, the topic became a subject of discussion. Internationally, the allocation of refugee responsibility is currently "regulated" primarily by the principle of non-refoulement: states may not deport migrants who have fled to their territory to areas where they are at risk of serious human rights violations. In the absence of significant and functioning redistributive systems for such groups, the costs for them ultimately remain with those close to the contexts that caused the forced migration. Aside from refugees from Russia's current war on Ukraine, this currently primarily affects the Global South, as noted above. In the eyes of Matthew Gibney (2015: 451–452), the situation represents an injustice on two levels, as not only is there no fair distribution of the corresponding costs, but the Global North, through its border and visa policies, compels the Global South to bear the costs alone.

The distinction between the two levels is important because, although injustice at the second level is currently the main reason for the unequal distribution of burdens between South and North, simply opening the borders would not itself lead to a fair distribution of costs. Indeed, if forced migrants were to decide for themselves where to seek protection, this could again lead to an unequal distribution of burdens—this time possibly to the disadvantage of a few countries in the Global North. Gibney (2015: 449–450) therefore argues that a globally just distribution would need to occur through a global resettlement program. It should take into account three factors for the respective distribution of groups: first, the capacity of states to integrate forced migrants, measured by population size, gross domestic product, and the number of forced migrants already admitted; second, the interests of forced migrants in the respective contexts of arrival; and third, the conditions under which forced migrants could lead new and better lives.

Gibney's proposal has led to an extensive debate on the allocation and distribution of forced migrants. On the one hand, it discusses how convincing the factors for distribution suggested by Gibney are and whether other factors, such as the willingness of the respective societies to accept forced migrants, should not be added. On the other, there is an examination of what would be just under the non-ideal condition that some states do not meet their fair share (Owen 2016; see also chapter "Refugee Ethics" in this Handbook). What hardly matters in these debates, though—surprisingly, given the goal of including forced migration as a topic in the field of theories of global justice—is how the corresponding distributional question relates to other dimensions of global injustice. For forced migration does not usually occur under circumstances that are unrelated to global conditions, but it is often at least also the result of trans- and international actions and structures. Against this background, we should ask whether the contribution to the causes of forced migration should not also be an important factor for the distribution of forced migrants or the costs they generate. This would give greater weight to the role that respective actors have played in creating unjust global conditions in determining the rights to migrate or of migrants. Migration would be seen less as a phenomenon or instrument in its own right and more as an essential feature of the dynamics of global injustice. Unfortunately, the current refugee debate is dominated by approaches that focus on overcoming the suffering of refugees and disregard the reasons for their suffering also in order not to create reasons for not assuming remedial responsibility.

3 Towards a Better Understanding of Migration in Light of Global Justice

Rainer Forst (2013: 155–160) has distinguished two different "pictures" of justice in theories of justice: a first "unpolitical" picture, which focuses on the recipients of opportunities and goods, and another "political" one, which centers on the question of who decides that and how opportunities or goods are generated and distributed. In the first picture, the theories look for principles that are translated into structures and institutions to implement a fair distribution of opportunities, etc. In the second picture, approaches look for distributions of opportunities to participate in those structures and institutions that make those structures and institutions themselves just. As far as the global situation is concerned, both pictures agree that it is unjust. However, their essential difference becomes apparent precisely when they consider migration as a reaction to the injustice: In the first picture, it is indeed difficult to see why the fact that some have better access to opportunities, resources, or goods through emigration should significantly change the injustice. Global justice means equality of opportunity, etc. for all people and not just for those who have the means and skills to emigrate. If, however, we assume in the second picture that globally everyone is entitled to globally just conditions, then the scandal of the current situation is not that some goods are not equally distributed, but that the Global North refuses to change the prevailing conditions. In this perspective, too, we should not overstate the importance of migration and assume a fundamental shift in global power relations just because some of those previously excluded from power may end up in contexts that control them. But the opening of borders, that is, a right to migrate for those who are exposed to a global situation in which some dominate others and prevent them from changing the basic global structure, would give everyone the opportunity to be heard and, if necessary, to participate politically where decisions are made. This claim is not invalidated by the fact that it cannot be used equally and directly by everyone in the Global South. Every individual can be subject to global domination and thus every individual can also claim non-domination and structures that secure it. In many cases migration is not simply a matter of a human right to freedom of movement, but a right of individuals or even collectives, arising from the claim to global justice, to resist global injustice (Niederberger 2021).

In the future debate on migration and global justice, we must focus on three questions in particular: First, it is necessary to examine whether and under what conditions subjection to global economic, political, and legal structures and institutions without significant means of controlling them leads to a right to migration. This includes the question of whether the consequence is a general right to migration or a qualified right to migrate to a specific context. Second, we should empirically study the impact of migration on changes in global structures and institutions, and what this means for the opportunities that migrants in contexts of arrival particularly need in order for them or their migration to contribute to desired changes. Third, and finally, beyond the pressing urgency of reforming existing dominating structures and institutions, we need to ask what the broader perspective is for developing a just global migration regime. To this end, we must seek comprehensive and inclusive procedures that recognize the different interests in sedentariness and mobility, exclusive self-determination, and a dynamic and transnational shaping of social and cultural contexts.

Literature

Brock, G. 2009. *Global Justice. A Cosmopolitan Account*. Oxford: Oxford University Press.

Brock, G. 2015. Part I. In *Debating Brain Drain. May Governments Restrict Emigration?*, 9–107. Oxford: Oxford University Press.

Brooks, Th., ed. 2008. *The Global Justice Reader*. Oxford: Blackwell.

Carens, J. H. 2013. *The Ethics of Immigration*. New York: Oxford University Press.

Cavallero, E. 2006. An Immigration-Pressure Model of Global Distributive Justice. *Politics, Philosophy & Economics* 5:97–127.

Forst, R. 2013. A Kantian Republican Conception of Justice as Nondomination. In *Republican Democracy. Liberty, Law and Politics*, ed. A. Niederberger and Ph. Schink, 154–168. Edinburgh: Edinburgh University Press.

Gibney, M. J. 2015. Refugees and Justice Between States. *European Journal of Political Theory* 14:448–463.

Goodin, R. E. 1992. If People were Money…. In *Free Movement. Ethical Issues in the Transnational Migration of People and of Money*, ed. B. Barry and R. E. Goodin, 6–22. University Park, PA: Pennsylvania State University Press.

Higgins, P. 2013. *Immigration Justice*. Edinburgh: Edinburgh University Press.

Juss, S. S. 2006. *International Migration and Global Justice*. Aldershot & Burlington, VT: Routledge.

Lenard, P., and Ch. Straehle. 2011. Temporary Labour Migration, Global Redistribution, and Democratic Justice. *Politics, Philosophy & Economics* 11:206–230.

Milner, J. 2014. Protracted Refugee Situations. In *The Oxford Handbook of Refugee & Forced Migration Studies*, ed. E. Fiddian-Qasmiyeh, G. Loescher, K. Long, and N. Sigona, 151–162. Oxford: Oxford University Press.

Niederberger, A. 2013. Liberales Tolerieren statt globaler Gerechtigkeit. John Rawls' Konzeption eines Rechts der Völker. In *Politischer Liberalismus und freiheitliche Ordnung. John Rawls und der Verfassungsstaat*, ed. M. Becker, 131–164. Baden-Baden: Nomos.

Niederberger, A. 2021. Migrationsethik in der Krise. Einige grundlegende philosophische Überlegungen. *Zeitschrift für Migrationsforschung* 1:97–123.

Niederberger, A. 2024. Ist Migration eine Herausforderung für den Kosmopolitismus? In *Migration – Herausforderungen im Spannungsfeld von Theorie und Praxis*, ed. O. Angeli, A. Niederberger and H. Vorländer, 49–76. Baden-Baden: Nomos.

Oberman, K. 2011. Immigration, Global Poverty and the Right to Stay. *Political Studies* 59:253–268.

Oberman, K. 2015. Poverty and Immigration Policy. *American Political Science Review* 109:239–251.

Owen, D. 2016. Refugees, Fairness and Taking up the Slack. On Justice and the International Refugee Regime. *Moral Philosophy and Politics* 3:141–164.

Pogge, Th. 1997. Migration and Poverty. In *Citizenship and Exclusion*, ed. V. Bader, 12–27. Houndmills, Basingstoke: St. Martin's Press.

Risse, M. 2012. *On Global Justice*. Princeton: Princeton University Press.

Sager, A. 2016. Methodological Nationalism, Migration and Political Theory. *Political Studies* 64:42–59.

Weber, L., and C. Tazreiter, eds. 2021. *Handbook of Migration and Global Justice*. Cheltenham & Northampton, MA: Edward Elgar Publishing Limited.

Migration and Historical Injustice

James Souter

Abstract In this chapter, I explore the implications of historical injustice for contemporary debates surrounding the ethics of migration, questioning how, if at all, states should respond to historical injustice in their various migration and asylum policies. In Part I, I map out some of the interrelations between migration and historical injustice during the various stages which migrants may pass through, including emigration, admission to another state's territory, migrants' acquisition of membership, and repatriation. I suggest that migration and the processes surrounding it may both compound historical injustice, and potentially offer an important form of reparation for that injustice. In Part II, I examine theoretical debates surrounding migration and historical justice, surveying work that applies principles of reparative justice to the ethics of migration and asylum. This work suggests overall that, while migration may play a potentially significant role in redressing historical injustice, and states often should use their migration and asylum policies to offer reparation to those affected by the legacies of historical injustice, there are various theoretical and practical complications surrounding this line of thought—for instance, surrounding issues of causality, responsibility, state capability, and the prioritisation of migrants for entry—which need to be addressed in order for these arguments to fully succeed. I conclude by identifying additional aspects of this controversy which merit further research and analysis, such as the need for greater engagement with empirical evidence to begin to ascertain the reparative duties of actual states towards migrants in practice.

Migration and historical injustice have a close yet complex relationship. While many injustices throughout history, such as conquest and persecution, prompted immediate migration, some injustices perpetrated generations ago still deeply shape contemporary migratory patterns. While *all* migration clearly occurs within a global system that generates unjust global inequality, which is often regarded as at least partly stemming from the legacies of European colonial rule, many migrants are

J. Souter (✉)
School of Politics and International Studies, University of Leeds, Leeds, UK
e-mail: J.Souter@leeds.ac.uk

A. Niederberger et al. (eds.), *Handbook of Migration Ethics*,
https://doi.org/10.1007/978-3-031-89877-8_13

arguably led to move by historical legacies more directly, whether they are at the sharp end of that inequality, or affected by conflicts resulting from these legacies. Some migration may itself constitute a vast historical injustice, such as the transatlantic slave trade which forcibly transferred millions of Africans over the course of several centuries. Migration is also deeply entwined with two cases of historical injustice, concerning indigenous peoples and African-Americans, that are generally focused upon by theorists and reparation advocates (Spinner-Halev 2012: 57), as the dispossession of the former resulted directly from the migration of settlers to their lands, while the plight of the latter stems from the slave trade, in both cases centuries ago.

However, although migration and historical injustice can be mutually constitutive, migration may also potentially act as an important means of making amends for such injustice, by creating opportunities for migrants previously denied to them due to that injustice. As I suggest in this chapter, while states have more frequently created historical injustice by generating migration and mistreating migrants, admission into another state may at times constitute a form of *reparation* for historical injustice, by going at least some way towards redressing that injustice. Migration, then, can play a reparative role, potentially contributing to a more just global future overall.

My concern in this chapter is to address the following questions. First, what are the interrelations between historical injustice and contemporary migration? Second, how—if at all—should states respond to historical injustice in their various migration policies, especially where they bear causal or moral responsibility for that injustice? I engage with these questions largely by drawing on work within contemporary analytic political theory, and especially on work concerning the ethics of migration. Although there are no major works in this field which explicitly aim to analyse the relationship between migration and historical injustice, this literature nevertheless contains important arguments that can be readily brought together in order to illuminate both this relationship and the responsibilities of states in this context.

Before proceeding further, it is important to clarify the understanding of historical injustice drawn upon here. While historical injustice has been defined as "those harms or wrongs committed by individuals, groups, or institutions against other individuals and groups who are now dead, but whose descendants live today" (Ivison 2008: 509), in what follows I give particular attention to two types of this injustice. First, I attend particularly to what Jeff Spinner-Halev (2012: 56) has termed "enduring injustice", which he defines as injustice that has "roots in the past, and continues into the present day". This is because, unlike some historical injustices which do not affect anybody currently alive, such enduring injustices are most morally and politically urgent. Second, I include discussion of what can be termed "past injustice", which may involve perpetrators and victims who remain alive, despite occurring some time ago. While it would be difficult to identify a precise cut-off point between past and recent injustice, much discussion of reparative justice and migration is concerned with past, as well as historical, injustice.

1 General Issues at Stake in the Controversy

Many contemporary political theorists view historical injustices as morally significant, and believe that at least some of these injustices, including those affecting migrants, should be repaired (e.g. Thompson 2002; Butt 2009; Spinner-Halev 2012; Lu 2017). While I examine arguments based upon principles of reparative justice in the following section, at this point I explore in more detail the major relevance of historical injustice for the various stages that migrants may pass through, both *causally* (concerning how migration and historical injustice may be related in practice) and *normatively* (concerning the ethical and political implications of this injustice for states' migration policies). Historical injustice, I suggest, is often closely intertwined with the causes of contemporary migration; processes of emigration across borders; migrants' admission to (or exclusion from) other states; their acquisition of membership; and their return to their states-of-origin, and it also significantly shapes states' responsibilities in these contexts. Although historical injustice can also be heavily bound up with internal migration, I largely focus here upon international migration, especially between the global "South" and the "North".

1.1 Causes

Although not every decision to migrate is directly affected by legacies of injustice, many of the major causes of contemporary international migration are nevertheless rooted in such legacies. A few examples can illustrate this point. Although the relationship between migration and global inequality is complex (e.g. see Castles 2010: 1568), much migration is driven by desire to access resources and opportunities in some regions that are scarce or non-existent in others. And, while much global inequality can be partly attributed to more recent injustices such as unfair international trade rules (Pogge 2002), it appears to have its roots in the domination and social exclusion that characterised European colonial rule. For instance, the group enmities which lead to persecution and violence, and in turn forced migration, may stem from historical grievances between groups, or themselves be at least partly the product of colonial policies, such the imposition of arbitrary colonial borders which did not reflect pre-existing social structures (Holton 2014: 59).

Furthermore, environmental drivers of migration can be strongly linked to historical injustice. Environmental displacement, set to accelerate given climate breakdown throughout the twenty-first century, is being spearheaded by the excessive carbon emissions largely of Western economies. Although the harms set in motion by Western emissions during the Industrial Revolution may not constitute culpable historical injustice*s* per se, given they were not foreseeable at the time (Caney 2005: 761), modern scientific understanding of the anthropogenic nature of climate change is such that the decisions of Western states to continue emitting at high levels in previous decades despite this does constitute a highly significant past (and ongoing) injustice. Given the more recent culpability of high-emitting states and their growing impact on patterns of forced migration, debate in political theory has developed concerning their duties towards the environmentally displaced (e.g. Risse 2009; Byravan and Rajan 2010; Draper 2018; Eckersley 2015; Buxton 2019).

1.2 Emigration

While migrants' decisions to move may be shaped by historical injustice for such reasons, their emigration also may have an important bearing on whether, and how, such injustice is addressed. For example, emigration can lead to the so-called 'brain drain', whereby the departure of highly skilled professionals such as medical personnel from the global 'South' to the 'North' often severely saps the ability of their states-of-origin to address developmental and human rights issues such as healthcare shortages, which may stem in large part from historical injustice (Shah 2010: 92).

Accordingly, issues of historical and past injustice feature in some theoretical debate concerning the ethics of brain drain. Some, such as Gillian Brock (2009: 287) have highlighted the active recruitment of medical professionals from the developing world by Western states in recent decades as a past injustice creating "duties of rectification", given that this policy "looks remarkably like raiding other countries of their (precious) human resources". Second, some responses to the brain drain—such as immigration restrictions imposed by Western states on skilled workers from the global "South"—may be illegitimate given the legacy of injustice. As Kieran Oberman (2013: 451) has suggested, Western states may not be able to enforce such restrictions without hypocrisy, given their partial responsibility for underdevelopment in the "South" and the lack of reparation offered for this.

1.3 Journeys Across Borders

Once migrants have emigrated, both their journeys across borders and the institutions of border control they generally encounter are deeply shaped by historical injustice. First, migrants' intended journeys can be affected by this injustice, as many migrants wish to move to states that colonised their states-of-origin, given the cultural or linguistic affinities they share with those states as a result (e.g. Robinson and Segrott 2002: 30). Second, the history of Western border controls is littered with unrepaired injustices that may well affect their current legitimacy. Many immigration policies have been overtly racist (Hayter 2000; Dummett 2001), and the dehumanisation of migrants and asylum seekers in Western policy has been presented as "an example of the coloniality of power" and shaped by continuing notions of racial hierarchy rooted in the colonial period (Mayblin 2017: 1). Strong parallels exist between past and present migration systems, with Phillip Cole (in Cole and Wellman

2011: 221) observing that "the global migration regime of the colonial period was a system of domination and exploitation, and the current global migration regime operates in the same way" with similar patterns of exclusion often continued by states behind a veneer of more "'acceptable' policy objectives" (Fine 2016: 132).

More generally, the unjust harms of border control throughout its history may create reparative obligations borne by the states that have operated it, whether to compensate those previously coerced unjustly by them, or to ensure that current and future migrants avoid such treatment (see Al Hashmi 2023). Even for those who see some form of border control as legitimate, states may bear causal and moral responsibility for the violence and harm that has been routinely involved in the enforcement of borders historically and currently (see Oberman 2020; Parekh 2020). Alternatively, if Hillel Steiner (2001) is correct to argue that anyone deprived of an equal share of resources cannot be excluded from a state without compensation, then it seems compensation is due to countless migrants denied entry throughout history. Moreover, the exclusion of those owed admission as reparation may constitute a *double* injustice, given that, as Butt (2009: 176) has observed more generally, "the ongoing failure to fulfil [...] compensatory obligations itself constitutes an ongoing act of injustice".

In addition to such active injustices, the previous moral failures of states in the context of migration may also constitute historical injustices. For instance, Joseph Carens (2013: 194) considers the refusal of Jewish refugees fleeing Nazi Germany by other states to be of great significance to their current responsibilities towards refugees, suggesting we take "this terrible failure [...] as a constraint upon our inquiry into the ethics of admitting refugees". Such failures to protect refugees can contribute significantly to concurrent historical injustices, given that some rejected Jewish refugees went on to swell the number of victims in the Holocaust.

Moreover, the origins of a great many borders in historical injustice, such as violent conquest, might be taken as a serious challenge to their legitimacy, rendering them morally arbitrary. However, various theorists have argued that at least some territorial borders can be justified despite their morally tainted histories. In addition to Jeremy Waldron's influential argument that the injustice of colonial conquest can be superseded by the legitimate expectations of settlers' descendants (Waldron 1992), Ryan Pevnick (2011: 43) has observed that, given the ubiquity of historical injustice, claiming that such injustice invalidates states' right to control their borders would also threaten any claim of ownership in other contexts, and Sarah Song (2018: 63) believes states' territorial rights can be justly maintained "if serious efforts are made to remedy the harms caused by historical injustice".

Given the realities of border control, a great many migrants are refused entry and are instead forced to languish in a protracted state of liminality, often within refugee camps (see Parekh 2020). These states of limbo are hardly unaffected by historical injustice either. For instance, one large-scale attempt to remedy historical injustice suffered by Jews, through the creation of Israel, has led to the protracted displacement and encampment of many Palestinian refugees. In Spinner-Halev's terms, mentioned earlier, long-term encampment and exile can become a clear example of "enduring injustice" (Spinner-Halev 2012: 7), given the lack of autonomy residence within them generally entails.

1.4 Admission

For those spared these predicaments and admitted to another state's society, their admission may either *deepen* or work to *rectify* historical injustice. On the one hand, admission into another state may simply usher in another round of injustices also requiring redress. Indeed, immigrants to Western states have very often faced a range of injustices following their arrival—from racial discrimination and social exclusion to formal segregation and violence—with the harassment of Chinese immigrants in the nineteenth-century United States being a more extreme example (Spinner-Halev 2012: 30). In the case of migrants owed admission as reparation for historical injustice, subjection to these further injustices adds insult to injury.

There may also be some cases in which an open entry policy into a territory may undermine a nation's capacity to rectify historical injustices they have suffered, or even deepen that injustice. Groups that have a history of subjection to invasion or colonisation, such as indigenous peoples, may feel this keenly. As Margaret Moore (2015: 200) has suggested, larger-scale immigration of members of a dominant group into areas historically occupied by indigenous peoples can potentially undermine their self-determination. In such cases, consideration of historical injustice may point to the restriction of mobility, rather than to the creation of opportunities for it.

On the other hand, various scholars have argued that the admission of migrants can potentially play an important reparative function in response to both historical and contemporary injustices (e.g. see Wilcox 2007; Achiume 2019; Kling 2019; Santi Amantini 2022; Souter 2022; Goldstone 2024). Although the philosophical rationale for these arguments will be explored in the following section, such theorists have contended that the losses many migrants suffer as a result of historical injustice can be meaningfully redressed by offering safety and opportunities within another state. It has also recently been argued that an inclusive refugee policy can help to rectify the injustice of discrimination suffered by minorities, such as Muslim citizens, in Western states (Lindauer 2021). More broadly, past failures to admit vulnerable migrants may motivate and justify the adoption of more generous admissions policies; for instance, the inclusion of a wide-ranging right to asylum within Germany's constitution following World War Two has been described by Joppke (1999: 122) as a "conscious act of redemption" in response to its Nazi past.

1.5 Membership

Although admission may potentially constitute a form of reparation, it is unlikely in itself to redress the impacts of historical injustice on migrants fully. The desperate plight of stateless persons within states' territories, such as members of the Rohingya minority following their flight from Burma to Bangladesh, demonstrates the insufficiency of mere admission, both as a means of meeting one's basic needs, let alone as a mechanism for rectifying historical injustice. In order to be properly compensated for harms stemming from historical injustice, migrants need access to a range of basic resources and opportunities within their host states afforded by "social membership" (Carens 2013: Chap. 8), if not formal citizenship. This may also need to be accompanied by multicultural policies aimed at protecting migrants' cultural practices, especially where they have been undermined through historical injustices. As Will Kymlicka (1995: 30) has suggested, "polyethnic rights"—involving, for instance, legal protections for certain cultural practices brought by immigrants—are at times required in order to ensure their equality of opportunity. This may especially be the case where historical injustices suffered by them, both abroad or in their host state, have previously denied them this equality.

After becoming members of other states, migrants acquire responsibilities as well as rights, but it is debatable whether they should be required to contribute to the fulfilment of their new states' reparative obligations. Although it might be thought unfair to expect immigrants to contribute to reparation for actions they themselves have not been associated with, Ivison (2008: 521) argues that "[a]s long as migrants are given equal opportunity to participate in and benefit from the collective practices and institutions of that community, then arguably they [...] form part of the intergenerational structure that underpins civic responsibility for the goods (and bads) produced by the community over time".

1.6 Repatriation

Through either choice or compulsion, many migrants later find themselves returning to their states-of-origin and, like admission, repatriation may also potentially deepen or help to redress historical injustice. On the one hand, processes of deportation can exacerbate historical injustice, if they return migrants to volatile situations still reeling from the effects of that injustice, and especially if the deporting state is failing to fulfil a reparative obligation towards a migrant by deporting her (see Espindola 2024). In such cases, deportation can constitute yet another stage in a long line of injustices, stretching from the original historical injustice, through to a series of failures to adequately rectify it through admission and membership.

On the other hand, voluntary repatriation may constitute a form of redress itself, both generally (Bradley 2013: 61–64) and specifically in relation to historical injustice. If, as is often the case, migrants' states-of-origin are primarily responsible for a historical injustice which led to their flight, then safe and voluntary return may be a way of compensating them. As Moore (2015: 141) has suggested, the right of return is the "appropriate corrective justice remedy for a violation of the right of residency or occupancy". For instance, the claims of Palestinians to an ongoing right of return may be justified even for the children of those expelled from their homes, given that their parents have not received reparation for this expulsion (Meyer 2004: 310). Although it is debatable how far back one should go, and whether this right transfers across multiple generations, Meyer (2004: 309) contends that "the normative relevance of past wrongs will depend upon their causal relevance to the well-being of currently living (and future) people".

Moreover, repatriation may be the optimal form of reparation for cultural groups in exile. While their individual dispersal across states may deny them communal autonomy, repatriation can "protect collective group identity and political agency" (Long 2013: 40), and their forms of "collective narrative" (Spinner-Halev 2012: 67) by offering opportunities for collective self-determination within their homelands. Return may also facilitate transitional justice and reconciliation processes which address the legacies of injustice, by enabling the participation of returnees who have been heavily affected by such injustice (see Duthie 2012).

Nevertheless, for all its reparative potential, repatriation often creates difficult ethical quandaries. For instance, the return of individual migrants can lead them to claim their original homes, although others may have become secondary occupants and have developed legitimate claims to that property in the meantime. In cases where the returnees are relatively affluent, while secondary occupants are worse off, property restitution claims can create a conflict between the demands of reparative and distributive justice (see Gerver 2018: Chap. 8).

2 Positions and Arguments in the Controversy

Having offered a short survey of some of the interrelations between migration and historical injustice, in this section. I give a deeper analysis of some of the principal normative arguments made in this context by contemporary political theorists. Before doing so, it is worth observing that, despite these interrelations, much normative theorising on migration takes what can be fairly described as a *ahistorical* focus, focusing largely on the current situations of migrants and states, rather than on pre-existing relationships between them (Souter 2022: 7). Consequently, much of this work has proceeded as if historical injustices such as "colonial exploitation" have "never occurred" (Cole 2000: 14).

Nevertheless, it would be mistaken to characterise *all* normative theory on migration as fundamentally ahistorical. In particular, as noted above, some theorists have contended that states may bear a special responsibility to admit certain migrants, by appealing to principles of reparative justice. Although, at the broadest level, *all* grants of asylum (and some wider migrant admissions) are reparative in the sense of repairing the wrongs of displacement, such theorists are concerned specifically with the reparative obligations of those causally or morally responsible for those wrongs. For instance, Shelley Wilcox (2007: 277) posits a global harm principle that asserts in part that "agents who have […] harmed others must compensate their victims", while Rogers Smith (2014: 384) applies a "principle of coercively constituted identities" to migration, claiming that states' duties to assist non-citizens are "roughly proportionate to the degree that they have coercively constituted those persons' identities". In my work on asylum as reparation, the guiding principle is that "[w]hen an agent bears […] responsibility for causing another agent unjustified harm, the first agent bears a special obligation to provide the second agent with the most fitting form of reparation for that harm that is available" (Souter 2022: 44).

In contrast, some recent theoretical work on migration has gone beyond this view of reparative claims as arising from discrete interactions between identifiable agents, examining the responsibilities that may flow from what Iris Marion Young (2006: 114) has influentially described as *structural injustices*, which she defined as injustices that occur "as a consequence of many individuals and institutions acting in pursuit of their particular goals and interests, within given institutional rules and accepted norms". Some theorists have recently characterised some injustices in the context of displacement and migration that are rooted in past and historical injustice in these terms (Owen 2020: 2587; Parekh 2020: Chap. 6).

The reparation states may owe to migrants can take various forms, whether that be *restitution* (which involves recreating the situation which obtained before a violation occurred); *compensation* (which attempts to replace what was lost with something of equal or approximate value); or *satisfaction* (consisting of apologies and acknowledgement). Moreover, such principles of reparative justice may take broad or narrow forms. While a broad approach understands reparation as being owed even in the absence of any wrongdoing, the narrower view sees reparation as owed *only* when the harm-doer is morally culpable. To use David Miller's terminology, some see reparative duties flowing merely from *causal* responsibility, whereas others believe *moral* responsibility is required (Miller 2007: 86–89).

Whatever the form reparation takes, work in this area recognises that reparation may be owed to migrants by a range of actors, depending on where responsibility for their plight lies. *Ceteris paribus*, in cases where their predicament is entirely or largely the responsibility of their states-of-origin, reparation may be owed in the form of voluntary repatriation, whereas the responsibility of external states may generate an obligation to admit them. In addition to admission as reparation, there may be a place for inter-state compensation for the costs of refugee protection: if a refugee finds herself in a state lacking any particular responsibility for her flight, then a monetary transfer to her host state may indirectly provide reparation.

As reparative claims are often advanced alongside various other moral arguments for admitting migrants, they need to be distinguished from one another. For instance, reparative arguments are similar to, but analytically distinct from, *affinity-based* or *associative* accounts, which claim that states may bear special responsibilities to admit migrants on the basis of certain kinds of affinities shared with them (Walzer 1983: 49). While, in some cases, there is overlap between reparative and affinity-based claims (for instance, where migrants harmed by a state also share affinities with it), each can have divergent implications for states' migration policies. Taking the example of duties arising from the colonial legacy, while an affinity-based view grounds states' duties to admit citizens from former colonies on the "intertwined histories" binding them together (Amighetti and Nuti 2016), a reparative approach looks more specifically to harms stemming from colonialism that contribute to current migratory patterns (see Collste 2015: Chap. 12). While the latter only justifies admissions from former colonies where causal links between colonial rule and migrants' plight can be established, the former justifies them on the broader basis of shared ties, potentially enjoining states to admit a larger number of migrants.

2.1 Complications and Challenges

Despite being based on intuitive moral principle, reparative arguments concerning migrants face various theoretical and practical complications. First, as noted above, there are questions concerning *causality*: how direct must be the causal links between a state's actions and a migrant's plight be for the former to owe reparation to the latter (Souter 2022: Chap. 4)? Second, there are questions of *justice*: what if the state's actions in creating an injustice were justified, as in potential cases of refugees unavoidably produced by necessary humanitarian intervention? Third, there are issues of *fittingness*: should we always assume that admission is always the most fitting form of reparation (Miller 2016: 114)? Fourth, there are matters of *capability*: how should we respond if it turns out that a state lacking any particular responsibility for their plight is better placed to accommodate migrants? Fifth, there are issues of *agency*: how is reparative justice to be served if migrants prefer residence in a state lacking that particular responsibility (Eckersley 2015: 495)? Sixth, there are problems of *prioritisation*: what should states do if the migrants to whom they bear reparative obligations are not those most in need (Souter 2022: Chap. 8)? Lastly, there are normative challenges surrounding historical injustice more generally, such as whether present generations should be held responsible for the actions of their ancestors (e.g. Miller 2007: Chap. 6).

While these are difficult questions, they point to the likelihood that there are some migrants with *prima facie* reparative claims who, all things considered, are not owed admission as reparation. For instance, those adhering to the narrow approach to reparation mentioned above resist calls to admit migrants for reparative reasons if they were displaced as part of a justified military intervention (Blake 2013: 114–119), while those who prioritise migrants' quality of life would argue for their admission to the most capable state, instead of a guilty one. Moreover, while refugees cannot be offered reparation while they remain in their homes, the optimal form of reparation for other migrants might well be *in situ* (Miller 2016: 114). Consequently, offers of admission are often not *fully* able to redress the effects of historical injustice suffered by migrants, let alone by people worldwide more generally. Other measures—such as the global redistribution of resources, transitional justice and reconciliation processes, and apologies—are also necessary to advance the cause of reparative justice overall.

3 Desiderata for Future Research

Although I have indicated how theoretical work has illuminated some of the normative issues surrounding migration and historical injustice, there is much that remains to be done to deepen our understanding further. As noted above, there is a lack of work within analytic political theory which explicitly takes migration and historical injustice as its main focus, and so we lack any unified and focused account of the interrelations between them. More specifically, however, there are various lines of enquiry to take forward. First, as Fine (2016: 148) has underlined, there remains a need for far greater engagement with the entanglements among race, racism and immigration controls that, as explored above, have been present throughout the history of such controls in this area of political theory. Second, some of the challenges surrounding reparative arguments in the context of migration would benefit from much more sustained attention. For instance, some work on migration presupposes a particular conception of reparation—such as the narrow approach, which requires culpability—without argument (Blake 2013: 114–119), and could be strengthened by deeper engagement with work on reparation more broadly. Third, the work outlined here largely focuses on international migration from the global "South" to the "North", leaving the implications of reparative justice for internal migration, and migration within particular world regions, largely obscure. This literature is also generally very state-centric, focusing on states' reparative obligations towards migrants, and glossing over the potential reparative obligations of non-state actors, such as private corporations, given the roles they may have played in past and historical injustice (see Buechel 2023).

Thirdly, there is a particular need for greater attention to the implications of abstract theoretical argument for states' real-life migration policies. Given that the conclusions reached depend on one's empirical assumptions, engagement with empirical evidence is important when undertaking this task. Currently, work in this area leaves it largely as an open question whether actual states owe admission as reparation to migrants only in fairly rare cases or much more frequently. Some suggest that reparative arguments cannot in themselves take us all the way to open borders. As Cole (in Cole and Wellman 2011: 221) observes, although "those nations that benefited from colonialism and slavery need to make amends", they "can do that in ways other than opening their borders, and so we cannot move directly from the injustice of colonialism to freedom of international movement" (see also Song 2018: 82). Others apply reparative principles in a fairly broad way, with Alise Coen (2017) suggesting that past failures of the United States' foreign policy in relation to Syria and the Middle East create reparative duties towards many refugees, while Luke Glanville (2020: 4) has argued that Western states' current immigration policies are hypocritical, given that one justification for European colonial rule was indigenous peoples' purported violation of a duty of hospitality, which former colonial powers now work to circumvent through such policies. In contrast, Miller (2017: 765) takes a narrower view, pointing to "exceptional cases", while cautioning that "allowing reparative concerns to dominate immigration policy" would be "a backward step". More thorough engagement with evidence on the causes of contemporary forced migration is needed to either vindicate or challenge Miller's judgement here, and to deepen our understanding of the implications of historical injustice for states' migration policies overall.

Literature

Achiume, E. T. 2019. Migration as Decolonization. *Stanford Law Review* 71:1509–1574.

Al Hashmi, R. 2023. Historical Injustice in Immigration Policy. *Political Studies* 71:1261–1276.

Amighetti, S., and A. Nuti. 2016. A Nation's Right to Exclude and the Colonies. *Political Theory* 44:541–566.

Blake, M. 2013. Immigration, Causality, and Complicity. In *Varieties of Sovereignty and Citizenship*, ed. S. R. Ben-Porath and R. M. Smith, 111–123. Philadelphia: University of Pennsylvania Press.

Bradley, M. 2013. *Refugee Repatriation. Justice, Responsibility and Redress*. Cambridge: Cambridge University Press.

Brock, G. 2009. *Global Justice. A Cosmopolitan Account.* Oxford: Oxford University Press.

Buechel, B. 2023. Beyond the State. The Moral Nexus between Corporations and Refugees. *Critical Review of International Social and Political Philosophy* 26:461–483.

Butt, D. 2009. *Rectifying International Injustice. Principles of Compensation and Restitution Between Nations*. Oxford: Oxford University Press.

Buxton, R. 2019. Reparative Justice for Climate Refugees. *Philosophy* 94:193–219.

Byravan, S., and S. C. Rajan. 2010. The Ethical Implications of Sea-Level Rise Due to Climate Change. *Ethics and International Affairs* 24:239–260.

Caney, S. 2005. Cosmopolitan Justice, Responsibility, and Global Climate Change. *Leiden Journal of International Law* 18:747–775.

Carens, J. H. 2013. *The Ethics of Immigration*. Oxford: Oxford University Press.

Castles, C. 2010. Understanding Global Migration. A Social Transformation Perspective. *Journal of Ethnic and Migration Studies* 36:1565–1586.

Coen, A. 2017. Capable and Culpable? The United States, RtoP, and Refugee Responsibility-Sharing. *Ethics and International Affairs* 31:71–92.

Cole, P. 2000. *Philosophies of Exclusion. Liberal Political Philosophy and Immigration*. Edinburgh.

Cole, P., and C. H. Wellman. 2011. *Debating the Ethics of Immigration. Is There A Right to Exclude?* Oxford: Oxford University Press.

Collste, G. 2015. *Global Rectificatory Justice*. Basingstoke: Palgrave Macmillan.

Draper, J. 2018. Responsibility and Climate-induced Displacement. *Global Justice: Theory Practice Rhetoric* 11:59–80.

Dummett, M. 2001. *On Immigration and Refugees*. London: Routledge.

Duthie, R., ed. 2012. *Transitional Justice and Displacement*. New York: Social Science Research Council.

Eckersley, R. 2015. The Common but Differentiated Responsibilities of States to Assist and Receive 'Climate Refugees'. *European Journal of Political Theory* 14:481–500.

Espindola, J. 2024. Compensatory Justice and the Wrongs of Deportation. *Critical Review of International Social and Political Philosophy* 27:536–563.

Fine, S. 2016. Immigration and Discrimination. In *Migration in Political Theory. The Ethics of Movement and Membership*, ed. S. Fine and L. Ypi, 125–150. Oxford: Oxford University Press.

Gerver, M. 2018. *The Ethics and Practice of Refugee Repatriation*. Edinburgh: Edinburgh University Press.

Glanville, L. 2020. Hypocritical Inhospitality. The Global Refugee Crisis in the Light of History. *Ethics and International Affairs* 34:3–12.

Goldstone, Z. 2024. Migration as Reparation for Colonialism. *Res Publica* 30:763–778.

Hayter, T. 2000. *Open Borders. The Case Against Immigration Controls*. London: Pluto Press.

Holton, R. J. 2014. *Global Inequalities*. Basingstoke: Palgrave Macmillan.

Ivison, D. 2008. Historical Injustice. In *The Oxford Handbook of Political Theory*, ed. J. S. Dryzek, B. Honig, and A. Phillips, 507–527. Oxford: Oxford University Press.

Joppke, C. 1999. Asylum and State Sovereignty. A Comparison of the United States, Germany and Britain. In *Challenge to the Nation-State. Immigration in Western Europe and the United States*, ed. C. Joppke, 109–143. Oxford: Oxford University Press.

Kling, J. 2019. *War Refugees. Risk, Justice, and Moral Responsibility.* Lanham: Lexington Books.

Kymlicka, W. 1995. *Multicultural Citizenship. A Liberal Theory of Minority Rights*. Oxford: Oxford University Press.

Lindauer, M. 2021. Righting Domestic Wrongs with Refugee Policy. *Critical Review of International Social and Political Philosophy* 27:206–223.

Long, K. 2013. *The Point of No Return. Refugees, Rights, and Repatriation*. Oxford: Oxford University Press.

Lu, C. 2017. *Justice and Reconciliation in World Politics*. Cambridge: Cambridge University Press.

Mayblin, L. 2017. *Asylum After Empire. Colonial Legacies in the Politics of Asylum Seeking*. London: Rowman & Littlefield.

Meyer, L. H. 2004. Historical Injustice and the Right of Return. *Theoretical Inquiries in Law* 5:305–315.

Miller, D. 2007. *National Responsibility and Global Justice*. Oxford: Oxford University Press.

Miller, D. 2016. *Strangers in Our Midst. The Political Philosophy of Immigration*. Cambridge, MA: Harvard University Press.

Miller, D. 2017. Migration and Justice. A Reply to My Critics. *Critical Review of International, Social and Political Philosophy* 20:763–773.

Moore, M. 2015. *A Political Theory of Territory*. Oxford: Oxford University Press.

Oberman, K. 2013. Can Brain Drain Justify Immigration Restrictions? *Ethics* 123:427–455.

Oberman, K. 2020. Border Rescue. In *The Political Philosophy of Refuge*, ed. D. Miller and C. Straehle, 78–96. Cambridge: Cambridge University Press.

Owen, D. 2020. Migration, Structural Injustice and Domination on 'Race', Mobility and Transnational Positional Difference. *Journal of Ethnic and Migration Studies* 46:2585–2601.

Parekh, S. 2020. *No Refuge. Ethics and the Global Refugee Crisis*. Oxford: Oxford University Press.

Pevnick, R. 2011. *Immigration and the Constraints of Justice. Between Open Borders and Absolute Sovereignty*. Oxford: Cambridge University Press.

Pogge, T. 2002. *World Poverty and Human Rights. Cosmopolitan Responsibilities and Reforms*. Cambridge: Polity.

Risse, M. 2009. The Right to Relocation: Disappearing Island Nations and Common Ownership of the Earth. *Ethics and International Affairs* 23:281–300.

Robinson, V., and J. Segrott. 2002. Understanding the Decision-Making of Asylum Seekers. *Home Office Research Study*, 243. London: H.M.S.O.

Santi Amantini, L. 2022. Reparative Responsibility for the Harms of Forced Migration. *Journal of Social Philosophy* 55:605–623.

Shah, R. S. 2010. *The International Migration of Health Workers. Ethics, Rights and Justice*. Basingstoke: Palgrave Macmillan.

Smith, R. M. 2014. National Obligations and Noncitizens. Special Rights, Human Rights and Immigration. *Politics and Society* 42:381–398.

Song, S. 2018. *Immigration and Democracy*. Oxford: Oxford University Press.

Souter, J. 2022. *Asylum as Reparation: Refuge and Responsibility for the Harms of Displacement*. Cham: Palgrave Macmillan.

Spinner-Halev, J. 2012. *Enduring Injustice*. Cambridge: Cambridge University Press.

Steiner, H. 2001. Hard Borders, Compensation, and Classical Liberalism. In *Boundaries and Justice. Diverse Ethical Perspectives*, ed. D. Miller and S. Hashmi. Princeton: Princeton University Press.

Thompson, J. 2002. *Taking Responsibility for the Past. Reparation and Historical Justice*. Cambridge: Polity.

Waldron, J. 1992. Superseding Historic Injustice. *Ethics* 103:4–28.

Walzer, M. 1983. *Spheres of Justice. A Defence of Pluralism and Equality*. Oxford: Martin Robertson.

Wilcox, S. 2007. Immigrant Admissions and Global Relations of Harm. *Journal of Social Philosophy* 38:274–291.

Young, I. M. 2006. Responsibility and Global Justice. A Social Connection Model. *Social Philosophy and Policy* 23:102–130.

Migration and Borders

Sarah-Lea Effert

Abstract Who has the right to control the border? Should borders be open or closed? What even is the border? And how should philosophy approach an issue which raises fundamental philosophical questions at the most general level whilst impacting the lives of so many in the most political and concrete manner? These are some of the questions at the heart of the contemporary debate on migration and borders, which this article follows. Section 1 presents the ways in which the philosophical debate on migration and borders has typically been framed, drawing out the distinction between the legitimacy of migration control and the right or wrong of closed or open borders. Section 2 explores the normative positions that have been taken in this debate, and includes a discussion of approaches that have looked more closely and critically at what the border actually is. This section passes fluently into Sect. 3, which explores areas of future research.

Between 2014 and at the time of writing in July 2022, the Missing Migrants Project of the International Organization for Migration has recorded the deaths of 49,353 migrants, refugees, and asylum-seekers who died at the border or en route to an international destination (International Organization for Migration, July 1, 2022). This fact (based on low estimates and excluding deaths in detention centres or refugee camps) is a reminder of what is at stake in migration.

Within about the same time-frame, the philosophical debate on migration and borders, earlier contributions to which date back to approximately the 1980s, has intensified significantly. In this article, I follow the contemporary debate, which has increasingly attempted to reflect upon the realities on way to, and at the border. In Sect. 1, I present the ways in which the philosophical debate on migration and borders has typically been framed, drawing out the distinction between the legitimacy of migration control and right or wrong of closed or open borders. Section 2 explores the normative positions that have been taken in this debate, and includes a discussion of approaches that have looked more closely and critically at what the border

S.-L. Effert (✉)
Department of Philosophy, University of Duisburg-Essen, Duisburg, Germany

A. Niederberger et al. (eds.), *Handbook of Migration Ethics*,
https://doi.org/10.1007/978-3-031-89877-8_14

actually is. This section passes fluently into Sect. 3, which explores areas of future research.

1 General Issues at Stake in the Controversy

The philosophical debate on migration and borders has centrally evolved around two questions: Does the state have the right to control its borders? And, to what extent should these borders be open or closed? In public debate, these questions are often treated as one and the same—in terms of the justifiability of restrictions to immigration. Philosophy has also been called out for occasionally conflating them within the overall debate on migration and borders (Yong 2017). Using Caleb Yong's language, we should differentiate between legitimacy-right claims and justification-right claims regarding states' border regimes. The first type of claim is a claim about legitimacy, taking a position on the extent to and the conditions under which the state may legitimately determine and enforce regulations at its borders. The second type of claim makes a moral judgement (or a justice claim) on what these regulations should look like. Making the first type of claim does not necessarily entail any claim of the second type. For example, we might believe the state has a legitimacy-right to control its borders, yet consider the particular regulations morally unjustified or unjust—in the way in which we might believe the state is legitimately enforcing a taxation system, despite considering its particular system unjust. As an example for this type of position, Sarah Song (2017) has provided an argument from democratic self-determination for the state's right to control its borders—yet emphasised she has not thereby provided a case for closed borders. In fact, she points towards strong moral (in particular liberal democratic) reasons for states to implement much softer border policies than many do in practice. Conversely, we might also believe a particular border policy to be morally justified, yet hold that the decision-making power regarding this policy does not legitimately lie with the state (generally, or in a particular instance). As an example for this type of position, Arash Abizadeh (2008) at least allows for the option of there being good moral reasons for limiting migration, whilst taking a strong position against states having the unilateral right to border control.

The reason for why these two questions or types of claims have been conflated in philosophy and public debate, presumably, lies in the fact that—as we will see in section two—the reasons for ascribing legitimacy-rights to the state in enforcing a border regime often resemble the reasons for valuing at least some restrictions on immigration (for example, a concern for political self-determination), and the reasons for calling into question legitimacy-rights of states sometimes resemble the reasons for valuing (more) open borders (for example, a concern for global justice). Further, it is reasonable to assume that legitimacy-rights of states to enforce immigration restrictions at their borders are conditional on at least a minimal standard of the moral justifiability of their actual border regimes, such as respect for the human rights of (potential) immigrants. Or, if you take a very strong position against any kinds of restrictions to immigration ever being justifiable, this can entail the position that the state can never legitimately control immigration. Often, this is what is meant when people argue for "open borders", and here, the answers to the two questions outlined in the beginning indeed boil down to one and the same.

The reason why we should still differentiate between positions on the legitimacy of state control of borders, and the moral justifiability of substantive immigration regulations—even whilst recognising these positions will not be entirely independent from each other—is not purely conceptual clarity or philosophical rigour. It might have important normative implications for example in terms of thinking about rights or duties to resist immigration regimes, a relatively new area of philosophical thinking about migration and borders, which will come up again in Sect. 3 of this article. It might further be relevant in terms of thinking about alternatives to the current system of border regimes, with some proposals focusing on decision-making procedures about immigration (e.g. Abizadeh 2008) and others focusing on the content of these decisions (e.g. Ypi 2008). There might even be some tension regarding policy interventions stemming from legitimacy and justice (or, depending on terms, justifiability) requirements. Peter Higgins (2021) for example, has recently argued that the legitimacy requirements Gilian Brock (2020) has defended actually undermine her suggestions for substantive migration policy change in terms of justice, because virtually no currently existing state could legitimately establish these changes. His proposal to focus exclusively on justice in migration stands in contrast to others who have argued that the migration debate indeed suffers from a heavy focus on justice rather than legitimacy, characteristic of contemporary analytical political philosophy (Sandven and Scherz 2022).

Another feature uniting debates on both justice and legitimacy in migration is their framing in terms of weighing up claims of the state, or the national political community, against claims of (would-be) immigrants. Put simply, the starting point of the debate is often the assumption that the state has a claim to determine the fate of its political community, including making decisions about membership, which competes against migrants' claims to freedom of movement. These claims are typically regarded as *pro tanto* claims, giving us good reasons for a particular institutional set-up or course of action, but leaving open the possibility that they might be overridden by competing claims. Even when presenting strong arguments for the state to control its borders, philosophers hardly ever argue that this claim overrides all others, making room at least for the claims of refugees fleeing political persecution to override the *pro tanto* claim of the state (Wellman 2008 is an exception here). While it is common to focus on the claims of the receiving state and migrants, informed by global justice theorising, philosophers have also argued there can be claims of the state of origin (see the discussion on brain drain, e.g. Brock and Blake 2015), family members (for example via the role of remittances, e.g. Cheneval and Rochel 2012), or environmental reasons to be considered (de Guchteneire et al. 2011). When looking at the philosophical literature on migration and borders, it is hence important to understand whether someone is making a *pro tanto* claim to be considered in the debate, or an overall normative evaluation considering all relevant claims they believe to be at stake.

The claims I have covered so far have all taken a position on different ways to envision a world with state borders. Note that even a very strong "open borders" position, including a rejection of unilateral state determination of immigration policies, does not mean envisioning a borderless world. Such a position would be more adequately described by "no borders", rejecting territorial state jurisdiction entirely, and many philosophers advocating for open borders would not want to go as far as advocating for a world without state borders (Sager 2020). There is an extended debate in philosophy also about states and their rights over territory in general, which cover "a bundle of claims" including rights to general jurisdiction, control over land and resources, rights to taxation, and rights to control movement and membership (Stilz 2009: 186). This debate can be seen in one sense as prior to the debate about control of immigration, given that state territorial rights are necessary for states' rights to control immigration. Yet whether they are also sufficient, that is whether state territorial rights must necessarily comprise the whole bundle mentioned above, and where the limits of these rights might lie, is up to debate.

As philosophical thinking about migration and borders has intensified over recent years, at least partly in response to witnessing the situation of migrants on their way to or at the border, and to heated controversies in public debate about migration, philosophers have started to criticise existing literature mainly from analytical philosophy for not adequately grasping what really is at stake when it comes to migration and borders, and for whom. For much of the philosophical debate, the border has been viewed simply as the line that demarcates state territory—not itself central to philosophical deliberation. More and more, theorists have attempted to move the border itself to the foreground of their theorising: What is a border? What, if anything, makes state borders special? What happens at the border, and in its periphery? What are experiences like at the border, and how does it affect different people differently? These philosophers have, in very different ways, argued that framing the debate on migration and borders in general terms of openness or closedness, or in terms of the state's right to restrict immigration, has in important ways masked how the border functions as an institution in and of itself, just as it has blurred how and who borders exclude—e.g. in terms of race or class. We could summarise this as an attempt to bring more empirically informed critical theorising to philosophical reflection on migration and borders. How this might impact normative arguments on migration and borders will be explored in the next section.

2 Positions and Arguments in the Controversy

This section will give a more detailed overview on normative arguments regarding the state's right to control immigration, open or closed borders, and other normative issues philosophers have been grappling with in terms of migration and borders.

The leading arguments for legitimate state border control ground said legitimacy in (a) the value of national public culture, (b) freedom of association, (c) property, and (d) democratic self-determination. David Miller has prominently presented a case based on the value of national public culture. He argues that each state requires a distinct national public culture in order to sustain the political community and the democratic and social goods the state provides, and that citizens have an interest in controlling or shaping this culture. In particular, whilst continuous change of the public culture is to be expected and indeed valued, with and without immigration, citizens have a claim to attempt to maintain a certain cultural continuity that allows them to "see themselves as the bearers of an identifiable cultural tradition that stretches backward historically" (Miller 2005: 370). Hence, citizens are entitled to control immigration to the extent necessary to preserve cultural continuity, which he admits will be an empirical question and dependent on circumstances. The argument can be challenged precisely on these grounds, in contesting that immigration poses a (sufficient) threat to the continuity of public culture. Alternatively, one could argue that citizens' interest in preserving their public culture is not sufficient to establish a very strong claim for the state to control immigration (perhaps because it is not as instrumental for the state to perform its functions or as valuable as Miller has outlined), particularly vis-à-vis the claims of migrants which Miller recognises as relevant.

The argument from freedom of association begins from the conviction that freedom of association is something we value very highly as an essential part of self-determination. Christopher Wellman (2008) has developed this point by drawing a comparison to marriage, religious association, or private sports clubs, emphasising our commitment to people choosing their marriage partners and religious associates freely. Building on the same examples, he continues to argue that freedom of association must include the freedom not to associate, or to exclude (i.e. to reject a marriage proposal). Transferring these deliberations to the state, Wellman has argued that a self-determining state must have the right to exclude immigrants based on its claim to freedom of association. Whilst in a first step, Wellman presents his argument as a pro tanto claim, potentially to be overridden by the claims of migrants, he ultimately argues for a far-reaching right of the state (which may choose to fulfil its global justice duties in a different manner), which includes the freedom to exclude even refugees. The argument has been criticised by questioning the analogy between marriage, religious associations, sports clubs, and the state. Sarah Fine (2010), for example, has argued that the former rest on being "intimate or primarily expressive" (352) associations or on exclusion being "reasonably 'harmless'" (352), neither of which she believes to be the case for the state excluding immigrants.

Philosophers have also defended the state's right to control its borders based on Lockean theories of property (Nine 2008; Pevnick 2014). Generally speaking, they have claimed that members of the state have transformed the land of the state and created and maintained political and other institutions. They have hence acquired something akin to a property right, or ownership to the state's institutions and territory (the link between the two is not straightforward to establish). As a result, they enjoy at least a certain level of discretion in who is allowed to benefit from these. Arguments here differ in who is thought to have acquired this (quasi) property right, be it individuals, individuals jointly in their role as citizens or simply the state treated as a moral agent itself. Critique of these accounts has been provided for example by Sarah Song (2017) who claims that they do not succeed in establishing a sufficient relationship between "the collective entity of 'the state' and individual members as 'owners'" (30). She also argues that they fail to show contributing to public institutions is sufficient for establishing ownership rights and implicitly rely on consent in order to account for the case of unauthorised migrants.

Along others like Frederick Whelan (1988), Song (2017) herself has provided an argument from democratic self-determination for the state's right to control immigration. On this account, a people's right to self-determination should be understood as a right to democratic governance over "significant aspects of its common life" (35), including immigration. An important step in the argument is to show that the demos must be bounded. Here, she lists the state's role in securing the conditions for democracy, the role of solidarity and trust in democratic governance, and the relation between citizens and political representatives and the accountability of the latter. Democratic reasons for assigning legitimacy-rights to border control to the state have been emphatically rejected by Arash Abizadeh (2008), who in turn has developed a democratic argument against unilateral state control of border regimes.

Abizadeh has presented the case that "to be democratically legitimate, a state's regime of border control must result from political processes in which those subject to it—including foreigners—have a right of democratic participation" (Abizadeh 2010: 121). Contrary to the theorists discussed above, Abizadeh considers democracy to be "in principle unbounded" (2008: 38). For him, democracy requires generally that coercion be replaced by systems of mutual justification respective of freedom and equality for all. Given that border regimes are coercive of both citizens and non-citizens, they suffer a substantial legitimacy gap in terms of the coercion of foreigners. In response, David Miller (2010) has questioned the assumption that not being admitted to a country constitutes coercion, at least in cases where potential migrants have adequate alternative options to live an autonomous life. For Miller, restricting entry does indeed limit freedom and is hence in need of justification, yet since it does not constitute coercion, this justification need not involve taking part in democratic decision-making.

Abizadeh's line of argument differs from the arguments more typically brought up against the state's right to control borders or against immigration restrictions (as outlined in Sect. 1, the distinction is not always clear and arguments often rely on the same values). The most common arguments have been made on liberal egalitarian grounds, with explicit reference to the values of (a) freedom and (b) equality respectively. One of the most prominent defenders of open borders, Joseph Carens, has made both types of argument. He has argued that freedom of movement is one of the most essential liberties individuals are due. "One has only to ask whether the right to migrate freely *within* [sic] a given society is an important liberty. The same sorts of considerations make migration across state boundaries important" (Carens 1987: 258). Constructing freedom of movement as a human right, of course, implies that the claims migrants have are very strong. Kieran Oberman (2016) takes this position, defending a human right to freedom of movement based on essential personal (accessing the full range of existing life options) and political (engaging freely in political activity) interests. Although he does not claim that the human right to freedom of movement is absolute, hence it might be overridden by severe other concerns, he is adamant that the standards for this to be the case are high and no lower for immigration than any other restrictions of human rights.

Returning again to Carens, he has also built on global egalitarian concerns, pointing out the arbitrariness of one's place of birth or citizenship. He has rather famously claimed that the system of restrictive citizenship "is the modern equivalent of feudal privilege [...] hard to justify when one thinks about it closely" (Carens 1987: 252). Generally speaking, concerns of equality, or, less ambitious, poverty, have been brought up by many in defence of (more) open borders. According to this line of reasoning, our world as it is is hugely unjust, unfairly denying people opportunities and an acceptable standard of life. Migration is seen as a way to rectify these injustices.

In terms of responses to these types of arguments, for the most part, even proponents of the permissibility of restrictions to immigration accept at least some aspects of them. Typically, they argue that these arguments do not succeed in establishing a strong case for open borders, for example, because they move too quickly from establishing interests to establishing rights, or because they neglect what is at stake for the receiving political communities or morally valuable about (restricted) citizenship (for a helpful overview and discussion, see Wilcox 2009).

At this point, it is worth re-iterating two observations made in Sect. 1 of this article. The first of these is that most of the arguments presented appear to be shared across the debate in being relevant pro tanto arguments for the claims of the state and immigrants respectively, which ought to be considered when making an overall judgement. And even where there is apparently stark disagreement, for example on the question of whether or not justice requires open borders, philosophers still agree on many of the values they consider relevant, including the value of belonging to a political community, making and realising one's own life plan, and so on. Joseph Carens's (2017) discussion of David Miller's positions—someone on the supposed other end of the debate—is illuminating in this regard. As Carens alludes to, and this is the second point I wish to emphasise here, at least some part of the disagreement appears to be contingent, relying on different empirical understandings, or interpretations of the requirements for realising certain values, like a political culture supportive of democratic institutions. And, despite their differences, Carens finds that he and Miller appear to share many convictions, if not on the most general question of open or closed borders, then e.g. on selecting immigrants (under the current system of state control of borders), integration, the rights of temporary workers and so on. This is to say that whilst the debate on migration and borders has often been conducted on a very general level, and presented in terms of starkly contrasting positions on the fundamental question of state control of borders and openness or closedness (an approach I have so far followed in this article), taking a stance on these questions does not necessarily seem to imply a position on many of the other normative issues regarding migration.

This observation gives rise to discussing contributions to the philosophical debate on migration and borders that have attempted to work outside—and often been rather critical of—the framing of a debate of open vs closed borders, the rights of the state vs the rights of immigrants. Although this literature is highly diverse and difficult to group together, a uniting feature of the approaches I will discuss below is their increased focus on the border itself and bordering practices.

Lea Ypi (2018) has argued that a problem of the philosophical literature on migration and borders is that it has treated migration as posing "a problem of justice per se", yet "whether or not it does, and to what extent, depends on who you are", because "borders have always been (and will continue to be) open for some and closed for others." (142) She claims that the focus on "the abstract value of freedom of movements, and its implications for border control" (142) represents a focus on a secondary issue not actually very relevant for the politics of migration and the people affected by it. She argues that many of the supposed conflicts migration might give rise to, be they regarding distributive justice within the receiving state or cultural values, rather stem from class-based oppression to which both lower income migrants and citizens are subjected to. Agreeing that there is nothing normatively special, so to say, about migration in terms of crossing state borders, Alex Sager has influentially argued that "instead of debating open and closed state borders, a more fruitful investigation addresses how borders and boundaries exclude people and hinder mobility more generally" (Sager 2020: 11). He pleads for a general ethics of mobility which does not normatively differentiate between movement across or within states, where boundaries that hinder free movement exist as well. Luke Maring (2019) delivers a supportive normative argument, aiming to show that if the right of the state to exclude is significant enough that it overrides the claims of would-be immigrants, it would be transferable to the claims of provinces, cities, and municipalities.

Taking a closer look at state borders, based also on his own empirical field-work, Matthew Longo (2017) has argued that political theory ought to recognise the "institutional existence" (2) of borders themselves. He describes borders as "*thick, multi-faceted* and *bi-national* [sic] institutions that have evolved greatly in recent decades" (2). Longo further draws attention to the practice he has termed "*co-bordering* [sic]" (5), where neighbouring or other states cooperate intensively in order to jointly police the border and control immigration, "*ceding classic markers of sovereignty for security* [sic]" (5). If this is the case, it lends strong support to the idea that framing the debate on migration and borders in terms of state sovereignty vs rights of immigrants is misleading, and that normative attention should be paid for example to the design and impact of bi- and multi-lateral agreements between governments in an attempt to control the movement of people. As they currently stand, Longo points out, these agreements often are instruments of domination and hierarchy. That cooperation between states in terms of migration regulation could also be seen as normatively desirable from a global justice perspective is something Longo briefly acknowledges as a theoretical option, and a point which Paulina Ochoa Espejo (2020) has explicitly defended. Ochoa Espejo joins some of the other authors mentioned in this section in criticising the way philosophy has modelled the border. Focusing in particular on the role of our natural environment, she contrasts a still dominating "desert island model" with a "watershed model" which views borders as “interconnected institutions, where people, the biota, and the land together create overlapping civic duties and relations” (abstract). Normatively, she claims legitimacy of state border control cannot be established from the internal legitimacy of the state, but would have to stem from international conventions and joint governance by neighbouring states (note that a similar, perhaps more ambitious option, is also suggested by Arash Abizadeh, despite him arguably writing within the “desert island model”, or at least without paying significant attention to the nature of the border). Ayelet Shachar’s conception is that of “the shifting border” (2020). She argues we should concentrate on the movement of borders rather than the movement of people and recognise how bordering practices and state control of migration have shifted far beyond state territory. Her normative intervention based on this analysis is that as the state moves the place of control, i.e. the place where it enacts its claims, the place where migrants’ claims are processed ought to move as well. If institutionalised, it would for example mean migrants’ claims would be assessed in their own location, without having to first arrive at the state’s territory, risking their life during the journey. Recalling Longo’s point about the realities of multilaterally shared systems of immigration control, we might be wary about the implications of such a system. Rather than increasing state responsibility, it might very well be seen as a way to keep migrants as far away from state territory as possible for governments restrictive on migration. Irrespective of our normative judgement of this and other proposals, Shachar and the previous authors mentioned have changed, or added significantly to, the questions philosophers and political theorists have asked when it comes to migration and borders. They urge us to ask not only what the claims of the state or the migrant might be, but where, and how, these claims ought best to be processed and enacted, considering the realities of the actual institution of the border.

3 Desiderata for Future Research

Given the fast development of the philosophical debate on migration and borders, it is hard to draw a line between established positions within the literature and areas of future research. A great many of the works discussed in second part of the section above could have well been included as an initial impetus for future research within this section (which in turn will draw on already existing scholarship that leads the way in identifying un- or underexplored questions). I have decided to discuss them under the heading of positions and arguments in the debate in order to acknowledge their foundational contribution in terms of re-framing the debate on migration and borders by ways of focusing on what the border actually is, and how it affects different people differently. Much work surely remains to be done to follow-up on the critical analysis of and normative reflection on bordering practices, in particular as they shift, change, and adapt before our eyes in an increasingly fragile political climate. Although many of the authors discussed above include, or better build their analysis on actual border regimes in particular places, more could be done in order to understand and evaluate bordering systems and their particular features all over the world.

An increasing such feature of bordering systems is their technological- and data-driven nature, including the role of big data, artificial intelligence, digital registration, biometric data collection, and communication techniques. Digital technologies are likewise used by migrants, activists, and the state, they change the character of the border and give rise to new questions at the intersection of migration and digital /AI ethics that are yet to be fully explored (Ajana 2015). Next to being more and more technologised, it has been pointed out that border control has increasingly been privatised—adding to the theme of a voluntary ceding of state sovereignty discussed above. Described by Tally Kritzman-Amir (2011) as a way to retain control whilst simultaneously evading accountability, the delegation of migration regulation to private actors demands further exploration. In their study of carrier sanctions, Tendayi Bloom and Verena Risse (2014) show that this line of research fruitfully interacts with the more general and abstract debate of border coercion, when they argue that carrier sanctions amount for "hidden coercion" that cannot be justified even when the state is seen as legitimate in controlling its borders. More could be done to clarify when and how, if ever, the delegation of border control to private actors can be justified.

Despite the many points of disagreement covered in this article, most philosophers agree that at least in practice, various features of migration control are either illegitimate, unjust, or both. This assessment is shared by many migrants, activists, and citizens generally, some of whom have attempted to evade or actively resist bordering practices, including through the help by smugglers. Their actions, as well as the actions of those who comply with state border regimes, merit philosophical discussion. When is resistance to state border control and disobedience of migration restrictions morally justified, when might it be morally required? What means can legitimately be used to evade, change, or resist border control, and by whom? To answer these questions, philosophers have called for and started to develop an ethics of migration that focuses on the individual and her actions vis-à-vis migration control (Yong 2018; Hildago 2018). Again, these are questions that merit further discussion, both on a general level as well as through an assessment of specific acts of resistance against specific border regimes (an example being Sandven and Scherz (2022) on the legitimacy of EU border control and unauthorised rescue missions in the Mediterranean). Studying resistance to border regimes might in this way even open a door for greater convergence or at least dialogue between the approaches covered in section two of this article.

Finally, let me point towards a lack of differentiation in this article the reason for which I hope to justifiably locate in a future research desideratum. When philosophers discuss migration and borders, particularly within the framework of legitimate state control, they typically envision the migrant who is here to stay (hence the discussion about political culture or democracy). Yet borders do not only impact long-term, but also temporary migrants, people who come to see family, celebrate weddings or attend funerals, make business contacts, spend time at university as a guest student or researcher, for religious pilgrimage or as a tourist. Within a time of heightened immigration restrictions and hostilities, visa-policies in general have in many places become more restricted as well (similarly impacted by technology and privatisation and structured along racial or class lines). This was further aided by the COVID-19 pandemic, with countries closing their borders partly or fully in response, with many heightened restrictions still remaining. The closure of borders not only for migrants but more generally, as well as the impact of the pandemic on practices of border control and visions of open borders (Owen 2020), are hence the final topics on a non-conclusive and open-ended list I wish to suggest for future philosophical research. They again highlight how urgent the philosophical debate on migration and borders can be.

Literature

Abizadeh, A. 2008. Democratic Theory and Border Coercion. No Right to Unilaterally Control Your Own Borders. *Political Theory* 36:37–65.

Abizadeh, A. 2010. Democratic Legitimacy and State Coercion. A Reply to David Miller. *Political Theory* 38:121–130.

Ajana, B. 2015. Augmented Borders. Big Data and the Ethics of Immigration Control. *Journal of Information, Communication and Ethics in Society* 13:58–78.

Bloom, T., and V. Risse. 2014. Examining Hidden Coercion at State Borders. Why Carrier Sanctions Cannot Be Justified. *Ethics & Global Politics* 7:65–82.

Brock, Gillian. 2020. *Justice for People on the Move. Migration in Challenging Times*. Cambridge: Cambridge University Press.

Brock, G., and M. Blake. 2015. *Debating Brain Drain. May Governments Restrict Emigration? (Debating Ethics)*. Oxford: Oxford University Press.

Carens, J. H. 1987. Aliens and Citizens. The Case for Open Borders. *The Review of Politics* 49:251–273.

Carens, J. H. 2017. "Why Do Political Philosophers Disagree? Reflections on David Miller's *Strangers in Our Midst*". *APSA Preprints* Working Paper, https://preprints.apsanet.org/engage/apsa/article-details/5d105b2995b36e00199fc135 (last checked November 19, 2023).

Cheneval, F., and J. Rochel. 2012. An Ethical View on Remittances and Labor Migration. *Global Justice. Theory Practice Rhetoric* 5:16–30.

Fine, S. 2010. Freedom of Association Is Not the Answer. *Ethics* 120:338–356.

de Guchteneire, P., A. Pecoud, and E. Piguet. 2011. Migration and Climate Change. An Overview. *Refugee Survey Quarterly* 30:1–23.

Higgins, P. W. 2021. Migration Justice and Legitimacy. *Res Publica*1–9.

Hildago, J. S. 2018. *Unjust Borders. Individuals and the Ethics of Immigration*. New York: Routledge.

International Organization for Migration. 2022. "Missing Migrants Project", https://missingmigrants.iom.int/ (last checked July 3, 2022).

Kritzman-Amir, T. 2011. Privatization and Delegation of State Authority in Asylum Systems. *Law & Ethics of Human Rights* 5:194–215.

Longo, M. 2017. *The Politics of Borders. Sovereignty, Security, and the Citizen after 9/11*. Cambridge.

Maring, L. 2019. Which Borders? *Moral Philosophy and Politics* 6:133–146.

Miller, D. 2005. Immigration. The Case for Limits. In *Contemporary Debates in Applied Ethics*, ed. A. I. Cohen and C. H. Wellman, 193–206. Malden: Blackwell Publishers.

Miller, D. 2010. Why Immigration Controls Are Not Coercive. A Reply to Arash Abizadeh. *Political Theory* 38:111–120.

Nine, C. 2008. A Lockean Theory of Territory. *Political Studies* 56:148–165.
Oberman, K. 2016. Immigration as a Human Right. In *Migration in Political Theory. The Ethics of Movement and Membership*, ed. S. Fine and L. Ypi, 32–56. Oxford: Oxford University Press.
Ochoa Espejo, P. 2020. *On Borders. Territories, Legitimacy, and the Rights of Place*. Oxford: Oxford University Press.
Owen, D. 2020. Open Borders and the COVID-19 Pandemic. *Democratic Theory* 7:152–159.
Pevnick, R. 2014. *Immigration and the Constraints of Justice. Between Open Borders and Absolute Sovereignty*. Cambridge: Cambridge University Press.
Sager, A. 2020. *Against Borders. Why the World Needs Free Movement of People (Off the Fence. Morality, Politics, and Society)*. Lanham: Rowman & Littlefield International.
Sandven, H., and A. Scherz. 2022. Rescue Missions in the Mediterranean and the Legitimacy of the EU's Border Regime. *Res Publica*1–20.
Shachar, A. 2020. *The Shifting Border. Legal Cartographies of Migration and Mobility. Ayelet Shachar in Dialogue (Critical Powers)*. Manchester: Manchester University Press.
Song, S. 2017. Why Does the State Have the Right to Control Migration? *Nomos* 57:3–50.
Stilz, A. 2009. Why Do States Have Territorial Rights? *International Theory* 1:185–213.
Wellman, C. H. 2008. Immigration and Freedom of Association. *Ethics* 119:109–141.
Whelan, F. 1988. Citizenship and Freedom of Movement. An Open Admissions Policy? In *Open Borders? Closed Societies? The Ethical and Political Issues*, ed. M. Gibney, 3–39. New York: Greenwood.
Wilcox, S. 2009. The Open Borders Debate on Immigration. *Philosophy Compass* 4:813–821.
Yong, C. 2017. Immigration Rights and the Justification of Immigration Restrictions. Immigration Rights and the Justification of Immigration Restrictions. *Journal of Social Philosophy* 48:461–480.
Yong, C. 2018. Justifying Resistance to Immigration Law. The Case of Mere Noncompliance. *Canadian Journal of Law & Jurisprudence* 31:459–481.
Ypi, L. 2008. Justice in Migration. A Closed Borders Utopia? *Journal of Political Philosophy* 16:391–418.
Ypi, L. 2018. Borders of Class. Migration and Citizenship in the Capitalist State. *Ethics & International Affairs* 32:141–152.

International Freedom of Movement

Andreas Cassee

Abstract Do states have a moral right to restrict immigration as they see fit? Or should individuals be free to cross international borders and settle in whatever country they choose? Those who defend a right to exclude base their arguments on principles of freedom of association, associative ownership, freedom from imposed obligations, communal independence and national self-determination. In the opposite camp, a right to international freedom of movement is defended by way of an analogy between international and domestic mobility, and on comprehensive theories of justice ranging from libertarianism to liberal egalitarian contractualism. Finally, there are views that criticise the right to exclude without endorsing a universal right to freedom of movement. These include the view that immigration restrictions should be evaluated with an eye to global distributive concerns, and the view that immigration restrictions would have to be sanctioned by a global *demos* to be legitimate. This chapter gives an overview of this controversy about the justifiability of restrictions on immigration. It closes with some desiderata for future research on the duties of various agents under non-ideal circumstances.

The public debate about immigration is often focused on the burdens and benefits various immigration regimes impose on the current citizenry of the country of destination. Some believe that a more liberal policy would be in the economic interest of the citizenry, while others argue for a more restrictive regime to prevent downward pressure on domestic wages or unwanted cultural change. Participants of this debate disagree about the preferred policy, but they typically share an important assumption: States (or their citizens) have a right to exclude non-citizens from their territory if they wish to do so, at least if those non-citizens do not qualify as refugees.

In the philosophical literature, the right to exclude would-be immigrants is itself an important point of contention. The contemporary debate about its justifiability was sparked in part by two publications defending opposite points of view in the 1980s. Michael Walzer's (1983) monograph *Spheres of Justice* defends a (qualified)

A. Cassee (✉)
Department of Philosophy, University of Mannheim, Mannheim, Germany
e-mail: cassee@uni-mannheim.de

A. Niederberger et al. (eds.), *Handbook of Migration Ethics*,
https://doi.org/10.1007/978-3-031-89877-8_15

right to exclude. Without immigration controls, Walzer argues, "there could not be *communities of character*, historically stable, ongoing associations of men and women with some special commitment to one another and some special sense of their common life" (Walzer 1983: 62). Joseph Carens's (1987) essay "Aliens and Citizens" defends the opposite view that the liberal commitment to freedom and equality entails a moral right to international freedom of movement. This chapter gives an overview of the most prominent positions and arguments in the ensuing debate.

1 Setting the Stage

Virtually all inhabitable parts of the surface of the earth are today claimed by states as their territories, demarcated by (sometimes disputed) international borders. Some may think that the existence of such borders is itself problematic, opting instead for a world state or a "no borders" position. In what follows, however, I will take the existence of international borders for granted to ask what normative principles should govern the mobility of people across such borders.

At first approximation, we may think of the responses to this question as falling on a spectrum between two extremes. At one end of the spectrum, states (or their citizens) have full discretion over the admission of non-citizens. At the other end of the spectrum, would-be immigrants have an absolute right to move and settle wherever they wish.

Current international law takes a position close to the former end of the spectrum. While there are established human rights to emigrate and to freedom of movement and residence within one's country of residence (International Covenant on Civil and Political Rights, article 12), there is no corresponding human right to immigrate or to international freedom of movement and residence. It is up to individual states to determine how they want to regulate immigration. The most important exception to this is the principle of *non-refoulement* in international refugee law, which prohibits the deportation of a refugee to a country "where his life or freedom would be threatened on account of his race, religion, nationality, membership of a particular social group or political opinion" (Convention relating to the Status of Refugees, article 33). But even refugees do not have a right to move to a safe country of their own choice, and would-be immigrants who do not qualify as refugees generally have no claim to be admitted in any country of which they are not already citizens.

In the philosophical debate, something akin to this legal status quo is defended by proponents of a "right to exclude". The bearers of this alleged moral right are states or, in the case of democratic states, ultimately their citizens. The right to exclude entitles them to deny non-citizens access to and/or settlement in the territory in question. Indirectly, this control over the territorial presence of non-citizens allows states to control access to other goods and institutions such as the labour market, social benefits, and ultimately citizenship.

Javier Hidalgo (2014: 225) helpfully distinguishes an "authority" aspect from a "permissibility" aspect of the right to exclude. The state's mere authority to restrict immigration is in principle compatible with the view that states themselves have a moral duty to respect a right to international freedom of movement (Angeli 2015: 106–109). However, most proponents of a right to exclude believe that states not only have the authority to control territorial access, but also enjoy wide discretion over how they want to exercise this authority.

This is not to say that the right to exclude is necessarily an absolute right. To the contrary, most proponents think that there are limits regarding the exclusion of specific groups of people such as the politically persecuted (e.g. Miller 2016: 76–93) or those escaping severe poverty (e.g. Pevnick 2011: 86–96). Furthermore, many authors believe that there are limits to the kinds of criteria states may apply to restrict immigration, the most commonly discussed case being criteria that openly discriminate on the basis of race (e.g. Miller 2005: 203–204; Wellman 2008: 137–141). The right to exclude is therefore best understood as a default: States are normally free to restrict immigration, except when there is special reason to think that the exclusion of a particular applicant or the employment of a particular criterion is morally problematic.

The opposite default is endorsed by those who argue for a (non-absolute) moral right to international freedom of movement: Individuals should normally be free to live in whatever country they wish, and restrictions are only justifiable in special circumstances, the most broadly accepted limitation being the "public order restriction", according to which states may restrict immigration if and to the extent that this is necessary to prevent a breakdown of public order (Carens 1987: 258–260).

International freedom of movement is typically understood to include both the freedom to travel across international borders (or freedom of movement in the narrow sense) and the freedom to settle where one has moved (freedom of residence). On most positions, the exercise of this right triggers a number of further claims that attach to territorial presence, such as a right to access the labour market, a right to access certain social benefits, and, perhaps after prolonged residence, a right to political participation and citizenship.

Not all arguments against a general right to exclude are arguments for a universal right to international freedom of movement. Some authors believe that the problem with current immigration restrictions is not (only) that they curtail individual liberty, but (also) that they have unacceptable distributive implications, for example because they violate a principle of global equality of opportunity (e.g. Moellendorf 2002: 61–67). The argument from equality of opportunity does not justify a universal right to international freedom of movement (we may imagine a world in which there are restrictions on mobility, but everybody enjoys the same overall level of opportunity). Nevertheless, this argument offers an important critique of the right to exclude as it is exercised in the real world.

Departing more profoundly from the usual framing of the debate around first-order individual rights to territorial access and second-order rights of states to control access, Arash Abizadeh (2008) argues that restrictions on immigration may in principle be legitimate, but only if they were sanctioned by a global *demos*. States have no right to *unilaterally* restrict immigration.

In what follows, the central arguments for each of these positions are discussed, starting with the arguments for a right to exclude.

2 Arguments for a Right to Exclude

Most proponents of the right to exclude base their case on some kind of claim to collective self-determination, though they disagree about its exact content and its justification. Communitarians and liberal nationalists believe that the relevant right to collective self-determination is held in the first instance by cultural groups or "nations". States have a right to self-determination because (and insofar as) they offer an institutional home to such cultural groups. In recent years, a number of authors (I will call them 'statists') have developed arguments for a right to exclude that abstain from such reference to national categories.

2.1 Freedom of Association

The most prominent defence of the right to exclude along statist lines is offered by Christopher H. Wellman (2008, 2011). His argument runs as follows: First, legitimate states have a right to political self-determination. Second, this right to political self-determination includes a right to freedom of association. Just like individuals are not truly self-determining if they are denied the ability to choose whom they want to marry, with whom they want to practice a religion or play golf, so states are not truly self-determining if they are not granted freedom of association. Third, freedom of association includes a right not to associate with specific others. Just like a golf club may refuse to admit new members, states have a right to refuse association with potential newcomers. Therefore, states have a right to exclude would-be immigrants.

Critics have objected to the freedom of association argument on a number of levels. First, freedom of association seems to pull both ways when it comes to immigration (Miller 2007: 210). Restrictions on immigration may be seen as a collective exercise of freedom of association at the level of the nation state, but they also restrict individual freedom of association, because states contain a whole range of private associations within them. Second, the freedom of association argument fails to address the territorial aspect of the right to exclude. As Sarah Fine (2010: 353–355) points out, a yoga club whose members meet in Central Park may refuse to admit new members, but the club does not therefore have a right to exclude non-members from Central Park. Third, the analogy between states and private clubs seems to overshoot, as freedom of association gives private clubs a right to exclude not only "outsiders" but also present members or their children (even if they are born on the club's premises). Wellman's argument therefore seems to imply a right to deport or de-naturalize current citizens and "newcomers by birth" (Brezger and Cassee 2016: 371–373). Finally, states and private clubs arguably serve different purposes. Clubs exist for individuals to pursue particular projects, and if clubs had no right to exclude, their members would have no means to ensure that their club remains committed to *their* particular project. States, by contrast, arguably serve the purpose of providing a framework within which different personal projects can be pursued. This calls into question Wellman's assumption that the same normative principles apply to both sorts of associations (Cassee 2016: 58–61).

2.2 *Associative Ownership*

Ryan Pevnick (2011) offers an alternative statist justification for the right to exclude that promises to avoid some of these problems. He bases his argument on a collectivist version of the Lockean theory of property (Locke 1980). Those who collectively employ their labour to produce a good have a claim to this good, regardless of whether membership in the relevant collective is chosen freely. Public institutions are not like "manna from heaven" (Pevnick 2011: 116). Rather, they are created and upheld by citizens' contributions and passed down from one generation to another. As co-owners, citizens have a right to control access to their institutions and thus a right to exclude would-be immigrants.

One worry about this position is that some citizens such as the severely handicapped may not count as contributors and may thus be excluded from political participation (Blake 2014: 531). Furthermore, if individual contributions are the basis for rights to collective self-determination, it is unclear why the upshot should be *democratic* self-determination rather than, for example, a shareholder model where those who contribute more have more of a say (Wilcox 2012: 620–621). Finally, given colonial history, the application of Pevnick's principles may run into problems of historical injustice, thus raising the question of whether former colonial powers have a right to exclude immigrants from the former colonies (cf. Amighetti and Nuti 2016).

2.3 *Freedom from Imposed Obligations*

A third statist argument for the right to exclude is put forward by Michael Blake (2013) on the basis of the structure of human rights obligations. While the duty to respect human rights is owed to everyone by everyone, duties to protect and fulfil are assigned on a territorial basis. Therefore, when an immigrant enters a country, she imposes new obligations to protect and fulfil her human rights on the inhabitants of that country. However, Blake insists, "we have a presumptive right to be free from others imposing obligations on us without our consent" (Blake 2013: 115).

Therefore, states have a right to exclude would-be immigrants, at least if their human rights are adequately protected in their country of origin.

One objection to this argument is that Blake overstates the new obligations immigrants impose on residents: Immigration does not lead to new types of obligation but only to new tokens of existing obligations (Kates and Pevnick 2014: 187–190). Another problem concerns the human rights obligations 'imposed' on society by those who have children. Blake acknowledges this problem and argues that the right to avoid unwanted obligation holds "except when we have an existing obligation to acquire such new obligations" (Blake 2013: 119), reproductive rights being a case where there is such a standing obligation. However, the argument then seems dangerously close to a petitio. By assuming that there is no standing obligation to incur new obligations towards immigrants, Blake seems to presuppose that there is no right to immigrate. Finally, it is important to note that in a world without international migration, *all* human rights obligations would be incurred in a non-consensual way. This raises questions about the strength of the presumptive case against imposed human rights obligations (Brezger and Cassee 2016: 373–375).

2.4 Communitarianism

The arguments discussed so far rely on universalist principles concerning a globe populated by abstract individuals, considered in isolation from their cultural attachments. Communitarians like Michael Walzer (1983) believe this approach to be mistaken. According to Walzer's particularist theory of goods, cultural communities lend meaning to goods, and those shared meanings are the source of principles of distributive justice (Walzer 1983: 6–10). Membership in a political community is itself a good that is distributed in line with the meaning membership has in this particular community (Walzer 1983: 31–63). Communities that understand themselves as democratic cannot tolerate the establishment of a category of "guest workers" who live permanently on a state's territory as second-class citizens. However, communities are largely free to restrict immigration. Drawing on analogies between the political community and neighbourhoods, clubs and families, Walzer develops a position that understands the right to exclude as a core component of "communal independence" (Walzer 1983: 62): "The restraint of entry serves to defend the liberty and welfare, the politics and culture of a group of people committed to one another and to their common life" (Walzer 1983: 39).

Liberal universalists of course object to the communitarian premises of Walzer's account. Furthermore, Carens argues that Walzer's account "must confront, as a methodological paradox, the fact that liberalism is a central part of our culture" (Carens 1987: 269). Even if culturally shared meanings are the only source of moral principles, *liberal* cultures are committed to the equal moral worth of *all* individuals, not just members of their own community. Therefore, Carens claims, liberal communities have to take the rights of outsiders more seriously than Walzer does, even on Walzer's own account.

2.5 Liberal Nationalism

Like communitarians, "liberal nationalists" believe that cultural attachments and national identities are normatively significant. However, they argue that this commitment to "nationalism" is consistent with a commitment to universalist liberal principles. The most prominent liberal nationalist in the contemporary debate is David Miller (1995, 2005, 2007, 2016). Miller's argument for taking national identity seriously is partly instrumental. National identity, he argues, is an important source of social trust, which is in turn a prerequisite for a strong welfare state and deliberative forms of democracy. Open borders would undermine this social trust and would thus have detrimental effects on political institutions (Miller 1995: 90–99, 2016: 9–10, 64–65; for a critique see Pevnick 2011: 154–161). However, Miller's nationalism runs deeper than that. He believes that national communities are intrinsically valuable. This thesis about the intrinsic value of the nation in turn supports a thesis about justified partiality among co-nationals (Miller 2007: 23–50). While the interests of would-be immigrants count for something, they do not count in the same way the interests of citizens do (Miller 2007: 222–224). And citizens have a weighty interest in national self-determination: "[T]he public culture of their country is something that people have an interest in controlling: they want to be able to shape the way that their nation develops, including the values that are contained in the public culture" (Miller 2005: 200).

Critics object that liberal nationalism paints too rosy a picture of benign and neatly separated national groups, ignoring both the violent history of nationalism and those who fall between the cracks, such as citizens who are not members of the dominant culture (Fine 2017). Furthermore, Miller is charged with jumping from the assumption that national belonging is *one* source of intrinsic value to the view that national belonging should be privileged vis-à-vis other sources of intrinsic value, such as the personal relationships of would-be immigrants (Owen 2010). Finally, the idea of collective cultural self-determination is criticised as being incompatible with a commitment to individual liberty. If we accept that freedoms of speech, religion or reproduction limit the nation's ability to collectively control its cultural future, we should resist the idea that freedom of movement may be curtailed on the basis of a claim to collective cultural self-determination (Freiman and Hidalgo 2016).

3 Arguments for International Freedom of Movement

While the right to exclude is typically defended by reference to a claim to *collective* self-determination, those who argue for a universal right to international freedom of movement typically base their claims on a commitment to *individual* self-determination or autonomy.

3.1 The Cantilever Argument

The most straightforward way to argue for a moral right to international freedom of movement is the so-called "cantilever argument", which starts from the broadly shared assumption that individuals have a moral right to move freely within any given country's territory. The same considerations that support this broadly accepted and legally enshrined domestic right, the argument claims, also justify a right to freedom of movement across international borders (Carens 1987, 1992, 2013; Oberman 2016).

As Carens points out, international freedom of movement is important for the realisation of some people's life plans for the same sorts of reasons domestic freedom of movement is important: "One might want a job; one might fall in love with someone from another country; one might belong to a religion that has few adherents in one's native state and many in another; one might wish to pursue cultural opportunities that are only available in another land" (Carens 1992: 27–28). Why, then, should freedom of movement be considered a fundamental right in the domestic but not in the international context?

One way to counter the cantilever argument would be to reject the assumption that there is a moral right to domestic freedom of movement. The more common response, however, is that there are valid grounds for a (duly limited) right to domestic freedom of movement, but that these grounds do not at the same time support a right to international freedom of movement. One version of this response is defended by Miller (2005: 194–196). On Miller's view, to be a (co-)author of your own life, you need an adequate range of options to choose between in important areas of life. However, there is no need for unlimited options, and what matters are not the idiosyncratic interests of a particular person (e.g. to be an opera singer) but generic human interests (e.g. in musical expression). As long as individuals enjoy a sufficient range of options (a number of places to live, jobs, potential partners, etc.) in their country of origin, there is no need for international freedom of movement to protect individual autonomy.

However, this "adequate range view" is not without problems. First, it raises questions about the substitutability of life options (Owen 2010: 108–109). For example, if what you really want is to live with a specific romantic partner from another country, it seems doubtful that your interest is reducible to a generic interest in romantic love which is served just as well by having a range of different partners to choose between. Second, the "adequate range view" justifies less than the domestic right to freedom of movement as it is commonly understood. Assuming that a relatively small state like Belgium provides its citizens with an adequate range of options, Miller's position seems consistent with the practice of dividing a larger state into Belgium-sized regions and restricting movement between those regions (Oberman 2016: 39; also see Torresi 2010: 30–31). Third, the purpose of liberty rights is arguably not just to guarantee individuals an adequate range of options. Rather, liberty rights exist to secure a sphere in which individuals can choose independently (without coercive interference) between all existing options (Oberman

2016: 38–44). If freedom of movement is understood in this way, the fact that a person already enjoys an adequate range of options in their country of origin does little to justify interference with that person's choice of a particular option abroad.

3.2 *The Libertarian Case*

Besides appealing to the analogy to domestic mobility, proponents of a right to international freedom of movement also try to ground this right in broader theories of justice, ranging from libertarianism to liberal egalitarianism.

According to libertarianism, the sole legitimate purpose of the state is to enforce individual property rights, including each person's "property" in their own body (Nozick 1974). On this theory, restrictions on immigration violate both the rights of resident citizens and the rights of potential immigrants (Carens 1987: 252–254; Steiner 2001). They violate the rights of "insiders" because each individual citizen has a right to employ foreign workers, rent apartments to foreigners, or host them as guests. The citizenship collectively has no right to interfere in such voluntary transactions. Restrictions on immigration also violate the rights of "outsiders", because state coercion is illegitimate except to enforce individual rights. Yet migrating to another country does not infringe anybody's rights. As long as immigrants do not trespass on private property or violate individual rights in some other way, there is no basis for keeping them out.

While international freedom of movement is hard to reject on libertarian premises, these premises themselves are highly controversial. Some will see migration as just one of many cases where libertarianism is at odds with broadly held convictions about the things states may legitimately do. However, international freedom of movement is also defended on liberal egalitarian grounds that allow for a justification of many of the other state functions libertarianism denounces.

3.3 *The Contractualist Case*

The most discussed liberal egalitarian argument for international freedom of movement is based on a cosmopolitan version of John Rawls's (1971) contractualist theory of justice. The basic idea of Rawlsian contractualism is to ask what principles of justice individuals would choose if they had to choose behind a "veil of ignorance", that is, without knowing any particularities of their own situation like their class, their talents or their conception of the good. Carens (1987: 255–262) extends this thought experiment to the international level and applies it to the question of international migration: What principles regarding human mobility would we accept if we did not know our country of birth, our economic situation or our idea of a life well lived? Carens contends that individuals behind a global "veil of ignorance" would agree to include a right to international freedom of movement in the list of basic liberties. These liberties may be restricted for the sake of liberty itself. Immigration may therefore be restricted to the extent that this proves necessary to uphold public order and thus preserve the overall system of liberties. But in the absence of such special circumstances, individuals should be free to move across international borders.

One obvious route for criticism of this argument is to resist the direct extension of Rawls's domestic theory of justice to the international level. Rawls (1993) himself rejects this extension and conceives of international justice as a contract between delegates of "peoples", rather than individuals. Individual interests of a transnational kind (such as the interest to migrate to another country) do not figure prominently in this account. Accordingly, Rawls's "Law of Peoples" affirms a "qualified right to limit immigration" (Rawls 1993: 48 Fn. 17).

4 Beyond Freedom of Movement and the Right to Exclude

4.1 Global Equality of Opportunity

Besides restricting individual liberty, immigration restrictions arguably play an important role in upholding global economic inequalities. According to one estimate, "location" explains more than two thirds of income inequality among individuals worldwide (Milanovic 2013). Against this background, Carens (1987: 252) describes citizenship in a Western liberal democracy as "the modern equivalent of feudal privilege—an inherited status that greatly enhances one's life chances".

This analogy to feudal privilege has often been interpreted as an implicit argument about equality of opportunity. If wealthy states close their borders to the globally disadvantaged, they violate a standard of global equality of opportunity. As not everybody is among the globally disadvantaged, this does not amount to an argument for a universal right to international freedom of movement. But it is nevertheless an important argument for *more* open borders.

Critics object to the equality of opportunity argument on three different levels. First, some authors dispute that admitting more immigrants is an effective means to better the situation of the globally disadvantaged. The poorest of the poor hardly have the financial means to migrate themselves, and they might even suffer from the emigration of skilled workers (e.g. Miller 2005: 198). The overall effects of emigration on those left behind in the country of origin remain controversial, with pessimists stressing the negative impact of "brain drain" (cf. Brock and Blake 2014), while optimists insist that the opportunity to emigrate can itself be an important driver for investment in higher education, and a skilled diaspora abroad can benefit the population of the country of origin through remittances and network effects (Rapoport 2016).

Second, there is normative disagreement about the scope of justice. Some authors argue that equality of opportunity and other demanding standards of distributive justice apply only in the domestic arena (e.g. Rawls 1993; Blake 2001).

Third, even if the demand for global equality of opportunity is justified and allowing more immigration is one means to fulfil this demand, it is hardly the only means. According to the "choice view" (Oberman 2011), states may choose whether they want to fulfil their duties towards the globally disadvantaged by admitting immigrants or by some other means, for example by transferring resources (e.g. Miller 2005: 198; Wellman 2008: 127). However, proponents of the equality of opportunity argument insist that the mere possibility to address global inequality of opportunity in some other way does not invalidate the argument for more open borders as long as those other means are not actually employed (Wilcox 2014: 131).

4.2 Democratic Theory

A final critique of immigration restrictions as they are in place today is based on democratic theory. Like other opponents of the right to exclude, Abizadeh (2008) thinks that immigration restrictions are in need of justification vis-à-vis potential immigrants. However, he understands this need for justification in a procedural way.

Abizadeh's argument runs as follows: First, for the coercive exercise of political power to be legitimate, all those subject to coercion need to have a say in the relevant political decisions. Second, immigration restrictions subject non-citizens abroad to coercion. Therefore, non-citizens abroad have a right to participate in decisions about immigration restrictions. A global *demos* could decide to delegate the power to restrict immigration to individual states. But in the absence of such delegation, unilateral restrictions on immigration are democratically deficient.

Some critics dispute Abizadeh's first premise that immigration restrictions are coercive in the relevant sense. Miller (2010) argues this point by introducing a distinction between coercion proper, where an agent is forced to take a specific course of action, and "prevention", where a threat is used to prevent an agent from taking a specific course of action. A second line of objection disputes the second premise that all coercion needs to be justified democratically (Saunders 2011). If there is a convincing substantive moral justification for a right to exclude, then perhaps there is no need to justify the exercise of this right democratically. This throws the discussion back to the arguments discussed in Sect. 2.

5 Future Developments and Desiderata

Whether there is a moral right to international freedom of movement is of great importance not only for what counts as a just policy of territorial admission, but also for other areas of the ethics of immigration. If there is a right to international freedom of movement, then all individuals worldwide are, at least in principle, eligible to access any of the other rights that come with (prolonged) residence in a state's territory. Furthermore, a right to international freedom of movement would plausibly impose constraints on the debate about the rights and duties of immigrants in their country of residence. For example, if the admission of immigrants is a discretionary act, then the host country may arguably place conditions on those admitted that would be unjustifiable if immigrants were owed admission in the first place (cf. Oberman 2017).

Given its importance for what counts as a just way of dealing with human mobility, the debate about international freedom of movement and the right to exclude is here to stay. However, there is often a considerable distance between the principles for a fully just immigration regime that are discussed in this debate and the moral questions faced by various agents in the real world.

One such question, which is especially relevant for proponents of a right to exclude, concerns the means of enforcing restrictions on immigration. If states may in principle control immigration, what may they legitimately do in practice to keep out unwanted migrants? Is there any way to effectively enforce restrictions without violating basic rights (Mendoza 2015)?

Proponents of a right to international freedom of movement face difficult questions about the duties of particular agents in a world in which this right is routinely violated. For example, does an individual state have to open its borders even if other states fail to do so (Bertram 2018: 105–118)? And what about would-be migrants themselves: Are they morally justified in evading unjust immigration restrictions (Hidalgo 2015)?

While the general debate about the justifiability of immigration restrictions continues, addressing these more applied issues is an important desideratum moving forward.

Literature

Abizadeh, A. 2008. Democratic Theory and Border Coercion. No Right to Unilaterally Control Your Own Borders. *Political Theory* 36:37–65.

Amighetti, S., and A. Nuti. 2016. A Nation's Right to Exclude and the Colonies. *Political Theory* 44:541–566.

Angeli, O. 2015. *Cosmopolitanism, Self-Determination and Territory. Justice with Borders.* Basingstoke: Palgrave Macmillan.

Bertram, C. 2018. *Do States Have the Right to Exclude Immigrants?* Cambridge: Wiley.

Blake, M. 2001. Distributive Justice, State Coercion, and Autonomy. *Philosophy & Public Affairs* 30:257–296.

Blake, M. 2013. Immigration, Jurisdiction, and Exclusion. *Philosophy & Public Affairs* 41:103–130.

Blake, M. 2014. The Right to Exclude. *Critical Review of International Social and Political Philosophy* 17:521–537.

Brezger, J., and A. Cassee. 2016. Debate. Immigrants and Newcomers by Birth. Do Statist Arguments Imply a Right to Exclude Both? *Journal of Political Philosophy* 24:367–378.

Brock, G., and M. Blake. 2014. *Debating Brain Drain. May Governments Restrict Emigration?* Oxford: Oxford University Press.
Carens, J. H. 1987. Aliens and Citizens. The Case for Open Borders. *The Review of Politics* 49:251–273.
Carens, J. H. 1992. Migration and Morality. A Liberal Egalitarian Perspective. In *Free Movement. Ethical Issues in the Transnational Migration of People and Money*, ed. B. Barry and R. E. Goodin, 25–47. Hemel Hempstead: Harvester/Wheatsheaf.
Carens, J. H. 2013. *The Ethics of Immigration*. Oxford: Oxford University Press.
Cassee, A. 2016. *Globale Bewegungsfreiheit. Ein philosophisches Plädoyer für offene Grenzen.* Berlin: Suhrkamp Verlag.
Fine, S. 2010. Freedom of Association Is Not the Answer. *Ethics* 120:338–356.
Fine, S. 2017. Migration, Political Philosophy, and the Real World. *Critical Review of International Social and Political Philosophy* 20:719–725.
Freiman, C., and J. Hidalgo. 2016. Liberalism or Immigration Restrictions, But Not Both. *Journal of Ethics & Social Philosophy* 10:1–22.
Hidalgo, J. S. 2014. Freedom, Immigration, and Adequate Options. *Critical Review of International Social and Political Philosophy* 17:212–234.
Hidalgo, J. 2015. Resistance to Unjust Immigration Restrictions. *Journal of Political Philosophy* 23:450–470.
Kates, M., and R. Pevnick. 2014. Immigration, Jurisdiction, and History. *Philosophy & Public Affairs* 42:179–194.
Locke, J. 1980. *Second Treatise of Government.* Ed. C. B. McPherson. Indianapolis: Hackett Publishing Company.
Mendoza, J. J. 2015. Enforcement Matters. Reframing the Philosophical Debate over Immigration. *Journal of Speculative Philosophy* 29:73–90.
Milanovic, B. 2013. Global Income Inequality in Numbers. In History and Now. *Global Policy* 4:198–208.
Miller, D. 1995. *On Nationality*. Oxford: Oxford University Press.
Miller, D. 2005. Immigration. The Case for Limits. In *Contemporary Debates in Applied Ethics*, ed. A. I. Cohen and C. H. Wellman, 193–206. Malden: Blackwell Publishers.
Miller, D. 2007. *National Responsibility and Global Justice*. Oxford: Oxford University Press.
Miller, D. 2010. Why Immigration Controls Are Not Coercive. A Reply to Arash Abizadeh. *Political Theory* 38:111–120.
Miller, D. 2016. *Strangers in Our Midst. The Political Philosophy of Immigration*. Cambridge, MA: Harvard University Press.
Moellendorf, D. 2002. *Cosmopolitan Justice*. Colorado: Boulder.
Nozick, R. 1974. *Anarchy, State, and Utopia*. New York: Basic Books.
Oberman, K. 2011. Immigration, Global Poverty and the Right to Stay. *Political Studies* 59:253–268.
Oberman, K. 2016. Immigration as a Human Right. In *Migration in Political Theory. The Ethics of Movement and Membership*, ed. S. Fine and L. Ypi, 32–56. Oxford: Oxford University Press.
Oberman, K. 2017. Immigration, Citizenship, and Consent. What is Wrong with Permanent Alienage? *Journal of Political Philosophy* 25:91–107.
Owen, D. 2010. Global Justice, National Responsibility and Transnational Power. *Review of International Studies* 36:97–112.
Pevnick, R. 2011. *Immigration and the Constraints of Justice. Between Open Borders and Absolute Sovereignty*. Cambridge: Cambridge University Press.
Rapoport, H. 2016. Who is Afraid of the Brain Drain? A Development Economist's View. *Law, Ethics and Philosophy* 4:119–131.
Rawls, J. 1971. *A Theory of Justice*. Cambridge, MA: Harvard University Press.
Rawls, J. 1993. The Law of Peoples. *Critical Inquiry* 20:36–68.
Saunders, B. 2011. Immigration, Rights and Democracy. *Theoria* 58:58–77.

Steiner, H. 2001. Hard Borders, Compensation, and Classical Liberalism. In *Boundaries and Justice. Diverse Ethical Perspectives*, ed. D. Miller and S. H. Hashmi, 79–88. Princeton: Princeton University Press.

Torresi, T. 2010. On Membership and Free Movement. In *Citizenship Acquisition and National Belonging. Migration, Membership and the Liberal Democratic State*, ed. G. Calder, P. Cole, and J. Seglow, 24–37. Basingstoke: Palgrave Macmillan.

Walzer, M. 1983. *Spheres of Justice. A Defense of Pluralism and Equality.* New York: Basic Books.

Wellman, C. H. 2008. Immigration and Freedom of Association. *Ethics* 119:109–141.

Wellman, C. H. 2011. Freedom of Association and the Right to Exclude. In *Debating the Ethics of Immigration. Is There a Right to Exclude?* ed. C. H. Wellman and P. Cole, 11–155. Oxford: Oxford University Press.

Wilcox, S. 2012. Ryan Pevnick, Immigration and the Constraints of Justice. Between Open Borders and Absolute Sovereignty. *Ethics* 122:617–622.

Wilcox, S. 2014. Do Duties to Outsiders Entail Open Borders? A Reply to Wellman. *Philosophical Studies* 169:123–132.

Ethics of Migration Policies

Desiree Lim

Abstract It is widely assumed that states have the right to exclude non-citizens through systems and mechanisms of border control. States routinely enact and enforce migration policies designed to control their borders. In recent years, there has been mounting pressure for states to make their immigration policies more restrictive than ever before. In this chapter, I discuss the ethical standards that states should be held to while exercising their right to exclude, especially with regard to migration policies that are widely regarded as uncontroversial, even desirable. While it may be argued that states do not have a right to exclude non-citizens, I will assume what Joseph Carens refers to as "the conventional view", under which states do have such rights (2008). In particular, I am concerned with the ethics of selection (how states select non-citizens for admission) and the ethics of enforcement (how states enforce the exclusion of non-citizens who have not been granted admission). In the first case, I consider the permissibility of various forms of discrimination, including racial, health, and skill-based discrimination, the prioritisation of family migrants, and citizenship-by-investment schemes. In the second case, I examine states' use of detention and deportation and how they stand in need of ethical justification, particularly because of, especially with regard to the coercive force they subject non-citizens to. I conclude by suggesting that theorists pay closer critical attention to concrete processes of immigration selection and enforcement.

It is widely assumed that states have the right to exclude non-citizens through systems and mechanisms of border control. Following Sarah Fine (2013), there are three possible modes of exclusion: firstly, the right to prevent non-citizens from entering the territory; secondly, the right to prevent non-citizens from settling on the territory (e.g. they may enter as tourists but not be allowed to work or own property), and thirdly, the right to prevent non-citizens from becoming citizens (e.g. a permanent resident being denied citizenship because they fail a mandatory language test. Some well-known justifications for the right to include the state's right to control its

D. Lim (✉)
Department of Philosophy, Pennsylvania State University, University Park, PA, USA
e-mail: dyl10@psu.edu

A. Niederberger et al. (eds.), *Handbook of Migration Ethics*,
https://doi.org/10.1007/978-3-031-89877-8_16

cultural composition (Miller 2013) or its right to the freedom of association (Wellman 2008). Even if there is no solid consensus on the exact philosophical basis of the right to exclude, what unites many theorists is a sense that existing citizens, at the very least, have a right to control who gets to become a member of their community. For this reason, states have been compared to families, clubs, or other similar associations (Walzer 1983), and this model of self-determination is largely mirrored in the everyday discourse on immigration and decisions made at the institutional level. States routinely enact and enforce policies designed to control their borders. In recent years, there has been mounting pressure for states, especially those of the wealthy Western variety, to make their immigration policies more restrictive than ever before.

We may, of course, express skepticism about the right to exclude. One argument is that it lacks a persuasive conceptual grounding. Alternatively, even if we agree that states have a *pro tanto* right to exclude, perhaps it is ultimately trumped by competing considerations, such as non-citizens' right to free movement or their interest in accessing better opportunities in wealthier countries. In this chapter, rather than casting doubt on its existence, I focus on how recognition of the right to exclude *does not* commit us to the further claim that states have absolute freedom over how they choose to control their borders. Rather, states' exercise of their right to exclude must be reined in by relevant ethical considerations in order to be legitimate. This is because having a right to X does not necessarily mean that our claim on X is *prior* to ethics. In at least some cases, my right to X is part and parcel of a broader system of ethical norms that, in turn, limits how I may exercise that right. Consider a simple analogy: my having a property right over my wallet does not, on its own, permit me to use deadly force against a thief. This is because my property rights are generally justified by the ethical norm of equal respect for all persons, which in turn endows them with robust protections against violations of bodily integrity. In short, the normative source of my property right over my wallet also gives the thief a right against being killed by me in my bid to retrieve it. Similarly, a state's right to control its borders does not license states to adopt egregious policies like shooting would-be migrants on sight, or of forcibly separating children from their parents at the border. At the same time, states have implemented policies that seem innocuous at first glance. For example, many take for granted that states should prioritise the admission of skilled immigrants who are economically productive for the receiving state, and that family migration should be limited only to one's immediate relatives. Equally, many take for granted that immigration laws can be enforced through the use of detention and deportation. Yet, as I will argue, these "commonsense" practices are surrounded by deep ethical questions about how migration policy ought to be designed.

1 General Issues at Stake in the Controversy

In this chapter, I focus on two aspects of migration policy: *the ethics of selection* and the *ethics of enforcement*. More specifically, I will be concerned with the ethically permissible grounds for selecting one non-citizen over another for the purposes of inclusion, and the means of enforcement that may be used to exclude non-citizens. In order to focus more sharply on these two aspects of immigration, my discussion will be limited in several significant ways. First, I will not be discussing migration policies pertaining to refugees, given the widespread consensus that refugees constitute an "exceptional" population that states have special duties to include. Instead, I am primarily concerned with economic migrants who move for work or investment-related pursuits, and family migrants who move to reunite with their kin. I also bracket the fact of historical injustice by assuming that these economic or family migrants have no prior political connection to the receiving state in question, as in the case of ex-colonial subjects. Finally, I will not be considering global distributive justice and how it may impose duties on wealthy states in the Global North to include citizens of the Global South in their polities, with the aim of equalising opportunities around the world. While I do not have the space to take them up here, it should be noted that these three topics also play a significant role in the philosophical discourse on the ethics of migration.

States routinely make bureaucratic decisions about who to select for the purposes of inclusion. But which criteria, exactly, ought to play a role in immigration selection? While some may argue that states can choose to admit immigrants on *any* basis they please, others find the idea of racially discriminatory selection policies troubling. They suggest that selection policies, at the very least, cannot fall afoul of anti-discrimination norms *even if* states reserve the right to control the numbers of immigrants they admit. The commitment to anti-discrimination, however, may have surprising consequences. Politicians who argue for a more restricted border frequently claim that immigration is bad for the economy; in response, their opponents tend to emphasise the valuable economic contributions of immigrants. Both these positions suggest that it is morally desirable for states to prefer family migrants who bring their skills and expertise with them. Yet I explain how prioritising the admission of highly-skilled immigrants may come into tension with anti-discrimination norms.

States also typically reserve spaces for family migrants admitted on the basis of their relationships to citizens or residents. In the existing political climate, the perception of immigrants as economic threats is frequently coupled with a dislike of family migrants who are not perceived as useful or productive. Many have expressed the concern that states are facing the opportunity costs of admitting too many family members instead of highly-skilled migrants. To respond to these concerns, we need to look more closely at the philosophical grounding for family migration, a task I will take up in the next section. Finally, despite their declaration to tighten up on immigration controls, some states continue to offer "citizenship by investment" schemes that provide wealthy investors with a relatively easy path to citizenship. I briefly comment on those schemes, especially with regard to how states tend to justify their modes of exclusion.

Apart from disagreements over immigrant selection criteria, today's mainstream immigration debates tend to revolve around border enforcement policies. States' inclusion of skilled or family migrants, in today's context, necessarily implies the exclusion of many other would-be migrants who fail to be admitted under these categories. Some non-citizens attempt to enter the state or extend their legal residency through conventional means, only to be rejected; others know there is no point in even trying. As a last resort, they enter or remain as *unauthorised* migrants. Political opinion on unauthorised migrants tends to be deeply divided. On one hand, it is believed that states can use harsh deterrent measures to obstruct or remove unauthorised migrants from the territory. Consider the practices of detention and deportation, which are often built into the basic machinery of immigration systems. While seemingly justifiable in theory, these practices have serious ethical ramifications: as I show, they subject migrants to arbitrary forms of coercion. On the other hand, many vigorously defend the rights of unauthorised migrants to remain in the host country, living and working without the constant fear of detection by border agents. Take, for example, "sanctuary cities" that intentionally refrain from enforcing immigration law against unauthorised populations. More than this, some argue that unauthorised migrants are be entitled to opportunities to regularise their status. I will expand on the philosophical basis of the right to remain.

2 Topics in the Controversy

Having provided an overview of two key issues, I now examine the philosophical controversies in more detail. Returning to the ethics of selection, recall that many theorists endorse the commitment to anti-discrimination norms: the belief that nobody should be disadvantaged on the basis of their gender, race, sexual orientation, or other historically stigmatised attributes. I will use the case of racial discrimination as an anchor point for my discussion. Racially discriminatory immigration policies, like the US's past Chinese Exclusion Act, or Australia's "White Australia" policy, are to be roundly condemned (see Walzer 1983 for an exception). Even if theorists agree that racially discriminatory immigration policies are wrongful, they differ on their explanations as to *why* racial discrimination is wrong. For example, David Miller has claimed that they are indefensible because they are *insulting* to would-be migrants to be excluded based on their race, given that it does not "connect to anything of real significance to the society they want to join" (2013). Miller's explanation is not entirely convincing. It seems like race *can* be of genuine significance to the receiving state in question: members of the community might value being around other people of the same race. In addition, the discriminating state might claim that insult is fully unintended. Rather, they simply treat race as a reasonably accurate proxy for "relevant" attributes like cultural similarity.

Against these views, Joseph Carens has argued that there is something *distinctively wrong* about racial exclusion. Focusing on the case of "White Australia", he contends that "moral meaning of the White Australia policy cannot be detached from its history" (1987). I take him to mean that racial selection policies were implemented in a particular social context: under a colonial racial hierarchy, non-white migrants were excluded on the basis of their inferiority. What is at stake is not mere insult, but familiar manifestations of racial subordination and disrespect.

Keeping this history in mind, selecting immigrants by race today "inevitably endorses the history of oppression from which it emerged" (ibid). The expression of endorsement is *inevitable* insofar as it does not rely on the presence of racist intentions. Regardless of their justification, racially discriminatory immigration policies *by their very nature* re-inscribe insidious forms of racism. However, some disagree that racial selection is necessarily off-limits. For example, Christopher Heath Wellman has proposed that racial selection policies are only unjust when they insult *citizens* of that race who, unlike non-citizens, have a right to be respected as "equal partners in the political cooperative" (2008). On Wellman's account, the US's recent move to exclude nationals of seven Muslim-majority countries is impermissible because it disrespects Muslim-Americans: it sends the message that they are less worthy than other citizens. However, even if the policy wrongfully subordinates Muslims as a whole, it is permitted to implement this policy so long as no existing citizens adhere to the faith.

To recap: for theorists like Wellman, who restrict the scope of equal respect, selection policies are only impermissible if they subordinate existing citizens. Alternatively, you might hold that *all* persons are owed equal respect by virtue of their humanity, and states have a responsibility to craft selection policies that do not fall short of that standard, even under circumstances where they will have no effect on citizens. The latter position may strike us as more appealing: racially discriminatory policies do not just subordinate citizens, but the *non-citizens* who are blocked from entry. However, it may be accompanied by some unexpected implications for immigration selection. In particular, the duty not to subordinate non-citizens may have consequences for skill-selective immigration policies. Many states seek to encourage the migration of highly-skilled migrants. For example, countries like Canada, Australia, and the United Kingdom utilise "points-based" immigration systems that assign migrants "entry points" based on factors like their education level, past experience in skilled occupations, or possession of a valid job offer. Australia admitted 128,550 migrants on skill-based visas in 2018–19, in comparison to 57,400 on family visas (see Australian Government 2018). Similarly, in 2017, Canada admitted 159,262 economic migrants, in comparison to 82,470 family migrants (see Government of Canada 2018). Under such systems, migrants must reach a threshold of points to be eligible for admission. In general, skilled migrants are more frequently admitted than migrants from other streams.

To be sure, receiving states do not completely exclude low-skilled migrants from entry. Yet relatively low admission quotas are not the only face of discrimination. Many low-skilled migrants are admitted on visas with disadvantaging conditions compared to visas for highly-skilled workers. Specifically, "guest workers" are admitted in order for them to perform necessary forms of labour (e.g. domestic work; construction; fruit-picking), often for significant lower salaries than citizens would command. They are "guests" precisely because they are effectively barred from taking up long-term residence in the receiving state. Guest worker programs have been long-established in countries like the Gulf States and Singapore, but also exist in liberal states like the US, Canada, and Australia. The terms of their admission tend to be harsh: typically, guest worker visas exclude their holders from entitlement to membership rights, including the right to vote, bring their families, or participate in social welfare schemes. Some internal critiques of guest worker programs may be that they are fundamentally inegalitarian or exploitative because they take advantage of guest workers' weak bargaining position and render them vulnerable to dominating relationships, but I will not address those arguments in detail here.

Rather, I want to situate the precarity of guest workers within the wider context of skill-selective immigration policies. Overall, there seems to be no question that states discriminate against low-skilled migrants. Yet unlike racial discrimination, skill is largely regarded as a reasonable basis for selection. Perhaps the relative exclusion of low-skilled migrants is no more problematic than the exclusion of non-qualified applicants from particular jobs (see MacKay 2016). However, we may have reason to believe that skill-based selection is not as innocent as it initially seems. Wrongful discrimination is often packaged in a seemingly legitimate guise. Consider, for example, the existence of states that restrict the admission of HIV-positive non-citizens (Australia, China, and Singapore, to name a few). Health-based discrimination against HIV-positive people may strike us as objectionable because it reinforces the severe stigma faced by people living with HIV. Though veiled in the language of preserving public health, it sends the demeaning message that HIV-positive persons as fundamentally dangerous and polluted. As this case illustrates, discrimination in the name of the collective interest can nevertheless be wrongful because it marks out a group as inferior or undesirable.

I want to argue, now, that skill-selective immigration policies are subordinating to low-skilled migrants. To start with, states' preference for highly-skilled migrants is difficult to disentangle from their disdain for low-skilled migrants. In populist rhetoric, the myth of the highly-skilled "model migrant", who productively contributes to society, tends to be juxtaposed against low-skilled migrants who commit crimes, steal jobs from citizens, and abuse public welfare. For example, then-President Donald Trump recently declared the government's intention to "encourage talented and highly skilled people to pursue career options in the US", despite campaigning on the premise that low-skilled Mexican workers were taking jobs from American citizens and killing them. These claims, which portray low-skilled migrants as dangerous and parasitic, are explicitly subordinating. As I have argued, skill-selective immigration policies cannot be detached from their present social context, which is highly disrespectful towards low-skilled migrants (Lim 2017).

I will pause to consider two objections. The first is that HIV-positive status is *socially salient*, while, being a low-skilled migrant is not. Following Kasper Lippert-Rasmussen's definition, "[a] group is socially salient if perceived membership of it is important to the structure of social interactions across a wide range of social contexts" (2006). I think this claim can be put aside easily. Being perceived as a low-skilled migrant can expose an individual to disadvantaging behavior across a significant range of contexts (see Olney and Cholewinski 2014). A second objection might be that, unlike discrimination against HIV-positive people, skill-selective immigration policies have a rational basis. HIV, unlike highly contagious health conditions like swine flu or SARS, is relatively difficult to contract, and medical interventions like PrEP can significantly reduce the likelihood of transmission.

Consequently, HIV-positive people do not pose a genuine threat to public health; it is irrational to single them out for exclusion. The economic threat posed by an influx of low-skilled migrants, however, has an empirical basis. One oft-cited claim is that low-skilled migrants depress the wages of low-skilled native workers (see, for example, Borjas and Katz 2007). Yet other considerations point to a more complex picture. Low-skilled industries like agriculture, construction, and healthcare may in fact be facing *shortages* of workers, with employers relying heavily on a workforce of unauthorised migrants. Low-skilled immigration may be necessary to fill such jobs. The point, here, is not to assert one position or the other. Rather, the widespread conviction that low-skilled workers are ultimately bad for states' economies and politicians' willingness to reinforce this belief seem rooted in prejudice, not fact.

The issue of family migration is closely related. As Stephen Macedo contends, the US's generous admission of family migrants, rather than the highly-skilled, may "worsen the relative standing of the American poor" (2009). Macedo's claim speaks to a broader national anxiety about the economic burden that disproportionately low-skilled and poorly-educated family migrants may pose to public coffers. Another concern is that large-scale demographic change, as a result of family migration, may undermine social trust and erode support for the welfare state (for a reply, see Pevnick 2009). It is perhaps no coincidence that EU member states have increasingly tightened immigration restriction towards family migrants, who may now "require the sponsor to have adequate accommodation, sufficient resources and health insurance, and impose a waiting period of no more than two years" (European Commission). The "sufficient resources" condition usually takes the form of a minimum income requirement that requires citizens to prove they can support the arriving family member without the state's assistance.

Despite its potential economic drawbacks, family migration may have strong non-instrumental value for citizens and non-citizens alike. Its primary purpose is not to maximise states' material success, but to support citizens and residents' capacity to engage in meaningful relationships that are best realised through territorial co-presence. Even in an age where technological innovation facilitates communication across borders, it can be challenging to care for one's parents, raise children with one's spouse, or even engage in an intimate romantic relationship from afar. Taking the ethical basis of family migration seriously, however, raises further questions about whether family members *in particular* ought to be regarded as having a special claim to inclusion. Turning this assumption on its head, Luara Ferracioli has argued that restricting citizen-sponsored migration to family members ultimately violates principles of state neutrality because it expresses a judgment about *which* special relationships are valuable (2016). What suffices is that the citizen *herself* regards the relationship as such. Ferracioli's intervention raises questions about the supposed unique rights of family members and if they reflect antiquated views about human relationships. However, against this radical approach, we might continue to advocate the more restrictive view that states should prioritise the admission of non-citizens who stand in *certain* special relationships with citizens, such as relationships of care that can in theory include non-family members (Yong 2016).

Up to this point, I have tried to lend some clarity to the philosophical debates on ethically permissible modes of immigrant selection. As we have seen, there is a tremendous emphasis on *skill* as a preferred criterion for admission. Decisions about selection seem to revolve around what would-be migrants are able to *contribute* to the state. Furthermore, states seem to have a narrow economic understanding of what "contribution" is, given that citizens may also stand to benefit greatly from low-skilled immigrant labour and the opportunity for reunion with their loved ones. The fixation on immigration as a means of attaining economic success is most concretely realised in "citizenship by investment" schemes. Countries like the UK, Australia, and Canada enable wealthy investors to "buy" citizenship and make use of its benefits, which may include a lower tax burden. Despite their substantial economic benefit, we may object to these schemes for several reasons. As one theorist has suggested, they devalue the meaning of citizenship by commodifying it and handing it out to individuals who treat citizenship as a mere business opportunity and have no true intention of fulfilling the civic obligations traditionally associated with it (see Shachar 2017). A similar point can be made about what Ayelet Shachar and Ran Hirschl have termed "Olympic citizenship" (2014). States have chosen to extend citizenship to "super-talented" non-citizens, including opera singers and star athletes, in a bid to heighten their global competitiveness. Expedited paths to citizenship, combined with states' oft-declared desire to admit the "best and the brightest" migrants, sit uncomfortably with states' hostility towards non-citizens who fail to demonstrate exceptional wealth, talent, or skill. In order to legitimate the exclusion of low-skilled family migrants, many have claimed that family migrants lack genuine commitment to the receiving state and have no desire to integrate into the existing community. Yet we rarely hear these complaints being made about investors, opera singers, and athletes.

I now turn to the ethics of enforcement. States' decisions to exclude non-citizens are, in many cases, accompanied by the use of coercive force. As I previously noted, non-citizens who have entered, or remain on the state's territory without authorisation (or are under suspicion of doing so) may be subject to detention and deportation. In the first case, detention can be defined as the "holding of foreign nationals, or non-citizens, for the purposes of realizing an immigration-related goal" (Silverman and Massa 2012). Detention involves primary and secondary deprivations of liberty. The primary deprivation concerns non-citizens' freedom of movement: they are held in designated facilities or informal holding areas in the custody of immigration officials (ibid). At the same time, secondary deprivations of liberty arise from the coercive measures deployed by detention facilities to observe control and order amongst detainees in their custody that closely resemble those used in prisons, including the handcuffing of detainees moving to and from various locations, the presence of guards, and even the use of solitary confinement. Deportation, which may follow from detention if the migrant's application to remain is unsuccessful, refers to the state-enforced departure of a non-citizen from the country. Like detention, it involves serious primary and secondary deprivations of liberty. The primary constraint on freedom, common to all cases of deportation, is the individual's removal from the territory and therefore their freedom to pursue the assorted interests and projects that are bound up with their presence. Secondary constraints, on the other hand, comprise the additional coercive measures that states may undertake to force out unwanted non-citizens who may resist expulsion: aside from detention centres, they may use "handcuffs, physical force, and sometimes even drugs, to effect departure" (Gibney 2008).

Detention and deportation have largely been treated as *administrative*, rather than punitive tools of enforcement. On the latter view, the goal is simply to monitor and control the movement of non-citizens as they are processed through immigration bureaucracy. However, even if they are not intended to punish migrants, these practices stand in need of ethical justification. One immediate concern might be deprivations of liberty they impose on non-citizens and their tendency to cause long-term psychological harm (see, for example, Dudley et al. 2012). It is widely recognised that, even if liberty-deprivations may be justified for particular important ends, such as placing criminals under arrest to fight crime, states must ensure that they are not arbitrary (e.g. ensuring that there is probable cause to hold someone under criminal custody). Proponents may argue that detention and deportation are non-arbitrary as long as they meet the state's collective interest of enforcing its immigration decisions.

We may, however, insist on a more stringent standard of non-arbitrariness. In this vein, Patti Lenard has argued that deportation must also meet the criteria of "publicity" and "justification" to be non-arbitrary: the rules and laws governing deportation must be "publicly debated and known, so that individuals are subject to these laws rather than the whim of individuals in positions of authority", and justifications for a particular decision to deport must be "offered in the public sphere and subjected to debate" (Lenard 2015). We may insist that detention meets similar criteria; for example, it is perhaps unjustifiable to detain migrants who have a low risk of absconding and can be tracked through other methods, such as regular check-ins. At the level of application, however, states are far from achieving these goals; as reports of death and abuse in detention centres continue to surface, there is serious need for robust accountability mechanisms designed to protect the interests of detainees. Worse still, the fear of facing detention or deportation may render unauthorised migrants more vulnerable to mistreatment at the hands of landlords, employers, or even their own spouses: they may be compelled to comply with the demands of persons who have the power to turn them over to border authorities.

It may also be wrongful to detain or deport at least some unauthorised migrants who have the *right to remain*. Carens has explored the notion that migrants can have a right to continue residing on the state's territory and eventually be granted citizenship, even if their initial entry was unauthorised (2013). Indeed, this sentiment is echoed by many immigration activists on the ground today. Suppose that an unauthorised migrant has lived and worked in a particular state for several years. In the process, they cultivate valuable relationships with other people on the territory and may even have children who are citizens. Alternatively, they may have arrived as children and have received their education and upbringing in the receiving state. Is it ethically permissible for states to detain and deport such migrants? On a hardline view, residence is irrelevant to the migrant's claim to stay: a violation of immigration law remains a violation even if it occurred many years prior and the appropriate penalties must be enforced upon discovery. Against this, Carens argues that "the longer the [unauthorised migrant's] stay, the stronger the moral claim to remain" (2013). He goes on to outline what he calls a theory of *social membership*, under which a migrant's length of residence should be interpreted as a proxy for the development of deep social roots. The moral weight of social membership may require that states gives unauthorised migrants the opportunity to regularise their status.

As a whole, my discussion has sought to sketch out the contours of the ethics of immigrant selection, as well as the ethics of immigration enforcement. To reiterate, there are grounds for questioning states' right to exclude non-citizens in the first place. Nonetheless, even assuming their right to exclude, it bears remembering that states do not have *carte blanche* over the immigration policies they craft; rather, their selection of immigrants and enforcement of immigration law must still be constrained by ethical principles.

3 Desiderata for Future Research

The previous section sought to reconstruct important philosophical contributions to the ethics of migration policy. Moving forward, how ought theorists build on these insights? Firstly, more attention must be paid to states' selection policies and whether they may inadvertently perpetuate wrongful forms of discrimination. As I have tried to show, supposedly innocuous or desirable criteria of selection may in fact be wrongfully subordinating. While my discussion has focused exclusively on direct discrimination, I believe that the *indirect* effects of selection policies merit closer examination. In a world rife with historical and structural injustice, policies that seem "neutral" with regard to race and ethnicity may have foreseeable, if not intended discriminatory consequences for certain groups (Fine 2016). At the same time, even if we successfully come up with a list of permissible selection criteria, we need to refine and clarify our understanding of who meets the criteria in the first place. While states might need to avoid overly narrow interpretations of "family", we may also require a more inclusive understanding of "skill" that takes into account the historical devaluation of various occupations (e.g. care work). More generally, our ethical assessment of a particular criterion of selection cannot take place in isolation; we must pay attention to how immigration systems are designed *as a whole*. For example, if points-based systems are wrongfully discriminatory, we need to investigate what a "fair" system of migration would look like, and to what extent existing systems could be re-designed.

The same critical approach ought to be adopted vis-à-vis existing enforcement policies. In the previous section, I sought to underscore the arbitrariness of enforcement policies like detention and deportation. It remains unclear *how* detention or deportation ought to be reformed, in order to be ethically permissible, or if they can even be reformed at all. However, there is a wealth of other enforcement policies that may, at first glance, appear to be ethically insignificant or even an unavoidable component of border control. Take, for example, the procedures through which unauthorised migrants are detected and apprehended, such as controversial "stop and search" measures that commonly operate on the basis of racial profiling, or the use of immigration raids. These enforcement policies stand in need of further justification. Aside their material effects on the persons they are used against, we may also be concerned about the identity of the *actors* who carry out border enforcement. Take, for example, the immigration industrial complex, which describes how states often stand to profit from their border enforcement policies through its use of the private sector. This may result in a scenario where "controls are exercised by multiple, overlapping, sometimes competing organizations", some of whom may lack democratic accountability altogether (Sager 2018: 13). Even more fundamentally, we may ask whether the right to exclude can ever be enforced in a way that does not end up wronging non-citizens. Overall, it is imperative that we pay close attention to the concrete details of how migration policy functions today.

Literature

Borjas, G. J., and L. F. Katz. 2007. The Evolution of the Mexican-Born Workforce in the United States. In *Mexican Immigration to the United States*, ed. G. Borjas. Chicago: University of Chicago Press.

Carens, J. 2013. *The Ethics of Immigration*. Oxford: Oxford University Press.

Dudley, M., S. Mares, L. Newman, and Z. Steel. 2012. Children and Young People in Immigration Detention. *Current Opinion in Psychiatry* 25:285–292.

Ferracioli, L. 2016. Family Migration Schemes and Liberal Neutrality. *Journal of Moral Philosophy* 13:553–575.

Fine, S. 2013. The Ethics of Immigration. Self-Determination and the Right to Exclude. *Philosophy Compass* 8:254–268.

Gibney, M. 2008. Asylum and the Expansion of Deportation in the United Kingdom. *Government and Opposition* 43:146–167.

Lenard, P. 2015. The Ethics of Deportation in Liberal Democratic States. *European Journal of Political Theory* 14:464–480.

Lim, D. 2017. Selecting Immigrants by Skill. A Case of Wrongful Discrimination? *Social Theory and Practice* 43:369–396.

Lippert-Rasmussen, K. 2006. The Badness of Discrimination. *Ethical Theory and Moral Practice* 9:167–185.

MacKay, D. 2016. Are Skill-Selective Immigration Policies Just? *Social Theory and Practice* 42:123–153.

Miller, D. 2013. Immigration, The Case for Limits. In *Contemporary Debates in Applied Ethics*, ed. A. I. Cohen and C. H. Wellman. Oxford: Blackwell.

Olney, S., and R. Cholewinski. 2014. Migrant Workers and the Right to Non-discrimination and Equality. In *Migrants at Work. Immigration and Vulnerability in Labour Law*, ed. C. Costello and M. Freedland. Oxford: Oxford University Press.

Pevnick, R. 2009. Social Trust and the Ethics of Immigration Policy. *The Journal of Political Philosophy* 17:146–167.

Sager, A. 2018. Private Contractors, Foreign Troops, and Offshore Detention Centers. The Ethics of Externalizing Immigration Controls. *APA Newsletter on Hispanic/Latino Issues in Philosophy* 17:12–15.

Shachar, A. 2017. Citizenship for Sale? In *The Oxford Book of Citizenship*, ed. A. Shachar, R. Bauböck, I. Bloemraad, and M. Vink. Oxford: Oxford University Press.
Shachar, A., and R. Hirschl. 2014. On Citizenship, States, and Markets. *The Journal of Political Philosophy* 22:231–257.
Silverman, S., and E. Massa. 2012. Why Immigration Detention is Unique. *Population, Space and Place* 18:677–686.
Walzer, M. 1983. *Spheres of Justice. A Defense of Pluralism and Equality*. New York: Basic Books.
Wellman, C. H. 2008. Immigration and Freedom of Association. *Ethics* 119:109–141.
Yong, C. 2016. Caring Relationships and Family Migration Schemes. In *The Ethics and Politics of Immigration. Core Issues and Emerging Trends*, ed. A. Sager. London: Rowman & Littlefield International.

Migration and Integration

Sophie Guérard de Latour

Abstract In the political domain, integration refers to the sociological process on which the social cohesion of a civic community depends. As such, integration has a functional dimension—i.e. the coordination of individual actions—as well as a normative component—i.e. the principles, values and culture that ground and motivate the citizens' sense of belonging. Migration raises specific issues with regard to integration, related both to the impact of immigration on social cohesion and to the political conditions of migrants' access to citizenship. In political theory, the connection between migration and integration therefore focuses on the fair terms of integration for immigrants in liberal democracies. Section 1 of this paper specifies these issues and places them in their intellectual background, going back to the nineteenth century's philosophical distinction between the "ethnic vs. civic" conception of the modern nation and to the twentieth century's sociological works on nations, nationalisms and interethnic relations. Section 2 analyses three prevailing normative models of integration in political theory: integration as assimilation, as multicultural recognition or as national civic inclusion. Section 3 questions the theoretical limits of such models that are focused on long-term immigrants and their possible naturalisation; it calls for future research that reassesses the fair terms of integration by addressing more precisely the changing features of international migration, the emergence of transnational citizenship and the issue of refugees.

1 Integration in or Beyond the Nation?

In its broad sense, "integration" means the unification of diverse elements in a coherent whole. Applied to the political domain, it refers more specifically to the social cohesion of a political society, which depends on the special bonds that unite its members into a single "people" or "nation" and that make them feel they belong to it. Integration thus implies two dimensions: At a factual level, some functional

S. Guérard de Latour (✉)
École Normale Supérieure de Lyon, Triangle (UMR5206), Lyon, France
e-mail: sophie.guerard@ens-lyon.fr

A. Niederberger et al. (eds.), *Handbook of Migration Ethics*,
https://doi.org/10.1007/978-3-031-89877-8_17

conditions are required to make social coordination possible and relatively stable; at a normative and ethical level, social cohesion also depends on the principles, values and identity that ground and motivate the citizens' sense of belonging to their political community.

Migrants are defined as foreign-born people who have crossed a national border to reside in a country. By contrast with tourists, they are not just visiting the country but have come there to study, work or join their family. By contrast with refugees, they are not fleeing political coercion but have come voluntarily to find better opportunities in another place where they may wish to stay. Therefore, migrants, who are more precisely designated here as immigrants, raise specific difficulties when it comes to integration. How do these outsiders enter "inside" the society they have come to? How do they shift from the status of non-members to the status of members? These questions are made complex by the various types of immigrant participation to the collective life of their host country, which are determined by the diversity of their situations. Immigrants are primarily contributing *members of economic and social activities*, as workers, students, parents or neighbours; if naturalised, they may also become *members of a political community* as citizens and possibly *members of a historical community*, in the case of naturalised immigrants who identify with the national culture and feel committed to their co-nationals. In all these situations, the relation between immigration and integration can be questioned both at the descriptive and normative levels: Is the economic, social, political or cultural participation of immigrants in the country where they live contributing to its social cohesion or is it undermining it? Do these various types of participation justify the immigrants' status as members of the nation, and which should prevail to legitimate it?

It is worth noting that, as regards the normative and ethical side of integration, the issue is made more complicated by the distinction between two points of view: the immigrant's opinion about their legitimate membership to the nation and the view held by the autochthonous citizen. While immigrants' participation to the various spheres of the collective life of a nation is likely to make them feel they belong to it and deserve to be acknowledged as members, their own appreciation about their membership tends to mismatch with the suspicious and hostile views that are often expressed in public opinions about them: Indeed, immigrants' economic participation is criticised by some as creating unfair competition for low-skilled national workers; others question their social participation as excessively burdensome for welfare states; others disqualify their cultural participation by accusing them of being disloyal members unwilling to conform to the national way of life and consider that they are importing foreign traditions instead. Hence the normative questions: What is morally required of immigrants to be integrated in their host society, i.e. to be considered as legitimate members? And what do these requirements imply in terms of what immigrants and autochthonous citizens legitimately expect from each other?

Political theory has been addressing these issues through the double intellectual heritage of philosophical debates on the "nation" and of the sociological works on integration, nationalisms and interethnic relations. As regards the former, two main positions have emerged in the late nineteenth century on the normative foundations of a "nation". On the one hand, the civic conception, famously upheld by the French philosopher Ernest Renan in his speech *What is a Nation?,* justified political membership through the active commitment to universal political principles that grounded the citizens' sense of belonging. This civic conception of the nation continues the rationalist tradition of modern contractualism in which the "people" is interpreted in artificial terms—by contrast with premodern naturalistic views—as made of individuals who consider each other as equals and voluntarily accept to form a self-governing community. On the other hand, the ethnic conception of the nation, as was defended by the German philosopher Johann Gottlieb Fichte in his *Speech to the German Nation,* emphasised the ethnocultural features that characterised the spirit of a people, namely its language, customs and folklore, arguing that the commitment to universal principles was conditioned by the sharing of an inherited national culture. This ethnic conception of the nation continues the romantic tradition that rejected the French revolutionaries' abstract view of humanity, anchoring it instead in the proper spirit and authentic genius of a people.

These competing views of the nation still exert significant influence on the contemporary debates about the model of integration that should be applied to immigrants. However, neither integration, nor migration, were the concerns of these philosophical approaches, which primarily addressed the issue of justifying political obligations to citizens—thus focusing on the normative dimension of social cohesion. "Integration" is instead a sociological concept that was originally designed to explain social cohesion in its functional dimension. It was theorised by European sociologists, such as Emile Durkheim and Max Weber at the turn of the twentieth century, in order to understand how democratic, industrialised and urbanised societies could create coherent and stable collectivities in spite of the decline of traditional modes of regulation and of the rise of individualism. It is worth recalling that the unprecedented development of migration within Europe and in the United States at the end of the nineteenth century, under the combined pressures of industrial development and of economic crises, weighed on the formation of sociology as a new scientific discipline. In many aspects, in Europe, the uprootedness of migrant workers echoed the experience of the national peasants who had left their villages to work in urban industrial zones, both populations facing the same problem of adapting to a modern way of life. Likewise, the massive arrival of European immigrants in the United States offered a rich field of investigation to the analysis of the migrants' adaptation to the environment of a modern society, as attested by the Chicago School's founding research. The main contribution of these works, and of their subsequent developments in the sociology of nationalisms and interethnic relations, has been to show how modern nations have managed to gather their members in a stable and prosperous society, and especially their new members, the immigrants. They have therefore provided tools to refine the philosophical binary approach about national membership by showing that it depended on a complex mix of ethnic and civic features.

It is from the 1990s onwards that political theorists started to explicitly address the issue of immigrants' integration in normative terms. The unexpected revival of nationalisms after the collapse of the communist bloc, combined with the growth of anti-immigrant discourses among European public opinions, raised issues rarely questioned so far: Is it legitimate to treat immigrants as invited workers who are temporarily contributing to socio-economic activities but should not be integrated in the national community, or does their residency in the host country require their naturalisation? Should they renounce their cultures of origin and fully conform to the national way of life to obtain citizenship, or should these new citizens have a right to preserve their cultural particularities as long as they comply with the legal and political norms? In political theory, a consensus soon emerged about the unfairness of disconnecting *social* and *political* memberships in liberal democracies. Indeed, denying immigrants the opportunity to become citizens submits them to the arbitrary power of "tyrant citizens" because it deprives them of a political voice and of the possibility to participate in the making of the laws that they must obey (Walzer 1983). Most of the political theorists therefore focused on the second question by asking which fair terms of integration were likely to settle the legitimate expectations of immigrants and autochthonous citizens, taking into account the functional and normative dimensions of social cohesion. The following section will expose the three main approaches that have dominated the debates: integration as assimilation, as multicultural recognition and as civic inclusion.

2 Assimilation, Multicultural Recognition or National Civic Inclusion?

The notion of *assimilation* rests on a communitarian perspective where the *people*, as a civic community, also forms a *nation*, considered as a homogenous cultural group whose members identify to a shared language, history and way of living. Concerning immigrants, the biological metaphor expressed in the word "assimilation" suggests that the differences related to their culture of origin are to vanish while they are incorporated in the national body. This approach of immigrant integration prevailed in Western societies from the late nineteenth century to the late 1960s. At first, it was related to a historical observation, namely to the fact that modern democracies were born as nation-states. The sociological works mentioned above have indeed shown that the pressure to assimilate in a common national culture involved all the members of a country because it was part of a broader process of social modernisation. The French sociologist Emile Durkheim analysed this process as a shift from a "mechanical" type of solidarity based on social "similarities"—i.e. on a same kinship and on shared traditions—to an "organic" type of solidarity based on "differences"—i.e. on the interdependence of specialised social functions in a complex society (Durkheim 2013). Durkheim's theory of organic solidarity thus allowed a challenging of the romantic and ethnic conception of the nation by arguing that in modern societies social cohesion no longer depended on a common origin and inherited culture, but rather on the social and economic functions performed by each individual in the social division of labour. As such, at a time when, in France, a large part of the workforce was composed of immigrants, his theory aimed more particularly at defeating the conservatives' thesis according to which immigrants were unassimilable to the French nation because of their foreign origins. Later, in the 1980s, theories of the nation-building process enriched the sociology of integration by specifying its cultural outcomes: Ernest Gellner argued that the politics of linguistic assimilation pursued in modern nation-states, which caused the marginalisation or disappearance of regional languages, was mainly a functional requirement of industrialisation, which implied the availability of a large and mobile workforce whose members could be trained and managed in the same language (Gellner 1983). Benedict Anderson explained how the circulation of cultural goods, such as novels and newspapers written in the official language, contributed to creating "imagined communities" in large-scale societies, thus causing the citizens' collective identification with the nation (Anderson 1983). Beside the theories of national integration, based on a macro-perspective approach, the sociology of interethnic relations completed the understanding of assimilation through a micro-perspective. The Chicago School's researchers richly documented the process of "acculturation", through which the successive waves of European migrants succeeded in integrating in American society. Robert Ezra Park's influential description of the "race relations cycle" replaced the biological view of race with a relational approach to culture which analysed the role of ethnic identities in the internalisation of the dominant way of life by immigrants: The cultivation of ethnic ties and cultural habits among communities of Irish, Polish or Italian immigrants equipped them with social and psychological resources that facilitated their transformation into American citizens, but which would eventually fade away; as such, migrants' ethnic identification was not an obstacle to their integration and rather functioned as a transitory step towards their fusion in the American "melting pot". It is worth noting here that national assimilation, even if it has been analysed by social sciences as a distinctive process of modern societies, was nevertheless associated with collective representations of the nation that fuelled ethnonationalism. Indeed, in order to give specific content to their common imaginaries and to foster identification among co-nationals, modern nations had to invent their own tradition, by celebrating founding myths and heroes, and nationalists therefore included ethnic features in the definition of the civic identity. Hence the conservative and nativist interpretation of the nation: By sacralising the origins of the nation and asserting the moral superiority of native citizens, ethnonationalism fostered hostility towards the foreign-born, deemed as ethnically unfit to become nationals.

In the sociological approach, integration understood as assimilation is a descriptive concept meant to explain how social cohesion works in modern nations. Nevertheless, such a description carries a normative dimension. It has indeed motivated a progressive assessment of assimilation until the 1960s because it suggested that differences of ethnic origins were no obstacle to the political inclusion of foreign-born individuals and that the merging of immigrants in the national culture was justified to include them as equal citizens. This normative view of integration as assimilation therefore grounded the adoption of the "Anglo-conformity model" of integration in the United States, Canada or Australia until the 1960s, when "immigrants were expected to assimilate to existing cultural norms, and, over time, become indistinguishable from native-born citizens in their speech, dress, leisure activities, cuisine, family size, identities and so on. This strongly assimilationist policy was seen as necessary to ensure immigrants become loyal and productive members of society" (Kymlicka 2001: 153).

The positive assessment of assimilation was totally reversed in the late 1960s under the pressure of the new social movements that struggled for the public recognition of racial, national and ethnic minorities. In the post WWII international context, dominated by the promotion of human rights and the progress of decolonisation, these movements powerfully contributed to disqualifying the racial hierarchies that motivated racism, antisemitism and ethnocentrism that were common sense before the war. Their contestation enabled the questioning of the alleged moral superiority of modern Western nations over the traditional cultures of postcolonial migrants or autochthonous peoples. Instead of accepting national assimilation as an emancipatory process towards civic integration, they reproached this political model for its oppressing and marginalising effects on the ethnic groups deemed as inferior. Moreover, the 1970s ethnic revival in the United States and United Kingdom, attested by the growing celebration of peoples' "roots" and foreign origins, evinced the powerful resilience of ethnic minorities in the face of pressures to conform to the Anglo-Saxon culture, which invalidated the sociological assumption about the expected obsolescence of ethnic identification in modern nations.

In the 1990s, political theorists articulated several reasons to denounce the illegitimacy of cultural assimilation in the case of immigrants, but this applied more largely to regional, linguistic and indigenous minorities. Some arguments focus on the moral harms inflicted by the loss of one's culture of origin: For Avishai Margalit and Joseph Raz, it unduly deprives someone of the "primary focus of identification" that warrants a secure and stable sense of oneself because identification with one's culture depends on belonging and not on achievement (Margalit & Raz 1990); for Charles Taylor, the dialogical nature of the self explains why the person's social recognition is a "vital need" that must be respected, by acknowledging both the person's dignity *qua* human and their authenticity *qua* person socialised in a specific culture (Taylor 1994). Other arguments emphasise the legal and political inconsistencies of assimilation, pointing to its incompatibility with the principles of non-discrimination and equal respect. For Will Kymlicka, the ethnic biases of the public sphere, which reflect the privileged position of historically dominant groups in the nation-building process, generate discriminatory practices that infringe on migrants' fundamental rights: For instance, when the framing of national laws, institutions and policies creates obstacles for the compliance of migrants to their religious duties (in terms of religious holidays, dress code, dietary requirements), the imperative to conform to the dominant culture restricts their religious freedom. Moreover, the public promotion of a national culture inevitably marginalises minority cultures and therefore corrupts the social bases of their members' self-respect: As long as the official culture privileges the dominant ethnic group(s), thus confirming implicitly the inferior status of ethnic minorities, their members tend to internalise the negative representations attached to their cultures and lose confidence in their ability to live a worthwhile life (Kymlicka 1995). Other critics, such as Iris M. Young and Tariq Modood, raised objections inspired by postcolonial studies to contest the cultural imperialism at stake in the liberal view of citizenship, which imposes the majority's practices and values on the minorities under the pretext of implementing universal principles (Young 1990; Modood 2013).

Multicultural recognition aims to overcome these normative limits. Acknowledging publicly the moral value of cultures, this normative approach does not require of immigrants that they fully conform to the national culture, but provides them instead with public protections against cultural assimilation. Drawing on John Rawls's contractualism, Kymlicka analyses membership in one's culture of origin as "a social primary good" that must be fairly distributed among citizens. He concludes from it that, as a matter of justice, cultural rights should be granted to immigrants whose minority status undermines the possibility to enjoy their culture. Cultural rights, defined as legal safeguards against assimilation, do not deny the legitimacy of political integration, but they rework it in an inclusive fashion to make it more hospitable to ethnocultural diversity; for instance, they allow the justification of multicultural policies directed at migrant populations, such as bilingual classes in primary schools, legal exemptions to accommodate religious duties and customary practices, or public subventions of ethnic festivals. This kind of policies, which are often accused of encouraging cultural separatism, should rather be considered as integrationist tools aimed at promoting immigrants' participation in social life and at redefining their political integration on equal terms.

Multicultural recognition interprets these equal terms in two ways. Firstly, it places the emphasis on anti-discrimination: Differences of origin should not be an obstacle to immigrants' access to education, employment, housing, politics, etc., as is the case with the assimilation model, where the privileged position of the national culture is taken for granted and unquestioned. By contrast, accommodating immigrants' cultural difference is justified because it compensates their social disadvantages by giving them public recognition; it is required to guarantee equality of opportunities for them. Secondly, multicultural recognition associates civic equality with the public endorsement of the moral value of different cultures. Consequently, public recognition of ethnic minorities is not limited to temporary measures designed to deconstruct racist and ethnocentric prejudices. Multicultural recognition does not aim "to bring us back to an eventual "difference-blind" social space but, on the contrary, to maintain and cherish distinctness, not just now but forever" (Taylor 1994: 40).

Thus defined, the equal terms of integration allow a specification of immigrants and nationals' expectations of each other. In this approach, integration should be considered as "a two-way street", grounded on a moral norm of reciprocity that demands mutual efforts of adaptation between ethnic minorities and the majority. On the one hand, immigrants who have come voluntarily to a liberal and democratic country have a "duty to integrate", which means that members of the majority can legitimately expect from them that they comply with the nation's laws and that they respect its political principles. Note that because these norms and principles have been historically implemented in a specific nation, immigrants' duty to integrate goes hand in hand with some requirement of cultural adaptation on their part so that they acquire the linguistic skills and institutional competence necessary to participate in social and political life. However, from the perspective of multicultural recognition, adaptation to the national culture is only justified for instrumental reasons, as a means to become a cooperating partner of a liberal and democratic society. This only retains that part of the national culture which is required for migrants to become functional workers, neighbours and citizens, but this rejects as illegitimate the expectation that immigrants fully conform to the dominant way of life in order to be considered as members.

Reciprocally, the majority has a duty to respect the immigrants' will to preserve some aspects of their cultures. This implies more than tolerating cultural differences that are compatible with the political principles of a liberal democracy; it also imposes on the state a positive obligation to correct the cultural biases that are expressed in the national institutions and laws due to the ethnic majority's historically privileged position. The majority's efforts to adapt the institutional framework to the immigrants' specific needs are required in order to prove its willingness to welcome the newcomers, by acknowledging publicly that they should be treated as equal citizens and full members of the political community. A new conception of national identity results from this norm of mutual adaptation. By contrast with assimilation, which interprets national identity as the "melting pot", i.e. the fusion of different ethnic groups in the dominant culture, multicultural recognition pictures it as a "mosaic" within which each group is allowed to cultivate its own culture while remaining attached together in colourful unity.

Multicultural recognition suffered a significant disgrace in the early 2000s both in theory and practice (Joppke 2004). In political theory, liberals have challenged its progressiveness, objecting that this political model weakens the emancipatory potential of liberal cultures for ethnic minorities. Susan Moller Okin criticised the oppressive effects of Kymlicka's defence of cultural rights for minority women because it encourages the preservation of traditional cultures which are more patriarchal than the modern culture of the majority. For her, Kymlicka's liberal clause according to which cultural rights forbid any restrictions of individual freedoms fails to protect minority women because sexual discrimination operates insidiously through girls' primary socialisation and because sexually oppressive practices, such as circumcision or forced marriage, are mostly performed in the private sphere, out of public scrutiny. Accordingly, assimilation to the majority culture seems fairer than the recognition of minority cultures because immigrant women "*might* be much better off if the culture into which they were born were either to become extinct […] or, preferably, to be encouraged to alter itself so as to reinforce the equality of women—at least to the degree to which this value is upheld in the majority culture" (Okin 1999: 23). Similarly, Brian Barry has questioned the "rule and exemption approach" that is typical of the legal accommodation of minority cultural practices, arguing that it deeply weakens the citizens' sense of equality. Instead of achieving a more inclusive form of equality, it endorses discrimination because it multiplies norms that give undue privileges to ethnic minorities—for instance when permitting British Sikh motorcyclists to ride without a helmet while non-Sikh motorcyclists have no right to do so. Moreover, it undermines the authority of the law, which is supposed to publicly acknowledge the general interest—for instance when legal exemptions about religious animal slaughtering contradicts the national laws forbidding animal suffering (Barry 2001).

In practice, these theoretical objections echo the doubts of politicians and citizens concerning the damaging effects of multicultural policies on social cohesion. In the post-9/11 era, dominated by the threat of Islamist terrorism and the revival of religious fundamentalism, many public discourses conveyed the idea that multiculturalism had failed because tolerance of ethnoreligious minorities had hampered their integration in the liberal culture of the majority; such discourses accused multicultural tolerance of being the Trojan horse that introduced illiberal values in liberal countries, thus weakening their moral fabric. As a consequence, in the late 2000s, many Western democracies undertook a "civic turn" by placing the emphasis on the "common values" constitutive of their political identities as liberal democracies. The civic turn led some European countries to adopt "citizenship tests" designed to make the immigrants' naturalisation conditional on their knowledge of the national history and institutions and on their adhesion to national values. These civic tests have been criticised for their discriminatory logic and normative ambiguity. On the one hand, they set a double standard between would-be citizens and actual citizens by requiring of the former an empirical knowledge and moral engagement that the latter may lack, which is incoherent or unfair: It is incoherent to test the immigrants but not the nationals if such civic skills are seen as necessary conditions of social cohesion; or it is unfair if immigrants are the only ones who are expected to plead allegiance to national values, while citizens would be free to reject them without being suspected of being disloyal to their nation. On the other hand, growing discourses on "common values" in liberal democracies are highly ambiguous inasmuch as they conflate universal principles with particular cultures. This ambiguity boils down to cultural imperialism, which comforts the majority's tendency to associate its own national culture with universal principles such as freedom of speech or gender equality and to denigrate accordingly minority cultures as inherently illiberal and undemocratic. Christian Joppke has concluded from this ambiguity that citizenship tests face an alternative that prevents them from achieving their objective of grounding social cohesion in liberal democracies on some degree of cultural conformity: Either they favour a thick interpretation of the national culture by valuing a specific way of life, which takes it back to the pitfalls of assimilation, or they adopt a thin interpretation by placing the emphasis on abstract political principles, in which case the question of identifying the distinctive cultural features of a nation is left open, given that liberal democracies profess the same principles (Joppke 2008). Moreover, the civic turn is problematic from the point of view of ethnic minorities to the extent that it exposes their members to strong performative effects of stigmatisation: the growing public discourses about the immigrants' alleged lack of will to integrate neglect the reality of their long-term acculturation, blame them for the discriminations from which they suffer and ultimately contribute to reinforcing their symbolic status of eternal strangers in the public opinion.

National civic inclusion offers an intermediate approach which takes the need for civic identification with a national culture seriously without reactivating the exclusive effects of assimilation. It has been developed by liberal nationalists and liberal republicans. Their normative position consists in justifying the promotion of nationality as a means to achieve liberal justice. Their instrumental justification differs from that of liberal multiculturalists as regards migrants: Cultural adaptation does not serve only the functional adaptation of newcomers to their host society, which conditions their ability to enjoy the full benefits of their rights on equal terms with the majority members; liberal nationalists and liberal republicans add a communitarian component to the process of acculturation and place the emphasis on the psychological identification and moral commitment that are needed to achieve the migrants' civic inclusion. Dominique Schnapper argues that a "community of citizens" cannot be motivated only by instrumental and interest-based connections—of economic type—nor by rational and abstract principles—of legal and political type. Citizens are more than producers and consumers because they are also committed to a common political project as co-nationals; migrants' integration should therefore not be limited to a citizenship primarily based on residency, according to which their long-term participation in socio-economic life suffices to justify their political integration after a certain time. Moreover, their civic sense of belonging cannot be based on abstract principles only but also depends on the symbolic and affective ties—of ethnic type—which make the citizens identify with the same historical collectivity (Schnapper 2017). Therefore, as David Miller has argued, national identification is a crucial condition of political solidarity: Liberal democracies need to cultivate a mutual commitment between their citizens in the absence of which the redistributive efforts required by welfare policies would lack social legitimacy and collective support. Such conditions are severely lacking in particular when ethnic minorities are not considered as equal members of the national community in public opinion and reciprocally when the former do not identify with the latter as co-nationals.

In spite of its communitarian underpinnings, national civic inclusion differs from assimilation because it interprets nationality as a dynamic identity prone to evolve under the influence of social and ethnic diversity. In this approach, national identity does not refer to a unitary and homogeneous culture, as in nineteenth century nationalist discourses. Instead, the legal and political principles institutionalised in modern nations allow migrants to express their cultural differences through the exercise of their individual freedoms (of religion, association, speech, etc.). This opens up the possibility for ethnic minorities to become involved in public deliberations and to make their own contribution to the discursive construction of the national identity. However, by contrast with multicultural recognition, national civic inclusion places limits on the public promotion of ethnic differences. It rejects cultural rights granted to ethnic groups *qua* groups because these rights carry the risk of fostering ethnic withdrawal and political balkanisation. In general, national civic inclusion "supports equal citizenship rather than a form of politics that is fragmented along group lines" (Miller 1995: 153). Schnapper argues that members of ethnic minorities should not claim for the public protection of their cultural specificities, but are rather expected to cultivate them in private and to identify, as citizens, with the national culture. She admits that this expectation is more demanding for immigrants and their descendants than for members of the historical majority because the latter's ethnic characters are part of the national culture (e.g. official language, religious public holidays), however, this is the price to pay to participate fully in the civic community. Such a cost is not an unfair burden for ethnic minorities because nationality is a specific type of cultural identification that differs from ethnicity: Its connection with the civic project makes it essentially inclusive, by contrast with the exclusivist logic typical of ethnicity (a person cannot decide to be Corsican if she was not born Corsican, but she can become French by naturalisation). In sum, nationality offers the trans-ethnic and common culture that is required to include people from diverse origins in the same political community.

However, the serious objections raised by theorists of multiculturalism about the capacity of national cultures to include ethnic minorities in the absence of legal and political recognition strongly suggest that national civic inclusion may well reproduce the shortcomings of assimilation in its own way. To avoid this, it must take the obstacles faced by ethnic minorities in the process of their integration seriously. Cécile Laborde's "critical republicanism" does so by offering a refined version of the third approach, enriched by the input of critical theory. It departs from multiculturalists' diagnosis and remedies, firstly by rejecting culture-based arguments about migrants' problems of integration, secondly by blaming multicultural policies for being at best ineffective or at worst counterproductive. The integration of immigrants is not hindered by a cultural misunderstanding with the majority but by the socio-economic obstacles they face due to their stigmatisation and racialisation as "people of a different origin". As such, racialised immigrants are not victims of a lack of recognition but rather of an "excess of the wrong kind of recognition" (Laborde 2008: 247). Then, by publicly granting a positive value to ethnic differences, multicultural recognition reverses the racial stigma without changing its causes, thus reinforcing the majority's prejudices about the minority instead of deconstructing them. Therefore, the legitimate strategy consists in radical reforms of "de-ethnicisation" designed to remove institutional, political, social, economic and cultural obstacles faced by members of ethno-racialised groups. It requires solutions that, contrary to programmes of affirmative actions, remain colour-blind as much as possible (e.g. anonymous resumes), even if, in special situations, the public promotion of visible minorities may be needed to pluralise the national imaginary, for instance in mass media. However, this kind of policy does not lead to a permanent acknowledgement of ethnic differences, by contrast with multicultural recognition. Rather, it is only justified as a temporary tool for mainstreaming ethnic minorities' differences and erasing their racial stigma, so that eventually hybrid and plural modes of identification with the national culture emerge.

3 New Questions for Research

Until the late 2000s, normative debates about integration mainly focused on long-term legal immigrants considered as would-be citizens and engaged as such in a process of cultural adaptation supposed to end up in their naturalisation. However, this dominant approach appears oversimplified in view of recent social and political changes, such as the growing complexity of migration fluxes, the development of transnational forms of sovereignty and recent refugee crises. These changes raise new questions for research.

Firstly, social scientists recently started to analyse the emergence of "super-diversity" in multicultural democracies (Vertovec 2007). Super-diversity refers to the unprecedented social configuration produced by the proliferation of migration channels and immigrant legal statuses, which challenges the classic scenario of their linear and progressive route towards naturalisation. In super-diverse societies, migrant populations have changed to the extent that migrants come from more diverse countries and tend to be more mobile or to be engaged in diasporic networks and transnational forms of identification. Besides, public authorities manage migrant populations in more complex ways than suggested by the standard view of integration as a process of progressive acquisition of full citizenship, with the whole bundle of rights associated with it. Instead, legal statuses related to migration are characterised by their growing diversity, which makes it necessary to reassess the fairness of these various statuses and the rights to which migrants are entitled. Finally, the situation of illegal migrants also needs to be taken into account because they form statistically significant population groups who contribute importantly to the economy of developed countries. Future normative research therefore needs to pay more attention than before to the difficulties raised by the great diversity of migrants' situations and explore in greater detail the fair terms of integration to which they are entitled (Carens 2013; Song 2019).

Secondly, transnational forms of legal and political integration, like in the case of the European Union, raise new issues with regard to migration. The emergence of European citizenship opens an important field of reflection about the articulation of legal norms at various scales (local, national, regional, European). Such an evolution inevitably challenges the dominant state centred perspective on migrants' integration. Intra-European migration is indeed interpreted by some as the source of post-national or cosmopolitan citizenship, in which individual rights and political solidarity gradually gain independence from national status (Sangiovanni 2013; Delanty 2018).

Thirdly, recent refugee crises also impact on debates on migrants' integration. So far, the standard view in normative political theory has distinguished these two categories on the basis of the assumption that migrants are permanent residents and participants who are likely to stay in the host country, unlike refugees who obtain asylum for a limited time in response to an immediate danger. According to this distinction, the issue of integration mostly concerns the former—and only marginally the refugees who are unable to return to their country. This differential approach derives from the distinct moral obligations that political theorists usually attribute to democratic states as regards refugees and migrants: They argue that humanitarian reasons require states to accept people whose life is in danger, while democratic states have no moral duty to open their borders to other migrants because their citizens have a fundamental right to self-determination, hence a right to decide who should enter their political community. However, the distinction between migrants and refugees appears to be empirically fragile and of little normative significance in view of global inequalities and of the current evolution of international migration. The distinction is blurred by the debates among international lawyers about including economic or environmental criteria in the legal definition of "refugees" in order to extend the category beyond the limits of political oppression fixed by the Geneva Convention. Moreover, at the political level, under the pressure of populist discourses and ethnonationalist governments, refugees and migrants are often conflated with one another and perceived as a common threat that should be contained. These observations require political theorists to pay more careful attention to the gap between their normative models and current political contexts (Miller and Straehle 2020).

Literature

Anderson, B. 1983. *Imagined Communities. Reflections on the Origin and Spread of Nationalism.* London: Verso Books.

Barry, B. 2001. *Culture and Equality. An Egalitarian Critique of Multiculturalism.* Cambridge, MA: Harvard University Press.

Carens, J. 2013. *The Ethics of Immigration.* New York: Oxford University Press.

Delanty, G. 2018. *Formations of European Modernity. A Historical and Political Sociology of Europe.* Cham: Springer International Publishing.

Durkheim, E. 2013. *The Division of Labor in Society.* Basingstoke: Palgrave Macmillan.

Gellner, E. 1983. *Nations and Nationalism.* Ithaca, NY: Cornell University Press.

Joppke, C. 2004. The Retreat of Multiculturalism in the Liberal State. Theory and Policy. *The British Journal of Sociology* 55(2):237–257.

Joppke, C. 2008. Immigration and the Identity of Citizenship. The Paradox of Universalism. *Citizenship Studies* 12(6):533–546.

Kymlicka, W. 1995. *Multicultural Citizenship. A Liberal Defense of Minority Rights.* New York: Oxford University Press.

Kymlicka, W. 2001. *Politics in the Vernacular. Nationalism, Multiculturalism, and Citizenship.* New York: Oxford University Press.

Laborde, C. 2008. *Critical Republicanism. The Hijab Controversy and Political Theory.* New York: Oxford University Press.

Margalit, A., and J. Raz. 1990. National Self-Determination. *Journal of Philosophy* 87(9):439–461.

Miller, D. 1995. *On Nationality.* New York: Oxford University Press.

Miller, D., and C. Straehle. 2020. *The Political Philosophy of Refuge.* Cambridge: Cambridge University Press.

Modood, T. 2013. *Multiculturalism.* A Civic Idea. 2nd ed. Cambridge: Polity Press.

Okin, S.M. 1999. *Is Multiculturalism Bad for Women?* eds. J. Cohen and M. Nussbaum, Princeton NJ: Princeton University Press.

Sangiovanni, A. 2013. Solidarity in the European Union. *Oxford Journal of Legal Studies* 33(2):213–241.
Schnapper, D. 2017. *Community of Citizens. On the Modern Idea of Nationality*. London: Routledge.
Song, S. 2019. *Immigration and Democracy*. New York: Oxford University Press.
Taylor, C. 1994. *Multiculturalism. Examining The Politics of Recognition*, ed. A. Gutmann, Princeton NJ: Princeton University Press.
Vertovec, S. 2007. Super Diversity and its Implications. *Ethnic and Racial Studies* 30(6):1024–1054.
Walzer, M. 1983. *Spheres of Justice. A Defense of Pluralism and Equality*. New York: Basic Books.
Young, I.M. 1990. *Justice and the Politics of Difference*. Princeton NJ: Princeton University Press.

Xenophobia in South Africa

Bernard Matolino

Abstract Post-apartheid South Africa has witnessed a number of xenophobia related incidents directed mainly against African immigrants. There have been as many attempts at explaining the underlying cause of these incidents. These explanations have ranged from those who seek to dispute that this violence is xenophobic to those who seek to explain it in terms of the complexities of migration and discrimination attached to it. An undertheorized aspect of xenophobia is how it represents Africa's failure to integrate economically, culturally, socially, politically as a unitary block with a shared identity, history and common future. Whatever the reasons for that failure, they must be understood to be a betrayal of African goals of political unity, material prosperity, and social stability. One way of overcoming this failure is to take seriously the founding principles of newly independent African leaders who sought to chart ways of finding a route to continental unity and co-operation. South Africa, beset with its own internal problems does, however, present itself as a viable alternative to the permanent state of failure of most sub-Saharan African countries. Young and old people who dare and wish to escape the trap of their countries see South Africa as a lifetime opportunity that will provide refuge to their troubled lives. South Africa has become a microcosm of the tensions caused by Africa's failed politics, economics, and social cohesion. Understanding xenophobia in South Africa requires going beyond the attitudes and behaviour of poor, frustrated, and disempowered South Africans who lash out at similarly disempowered fellow Africans.

Migration as a human phenomenon has been with us for as long as we have been around planet earth. The reasons for it vary from very basic such as foraging for food or fleeing from enemies to more complex ones such as a desire to experience the culture and prosperity of another place (see Adesina 2019: 112). However, since migration is a human phenomenon it is certainly going to elicit human reactions by hosts to emigres. While humans have the capacity to be hospitable, nice, and even accommodating to the stranger, they also have the capacity to be nasty, hostile, and

B. Matolino (✉)
Department of Philosophy, University of Pretoria, Pretoria, South Africa
e-mail: bernard.matolino@up.ac.za

A. Niederberger et al. (eds.), *Handbook of Migration Ethics*,
https://doi.org/10.1007/978-3-031-89877-8_18

even hateful of the stranger. In this chapter I will seek to explain the hatred of the foreigner, a very specific foreigner, in recent times in post-apartheid South Africa. My explanation for the emergence of that hatred will point to a number of factors that have combined to create an enabling environment for xenophobia to show itself in the way it has. While there could be many reasons and contestations of what the real cause is for the outburst of xenophobia, I will seek to present what I consider to be two most important background reasons to that outburst. The first reason is the general failure of the African continent to integrate economically, politically, and socially. The second reason is connected to the internal problems of South Africa as a country with a long history of division and othering. I will seek to show that when these factors, and others similar to them or even different from them, combine they produce an environment that enables incidents of xenophobia to rise amongst certain sections of the population.

1 Background

The violent incidents of 2008, 2015, and 2019 in some part of South Africa that saw black African foreigners being targeted and attacked caused worry among Africanists about the direction South Africa was taking as a fellow African country. These episodes saw the destruction of foreign owned shops and other property, murder, and demands for foreigners to leave South Africa and return to their countries of origin. While the violence was gruesome and indeed shocking; for anyone familiar with the historical and everyday experience of violence in the country, the violence directed at foreigners would just have been a different target but totally in keeping with occasional violence that some communities are known to be capable of. This violence may range from instantly delivered mob justice where alleged perpetrators of crimes are shot, castrated, stoned to death or burnt alive; all the way to communities taking part in the famous "service delivery protests". The latter is normally about complaints on the poor performance of local government authorities in executing their mandate to deliver services such as water, electricity, roads, houses or whatever the local community claims is due to them (see Tirivangasi and Mugambiwa 2016). Service delivery protests are notorious for members of society targeting and destroying social amenities such as clinics, schools, local government councillors' houses, and the occasional police station to register their unhappiness. Besides this, South Africa is a very violent country with a high murder, rape, carjacking, house invasion, and cash in transit heist rate. It is not an exaggeration to say that almost every South African adult knows someone who was either stabbed or shot to death in a criminal act.

Behind this context is the grotesque reality of apartheid deeply steeped in discrimination, categorisation, and, ultimately, dehumanisation of black people. The effects of apartheid did not end with the official arrival of democracy in 1994 or with the release, from prison, of Nelson Mandela in 1990. The physical structures that separated races into different zones of existence and the accompanying benefits and disadvantages of such discrimination continue to the present day. Formerly white areas remain largely white with the privilege that comes with it while black people are largely confined to township, rural, and informal dwellings. There is a clear division, as Thabo Mbeki once observed, between the South Africa that is white and fairly rich and the South Africa that is black and underprivileged. This material distinction is a continuation of the apartheid policies based on racial classification and blatant racism. Racism itself remains a significant social problem with black people still being discriminated against by certain sections of the white population. In education, there are shameful differences between private and public education and how students are subsequently skilled in these schools. At the university level, there is difference between what are called formerly white and black institutions. Politically, parties are divided between predominantly white and black parties. It is on rare occasion such as in the sporting arena or some other event that South Africans are joined together to celebrate something across racial, social, class, and political lines. Otherwise, the reality of South Africa is one of very public difference between races, the rich and the poor, and ultimately the local and the foreigner.

The point that must be appreciated by this brief sketch of South Africa is that both its history and present is one of divisiveness. Historically, South Africa has just emerged from the brutalities of the most total and evil system of white domination on the continent. Its present remains problematic as it has become the most unequal country in the world (with the majority of the black population severely impoverished). Yet at the same time South Africa has the most developed infrastructural system on the continent, has a vibrant economic system, a progressive constitution and independent courts, a genuine democracy (that shames most African tin pot dictatorships), and opportunities that do not exist for most Africans in their own countries. It is this bit that attracts all sorts of African immigrants.

Many Africans come to South Africa to eke out a living, pursue a dream, find security and safety, and even experience the cultural vibrancy of South Africa. All these are elements that are absent, for many migrants, in their home countries. And South Africa is the only hope they have to lead a decent life. Any African who is motivated to do so, and who can, then attempts to come to South Africa. Thus South Africa is unusual in that it has attracted African migrants who are wide ranging in their social standing. You will find, from the same African country, professional migrants pursuing their professions in ways that could not have been possible in their home countries alongside their kinsmen doing menial work. You can get a university professor and fruit vendor from the same village swearing that their decision to come to South Africa was the best, or a tacit admission that South Africa gives them privilege that they could not find back home. Thus, the distribution of migrants from the African continent is quite widespread in all sectors of South African society.

It is precisely this spread that causes problems in certain sections of South African society. When incidents of xenophobia sprung up in 2008, 2015, and 2019 they were to affect only certain African migrants. The migrants who were affected were township and informal settlements and, at times, inner city dwelling Africans. Affluent, educated, or Africans who lived and worked in 'more refined' areas were unaffected by xenophobia. Also not to be affected were other myriads of races and nations that have migrated to South Africa. What is clear is that xenophobic attacks had a class element to them. It was, as Chris Rock would joke, "broke on broke" violence.

Jonathan Crush and Sujata Ramchandaran have characterised these events as extreme xenophobia. They conceptualise it

> "... as a heightened form of xenophobia in which hostility and opposition to those perceived as outsiders and foreigners is strongly embedded and expressed through aggressive acts directed at migrants and refugees. The antipathy to migrants and refugees is acute and the threat perception attached to their presence is glaring and intense, influenced by numerous myths and biases. Citizens hold migrants responsible for the crime rate, bringing disease, and 'stealing' jobs, services, and resources and view them as being 'illegally' in the country. Moreover, perceptions of a rapid increase in the number of migrants intensify the levels of threats attached to them. Rights and entitlements for residents are directly and in a discriminatory fashion linked to citizenship, drawing the boundaries between those who are seen to belong and others who are not. High levels of antipathy lead to recurrent episodes of violence." (Crush and Ramchandaran 2015: 74–75)

Their assessment is that xenophobia is a well-entrenched attitude that is supported by a number of private and public views and commitments to how foreigners are to be understood. Such attitudes are found in ways that foreigners are treated by government officials responsible for migration, as well as the wording in government statutes that describes immigrants as aliens and illegal foreigners (Crush 2002: 105). These views also extend to locals' attitudes which tend to treat foreigners from countries such as Botswana, Lesotho, and Swaziland with leniency bust show hostility to Zimbabweans, Nigerians, and Somalis. Curiously, McDonald and Jacobs (2005: 130–131) state that Southern African countries of South Africa, Namibia, and Botswana are the troika of xenophobia in Southern Africa.

Crush and Ramchandran divide reactions to xenophobia into two camps. The first camp, of denialists, especially during the presidency of Thabo Mbeki seeks to downplay the violence against foreigners as anything but xenophobia. The violence was explained as ordinary acts of criminality or as isolated incidents. This became the official government position, in part as a response to the criticism by the African Peer Review Mechanism of xenophobic attacks on foreigners. Others blamed the xenophobic attacks on what they considered to be to be South Africa's porous borders that allowed anyone to come and go as they please. The argument was that if tighter border controls were implemented, then there would be low numbers of foreigners stemming the xenophobic flare ups. Crush and Ramchandran regard the latter point as unconvincing since South Africa has draconian border control laws and its deportation rate is comparable to that of Western countries such as the United States.

The second camp is that of minimalists who argue that violence against foreigners must be seen in the context of the general nature of violence in South Africa. Thus, the violence against foreigners, it is argued, is also witnessed against South Africans. Crush and Ramchandran reject this view on the grounds that it fails to properly assess processes involved in xenophobic attacks. They also reject any version of minimalism like ones that attempt to explain xenophobia in terms of the government's failure of transition post 1994—what is in crisis is the project of equality and access to scarce resources. However, Crush and Ramchandran are not convinced by this argument as they ask why xenophobia erupted in only certain (poor) areas and why certain groups were targeted. Further, they also ask why South Africans did not attack each other with the ferocity seen in the way foreigners were attacked. What they find from their research which they describe as xenophobia realism, is that South Africans actually operate under a number of myths such as the belief, by 90% of their respondents, that the country is overrun by a large number of foreigners (such as the number of Zimbabweans believed to be in the country), about 44% of respondents believed that most foreigners in the country were illegal immigrants, 63% believed that the presence of foreigners diminishes opportunities for citizens, 60% believed that the presence of foreigners leads to unemployment for locals, one in two people believed that immigrants contributed to crime, and only one-third believed that foreigners could benefit South Africa's skills shortage. Relying on a survey of 2008 Crush and Ramchandran find that the respondents were almost unanimous in their shared negative stereotyping of foreigners going as far as using apartheid era type of language to create a distinction between themselves and foreigners (2015: 83–84). In addition, South Africans are shown not only to have negative stereotypes against foreigners but to be prepared to either use violence in removing illegal foreigners from their midst or to turn them in to authorities.

My purpose in citing Crush and Ramchandran is to secure two points. The first is that xenophobia is only too real and should not be confused with other events that may be seen as problematic or synonymous with South Africa's social, economic, and political failure as a transitioning society. The point is that there actually exists concrete attitudes and opinions, both among the general population and officials, against the presence of foreigners (Neocosmos 2008: 588–589). The second point is meant to show that there is not a single authoritative way of interpreting and accounting for events that have been described as xenophobic. For example, Bronwyn Harris identifies three dominant ways of accounting for xenophobia which are the scapegoating hypothesis, the isolationist hypothesis, and the biocultural hypothesis. However, she argues that these hypotheses are not sufficient to understand the origins of xenophobia. What is needed, rather, is a new postulation that would take account of South Africa's transition from a past of racism into a future of nationalism (Harris 2002). On the other hand, Harris et al. (2018) provide formidable evidence to show that economic factors are behind anti-immigrant attitudes.

I will not seek to adjudicate over these differences or add to the strength of any side in this debate. What I think productive is to offer an attempt at understanding how disparate factors have combined to enable xenophobic attacks to occur. I think that there is a serious danger in either trying to show the inadequacy of each side or the strength of each side as the matter of xenophobia is complex and the place in which it occurred is also complex. For example Crush and Ramchandran ignore that in the course of xenophobic attacks, very quickly there developed a distinction amongst ethnic groups in South Africa. Nguni speakers launched attacks on fellow South Africans who did not belong to the Nguni groups not because they were genuinely mistaken for foreigners but because they were not seen as South African enough. Hence Darlington Mutanda writes: "When xenophobic violence in South Africa erupted, the victims were not only foreigners in terms of a different nationality but were, in fact, everybody who did not belong to the dominant ethnic groups like Zulu or Xhosa in the main cities. As a matter of fact, members of smaller ethnic groups in South Africa are also viewed as foreigners by fellow South Africans. There have been attacks on South Africans who 'looked foreign' because they were 'too dark' to be South African…" (2017: 280).

In what follows, then, I seek to offer an explanation of what I consider to be some of the most important reasons behind the occurrence of xenophobia. While I do not claim these reasons to be exhaustive, I think it is worthwhile to consider them key to fanning the flames. Further, I think it is unproductive to contest any reason that is forwarded as an explanation for xenophobic attacks because of the complexity of xenophobia which is accompanied by its suddenness in appearance and intensity in violence. What may be true from one perspective may be startling for another perspective. For example, my first reason takes the cause of xenophobia to places that are far from South Africa. The failure of these places directly contribute to the phenomena of xenophobia and maybe those places must be bold enough and take responsibility for their failure and the ripple effect caused by such failure.

The first explanation for xenophobia lies in the history of postcolonial sub-Saharan Africa as a remarkable place of failure. Post-independence political, economic, and social instabilities that have become defining characteristics of many states, have contributed to the so-called migration wave that South Africa has seen from very specific countries. For example, Zimbabwe's economic and political meltdown from 2000 onwards is said to have caused between one and three million of its citizens to move to South Africa (Serumaga-Zake 2017: 2) There is a simple fact that must be acknowledged if the discussion is going to be sincere and productive. There are many Africans who are impatient with their countries and are set on leaving. Whatever their actual final destination, it depends on logistics and easy access. It may also depend on other factors such as what one intends doing when they arrive at their intended destination and what they envisage themselves becoming. It is quite tragic to have to admit that for a very large number of immigrants in South Africa, who ultimately became victims of xenophobia, they were people who fled their homes to get somewhere they could just about do anything worthwhile with their lives. The property that was destroyed is not stock exchange kind of wealth but little corner shops, front house stalls, downtown small time enterprises. It is indeed a strange state of affairs which leads one to ask this unusual question: Do Somalians, Nigerians, Zimbabweans and whoever whose goods were destroyed—dream of leaving their familiar surroundings and their home so that they can sell loaves of bread, tinned fish, washing soap, and cigarettes in South Africa? Is the ultimate dream of the little hair salon shop owners and downtown sidewalk barbers to leave their homes to either braid or cut South Africans' hair in Johannesburg?

The story, the real story, is the long history of the failure of the post-independence African state and the failure of the continent to integrate as a block. The idea that

Africa must be one has received a great deal of attention from the time of Ghana's independence. There are two ways we can understand that desire. The first is political and the second is economic. The political desire is to be seen in Nkrumah's declaration that Ghana was not free unless the entire continent was free. This desire was based on the view that emphasised on colonialism's power to impede self-determination and political development. Nkrumah was aware of the natural resource potential spread all over Africa. However, in order for those resources to be transformed into viable economic activity that would benefit all Africans, there would be need for capital to be pooled together, for all economic barriers to be removed, and for the development of a single currency. This would lead to the continental economic development of Africa (Nkrumah 1964: 151–157). However, Nkrumah's dream of a free and united Africa never materialised. While all African countries eventually gained political freedom from their former oppressors, the problems that Nkrumah had feared would haunt Africa, did actually come around to haunt Africa. Neo-colonialism and lack of political and economic unity have kept Africa far from realising the potential Nkrumah had thought possible.

I will steer clear of articulating the reasons behind this failure. On the contrary, what I wish to pursue are the consequences of this failure on the general population. The most visible consequence of this failure has been Africa's inability to use its natural and human resources to modernise and develop. The reality is that the majority of African people live in rural areas where they are mainly engaged in subsistence farming. This lifestyle and its attendant economic activity is not suited for development or empowerment. It just ensures that, if the rain season is good and there are no diseases, they have enough to eat and a surplus to either sell or save as well as a few beast which are a symbol of wealth. Land is not properly utilised to advance a developmental agenda. At times, there are pseudo romanticised patrimonial attachments to the land which lead to restrictive views of the utility of the land. The very fact that the majority of the African population live in rural areas where there is largely no running water, electricity, decent roads, mechanised systems of production, and economically sound systems of using available resources, simply shows Africa to be an underdeveloped continent. The other part of the population that lives in cities and towns, is largely an impoverished lot with neither resources to empower their lives nor the means to escape widespread poverty. While there are pockets of development and displays of ostentatious wealth, in general, African populations are underprivileged as they live in depressed or stagnant economies. Even high annual economic growth rates for poor countries such as Mozambique and Angola have done nothing to transform the living standards of ordinary citizens. African lives are vulnerable to a lot of avoidable tragedies which are generated from the poor economic indicators of their countries; joblessness, limited financial rewards in the informal economic sector, and unrewarding hard labour.

The failure to integrate African economic activities that would have strengthened the entire continent's economic standing also means that most member states of this continent are economically weak and open to exploitation by their former colonisers. This in turn has a direct effect on citizens of these economically weak nations. These citizens then, particularly youths, are left with little option besides searching for better opportunities. This may mean that they move to the capital of their respective countries or other urban areas or take a step further by leaving their country. Thus, the movement of peoples is mainly for purposes of finding better opportunities. This has resulted in congested cities, neglected rural areas (as the most able bodied leave), and indeed attempts to leave for a country such as South Africa.

Since economic development has been uncoordinated and patchy, large areas and a significant number of people have been left living under conditions that are close to poverty or are actually impoverished. In reaction to these experiences, it is in human nature to seek better opportunities. However, these opportunities happen to be concentrated in already overpopulated zones that have their own internal problems to do with creating opportunities for locals. The new arrivals only add to problems that already exist. If it is internal migration, the new arrivals add to the housing crisis, overpopulation, the jobless, the army of informal traders and large numbers of people just waiting for something good to happen to them. This situation drives up social problems. If the migration is externalised, the new arrivals also add to the problems that already exist in the host countries. As the local population's excitement of hosting a few foreigners begins to wear off, they become more impatient with what they perceive to be an invasion. The invasion is nothing more than competition for survival by accessing menial jobs and unrewarding opportunities.

The economic meltdown and political problems of many African countries have been the major drivers of an upswing of African immigrants' arrival in South Africa. These immigrants range from refugees fleeing wars to ordinary people who are frankly fed up with countries of their birth. The plain truth is that had their countries not failed, but had forged a common path to genuine political freedom and economic empowerment of their citizens, then there would have been no need for one country in the entire sub-Saharan region to appear as the desired paradise that those in search of improved chances would automatically flock to. As David N. Field writes: "Migrants from the rest of Africa in South Africa can be found in all categories, with significant numbers falling into the category of survival migrants. Many Zimbabwean migrants can be categorised as survival migrants. In this case, the government has at times gone beyond its obligations to international law to make provisions for them. Other migrants may be understood as coming from the bottom end of the category economic migrants" (2017: 3).

The failure of Africa to integrate has not only meant that politically and economically, there are inequalities and problems of varying degrees right across the continent. It has also meant that Africans have remained restricted to their narrowly defined national identities and probably are as ignorant of each other, if not resentful of each other in damaging ways. The intention of the colonial mission was to divide Africa, to weaken its possibilities and to arbitrarily assign nationalist identities on citizens of this continent. This did succeed and has been perpetuated by Africans themselves. Some countries, are frankly not worth that title. With GDPs/GNPs that are less than some doughnut companies, and budgets that are dependent on aid, what makes those countries believe that they are worthy being called so? Could it be the case that political independence has turned out to be a trap that Africa can't escape from? Could it be the case that colonialism traumatised Africans to a point where they see co-operation or giving up a bit in order to gain another bit—to be effectively ceding their independence? I do not know the answers to these questions but what I observe are the contradictions of being African and how those contradictions have manifested themselves in how African countries operate as if they were not faced with the same fate that demands that they band together in confronting and overcoming this situation.

South Africa, like any other formerly colonised African country has its own internal problems of social, political, and economic transformation into a post-apartheid state that seeks to serve its citizens as equals. Some of these problems are commonly shared with other formerly colonised territories but others are peculiar to South Africa. Xenophobia, as a political and social problem, is just another layer to an already combustive cocktail of inequality, scarce resources, anger, violence, and at times resentment. What xenophobia signifies is a summary of what at first appears as senseless and unjustified targeting of certain groups in Africa. Yet if one looks beneath that veneer of senselessness one encounters layers upon layers of frustrated and disempowered citizens of a continent divided politically, socially, and economically. This division and lack of co-ordination shows itself in a variety of ways in different spots on the continent, and xenophobia is just one of them.

2 Xenophobia and Violence

Another possible reason for the explanation of xenophobia and the violence that came with it might be found in the social and political history of South Africa as a society that has a long history of structural division and discrimination. This discrimination has mainly been between whites and blacks but also between blacks (Nyamnjoh 2010). South Africa has had a long contest of insider and outsider identities, of who properly belongs and who does not. Even amongst the white population there is a long-standing history, traced to the Anglo-Boer war of the narrative of outsider and insider. Thus, the Afrikaans fighters saw themselves as defenders of the land, their land, their country against the British colonialists and invaders. When apartheid was instituted, English speaking descendants watched this spectacle as an Afrikaner aberration and the Afrikaner enforcers of apartheid were very suspicious of English-speaking white folk. Indeed, liberalism as a political tradition, became associated with English speaking whites, while Afrikaner officialdom defended the idea of the *volk*. The division was so prominent to the extent that there were provinces that were distinctly Afrikaner while certain provinces were viewed as English.

Since victims of xenophobia are black people, it is important to linger around the divisions between and conflicts among black South Africans both during and after apartheid. If we linger long enough around this conflict we might begin to get a grasp of the background in which xenophobia occurs. During apartheid, the violence perpetuated by black people on black people was quite extraordinary in both its scale and nature. The Inkatha Freedom Party (IFP) and the African National Congress (ANC) emerged as the main players. David B. Ottoway writing for the

Washington Post on the second of May 1991 reports the violence as follows: "At least 70 people have died in the fighting since last weekend, more than 600 since the start of the year and more than 6000 since the ferocious struggle began five years ago in Natal province, the ancestral Zulu homeland." Stuart J. Kaufmann states that while the transition from apartheid is rightly applauded, what is ignored is the violence that happened between 1985 and 1995. He divides this decade of violence into two phases, the first phase occurs between 1985 and 1986 pitting the ANC and its surrogate the United Democratic Front (UDF) against the apartheid government and its army. The second phase was between 1987 and 1995 pitting Inkatha, on one hand, against the ANC on the other hand. Kaufmann argues that Inkatha enjoyed the protection of the apartheid state machinery, which meant that the government of the day was not an innocent bystander in the violence that ensued. The major parts of the country where deaths occurred, which Kaufmann describes as civil war, were parts of KwaZulu-Natal and parts of Gauteng. In total, 23,888 people were killed between 1985 and 1995 (Kaufmann 2017: 509).

The details of events that occurred in this civil war are truly gripping for the nature of their violence and revenge cycles. However, what is even more disturbing is the division that black South Africans created between themselves to cement practices of othering. For example, Kaufmann points out that Inkatha was a Zulu based party that saw its attacks on ANC as an attack on a Xhosa party. What is even frightful is that the perpetrators and victims were both black people who were only divided, particularly in the then Natal midlands, by party affiliation. Since Xhosas and Zulus come from the same Nguni-speaking group one would have expected tolerance if not recognition of each other. With the end of apartheid, attempts were made to both understand and right this situation through the Truth and Reconciliation Commission (TRC), but the outcome of that process was neither universally accepted nor did it succeed in ending the culture of violence. Violence has continued to this day, with KwaZulu-Natal outstanding in its murder rate of local councillors. Sadly, this culture of violence has become embedded in everyday dangers of living in South Africa as the country records a daily murder rate of more than 50 victims. Most of these victims are shot or stabbed.

What this brief presentation serves, is to show three things; violence has a long and uninterrupted history, black on black violence did not start with South Africans targeting foreigners, and daily victims of violence are black South Africans. While it is tempting to categorise this violence according to period and purpose, I suggest that it is more useful to see violence in terms of its frequency. For whatever political and social forces that may point to foreigners as legitimate targets of attack, violence seems to be an easily and readily available form of either registering discontent or an expedient way of doing things. Xenophobia, then, is not necessarily planned or thought of as a violent act—but a continuity of the very violence on which society is constructed and which is always a trusted form of settling disputes. This is not meant to excuse the violence that has happened against foreigners, on the contrary, my point is that when we compare incidents of xenophobia and calculate the numbers of victims against the sort of violence that black South Africans have suffered from and that they continue to suffer from, we might get a better grasp why xenophobia always ends up in murder.

To fully grasp the nature of violence, we must also look at its occurrence in other forms besides murder of foreigners and pillaging of their stalls. South Africa has the most liberal and progressive constitution in the world. It is a constitution that guarantees human rights and actively protects claims to difference amongst citizens. South Africa recognises 11 official languages, actively protects indigenous practices, and recognises every person's right to cultural or religious identity. However, the reality of the matter is that not all 11 languages receive the same privileges and protection. English and Afrikaans remain dominant languages and indigenous cultural practices are still confined to the margins. There is violence against indigenous people's languages and cultural practices through marginalisation and exclusion.

Another example can be seen in the treatment of same-sex identifying individuals. While South Africa is one of the few countries that constitutionally recognises individuals' rights to sexual orientation, identity, and expression; the reality is that heteronormativity is commonly upheld as a standard and any other identification is seen as deviance which might be harshly punished. This is particularly so with females who identify as lesbians. A regrettable number has been victims of the violence of what is described as corrective rape (Mwambene and Wheal 2015: 59–61). The practice is carried out by black males who sexually force themselves on openly lesbian females in the belief that these women are in need of a real experience with men.

Whichever topic we choose, therefore, be it a soft topic of language and cultural rights or a hard topic of gay rights; there is always going to be an aspect of violence. The social and political landscape in its present day formulation and historical appearance, is littered with acts of extreme violence. An underlying theme and enabler of violence is othering. In that othering a narrative is created of what is authentic and what is not, who is closer to the norm and who is not, and those who happen to be identified as outside the norm are targeted.

Yet the strangest thing, as Njabulo Ndebelo correctly notes in his *Cry of Winnie Mandela*, if one looks at violence as something that is executed by the South African public towards any target they have chosen, one must also notice and be frank to acknowledge the warmth that South Africans extend towards the other. Thus, Ndebele asks what is it that makes the South African able to embrace and kill at the same time. Winnie Mandela, the protagonist of his novel, who stands accused of heinous crimes is then able to stare down South Africa and pronounce: I am you. Boldly, she states:

> "I am your pleasure and your pain, your beauty and your ugliness. Your solution and your mistake. Your hell and your heaven. I am your squatter camp shack and your million rand mansion. I am all of you who maim and rape. I am all of you who give love and succour. I am your pride and your shame. Your honour and your humiliation" (Ndebele 2003: 113).

This pronouncement indicates the contradictions of South Africa. Indeed, the foreigners who were attacked expressed surprise that they were even attacked. They had happily lived in the communities that had adopted but had later turned against them by demanding that they must leave. This was in complete contrast to those communities' famed humanity.

My own guess is that the speed with which South Africa moves from a warm embrace to violence on anyone, foreigner included, has to do with the trauma of its own past. Attempts by the TRC to deal with this trauma could have been inadequate as the wounds caused by the history of violence may have, to an unusual degree, devalued human life. It must be borne in mind that in some of these places where xenophobia occurred, other ghastly incidents occurred during apartheid such as necklacing (the placing of a tire around a victim, dousing them with fuel, thus burning them to death). It could be the case that what needs to be understood in greater detail is what the cumulative and continuing impact of apartheid's violence has been on the nation. It could be the case that this is one of the most subtle effects of apartheid which rears its ugly head in different situations for different sections of the population.

3 Imagining an African Future Without Violence

Xenophobia is just but one of the many instances of Africa's capacity to self-destruct through acts of counterproductive violence. What I hope to have provided is an explanation of how this particular violence may be contextualised. If there is any hope of forestalling this occurrence in the future, there is need to address some of the real issues that have directly and indirectly contributed to the emergence of xenophobia. Since I have divided the problems into two causes; the internal and external—I will also propose two solutions that may contribute to the alleviation of this problem.

At the continental level, there exists a dissonance between citizens of this continent, as political subjects and the continental body, the African Union. While, no doubt, the Union has done some important work its achievements are far from significantly serving the needs of the ordinary citizen. There are many issues of a very mundane and existential form that need to be fixed as a matter of urgency. While there are some advantages that may play to Africa's favour, these advantages need to be turned from mere potentials into actual gains. For the continent I will divide my assessment into three considerations; the rule of law and democracy, economic co-ordination, and healing wounds of the past. For South Africa I will only concentrate on one aspect: transformation from apartheid to democracy.

Africa's diversity in the application and respect for the law is well-documented. We see at one extreme functional democracies that take their constitutions seriously such as South Africa, and extremes where the rule of whatever envisaged law depends on the power of one person. In between these extremes are dictatorships, part time democracies, aspiring democracies, recovering warzones, aspirant dictatorships, and brutal oligarchies. Yet for Africa to function it needs to have stable countries that will contribute to the idea of a viable Africa as a place of the rule of law. If we have such divergent approaches, in terms of the application of the law and subsequent respect for it, the possibility of a well-coordinated continent disappears. I am not arguing for a single legal framework, rather I am interested in the development of a standard. The sort of standard I have in mind is one that emphasises predictability of the operation of the law and respect for such basic principles as no party will be a judge in his own case. When so many countries are basically permitted to recede into spaces that see the rule of law as tool that can be disregarded then all forms of problems begin to arise. Issues of corruption and dysfunctionality which lead to despair are primarily caused by disrespect for the law. The law as an instrument that governs society, that ensures justice, fair distribution of goods and opportunities must be binding and must bind each country together which in turn will have a positive effect on the continent. The failure of the continent to integrate and be of service to its people is in a very significant way due to the fact that the different regions of the continent have different approaches to how the law must be allowed to operate as the law. The law is related to the idea of democracy as a political principle since the greatest defender of democracy is the stipulation given by the law of what it is for a dispensation to be democratic. If the law is undermined then democratic principles, grounded in the law, are also undermined. The collapse of the law is the collapse of any chance at democracy. The second consideration is how Africa needs to co-ordinate its economic function so that it operates as a block in dealing with the rest of the world. This will enable maximisation of utilisation of all resources (human and natural). This can easily be achieved by promotion of co-operation between countries in the provision of resources and services to the global market. The myriad calls for the development of a single currency and stabilisation of economies has been repeated so many times, and those responsible for the realisation of such imperatives need to be seen to be working. And the last consideration is the need for Africa to heal. Africa is a place that has suffered so many human-made tragedies and its major complaint has been against the history and politics of colonialism. This event has changed the course for Africa for the worst. Overcoming the effects of this commonly shared tragedy has proven to be a brutalising process of its own that has produced violent spin-offs such as wars and xenophobia. As long as Africa remains trapped in the pain of the past, as long as Africa is unable to deal with that pain, and as long as its relationship with its former colonial powers is premised on the facts of colonial experiences, Africa will remain trapped. The third aspect relates to how Africa has to heal from the wounds of its colonial past. These wounds are so deep that they cause hurt every time their scars are made public. To heal from these wounds Africa needs to confront its colonial past with the sincerity it deserves. Such sincerity has to involve quantification of the gains of independence and benefits from such attainment. It will have to involve a frank assessment of whether Africans themselves are truly self-determining or are still bound by the reality of the colonial era. That episode, I think, has not been frankly dealt with beyond the promises of the rhetoric of independence.

South Africa, as an analysand, has contributed to the pains of the continent by providing senseless contradictions. For such a developed country to continuously sink to dehumanising vagaries from time to time as if it was seeking to confirm its own African standing has discredited its standing. Xenophobia, alongside other internal contradictions such as being the global leader in the massification of poverty of its citizens, racism, and violence points to an unfinished project of transformation. While the peaceful nature of transformation in 1994 was praiseworthy, it could be the fact that it was not complete. It is only South Africa that can face its deep-seated othering.

4 Conclusion

Xenophobia, I have sought to argue, is not a problem that arises as a stand-alone problem. It occurs as a result of the consequences of issues that are both near and far the site of violence. In order to overcome the flame of xenophobia, there is need to adjust the whole of Africa, not just South Africa, to get rid of all factors that lead to most vulnerable citizens of the continent turning on each other.

Literature

Adesina, O. S. 2019. Globalization, Migration and the Plight of Nigerians in South Africa. In *Nigeria-South Africa Relations and Regional Hegemonic Competence*, ed. O. Taylor, 109–127. Cham: Springer.

Crush, J. 2002. The Dark Side of Democracy. Migration, Xenophobia and Human Rights in South Africa. *International Migration* 38:103–133. https://doi.org/10.1111/1468-2435.00145.

Crush, J., and S. Ramachandran. 2015. Migration Myths and Extreme Xenophobia in South Africa. In *Global Migration. Old Assumptions, New Dynamics*, ed. D. Arcosta Arcarazo and A. Wiesbrock, vol. 3, 71–96. Santa Barbara: Praeger.

Field, D. N. 2017. God's Makwerekwere. Re-imagining the church in the Context of Migration and Xenophobia. *Verbum et Ecclesia* 38:1–8. https://doi.org/10.4102/ve.v38i1.1676.

Harris, B. 2002. Xenophobia. A New Pathology for a New South Africa? In *Psychopathology and Social Prejudice*, ed. D. Hook and G. Eagles, 169–184. Cape Town: University of Cape Town Press.

Harris, A. S., et al. 2018. The Economic Roots of Anti-immigrant Prejudice in the Global South. Evidence from South Africa. *Political Research Quarterly* 71:228–241. https://doi.org/10.1177/1065912917734062.

Kaufmann, S. J. 2017. South Africa's Civil War, 1985–1995. *South African Journal of International Affairs* 24:501–521. https://doi.org/10.1080/10220461.2017.1422012.

McDonald, D. A., and S. Jacobs. 2005. (Re)writing Xenophobia. Understanding Press Coverage of Cross-Border Migration in Southern Africa. *Journal of Contemporary African Studies* 23:295–325. https://doi.org/10.1080/02589000500274050.

Mutanda, D. 2017. Xenophobic Violence in South Africa. Mirroring Economic and Political Development Failures in Africa. *African Identities* 15:278–294. https://doi.org/10.1080/14725843.2017.1291325.

Mwambene, L., and M. Wheal. 2015. Realisation or Oversight of a Constitutional Mandate? Corrective Rape of Black African Lesbians in South Africa. *African Human Rights Law Journal* 15:58–88. https://doi.org/10.17159/1996-2096/2015/V15N1A3.

Ndebele, N. S. 2003. *The Cry of Winnie Mandela*. Claremont: David Philip.

Neocosmos, M. 2008. The Politics of Fear and the Fear of Politics. Reflections on Xenophobic Violence in South Africa. *Journal of Asian and African Studies* 43:586–594. https://doi.org/10.1177/0021909608096655.

Nkrumah, K. 1964. *Africa Must Unite*. New York: International Publishers.

Nyamnjoh, F. B. 2010. Racism, Ethnicity and the Media in Africa. Reflections Inspired by Studies of Xenophobia in Cameroon and South Africa. *Africa Spectrum* 45:57–93.

Ottaway, D. B. 1991. "ANC Vs Inkatha. Anatomy of a Slaughter". *Washington Post* May 2, 1991, https://www.washingtonpost.com/archive/politics/1991/05/02/anc-vs-inkatha-anatomy-of-a-slaughter/18376419-7fe7-4539-8a6d-4c7012be8793/.

Serumaga-Zake, P. A. E. 2017. Migration and Tourism. The Challenges of Zimbabwean Diaspora in South Africa. *African Journal of Hospitality, Tourism and Leisure* 6:1–20.

Tirivangasi, H. M. and Mugambiwa, S. S. 2016. "Citizenship Crisis or Xenophobia? A Critical Analysis of Service Delivery Protest in South Africa". *SAAPAM Limpopo Chapter 5th Annual Conference Proceedings*, https://pdfs.semanticscholar.org/8e14/5f0cf150a1bda0e46aba46efe8a6c7205851.pdf (last checked May 30, 2020).

Gender and Migration

Peter W. Higgins and Amy Reed-Sandoval

Abstract In this article, we argue that gender is relevant to the moral evaluation of migration policies concerning (a) immigrant admissions, (b) the status of migrants (documented and undocumented) in receiving countries, (c) refugee admissions, and (d) border enforcement and undocumented migration. We draw on existing feminist and other philosophical work to make the case that gender ought to be treated as a significant category of analysis by theorists writing on migration ethics in each of these four areas. We conclude by considering what policies gender justice requires in the area of international migration and propose several questions we believe philosophers are especially well equipped to address.

Mainstream philosophical work on international migration typically does not treat gender as an important category of analysis. Migrants are often theorised about as though they have no gender, with the consequence (given the tendency to regard the standard human person as male) in many cases that all migrants are conceived implicitly as having the social, political, and economic wherewithal of cisgender, heterosexual men. However, virtually all aspects of international migration on which philosophers write have gendered dimensions; that is to say, the conditions and consequences of international migration vary, sometime significantly, for people on the basis of their gender (by which we mean to include their self-identified gender, their externally ascribed gender, and other gender-related aspects of their social identity such as their sexual orientation). As expressed by the International Organization for Migration,

> "Gender influences reasons for migrating, who migrates and to where, how people migrate and the networks they use, opportunities and resources available at destinations, and relations with the country of origin. Risks, vulnerabilities and needs are also shaped in large

P. W. Higgins (✉)
Department of History and Philosophy, Eastern Michigan University, Ypsilanti, MI, USA
e-mail: phiggin1@emich.edu

A. Reed-Sandoval
Department of Philosophy, University of Nevada, Las Vegas, NV, USA
e-mail: amy.reed-sandoval@unlv.edu

A. Niederberger et al. (eds.), *Handbook of Migration Ethics*,
https://doi.org/10.1007/978-3-031-89877-8_19

part by one's gender, and often vary drastically for different groups. The roles, expectations, relationships and power dynamics associated with being a man, woman, boy or girl, and whether one identifies as lesbian, gay, bisexual, transgender and/or intersex (LGBTI), significantly affect all aspects of the migration process, and can also be affected in new ways by migration". (International Organization for Migration 2020)

In this article, we argue that gender is relevant to the moral evaluation of migration policies concerning (a) immigrant admissions, (b) the status of migrants (documented and undocumented) in receiving countries, (c) refugee admissions, and (d) border enforcement and undocumented migration. In the section that follows, we draw on existing feminist and other philosophical work to make the case that gender ought to be treated as a significant category of analysis by theorists writing on migration ethics in each of these four areas. In the concluding section, we consider what policies gender justice requires in the area of international migration, proposing several questions we believe philosophers are especially well equipped to address.

1 The Relevance of Gender in Several Areas of Migration Policy

1.1 Immigrant Admissions Policies

Immigrant admissions policies are those by which non-refugee migrants are permitted (or denied permission) to enter a country of which they are not a citizen for the purpose of permanent residence or temporary work. States' immigrant admissions policies often favor applicants for admission on the basis of criteria such as economic potential (e.g., formal training or tertiary education in particular fields of work), immediate family relationships with citizens of the receiving country, and cultural affinity. In contrast, applicants for admission with a criminal history, certain health conditions, or certain religious or political affiliations (which are sometimes assumed on the basis of their country of origin) are often denied admission. Given that most states limit the total number of immigration visas they grant annually, even prospective immigrants who satisfy the relevant criteria of admission may not be granted admission in many cases.

Gender is relevant to the moral assessment of immigrant admissions policies in at least two ways. First, (cisgender, heterosexual) men usually—simply by virtue of having greater control over or access to the resources that international migration typically requires—have greater means than others by which to exercise freedom of international movement, regardless of any particular state's admissions policies (Higgins 2013, 2017). Ayelet Shachar comments that freedom of international movement requires, on the part of migrants

"sufficient knowledge of where to move in order to improve one's economic prospects and to gain greater political freedom, in addition to basic linguistic and occupational skills, the willingness and ability to leave family members and a familiar cultural context behind, and related factors that deeply, and often unequally, shape the potential mobility of men and

> women in international patterns of migration. That is, there is a problem of gender, resource, and information asymmetries, which could potentially create a bias against those who may be most in need of the redistributive (and arguably also the liberating) dimension of global migration" (Shachar 2009).

This suggests that (changes to) immigrant admissions policies, even when formally gender neutral, will tend to leave (cisgender, heterosexual) men with greater options than they leave others. In examining the justice of immigrant admissions policies, it is thus necessary to apply a gendered lens to ensure an equitable distribution of opportunities.

Second, the criteria by which states admit or exclude prospective migrants may be biased in favor of (cisgender, heterosexual) men. Indeed, many countries have a long history of gender bias in immigrant admissions. Shelley Wilcox catalogues instances of gender bias in immigration policy, focusing in particular on the tendency to regard labour traditionally performed by women as "unskilled" and thus limiting women's options for migration or (if admitted) permanent residence (Wilcox 2005). While Wilcox focuses on the US, gender bias in judgments of the skilled-ness of work is the norm in countries of the global North, where increasingly large shares of visas allotted annually are reserved for "the skilled" (Kofman 2007). Similarly, Desiree Lim demonstrates that admissions criteria that favour the "highly-skilled" tend to favour men and thus indirectly discriminate against women migrants. She argues that policies that discriminate in this way are unjust because they perpetuate stereotypes of women as unskilled and as dependent on men, thus worsening existing disrespect toward women (Lim 2019).

In contrast to immigrant admissions policies that exclude on the basis of race, ethnicity, religion, sexual orientation, or political values, those that favour applicants for immigration on the basis of skilled-ness are commonly regarded as morally acceptable by liberals (see Carens 2013). However, attention to the gendered effects of such policies calls this conventional assumption into question. The gender-disparate effects of skills-based admissions policies (combined with the fact that, as Wilcox notes, women are the majority among the categories of immigrants with arguably the strongest moral claims to admission—refugees and immediate family members of current residents (Wilcox 2005)) reveals that questions about the justice of immigrant admissions policies cannot be thoroughly addressed in a "gender-neutral" analysis.

One of the most common channels through which women migrate internationally (both with and without the receiving country's authorisation) is employment in "carework" fields: as nurses, home health workers, maids, nannies, etc. Migration for carework employment is highly gendered not only because of gender roles and gender socialisation in sending countries but also because of the pervasive assumption in receiving countries that women are uniquely, even naturally, suited to such work. Migration for carework employment is not a specifically South-North phenomenon: a large percentage of carework migration is to Persian Gulf states, Singapore, and Hong Kong (Silvey 2009; Ghaeus 2013; Dumitru 2016; Aragon and Jaggar 2018). We discuss the consequences of carework migration for migrant women and receiving countries in the following subsection.

In this subsection, we discuss the consequences of women's carework migration for sending countries. The consequences of emigration for sending countries are widely discussed in the philosophical literature on the justice of states' immigrant admissions policies, which focuses on the phenomenon called "brain drain": the large-scale migration of highly-skilled professionals from, typically, countries of the global South to countries of the global North, and the effect this has on economic and human development in sending countries (see Brock 2009; Brock and Blake 2015). While the migration of women professionals (including nurses) is accounted for in these discussions, feminist theorists of international migration have sought to bring attention to another kind of drain caused by women's international migration: "care drain" (Hochschild 2000, 2002; Parreñas 2001). In general, women are more likely than men to have significant (socially assigned) responsibility for the care of others, including both children and aged parents. When women migrate internationally for employment purposes, they are unlikely to (be permitted by the receiving country to) bring those for whom they care with them. This results, either directly or indirectly, in the phenomenon controversially referred to as care drain. (The controversy stems from the fact that conceiving of this drain as distinct from that of the brain suggests that the labour involved is unskilled or that it does not exercise one's intellectual capacities—see Dumitru 2014, 2016.)

When men migrate internationally, they are significantly less likely than women to leave behind children, aged parents, or others who rely primarily on them for care. Regardless of whether others step in to care for dependents left behind, one consequence of women's international migration is, in some cases, a disruption to deeply meaningful relationships and difficulty maintaining important emotional ties (Kittay 2008, 2009; Weir 2008; Ghaeus 2013; Eckenwiler 2014). When women migrate internationally for carework employment, the effect is therefore sometimes referred to as the "global heart transplant" (Hochschild 2002; Kittay 2008). In many cases, others do step in to care for dependents of women migrants; typically they are poorer women in the migrant's country of origin for whom international migration is economically infeasible. Even in such cases—which instantiate what is called a "global care chain"—a care drain nevertheless occurs, as the children or aged parents of the woman who cares for the migrant's dependents resultingly experience a "care deficit" (Hochschild 2000, 2002; Parreñas 2001; Kittay 2008, 2009; Weir 2008; Ghaeus 2013; Eckenwiler 2014; Meghani 2016).

Discussions of the so-called care drain can misleadingly suggest the moral blameworthiness of migrant women. However, an examination of the "push" and "pull" factors that lead to women's international carework migration reveals complexity that stands in the way of assignment of blame to individual actors. Regarding the "pull" factors, there are at least three contributors: the increasing paid workforce participation of women from middle-class and wealthy households (those that can afford to hire others to perform labour in their home) over the last several decades; the reluctance of men from such households to respond by performing an equitable share of domestic labour (Weir 2008; Jaggar 2009; Ghaeus 2013; Meghani 2016); and declining public support to provide for the care needs of the residents of relatively affluent countries, often driven by the application of neoliberal principles that oppose state responsibility for social welfare and favour the privatisation of public goods (see Jaggar 2002, 2009). These "pull" factors create demand for migrant labour to perform the essential work of social reproduction. Simultaneously, lack of economic opportunities in countries of the global South—itself generally exacerbated by application of the same neoliberal principles in the form of Structural Adjustment Programs (Jaggar 2009; Kittay 2009; Eckenwiler 2014; Aragon and Jaggar 2018)—combine with (a) the actions of international labour recruitment agencies and (b) the creation of programs, in some countries that count migrant remittances as a major source of income (most prominently, the Philippines), whose purpose is to train residents to perform work (e.g., nursing) in other countries, to "push" people toward international migration (Eckenwiler 2009, 2011, 2014; Jaggar 2009; Kittay 2009; Crozier 2010; Dumitru 2016; Meghani 2016; Aragon and Jaggar 2018).

Theorists of migration ethics disagree about how receiving countries should respond to harm caused by brain drain—for example, whether it provides justification for excluding prospective migrants in some cases (see Higgins 2015). Nevertheless, even those who deny that considerations of brain drain should influence whom states ought to admit agree that the economic and human development consequences of brain drain are relevant to the moral duties of receiving countries. For example, Christopher Heath Wellman argues that wealthy receiving countries ought to compensate poor countries whose highly skilled emigrants they receive, even though they have a right to admit them (see Wellman and Cole 2011). Similarly, the harms associated with "care drain" are surely relevant to the moral assessment of the policies of countries to which people with (socially assigned) primary responsibility for the care of others migrate (even if one denies that considerations of "care drain" justify exclusion).

1.2 The Status of Migrants in Receiving Countries

In the previous section, we discussed women's opportunities for international migration, its conditions, and its consequences for sending countries. In this section, we turn to the effects of international migration for migrant women and for receiving countries.

Women who migrate internationally, and particularly women of colour, are directly targeted by intersecting racist and sexist stereotypes which often "spill over" to citizen and legal resident members of the same ethnoracial group as the immigrants in question. Some such stereotypes refer to pregnancy, childbirth, and child-rearing on the part of immigrant women of colour and other members of the same ethnoracial group. Feminist philosophers have explored, for instance, how the gestating bodies of immigrant women of colour are often considered "always, already perverse" (Cisneros 2013) in the United States, where attitudes and stereotypes often lead to a pernicious form of social "illegalization" (Reed-Sandoval 2020). Meanwhile, in the U.S., this harmful stereotyping often impacts the children of Latin American women and other immigrant women of colour, who may be referred to with the slurring term "anchor babies" (Chavez 2017). Meanwhile, Muslim women who veil are often directly targeted by intersecting racist and sexist anti-immigrant prejudices. The Canadian province of Québec recently went so far as to ban veils for certain members of the population, thereby centring Muslim women in a larger debate about the extent to which white Québec society ought to accommodate the religious practices of immigrants (Velji 2019). Falguni H. Sheth has argued that Muslim women who wear the hijab are construed by Western liberal regimes as "unruly subjects," or those who "are perceived to actively violate the ideal of the good (liberal) citizen" (Sheth 2006, 2019).

Historically, women's migration has been associated with family migration: prior to recent decades, most women who migrated internationally did so following their husband, who may have migrated for work or returned home after foreign military service. Women have also migrated as "mail-order brides", travelling abroad to join a husband they have known only for a short time through postal correspondence or (as is typical now) electronic communication. Many countries' immigration laws privilege (heterosexual) marriage, but women have been much more likely than men to migrate internationally for marriage (Jaggar 2009). One consequence of this is that many immigrant women find themselves in situations of legal and economic dependence on their citizen or legal permanent resident husband and disproportionately vulnerable to domestic abuse. As Uma Narayan argues, male citizens of receiving countries sometimes seek a foreign wife because they believe (in accordance with often racialised gender stereotypes) that they will find a submissive partner and because they can require such submissiveness: failure of the marriage undermines the basis of her visa and thus makes her likely to be deported (Narayan 1995). Immigrant women's vulnerability to domestic abuse is compounded by "gender-neutral" immigration laws that increase immigrant women's dependence (e.g., they are often forbidden to work legally for wages) and makes exit from the relationship without deportation difficult to achieve, given the legal, economic, and social resources of many immigrant women. As Narayan argues, these laws governing the rights of immigrants who enter the country on the basis of marriage are *falsely* gender-neutral because they are based on the tacit, unreflective assumption that those who must navigate them possess the legal, economic, and social wherewithal of white male citizens.

Women increasingly migrate internationally for employment and, as noted in the previous section, this employment is often in the carework sector. Migration for employment in the carework sector of wealthy economies enables many women to earn an income (much of which is remitted to non-migrating family members) that is substantially greater than what they can earn in their country of origin. However, migration for carework employment is attended by a number of serious harms, the analysis of which is a major theme in feminist writing on international migration. Among the first of the harms migrant women experience is the loss of a substantial portion of their income to various parties (recruitment and employment agencies, moneylenders, wire transfer companies, and others) immediately preceding and following migration (Rosewarne 2016). Further, in many receiving countries, migrant workers' visas are conditioned on ongoing employment with a specific employer, which predictably results in the wrongful exploitation of migrants' labour (being overworked or underpaid) and, in some cases, physical abuse (Kittay 2009; Jaggar 2009; Sager 2012; Straehle 2013). Due to employer expectations and migration restrictions, they often cannot take time off or return home to visit family (Kittay 2008; Eckenwiler 2011). Finally, migrants who work in family caregiving are among the most at risk for chronic, elevated stress, poor physical health, and depression (Eckenwiler 2011).

Among migrants, women carework migrants are particularly vulnerable to exploitation, abuse, and health risks for several reasons. First, they often perform labour in the homes of their employers, where privacy makes the exposure of exploitation and abuse less likely. Second, carework is generally an unregulated employment sector and unionisation for the purpose of collective bargaining is near-impossible to achieve (if it is even legally permitted). Third, migrants are expected to form a relationship of care with their employers or charges, which makes tough negotiations over employment conditions difficult (Crozier 2010). Fourth, most migrants admitted on the basis of employment lack the civil and political rights by which they might be able to challenge their exploitation or abuse (Eckenwiler 2011).

Feminist philosophers have conceived the moral wrongness in the circumstances that give rise to the harms women migrant workers experience in different ways. Given the nature of the work and what is often lost by women migrant careworkers, some feminist philosophers argue that the moral significance of the harms involved is captured well by the ethics of care. Eva Feder Kittay argues that we ought to understand the wrongness of harms women migrant workers experience in terms of the violation of the right to (give and receive) care (Kittay 2008, 2009). Other feminists have conceived of the harms women migrant workers experience in terms of the notion of structural injustice (Eckenwiler 2011; Aragon and Jaggar 2018), using Iris Marion Young's definition. According to Young, structural injustice occurs when "social processes put large groups of persons under systematic threat of domination or deprivation of the means to develop and exercise their capacities, at the same time that these processes enable others to dominate or to have a wide range of opportunities for developing and exercising opportunities available to them" (Young 2011). The circumstances surrounding migrant domestic labour clearly instantiate structural injustice as conceived by Young.

1.3 Refugee Admissions

The kinds of circumstances that lead individuals to seek asylum outside their country of origin vary significantly in relation to their gender (identity) and sexual orientation. Due to this fact, gender is an important category of analysis in normative inquiries concerning refugees. Neglect of gender (identity) and sexual orientation is likely to systematically obscure injustices faced by asylum-seeking individuals.

A clear example of how disregard for gender biases thought concerning refugees is the omission of gender and sex in the traditional definition of "refugee": a person who "owing to a well founded fear of being persecuted *for reasons of race, religion, nationality, membership in a particular social group, or political opinion*, is outside the country of his nationality, and is unable to or, owing to such fear, is unwilling to avail himself of the protection of that country" (UN General Assembly 1951, Article 1a; emphasis added). As Serena Parekh notes, when the UN Refugee Convention was adopted in 1948, it was seen as absurd that gender or sex would be included in the definition. Only since the 1990s have states begun to consider that women could be persecuted as women (Parekh 2012).

The persecution of women may be gendered in its motive, form, or both; that is, a woman may be persecuted because she is a woman, or her persecution may take a particular form because she is a woman (Freedman 2007; Parekh 2012). Clear examples of gender-related persecution include female genital cutting; forced pregnancy, abortion, and sterilisation; domestic and sexual violence; and prohibitions on lesbian relationships (Parekh 2012; also see Janssens et al. 2006; Lister 2016; Raghuram 2016; Sertler 2018).

In practice, Parekh argues, the individual being persecuted by the state for their political views or activism has been treated as the paradigmatic refugee. This paradigm privileges forms of persecution more likely to be suffered by men. In much of the world, women's persecution occurs in the private sphere and thus is seen as apolitical. Further, women are more likely than men to be persecuted by non-state (private) actors (e.g., their husbands, fathers, or brothers) (Parekh 2012). As this reveals, disregard for how gender and sex differentially shapes individuals' lived experience makes it more likely that those who ought to be recognised as refugees will not be. Whereas cisgender (heterosexual) women's refugee claims have been denied either because gender and sex were not regarded as bases for persecution or because the persecution women more typically suffered did not conform to the male paradigm and so was not recognised as persecution, LGBTQ migrants seeking refugee status have experienced distinct but parallel obstacles.

Through most of the twentieth century, prospective LGBTQ migrants were denied entry to the United States on the basis of their supposed psychopathy, sexual deviance, or perversion (Pitts 2018), regardless of claims of persecution. During this time, homosexuality and gender nonconformity were treated as tolerable, if not justified, bases of persecution. The US ban on homosexual migrants was lifted in 1990, in which year US immigration courts also recognised persecution for homosexuality as a basis for refugee admission (Morgan 2006).

Subsequent adjudication of LGBTQ migrants' claims for refugee status in the US has, despite being more permissive, treated gay white men (as they are stereotyped in mainstream US culture) as the paradigmatic LGBTQ refugee claimants (Morgan 2006). In an early case, a gay man's application for refugee status was denied because he was not feminine and, the court reasoned, he could easily conceal his "alleged" homosexuality should he be deported to Iran (Morgan 2006; Ferracioli 2014). Because immigration courts treat stereotypically effeminate gay men as the model of an LGBTQ person, the persecution claims of lesbians and closeted

LGBTQ persons are less likely to be regarded as credible (Morgan 2006). Stereotypes of men of colour as macho and aggressive also impede the credibility assigned to their claims to be persecuted for homosexuality (Morgan 2006). Similarly, the first transgender person admitted to the US as a refugee was classified as a "gay man with a female sexual identity" (Hipskind 2018). The United States did not recognise transgender or transsexual identity as a basis in itself for persecution until 2007.

1.4 Border Enforcement and Undocumented Migration

Gender and sexuality are also highly relevant to the evaluation of border enforcement. In the United States, visibly pregnant migrants—including those who have visas enabling them to legally enter the United States—can be denied entry at the discretion of immigration enforcement officials who deem them "Liable to Become a Public Charge". As Eithne Luibhéid explains, in the United States the Public Charge Rule "has been used against poor and minority women for over a century" (Luibhéid 2002). As a result, pregnant migrants entering the United States may seek to hide their pregnancies, even when doing so poses health risks. Those who are crossing the U.S.-Mexico border to obtain prenatal care in the United States frequently present "doctor's notes", or notes from midwives, to immigration officials stipulating that their medical bills for labour, delivery, and pre-natal care in the United States were paid (Reed-Sandoval 2020). Such practices should be contextualised within a broader history of immigration laws that specifically targeted women (particularly poor women and Asian women) and their sexuality for the purposes of exclusion and more extreme interrogation at ports of entry. Examples of such laws include the 1875 Page Law, which denied entry to Asian women brought to the United States supposedly for "lewd and immoral purposes" (Luibhéid 2002), the Immigration Act of 1903, which served to exclude prostitutes and pregnant people (reiterated in the Act of 1907), and the 1917 Act, which once again targeted women entering the U.S. for "immoral" purposes, and pregnant people.

If we treat gender as an important category of analysis in the ethics of migration, it becomes clear that we also need to consider borders and bordering practices that are *internal* to states, as opposed to working exclusively within a methodological nationalist paradigm that focus disproportionately, or even exclusively, on the ethics of national borders and questions of immigrant admissions. That is, we need to focus on internal/state borders and "bordering practices" and migrant journeys throughout the territories of two or more states, all of which can offer particularly strong barriers to migrant women, people with gestational capacity, and children. As Alex Sager argues, "the insistence on using the nation-state as the unit of analysis also obscures how mobility restrictions limit access to opportunities within nation states", and that "one upshot of recognizing how methodological nationalism has distorted thinking on the ethics of migration [...] is that this illuminates how borders enacting class-stratification, racialisation, and sexism affect mobility and access to opportunities within territories [...] for both migrants and citizens" (Sager 2018).

Indeed, expanding our understanding of "borders" and "bordering practices" beyond clearly demarcated ports of entry at national borders creates space for analysis of highly gendered aspects and consequences of border enforcement. For instance, North African migrants attempting to reach Italy and other parts of Europe are increasingly taking the dangerous and deadly Central Mediterranean route, which poses heightened risks to women and children of "detention, extortion, gender-based violence, abuse, exploitation, and drowning at sea" (UNICEF 2018) well before a state border is, in fact, reached. Central American women and girls experience sexual assault and trafficking at alarming rates while undertaking their "migrant journeys" through Mexico and the Sonoran Desert into the United States (Reed-Sandoval 2020). Meanwhile, citizens and non-citizens who enter the United States, or who cross the United States, to seek abortion care must regularly contend with enhanced interrogation at various internal and external immigration checkpoints—often taking steps to hide their abortion plans from potentially anti-choice immigration and police officers (Reed-Sandoval 2021). At the same time, "internal migration" can be a source of empowerment and transformation. Iván Sandoval Cervantes has explored how Zapotec women who migrate internally from rural Oaxaca to Mexico City (often due to social pressures to stay relatively "close to home" in case they are needed—pressures that are not equally applied to their male counterparts) often move and relocate within female networks of support that enable them to challenge oppressive gender roles in their communities. This is often in contrast to women who migrate internationally with their husbands, and therefore may not receive similar opportunities to question and challenge gender roles (Sandoval Cervantes 2017).

Meanwhile, sexuality—and, in particular, perceived sexual "deviance"—have long been a basis of exclusion, interrogation, physical harms, and even death at borders and throughout migrant journeys. Luibhéid has explored the history of sexuality-based exclusions in United States history, from the Immigration Act of 1891, which excluded, among others, those who had been convicted of adultery, bigamy, polygamy, and sodomy, and instructed physicians to "watch for signs of 'abnormal' sexual appetites and behaviour and for venereal disease" (Luibhéid 2002). Most of these grounds for exclusion were swiftly reinstated in the Immigration Acts of 1907 and 1910. Luibhéid also traces U.S history of lesbian and gay exclusions from 1917, when grounds were established for excluding those with so-called "abnormal sexual instincts" (Luibhéid 2002), and reinforced through the 1952 McCarren-Warren Act, which laid groundwork for excluding "homosexuals and sex perverts" (Luibbéid 2002), and established precedent for contemporary demands for "sexual confession" at borders and through other forms of immigration monitoring (Luibhéid 2002). Of course, sexuality-based oppression at borders, and through bordering practices, occur beyond the United States. For instance, along the "migrant" journey from Central America to the United States, LGBTQ migrants are often at heightened risk of sexual assault and murder, and are sometimes denied access to migrant shelters in northern Mexico (Reed-Sandoval 2020).

Along lines similar to those drawn above, understanding the relationship between sexuality and border enforcement also requires an expanded conception of where "borders" lie. For instance, Andrea Pitts expands upon our understanding of borders and bordering practices through arguing that sanctuary cities and gender-based asylum in the United States serve to uphold the carceral state by "reifying the state as a site of security and stability" and as an investment "in the state's carceral power to decide who is deserving of care, trust, health, and mobility [which] invariably leads to increased harms for trans, disabled, Black, indigenous, and any other people vulnerable to the state's refusal" (Pitts 2018). Meanwhile, empirical scholars have called attention to the heightened vulnerability of LGBTQ migrants in processes of arrest, detention, and deportation, and the ways in which Cisnormativity, in particular, often renders trans immigrants "hypervisible" and thus extremely vulnerable to policing and exclusion (Collier and Daniel 2019).

Meanwhile, legally undocumented status can make one's experiences of immigration enforcement at various levels considerably more challenging, particularly if one is a migrant of colour, a woman, and/or a sexual minority. For example, Stephanie J. Silverman and Petra Molnar have explored how the Canadian immigration system generates violence against women and children by subjecting them to unjust incarceration, failing to shield them from domestic abuse after deporting them, and sometimes contributing to their poverty and other vulnerabilities when men from their families are deported and detained (Silverman and Molnar 2016, 2018). In the United States, detained undocumented migrants have struggled to receive abortions upon request, while others who must travel long distances for abortion care (due to targeted-restriction-on-abortion-provider (TRAP) laws and a consequent lack of providers in their areas) are vulnerable to arrest and deportation as they cross state lines and confront immigration checkpoints (Reed-Sandoval 2021). People who live in mixed-status families—families in which some members have legal authorisation to live in their current country of residence, while others do not—often "experience illegality as a family", with children and so-called "legally documented" partners enduring fear, anxiety, and trauma associated with the arrest and deportation (real or imagined) of a family member (Rodriguez 2016). Finally, undocumented migrants who perform domestic care work are particularly vulnerable to workplace exploitation and sexual violence (Wilcox 2008).

2 Questions for Future Research: What Policies Are Required for Gender Justice in International Migration?

Despite a relatively recent proliferation of research on various dimensions of gender and migration, there is urgent need for significantly more normative and conceptual work in this area. Philosophers are uniquely well positioned to address questions of justice that arise from considering the gendered dimension of migration. Concerning immigrant admissions, these questions include: how must admissions policies be designed in order to ensure that they provide equitable opportunities among people of different genders and sexes? If skills-based admission policies are morally acceptable, can they be articulated in ways that do not disproportionately advantage male migrants? Should countries of immigration adopt policies for mitigating the harms of "care drain", and if so, to what extent should they parallel policies defended in philosophical literature for addressing "brain drain" (e.g., restrictions on skilled migration, compensation by receiving countries, emigrant taxation)?

With regard to carework migration, questions for which greater philosophical attention is needed include: should migrants with primary responsibility for the care of family members (even temporary workers or the "unskilled") be legally entitled to bring those family members with them, or be permitted to return home more frequently? What other policies to address harms associated with carework migration (such as collective bargaining rights, greater enforcement of migrants' rights against employers, protections against deportation in the event employment separation, greater state support for carework) should be adopted? Concerning the gendered aspects of refugee admission: can "persecution" be conceived in way that is sensitive to gendered differences in how it manifests, and yet distinguishes it from oppression? How should "social group" be defined for the purposes of refugee claim adjudication?

Furthermore, philosophers have a great deal to contribute to topics that have traditionally been explored exclusively by empirical scholars, such the experiences of mixed-status families, sexuality-based exclusions at state borders, and first-hand and phenomenological accounts of gendered migration experiences. (An exception to this, in the realm of philosophy, is that of Latina feminist scholarship; see, for instance, Ortega 2016.) Some questions to be explored include: are mixed status families, *qua* family units, deserving of special rights and protections—such as the "legalisation" of their undocumented members? Is the public charge rule just? How do intersecting racist, sexist, Cisnormative and heteronormative norms lead to acute anti-immigrant sentiments—along the lines of those that deem veiled Muslims "unruly subjects" and Latin American women highly "illegalised" regardless of their immigration status? Considerably more research should also be done on the various borders and bordering practices that coerce women and LGBTQ border-crossers at alarming rates.

Literature

Aragon, C., and A. Jaggar. 2018. Agency, Complicity, and the Responsibility to Resist Structural Injustice. *Journal of Social Philosophy* 49:439–460.

Brock, G. 2009. *Global Justice. A Cosmopolitan Account*. Oxford: Oxford University Press.

Brock, G., and M. Blake. 2015. *Debating Brain Drain. May Governments Restrict Emigration*. Oxford.

Carens, J. 2013. *The Ethics of Immigration*. Oxford: Oxford University Press.

Chavez, L. 2017. *Anchor Babies and the Challenge of Birthright Citizenship*. Stanford: Stanford University Press.

Cisneros, N. 2013. 'Alien' Sexuality. Race, Maternity, and Citizenship. *Hypatia* 28:290–306.

Collier, M., and C. Daniel. 2019. The Production of Trans Illegality. Cisnormativity in the U.S. Immigration System. *Sociology Compass* 13:1–15.
Crozier, G. K. D. 2010. Careworkers in the Global Market. Appraising Applications of Feminist Care Ethics. *International Journal of Feminist Approaches to Bioethics* 3:113–137.
Dumitru, S. 2014. From 'Brain Drain' to 'Care Drain'. Women's Labor Migration and Methodological Sexism. *Women's Studies International Forum* 47:203–212.
Dumitru, S. 2016. 'Care Drain'. Explaining Bias in Theorizing Women's Migration. *The Romanian Journal of Society and Politics*7–24.
Eckenwiler, L. 2009. Care Worker Migration and Transnational Justice. *Public Health Ethics* 2:171–183.
Eckenwiler, L. 2011. Women on the Move. Long-term Care, Migrant Women, and Global Justice. *International Journal of Feminist Approaches to Bioethics* 4:1–31.
Eckenwiler, L. 2014. Care Worker Migration, Global Heath Equity, and Ethical Place-Making. *Women's Studies International Forum* 47:213–222.
Ferracioli, L. 2014. The Appeal and Danger of a New Refugee Convention. *Social Theory and Practice* 40:123–144.
Freedman, J. 2007. *Gendering the International Asylum and Refugee Debate*. London: Palgrave Macmillan.
Ghaeus, A. 2013. Care Drain as an Issue of Global Gender Justice. *Ethical Perspectives* 20:61–80.
Higgins, P. 2013. *Immigration Justice*. Edinburgh: Edinburgh University Press.
Higgins, P. 2015. The Ethics of Immigration and the Justice of Immigration Policies. *Public Affairs Quarterly* 29:155–174.
Higgins, P. 2017. A Feminist Approach to Immigrant Admissions. *Hypatia* 32:506–522.
Hipskind, F. 2018. The Persecution of LGBT Individuals Fleeing to America from Muslim-Majority Nations. Has the Door to Refuge Been Closed? *Tulane Journal of Law and Sexuality* 27:53–74.
Hochschild, A. 2000. The Nanny Chain. *American Prospect* 11:32–37.
Hochschild, A. 2002. Love and Gold. In *Global Women. Nannies, Maids, and Sex Workers in the New Economy*, ed. B. Ehrenreich and A. Hochschild, 15–30. New York: Metropolitan Books.
International Organization for Migration. 2020. "*Gender and Migration*", https://www.iom.int/gender-and-migration (last checked December 30, 2020).
Jaggar, A. 2002. A Feminist Critique of the Alleged Southern Debt. *Hypatia* 17:119–142.
Jaggar, A. 2009. Transnational Cycles of Gendered Vulnerability. A Prologue to a Theory of Global Gender Justice. *Philosophical Topics* 37:33–52.
Janssens, S., M. Bosmans, E. Leye, and M. Temmerman. 2006. Sexual and Reproductive Health of Asylum-Seeking and Refugee Women in Europe. Entitlements and Access to Health Services. *Journal of Global Ethics* 2:183–196.
Kittay, E. 2008. The Global Heart Transplant and Caring Across National Boundaries. *The Southern Journal of Philosophy* XLVI:138–165.
Kittay, E. 2009. The Moral Harm of Migrant Carework. Realizing a Global Right to Care. *Philosophical Topics* 37:53–73.
Kofman, E. 2007. The Knowledge Economy, Gender, and Stratified Migrations. *Studies in Social Justice* 1:122–135.
Lim, D. 2019. The Indirect Discrimination of Skills-Selective Immigration Policies. *Critical Review of International Social and Political Philosophy* 22:906–928.
Lister, M. 2016. The Place of Persecution and Non-State Action in Refugee Protection. In *The Ethics and Politics of Immigration. Core Issues and Emerging Trends*, ed. A. Sager. London: Rowman & Littlefield International.
Luibhéid, E. 2002. *Entry Denied. Controlling Sexuality at the Border*. Minneapolis: University of Minnesota Press.
Meghani, Z. 2016. Women on the Move. In *Women Migrant Workers. Ethical, Political, and Legal Problems*, ed. Z. Meghani. New York: Routledge.

Morgan, D. 2006. Not Gay Enough for the Government. Racial and Sexual Stereotypes in Sexual Orientation Asylum Cases. *Law and Sexuality. A Review of Lesbian, Gay, Bisexual, and Transgender Legal Issues* 15:135–162.
Narayan, U. 1995. 'Male-Order' Brides. Immigrant Women, Domestic Violence and Immigration Law. *Hypatia* 10:104–119.
Ortega, M. 2016. *Latina Feminist Phenomenology, Multiplicity, and the Self.* New York: State University of New York Press.
Parekh, S. 2012. Does Ordinary Injustice Make Extraordinary Justice Possible? Gender, Structural Injustice, and the Ethics of Refugee Determination. *Journal of Global Ethics* 8:269–281.
Parreñas, R. 2001. *Servants of Globalization. Women, Migration, and Domestic Work.* Stanford: Stanford University Press.
Pitts, A. 2018. Embodied Thresholds of Sanctuary. Abolitionism and Trans Worldmaking. *APA Newsletter on LGBT Issues in Philosophy* 18:3–10.
Raghuram, P. 2016. Migration and Feminist Care Ethics. In *The Ethics and Politics of Immigration. Core Issues and Emerging Trends*, ed. A. Sager. London: Rowman & Littlefield International.
Reed-Sandoval, A. 2020. *Socially Undocumented. Identity and Immigration Justice.* Oxford: Oxford University Press.
Reed-Sandoval, A. 2021. "Travel for Abortion as a Form of Migration". *Essays in Philosophy*, https://doi.org/10.5840/eip20212236.
Rodríguez, C. 2016. Experiencing 'Illegality' as a Family? Immigration Enforcement, Social Policies, and Discourses Targeting Mexican Mixed-Status Families. *Sociology Compass* 10:706–717.
Rosewarne, S. 2016. Transnationalization and the Capitalization of Labor. Female Foreign Domestic Workers. In *Women Migrant Workers. Ethical, Political, and Legal Problems*, ed. Z. Meghani. New York: Routledge.
Sager, A. 2012. The Implications of Migration Theory for Distributive Justice. *Global Justice. Theory Practice Rhetoric* 5:56–70.
Sager, A. 2018. *Toward a Cosmopolitan Ethics of Migration*. Cham: Palgrave Macmillan.
Sandoval Cervantes, I. 2017. Navigating the City. Internal Migration of Oaxacan Indigenous Women. *Journal of Ethnic and Migration Studies* 43:849–865.
Sertler, E. 2018. The Institution of Gender-Based Asylum and Epistemic Injustice. A Structural Limit. *Feminist Philosophy Quarterly* 4:1–24.
Shachar, A. 2009. *The Birthright Lottery. Citizenship and Global Inequality*. Cambridge, MA: Harvard University Press.
Sheth, F. H. 2006. Unruly Muslim Women and Threats to Liberal Culture. *Pace Review. A Journal of Social Justice* 18:455–463.
Sheth, F. H. 2019. The Production of Acceptable Muslim Women in the United States. *The Journal of Aesthetics and Art Criticism* 77:411–422.
Silverman, S., and P. Molnar. 2016. Everyday Injustices. Barriers to Access to Justice for Immigration Detainees in Canada. *Refugee Survey Quarterly* 35:109–127.
Silverman, S., and Molnar, P. 2018. "How Canada's Immigration System Spurs Violence Against Women", *toronto.com* (last checked December 26, 2020).
Silvey, R. 2009. Transnational Rights and Wrongs. Moral Geographies of Gender and Migration. *Philosophical Topics* 37:75–91.
Straehle, C. 2013. Conditions of Care. Migration, Vulnerability, and Individual Autonomy. *International Journal of Feminist Approaches to Bioethics* 6:122–140.
UN General Assembly. 1951. Convention Relating to the Status of Refugees. *United Nations, Treaty Series* 189:137.
UNICEF. 2018. "*UNICEF Refugee and Migrant Crisis in Europe. Regional Humanitarian Situation Report #20*" (last checked December 26, 2020).

Velji, M. 2019. “Legislative Catharsis, Part One. A Primer on Québec’s Veil Bans for Europeans”. In “*The Religion Factor*” blog, University of Gronigen, https://www.rug.nl/research/centre-for-religious-studies/religion-conflict-globalization/blog/legislative-catharsis-part-one-a-primer-on-quebecs-veil-bans-for-europeans-22-05-2019?lang=en (last checked December 26, 2020).

Weir, A. 2008. Global Care Chains. Freedom, Responsibility, and Solidarity. *The Southern Journal of Philosophy* XLVI:166–175.

Wellman, C., and P. Cole. 2011. *Debating the Ethics of Immigration. Is There a Right to Exclude?* Oxford: Oxford University Press.

Wilcox, S. 2005. American Neonativism and Gendered Immigrant Exclusions. In *Feminist Interventions in Ethics and Politics. Feminist Ethics and Social Theory*, ed. B. Andrew, J. Keller, and L. Schwartzman. Lanham: Rowman & Littlefield Publishers.

Wilcox, S. 2008. Who Pays for Gender De-Institutionalization? In *Gender Identities in a Globalized World*, ed. A. M. Gonzalez, 53–74. Amherst, NY: Humanity Books.

Young, I. M. 2011. *Responsibility for Justice*. Oxford: Oxford University Press.

Expatriation/Denationalisation

Patti Tamara Lenard

Abstract The following entry will outline the central arguments in favour of and against the right of the state to revoke citizenship. According to those who defend the right of states to revoke citizenship, we must understand the relation between citizens and their state in terms of a contract, according to which each party has particular duties to carry out. Where someone has committed a particular set of crimes, they can be understood to have violated the terms of the citizenship contract, and are thereby no longer entitled—by their own actions—to the protection offered by the state. A second defence recommends revocation as an appropriate and *fitting* punishment for a particular subset of especially egregious crimes, including terrorism, since anyone who commits these crimes attacks the foundations of the state and correspondingly deserves to be excluded from its protection. There is a much wider range of arguments against the right to revoke, perhaps reflecting that none seems to effectively rule it out absolutely. According to those who oppose the right of states to revoke, the laws are problematic because they are likely to generate statelessness; because they are a violation of the equality to which democracies are committed; because they are likely to be arbitrarily applied; because they are a violation of democratic principles more broadly; and because they represent an abandonment of a collective, global, responsibility to fight terrorism.

Until the end of World War II, no one questioned the right of a state to denationalise its citizens, that is, to revoke their citizenship without or against their expressed consent. State sovereignty meant, among other things, that states were permitted to grant and retract citizenship from individuals, for all kinds of reasons. States revoked the citizenship of individuals who had resided abroad for extended periods of time; of women who married foreigners; of individuals whose allegiance was believed to be suspect, for example, because they participated in the politics of other states; of individuals who participated in foreign wars; and so on. Moreover, the practice of denationalising individuals as punishment—or perhaps more accurately banishing

P. T. Lenard (✉)
Graduate School of Public and International Affairs, University of Ottawa, Ottawa, Canada
e-mail: patti.lenard@uottawa.ca

A. Niederberger et al. (eds.), *Handbook of Migration Ethics*,
https://doi.org/10.1007/978-3-031-89877-8_20

or exiling them from the state's territory—was widespread. Of the latter, for much of human history, exile as a punishment was understood as considerably less severe than many others, including state-sanctioned torture and execution (Kingston 2005).

1 General Issues in the Controversy

However, the horrors of World War II forced states to rethink this easy equation of state sovereignty with the right to confer *and* retract citizenship: the genocide of Jews by Nazi Germany began in earnest with their denationalisation, rendering them stateless, and thus without the protection of any state (Arendt 1963). Subsequent political deliberations among states acknowledged the dangers faced by stateless individuals, which Arendt highlighted—no one stepped in to protect the human rights of Jews, once they had been rendered stateless, to protect their supposedly human rights—and international law, starting with the 1951 Geneva Convention Relating to the Status of Refugees, now recognises states' obligation to do their part to protect people from this status. According to Article 15 of the Universal Declaration of Human Rights, adopted in 1948: "(1) everyone has the right to nationality. (2) No one shall be arbitrarily deprived of his nationality nor denied the right to change his nationality" (United Nations 1948). From then, the right to nationality or citizenship (these terms will be used interchangeably in what follows) has been understood as a basic human right, rather than (merely) a privilege handed out by states. The essential role that citizenship status plays in protecting individuals from all manner of harm in part explains why its revocation is understood by some to be troubling. Certainly, the precise obligations that this right places on states have been the subject of much dispute, focused in particular on what the Declaration meant by referring to the dangers of its "arbitrary" removal, as will be detailed in what follows. But the general principle that, to the greatest extent possible, states must respect the right of individuals to hold citizenship status in at least one country, is widely recognised in international law.

The prerogative of the state to revoke citizenship has returned to public deliberations, however, in the face of its possible use as a punishment and deterrent of terrorist activities. As Sandra Mantu describes it, there is a revised interest in defending the right of states to revoke citizenship, especially as states expand their available toolkit to combat terror threats: "citizenship deprivation powers are inscribed within national security and counter-terrorism policies" (Mantu 2018: 38). Across many democratic states, there have been calls to adopt laws that permit the revocation of citizenship, to punish and deter terrorism; in some they have been adopted, for example, as in Australia, and in others they were briefly adopted but then rescinded, as in Canada; others, including many European countries, have reinvigorated laws that had remained on the books, but had been in disuse for decades. The use of revocation has expanded most dramatically in the United Kingdom, where the language of the law permitting revocation notes its applicability in cases where an individual and her actions are "seriously prejudicial to the vital interests" of the United Kingdom (Gibney 2013a).

The following entry, which uses the terms revocation and denationalisation interchangeably, and which reserves the term denaturalisation for the removal of citizenship acquired by naturalisation, will begin with an outline of the central arguments in favour of adopting these laws. According to those who defend the right of states to revoke citizenship, we must understand the relation between citizens and their state in terms of a contract, according to which each party has particular duties to carry out. Where someone has committed a particular set of crimes, they can be understood to have violated the terms of that contract, and are thereby no longer entitled—by their own actions—to the protection offered by the state. A second defence recommends revocation as an appropriate and *fitting* punishment for a particular subset of especially egregious crimes, including terrorism, since anyone who commits these crimes attacks the foundations of the state and correspondingly deserves to be excluded from its protection. There is a much wider range of arguments against the right to revoke, perhaps reflecting that none seems to effectively rule it out absolutely. According to those who oppose the right of states to revoke, the laws are problematic because they are likely to generate statelessness; because they are a violation of the equality to which democracies are committed; because they are likely to be arbitrarily applied; because they are a violation of democratic principles more broadly; and because they represent an abandonment of a collective, global, responsibility to fight terrorism.

2 Arguments in Favour and Against Revocation of Citizenship

2.1 Arguments in Favour of Revocation

Revocation is permissible for "contract" violations: One treatment of revocation situates it in an understanding of citizenship as a kind of contract between citizen and the state, according to which each party has a set of duties that it is charged with carrying out. This 'social contract' understanding of the state, the history of which is located in the works of both Thomas Hobbes and John Locke, suggests that where either party fails to carry out its duties, the contract is dissolved. What sorts of actions count as sufficient to dissolve the contract, on this view? Not all criminal actions are sufficient to do so. But, says Shai Lavi, citizens of democracies have a special duty to "uphold the foundations of a democratic order", which in turn requires that they refrain from acting "in a way that would intentionally undermine the power of the political community to self govern" (Lavi 2010: 796). Following from this logic, for Lavi, crimes that are terrorist in nature (as distinct from other types of crime) dissolve the contract and release the state from its duty to protect the offender from harm.

A related defence of revocation draws its inspiration from the ancient Athenian practice of ostracism. The ancient practice permitted citizens of Athens to ostracise an individual who, according to his co-citizens, had become powerful enough to subvert the democratic process (Forsdyke 2009). Ostracised citizens were banished from Athens for a period of 10 years, after which they could return to Athens, to their property and the full rights of democratic citizenship. This sort of temporary exile is not available as punishment in contemporary democracies, since one cannot cross borders without citizenship (and the passport to attest to it), however, and so translated to the present, Athenian practice appears to suggest more generally that where the actions of a citizen are such that they threaten to undermine the stability of democratic institutions, it may be well be appropriate to denationalise them (Miller 2016). For example, in the case of crimes in which criminals aim to harm citizens because of the democratic values they hold and which the criminal wishes expressly to repudiate through their actions it is reasonable for a state to respond by excluding them from the state.

Revocation for breach of allegiance to the state: This desire by the state to exclude those who would do it harm is, says Ben Herzog, of particular worry to states in an era in which dual citizenship is becoming increasingly normalised. In a political space in which states can never be sure of the loyalty of their citizens, but where nation states are still protective of their sovereignty, the practice of citizenship revocation gives democratic states a way to "regulate national allegiance" (Herzog 2016: 261). Precisely the anxieties that prompt some scholars, who will be considered below, to disavow revocation on the grounds that it targets dual citizens problematically, suggest to Herzog that there is a genuine legitimacy to states demanding and expecting allegiance from their citizens. Citizenship is understood and treated as retractable as a way to secure allegiance; and can thereby permissibly be retracted in cases where citizens choose against offering this allegiance by participating in the kinds of terrorist activities that undermine, or at least are aimed at undermining, democratic institutions (Herzog 2010).

Revocation is a fitting punishment for some (mainly terrorist) crimes: The defences of revocation reported above treat the state's job as one of protecting its citizens and considers if and when the state can reasonably be understood to be released from that job. They imply that permanent banishment is an appropriate action in these kinds of cases. Other defences of revocation focus not directly on the nature of the state-citizen relationship, and what severs it, but rather on the *fittingness* of banishment as a punishment for certain categories of crimes, including but not limited to crimes against the state. What does it mean to say that a punishment *fits* the crime? To say that it fits is to emphasise that there is some strong, logical, connection between the crime committed and the punishment that is thereby warranted. This view derives its strength from an expressivist theory of punishment, according to which one main goal of punishment is to express a disapproval or resentment of a particular action (Feinberg 1965). Often, the audience for the disapproving expression is not (solely) the perpetrator, but rather is the society at large who witnesses the punishment, and understands it in part as intended to protect it.

Operating with this basic framework in mind, terrorists, says Christian Joppke, operate to destroy the legal order: "Terrorists [...] posit themselves outside this order, even want to destroy it, so that reintegration—the usual rationale of the penal process—is not the obvious state response" (Joppke 2016: 731). Revocation of their citizenship is correspondingly a fitting response to their crime. Indeed, says Joppke, the legal revocation of citizenship merely formalises a choice made by the terrorist to renounce citizenship through a particular, violent, behaviour (Joppke 2016). It is notable that public opinion across many states endorses the revocation of citizenship for terrorists, broadly on the grounds that Joppke defends (for just one example, see Bell 2012). Having noted that, this kind of view was expressly rejected by the American Supreme Court, which considered whether a naturalised citizen, who had voted in an Israeli election, had by doing so renounced his American citizenship. The Court ruled that no action, other than the express renunciation of citizenship, should be understood in this way (Weil 2012).

Joppke is prepared to defend the additional claim that there is something special not simply about terror, but about Islamist terror, in its perpetrators' religious allegiance, and their supposed commitment to a "community of believers" rather than the political community of citizens who are treated as "enemies" (Joppke 2016: 732). Joppke is largely insensitive to worries that the adoption of revocation laws is part of a counter-terrorism effort that permits the discriminatory treatment of Muslim citizens (again, a claim that will be considered below). Rather, he says, it is the importance of citizenship that justifies revoking it from individuals who cannot see its value, and who seek to undermine its value for others. Citizenship is a privilege, which must be earned and deserved by its holders over time.

Displaying some discomfort with the kinds of arguments that offer an expansive space for revocation laws, Luara Ferraciolli and Christian Barry propose that the revocation of citizenship can be justified, but only under very constrained conditions. The first condition they specify is the avoidance of statelessness; however grievous the crime committed, an individual must be able to access citizenship status somewhere, and correspondingly where revocation would render the perpetrator stateless, it must be rejected. The second condition requires that the crime being punished is a severe *political* crime, which includes those crimes where "the perpetrator intentionally violates (or actively conspires to violate) the most basic rights of civilians (i.e. rights to life, liberty, freedom from torture and degrading treatment, etc.) in order to achieve political goals or communicate political messages" (Barry and Ferracioli 2015: 1065). So, revocation is for them reserved for individuals whose political violence amounts to a renunciation of their citizenship, just as Joppke proposed. The third condition requires that there be a state, of second citizenship, that is itself causally implicated in the particular terrorist crime that has been conducted. So, a dual citizen of X and Y, who commits a crime against Y with the support of X, can be denationalised by Y to X (Barry and Ferracioli 2015; for a thorough analysis and rejection of their claims, see Carey 2018). Although this defence of revocation leaves in principle room for revocation under the right conditions, the authors suggest that in practice they may be hard, if not impossible, to achieve.

2.2 Arguments Against Revocation

Revocation should not be permitted because it renders people stateless: The most powerful argument against granting states the right to revoke citizenship is that it will render someone stateless. Citizenship is typically understood to be a legal status, that is accompanied by a robust package of civil, economic, political and social rights. To be a citizen is not simply to possess these rights. It is also to be a member of a state that is responsible for protecting these rights for all of its citizens; democratic states are understood to be legitimate to the extent that they are able, and willing, to protect these rights for all of their citizens. So, rendering someone stateless deprives an individual of the formal protection of their rights. Thus, statelessness is widely agreed to be a severely insecure status, where one's human rights are not guaranteed to be protected: it is, says Matthew Gibney, "unjust and cruel" (Gibney 2013b: 651). It is unjust because it deprives people of a right to membership, and moreover puts individuals in the position of being subject to a coercive power—the state—without the right to have a say in how that power is exercised. It is cruel, Gibney continues, "because it may be a recipe for exclusion, precariousness, and general dispossession" (Gibney 2013b: 651). As a matter of practice, states that permit some forms of revocation generally recognise the need to avoid statelessness.

Revocation is a violation of equality: In most states that permit revocation, it is permitted by law only if it will not render someone stateless. The practical effect of this rule is that only individuals who possess two (or more) citizenships are eligible to have their citizenship revoked (there are exceptions, however, for example in cases where states permit the revocation of fraudulently-acquired citizenship, even where it will render someone stateless). Gibney terms this the "invidiousness" objection (Gibney 2013b: 652), which proposes that the introduction of an inequality in the possible application of revocation, only to dual-citizens, is a violation of the equality to which democratic states are committed (see also Lenard 2018; Kaya 2017). In particular, it looks like permitting revocation for dual citizens only renders their citizenship less secure in a meaningful way. What specifically is rendered insecure? On this view, the right to citizenship is protecting something like the secure residence of individuals. So, if only *some* citizens are liable to have their citizenship revoked, this renders their citizenship less secure than it is for single-citizenship citizens. Paradoxically, in the name of protecting security, the citizenship of some citizens is rendered less secure than the citizenship of others (Macklin 2014: 53).

Some respond to this objection by proposing that dual citizens are recipients of goods—to which single nationality citizens do not have access—that render them already *unequal* in relation to their co-citizens, including more residential, educational and travel opportunities (Ferracioli 2017: 6; Herzog 2016: 261). So, say some critics, permitting the revocation of citizenship for dual nationals can be understood as *returning* citizens in a state to a rough equality. One reason to resist this objection, however, is that while residential and travel opportunities are valuable, they are not nearly as valuable as residential security; the trade-off does not render an unequal situation equal.

Another equality-based objection to revocation focuses not simply on the inequality that is generated as a result of permitting the revoking the citizenship of only dual nationals, but emphasises in particular that, as a matter of fact, most dual nationals, especially in democratic states, are immigrants. This objection is especially strong where states permit the revocation of citizenship in the case of naturalised, but not natural-born, dual citizens. On this view, the reason to object to revocation laws on the grounds of inequality is that it is explicitly, or implicitly, intending to target immigrants in invidious ways. There is, observe these scholars, an extended history of scepticism towards immigrants, and dual nationals especially, focused on their potential disloyalty to their state of second citizenship. In light of this history of scepticism towards immigrants, including with respect to the ways in which many putatively disloyal migrants were treated—for example, both Canada and the United States interned citizens with Japanese ancestry during World War II—democratic states must be explicitly focused on protecting the equal treatment of immigrants rather than emphasising their differences. Thus, revocation laws are to be rejected because they target dual citizens, especially immigrant dual citizens, and are problematic for propagating the myth that immigrants may be disloyal, and for permitting inequality to flow from the method of one's citizenship acquisition.

Multiple scholars add an additional layer to this objection to revocation, in the case of dual citizens, noting the historical context in which these laws have gained renewed prominence. This context is noted in the introduction, in which democratic states are pre-occupied with combatting terrorism. Counter-terrorism measures of all kinds appear, to many scholars, to disproportionately target Muslims, and this targeting is co-incident with repeated demands that Muslims convince their fellow-citizens of their loyalty. Revocation policies are just one dimension of counter-terrorism efforts, which produce "the development of citizenship hierarchies whereby deprive-able citizens seem to occupy an intermediate position between 'real citizens' and 'foreigners'" (Mantu 2018: 39). One consequence is that although revocation laws are supposedly neutral among dual citizens (especially those that do not distinguish among naturalised and natural-born dual citizens), it remains clear to some scholars that those most likely to be affected by revocation laws are not dual citizens in general, nor immigrants in general, but rather specifically dual-citizen immigrants from Middle Eastern countries most of whom practice Islam, and whose origin states are tied to or seen as endorsing terrorism (Choudhury 2017). Moreover, empirical research suggests that in states where the citizenship of immigrants is rendered less secure via their liability to revocation, "majority members are significantly more likely to have negative attitudes towards immigrants. These results hold for immigrants in general, and Muslim immigrants in particular" (Kaya 2017: 43). The equality that is central to democratic principles, and which demands for many a special attention to those who are racially and religiously marginalised in a community (as is certainly the case with respect to Muslim citizens in many democratic states), requires that a stand against revocation be firmly taken, in part as a commitment to *them* and their equal standing in democratic states.

Revocation is arbitrary in its application: This formulation of the equality objection to revocation is interlinked with worries about its arbitrary application. In general, one worry about revocation laws is that, in practice, they are likely to be applied in an arbitrary way, targeting those who are already marginalised in a particular state—certainly, the list of individuals who have been denationalised is reason for thinking this worry is founded; the vast majority are of Middle Eastern or Muslim background. It is perhaps no surprise that citizens with these origins believe, moreover, that they are the intended target, and considerably be more likely to be their victims (Lenard et al. 2016). The application is arbitrary, for example, if it is applied in a discriminatory way, or if it results in de facto or even de jure statelessness, as well as when it appears that it is being applied simply to avoid protecting someone's rights, or to avoid taking the responsibility for punishing them (Mantu 2018: 30). One strategy revoking states have adopted to avoid charges of arbitrariness is to mandate its automatic application in the case of conviction for certain crimes, as has been done in Denmark, rather than by ministerial discretion, as has been done in the United Kingdom. The thought here is that requiring revocation as an automatic penalty for certain crimes removes from judges, or ministers, the discretion to make a choice about whether to revoke citizenship, where there is reason to suspect that such individuals may implicitly or explicitly hand down tougher punishments to marginalised individuals. Even here, however, the possibility of discrimination in identifying who can be charged of the relevant crimes remains a real danger, as has been demonstrated in other cases.

Arbitrariness can enter in another, distinct, location, having to do with the fate of individuals whose citizenship is revoked. In general, it is reasonable to assume an intimate connection between expatriation and deportation (Carey 2018). However, this connection is not necessary. That is, it is in principle possible that a citizen is denationalised but, because she is stateless as a result, remains permitted to reside in her state of prior citizenship as a resident. As a resident, she would be entitled—as are other non-citizen residents—to have a wide range of her rights protected by the state, although the rights of democratic political participation would largely be denied her. Additionally, she would be denied the right to cross borders, since revoking citizenship results in a revoked passport, without which one cannot gain entry into another country. Even a denationalised dual citizen may find herself resident in a country that has denationalised her, if for example her country of (remaining) citizenship is unsafe and unwilling or unable to protect her human rights; democratic states possess the obligation of non-refoulement, a duty that is mainly invoked in the case of asylum seekers, who cannot (by international law) be returned to their country of origin if that country is unsafe. Although in many recent cases of revocation—especially by the United Kingdom—the action is taken against citizens who are abroad, the practical effect of which is that they are denied the right to return to the UK, democratic states that denationalise an individual following a trial and conviction for relevant crimes generally recognise the obligation of non-refoulement and thus may nevertheless offer residence to denationalised individuals, and correspondingly protect the set of rights that residents can normally expect to have protected.

A complication is that denationalised individuals may find that their states of *alleged* second citizenship are unwilling to receive them—perhaps because they do not recognise the citizenship of the individual in question. This sort of disagreement can happen where denationalising states permit the denationalisation not only in cases of actual dual citizens, but also in cases where denationalising states have the rule that denationalising is permissible in the case not only of dual citizens, but also in the case of single citizens who are believed to have access to a second citizenship. Shamima Begum, a young British woman who left the UK when she was 15 years old to marry an ISIS fighter, and who has requested permission to return to the UK, has, as at the time of writing, had her British citizenship revoked. The British state alleges she has access to Bangladeshi citizenship through her parents, although she has never taken it up, and has never lived there (John 2019). These kinds of uncertainties that face denationalised individuals render the practice arbitrary.

Revocation is a violation of democratic principles of punishment: Another set of objections to revocation focuses on additional commitments that democracies possess in order to sustain their legitimacy. One such argument makes a claim about the nature of punishment in democratic states. On this view, one assumption that we must make about citizens in a democratic state is that they can and do change over time; applied to punishment, this means that we must treat citizen wrong-doers as capable of rejoining the democratic community after amends are made for their specific wrongdoing. We must in other words believe in something like the rehabilitative power of punishment, according to which punishment can, if effectively carried out, result in wrong-doers rejoining the democratic state as citizens with their full set of rights intact. The result must be a rejection on democratic grounds of any punishment that denies this possibility, including most obviously capital punishment, and in practical terms, citizenship revocation (Cohen 2016).

Revocation does not meet democratic standards of justification: Alternatively, say some objectors to revocation, democracies are committed to adopting laws that meet certain standards of justification, and these are not met in the case of revocation, including with respect to its necessity in the fight against terror. It is important to note that the demand for revocation policies, where it has arisen, is made in terms of their essential role in fighting terrorism, both by preventing terrorism (since terrorists might be deterred from carrying out attacks if they were genuinely worried about losing their citizenship), and by punishing it as severely as possible. Yet, according to some, very little evidence has been provided to support the claim that the revocation of citizenship can accomplish this sort of deterrence. In other contexts, there is evidence on the contrary that the prospect of harsh punishment—in the US context, for example, the prospect of capital punishment—has little impact on the choices that would-be criminals make. This observation is connected to the fact that democracies are ordinarily equipped with a robust criminal system that enables the effective punishment of citizen wrong-doers, so arguments that an entirely new form of punishment is required, especially one that is especially harsh, demands an explanation that contains an account of why the existing institutional framework for punishing wrong-doers is inadequate to the task (Lenard 2016).

Revocation undermines the global nature of the fight against terror: A final argument against revocation focuses on the global obligation to combat terror. On this view, every state is equally committed to the task of fighting terror, since the threat is global and criss-crosses borders. So, this argument proposes, we should worry that if we permit states to revoke citizenship, the rush to do so in the event of a terrorist attack committed by a dual national generates a kind of "rush to the bottom", where states who are both in some sense responsible for a terrorist work as quickly as possible to abandon responsibility for him. A better strategy is to avoid this race to denationalise, by accepting that fighting terror is best done under the condition that no states can revoke, and correspondingly by generating rules that guide us, as a global community, in determining which state is responsible for the prosecution of terrorist crimes, and the carrying out of the punishment when a fair one is determined.

3 Future Research Directions

The deliberations above, at their heart, are about how the right to citizenship should be understood. For those who are vocal against the right to revoke, it is proper to understand citizenship as a right, just as the Universal Declaration of Human Rights demands. For those who endorse the right of states to revoke citizenship, it is best understood as a conditional right, that is, a right that is conditional on certain kinds of behaviours and values. Citizenship, on this view, is a privilege to be earned, and which one can deserve or not. Correspondingly, the privilege can be rescinded where individuals do not behave in ways that are worthy of the status, and continuing to hold the status is conditional on behaving (or not behaving) in a particular way.

For those who argue that the right to revoke is a violation of the right to citizenship status, much is taken for granted about what this right is, and especially, why it is a distinct kind of right. It is not enough to say, simply, it is an unconditional right—it is certainly conditional in some sense, for example, for individuals who naturalise into the status after they fulfil the appropriate set of conditions. To say, however, that it is unconditional because it is a human right—which it may well be—is insufficient as well, since all human rights are conditional in some way (for example, on the condition that exercising them does not harm others' ability to do the same). Current rejections of revocation rely on the truth of there being a codified right to nationality, but do little to defend the features that distinguish it from other human rights.

What might the right to citizenship be grounded in? In earlier work, Lenard has grounded the right to citizenship in the right to a secure expectation of residence (Lenard 2018). What the right to nationality does is ensure that a citizen can expect to reside in her country of citizenship in the present and in to the future, and correspondingly that she is free to make her life in that place. As others have observed, this generates some uncertainty in the case of dual citizens who, on this view, thereby seem to have the right to secured residence in more than one location. Why, sceptics might ask, should a dual citizen be protected by a state of second citizenship, in which she has never lived, and to which she has no apparently genuine links (Bauböck and Paskalev 2015), if she commits a grievous act of terrorism on its soil? Lenard's answer, that the right to citizenship once granted cannot lapse, even in these kinds of cases (though she concedes that the rights of citizenship themselves can lapse, and may require sustained residence in order to ensure their protection), is unsatisfying to some who believe the account is inadequately robust to defend a fully unconditional right to citizenship. This work remains to be done well, and may be at the foundation of an argument against revocation that is able to withstand naysayers.

Finally, those who defend the right of states to revoke citizenship do widely acknowledge that an in-principle defence of revocation does not translate easily into the fair and justified application of revocation. Even among most of those who defend the right of states to revoke, there is no great appetite to see it revived and implemented on a regular basis. Thus, there exists considerable space, among those who believe that it can be applied fairly, to flesh out the procedures for doing so, as well as an account of where they can go wrong, and what sort of appeals procedure can protect those who are subject to such a significant penalty (Lenard 2022). Could a fairly applied procedure serve to assuage the anxieties of those who continue to seek a fully robust argument against revocation? It is unlikely, but that does not make the project one that should be abandoned. Efforts to flesh out the right to citizenship and its robustness can and should travel alongside efforts to make sure that, if it must be removed, it is done as fairly as possible.

Literature

Arendt, H. 1963. *Eichmann in Jerusalem*. New York: The Viking Press.

Barry, C., and L. Ferracioli. 2015. Can Withdrawing Citizenship Be Justified? *Political Studies* 64:1055–1070.

Bauböck, R., and V. Paskalev. 2015. Cutting Genuine Links. A Normative Analysis of Citizenship Deprivation. *Georgetown Immigration Law Journal* 30:47–104.

Bell, S. 2012. "Overwhelming Majority Supports Revoking Citizenship for Canadians Who Commit Treason". *National Post*, October 30, 2012, https://nationalpost.com/news/canada/overwhelming-majority-supports-revoking-citizenship-for-canadians-who-commit-treason-poll.

Carey, B. 2018. Against the Right to Revoke Citizenship. *Citizenship Studies* 22:897–911.

Choudhury, T. 2017. The Radicalisation of Citizenship Deprivation. *Critical Social Policy* 37:225–244.

Cohen, E. F. 2016. When Democracies Denationalize. The Epistemological Case against Revoking Citizenship. *Ethics & International Affairs* 30:253–259.

Feinberg, J. 1965. The Expressive Function of Punishment. *The Monist* 49:397–423.

Ferracioli, L. 2017. Citizenship Allocation and Withdrawal. Some Normative Issues. *Philosophy Compass* 12:1–9.

Forsdyke, S. 2009. *Exile, Ostracism, and Democracy. The Politics of Expulsion in Ancient Greece*. Princeton: Princeton University Press.

Gibney, M. J. 2013a. 'A Very Transcendental Power'. Denaturalisation and the Liberalisation of Citizenship in the United Kingdom. *Political Studies* 61:637–655.

Gibney, M. J. 2013b. Should Citizenship Be Conditional? The Ethics of Denationalization. *The Journal of Politics* 75:646–658.

Herzog, B. 2010. Dual Citizenship and the Revocation of Citizenship. In *Democratic Paths and Trends*, ed. B. Wejnert, 87–106. Bingley: Emerald.

Herzog, B. 2016. The Democratic Roots of Expatriations. *Ethics & International Affairs* 30:261–264.

John, T. 2019. "Shamima Begum Will Not Be Allowed into Bangladesh, Foreign Ministry Says". *CNN*, February 22, 2019, https://edition.cnn.com/2019/02/20/uk/shamima-begum-uk-citizenship-stripped-gbr-intl/index.html.

Joppke, C. 2016. Terror and the Loss of Citizenship. *Citizenship Studies* 20:728–748.

Kaya, S. 2017. Social Consequences of Securitizing Citizenship. Two-Tiered Citizenry and Anti-Immigrant Attitudes. *Canadian Ethnic Studies* 49:27–49.

Kingston, R. 2005. The Unmaking of Citizens. Banishment and the Modern Citizenship Regime in France. *Citizenship Studies* 9:23–40.

Lavi, S. 2010. "Punishment and the Revocation of Citizenship in the United Kingdom, United States and Israel". *New Criminal Law Review. An International and Interdisciplinary Journal* 13:404–426.

Lenard, P. T. 2016. Democracies and the Power to Revoke Citizenship. *Ethics & International Affairs* 30:73–91.

Lenard, P. T. 2018. Democratic Citizenship and Denationalization. *American Political Science Review* 112:99–111.

Lenard, P. T. 2022. Constraining Denaturalization. *Political Studies 70*(2):367–384.

Lenard, P. T., Nagra, B., Vucetic, S., Wark, W. and Winter, E. 2016. "*Securitizing Muslim Canadians. Evaluating the Impact of Counter-Terrorism, National Security and Immigration Policies since 9/11*". Working report prepared for the Department of Public Safety and Emergency Preparedness, http://www.academia.edu/24765979/Securitizing_Muslim_Canadians_Evaluating_the_Impact_of_Counter-terrorism_National_Security_and_Immigration_Policies_Since_9_11.

Macklin, A. 2014. Citizenship Revocation, the Privilege to Have Rights and the Production of the Alien. *Queen's Law Journal* 40:1–54.

Mantu, S. 2018. 'Terrorist' Citizens and the Human Right to Nationality. *Journal of Contemporary European Studies* 26:28–41.

Miller, D. 2016. Democracy, Exile, and Revocation. *Ethics & International Affairs* 30:265–270.

United Nations. 1948. "*Universal Declaration of Human Rights*", http://www.un.org/en/universal-declaration-human-rights/.

Weil, P. 2012. *The Sovereign Citizen. Denaturalization and the Origins of the American Republic*. Philadelphia: University of Pennsylvania Press.

Reverse Migration

Afe Adogame

Abstract Reverse migration may refer to the systematic act or process of relocation and movement of migrants or refugees, returnees who decided to go back or are being forced back to the initial place of departure. It should be treated as involving both stream and counter-streams, sending and receiving ends, as well as its translocal, transnational, multidimensional strands. This article looks at the various ethical and political challenges posed by such acts, processes and systems.

The increasing volume, dynamism, trends and complex patterns of international migration continue to generate wide ranging interests within academic and public arenas. As migration trajectories burgeoned with regards to human migration corridors and policies, the focus on reverse migration has gained more traction. However, as compared with the extensive literature on migration, the treatment of reverse migration generally lacks adequate theoretical and empirical knowledge. Reverse migration, also called reverse misorientation is a phenomenon recognised in the scientific study of bird migration (Lewis 1939; Alerstam 1991; Thorup 2004; Nilsson and Sjöberg 2016). Reverse migration of birds is widespread around the world and occurs for many species migrating or flying in opposite direction of what is typical of its species during the migrating time. The overall migration pattern (direction, path and timing) is relatively well described for many bird species. As Thorup (2004) vividly demonstrates, birds that migrate eastward in autumn from their breeding grounds, mostly in eastern Europe, are more likely to consistently reverse migrate than those species migrating southward mostly from southern Europe.

Until recently in the 1980s, research and publications devoted exclusively to human return migration are scant. The "terminological sloppiness" (Bovenkerk 1974) found in the relevant literature presents us with considerable difficulties in the study of reverse migration. Data on the sociodemographic characteristics of return migrants are grossly limited. Guzzetta (2004) points to the issue of limited data gathering by the U.S. on reverse migration despite the sizeable amount and quite

A. Adogame (✉)
Princeton Theological Seminary, Princeton, NJ, USA
e-mail: afe.adogame@ptsem.edu

A. Niederberger et al. (eds.), *Handbook of Migration Ethics*,
https://doi.org/10.1007/978-3-031-89877-8_21

reliable demographic information on arrivals. He aptly observes, "Immigration to the United States (U.S.) is made for many different reasons, which may be economic, political, or social or any combination of them, and subsequent reverse migration may occur for any of the same reasons. Discussion of immigration and the variety of circumstances attendant to it are wide-ranging, but usually such discussions rely on figures related to migration into the country and overlook, either purposefully or accidentally the fact that return migration has always been a significant movement in this country" (109).

One possible reason for the neglect of research on reverse migration is that "return is the most difficult aspect of the migration cycle to quantify" (Gmelch 1980: 135). While numerous migrants return voluntarily or forced to their countries of origin, original homelands, rural-urban nexus, these trends and mobilities are hardly noticed nor recorded. The lack of a definitional consensus impedes on needs assessment, intervention strategies and quantifying size and volume of reverse migration. Reverse migration is characterised with inherent conceptual difficulties and policy implications that could facilitate intervention within the contexts of origin and destination, as well as countries affected by irregular migration flows, temporary or permanent resettlements (Carling et al. 2011; Carling 2015). The trajectories of reverse migration, implications for migrants, families and contexts, migrants' vulnerabilities with regards to homecoming demand basic principles and policy frameworks for governing reverse migration and reintegration. However, reverse migration should not be treated as a homogenous phenomenon. This could result in common misunderstanding in policy framework, planning and intervention.

Reverse migration is the *afterlife* of an enduring migration process whose duration depends on the motivations, circumstances, process and contexts. Reverse migration may refer to the systematic act or process of relocation and movement of emigrants, immigrants, migrants, refugees, returnees who decided to go back or are being forced back to the initial place of departure, original homeland, country, village, town, thus crossing geo-cultural and territorial boundaries to resettle, albeit temporary or indefinite. Iterations of the concept in literature include return migration, return flow, return movement, back migration, second time migration, domestic return, chain migration. Similarly used are countercurrent, counterflow, repatriation, re-emigration, reflux migration, homeward migration, circular migration, remigration, retro-migration, reverse mission, reverse diaspora, reverse skilled migration, reverse remittances or the reverse migration of labour.

However similar or different meanings each of these terms might convey, they imply something more than the simple fact that the migrant goes back to his initial point of departure—international, cross-continental, cross-country, internal migration within country of origin, inter-state, rural-rural, urban-rural, urban-urban or rural-urban influxes. Returnees are referred pejoratively as "*Been-Tos*" (Ghanaian slang for those born or lived abroad and have returned), or named *Tokunbo* (Nigeria-Yoruba phrase for those who have come from abroad or beyond the seas). Minta (2007) employed *The Reverse Diaspora* to characterise "African Immigrants and the Return Home", in a study aimed at ascertaining why there was an urge in the African immigrant community to return to their country of birth; or to "return home".

The Great Reverse Migration in the US refers to the rapid growth of the black African-American migration to the South. Blow's (2021) *The Devil You Know. A Black Power Manifesto* recounts the story of the Great Migration and the impact it had on the South. He calls for a reverse migration, reimagining the return of Blacks to the South in order to claim political heritage lost during the migration in the Jim Crow era. Black suburbanisation was another driving force in the development of black-majority cities (Cunningham 2019). As Drummond notes, "Driven out by racism and lack of economic opportunity, Southern blacks have been migrating North in huge numbers for more than a century, [...] But now, with the promise unfulfilled in many cases, significant numbers of blacks are returning to the South, coming back down home" to a region that seems, at last, to offer as many economic opportunities and as much brotherhood as any other section of the United States" (1974).

Migration and reverse migration should be treated as processes or systems that examine both stream and counter-streams, sending and receiving ends, as well as its translocal, transnational, multidimensional strands. Earlier perceptions about migration in terms of a single individual action or mass movement of people in their quest for greener "pastures", better-life chances from home of origin (sending) and settlement in host (receiving) societies have become rather tenuous. Migration and reverse migration are now thought as multidirectional. It is not simply a process of mobility from poor, less developed, war-stricken, crisis-prone, overpopulated contexts to affluent, developed ones, but vice versa. It encompasses undocumented, asylum seekers, refugees, economic migrants, highly and less-skilled migrants, students, diplomats, businesspersons, missionaries, tourists, temporary, long-term and circular migrants.

Central to migration and reverse migration dynamics is the discourse on brain drain, brain gain, brain circulation, brain waste, reverse brain drain/gain viewed as catalysts for economic and social development (Iredale et al. 2003; Saxenian 2005; Chacko 2007; Dhar and Bhagat 2021). Iredale et al. (2003) provide timely insights into recent developments in return skilled migration in four regions in the Asia Pacific—Bangladesh, China, Taiwan and Vietnam, demonstrating how the mobility of skilled migrants, and the tacit knowledge they bring with them, is a vital component in the process of globalisation. With the recent changes in the global economy, growing unemployment rates in the developed countries and rapid growth of Indian economy there is a likelihood that the reverse brain drain has also been occurring. This is equally applicable for internal migration as well (Dhar and Bhagat 2021). In the absence of systematic studies on the magnitude and characteristics of return migration in India, their work fills an important gap examining the return migration, both internal and international, and the socio-economic and household characteristics of return migrants.

Returning high-skilled migrants bring with them knowledge, expertise, access to global networks and capital, but also international sensibility that influences where they work, live and their expectations and vision of life in the home contexts in which they settle. China and India have significant overseas populations that constitute global talent, and both countries recognise that reverse migration of diasporic talent can contribute to domestic economic development (Zweig et al. 2021). In their study, Zweig, Tsai and Singh compare China and India's capacity to encourage reverse migration with a focus on overseas entrepreneurs who return to start new firms in their homeland. In this way, they underscore how the world's two largest countries have significant overseas population that constitute "a global talent pool". As they remark, "So today, leaders in China and India recognize that reverse migration of diasporic talent can facilitate technological transfer from abroad" (2).

Recent research points to migration and reverse migration during the COVID-19 pandemic era (Devaki and Vanaja 2021; Dandekar and Ghai 2020), also identifying reverse migration from urban to rural contexts (Sen 2020; Sengupta 2020) or from rich to poor countries (Warren 2020). Unprecedentedly, reverse migration is assuming a buzz word in describing how COVID-19 pandemic has triggered remigration from metro cities towards towns and villages, urban to rural areas of origin, and from so-called developed to developing countries. Human mobility has been severely limited due to the novel coronavirus pandemic. Notably, it has forced many migrants to return to their home countries, impacting migration dynamics around the globe and creating a reverse migration paradox amid COVID-19 (Friborg 2020). Devaki and Vanaja (2021) demonstrate how the United Arab Emirates witness an exodus of long-term non-resident Indians, especially Malayalees, due to unforeseen impacts of the COVID-19 pandemic in 2020. Through a means-end approach, their study reviews respondents' attributes, consequences, and values at the time of migration and compares it with their perceptions during the sudden COVID-19 related reverse migration from the United Arab Emirates.

Sen observes, "We can point out here that barring abnormal times (as at present) the major part of the migrant flow in India has been from rural to urban areas, […] The current flow of reverse migration in India, which is from urban to rural, however, falls into none of the above categories describing the usual patterns in the movements from rural to urban centres" (2020: 1). Sengupta also notes, "Post COVID19 lockdowns in India, thousands of daily wage labourers belonging to the informal sector have been leaving big cities in droves. They have no other alternative but to return to their villages though their future there is also bleak […] Indeed, there has been a trend for reverse migration over the last few years, and Kolkata and Hyderabad have, in the recent past, received less number of migrants than before. Workers have been forced to go back because of the lull in construction activities due to environmental reasons. Some have gone back because they could not live in the horrible conditions anymore. But voluntary reverse migration has been less than forced migration" (2020: 1). Warren (2020) presents estimates of the undocumented population residing in the United States, highlighting demographic changes since 2010 and showing how the total undocumented population in the United States continued to decline in 2018, primarily because large numbers of undocumented residents returned to Mexico.

Female migration is evidently a key constituent of global migration and reverse migration processes. While a gendered understanding of migration is quintessential, only recently have efforts been exerted to incorporate gender into theories of international migration. Migration theories are increasingly becoming gender-sensitive, thus marking a shift from perception of female immigrants simply as "dependents", "wives" and "children" of male immigrants to incorporating explanations of the unique experiences of women migrants themselves. The complex dynamism of contemporary migration is partly reflected in the growing feminisation of migration (Adogame 2013: 9–13). Women and children's reverse migration experiences and expressions in their everyday lives vary when located in time and space. Women have not only accompanied or joined their husbands abroad through family reunification; professional women are sometimes leaving their spouses temporarily to care for their children while they take advantage of greener pastures or fall prey to the brain drain syndrome. The access to a variety of educational and employment opportunities acts as a magnet for women wishing to be economic and social actors in their own right. In such cases, their husbands and children have subsequently reunified with them. A gendered perspective of migration is fundamental to understanding the motives for migration, but also reverse migration (Adogame 2013; Kuschminder 2013, 2017).

Migration may also alter the status, attitudes and gender relations of men and women. New economic roles and new responsibilities affect spousal relationships, in some instances leading to considerable negotiations and resistance to change by both men and women. Migration represents a challenge in women's economic and familial roles, as well as in their usual pattern of relations with husbands and children. For some women, migration means an increase in social mobility, economic independence and relative autonomy. For others, getting better-paid employment is only a complementary medium improving the family's economic base. The impact of moving from one form of gender stratification system to another can be understood through the interaction between societal and family contexts. All stages of the migration process are important for considering the influence of gender, but also essential is the reverse migration process.

Children of migrants born and bred abroad face huge complexities in reverse migration. Thus, a specific aspect in the reintegration process concerns the children of migrants and intergenerational dynamics. Children born abroad face return as a migratory experience (Battistella 2018: 8). This is especially traumatic for children when they are forced to return with their parents, even though they may be citizens of the host country and have adapted to the language and culture of the host society. Reintegrating children in the home context or country of the parents is not unproblematic, especially with regards to language challenges and incompatibilities of educational systems. Similar difficulties in reintegration can apply to refugees who have the opportunity to return because the conditions that motivated their search for refuge have ceased. In the worst case, return is complicated by the lack of infrastructure, facilities, and economic conditions to support the reintegration of a large number of refugees (Muzhary 2017; Arowolo 2000).

There is the assumption that migrants settle, adapt or assimilate in and to their "new homes", and no sooner sever ties, links with their original homelands. There is now an increasing tendency to think of the longue durée of the migration process, (non-)settlement as a transient, life-long process encapsulating crucial life stages and sometimes transcending death. While Castles and Miller (2009: 21) aptly describes migration as "a collective action, arising out of social change and affecting the whole society in both sending and receiving areas"; even this idea of "sending" and "receiving" have now shifted from a two-way, bi-polar node to multiple, fluid nodes of "home" or "contexts". Thus, the conceptual "goal-post" of migration is increasingly becoming elastic, mutable, experiencing constant shifts just as the theoretical grid-making of migration is in flux (Adogame 2013: 2). This necessitates a revisiting of existing paradigms, whether and how they can withstand the enigmatic nature of the migratory process in the new global era. A broader treatment of reverse migration phenomenon should include typologies, motives (reasons) for return, theories, dynamics of adaptation and readjustment, corresponding impact on home and host contexts.

1 Understanding Reverse Migration

1.1 Typologies

Migration is a life-long journey, an enduring process involving several strands including reverse migration. "Return migration is the great unwritten chapter in the history of migration" (King et al. 1983, King 2000: 7). Typologies of reverse migration include dimensions, each of which has considerable variety and diversity. Migration trajectories and the explanations behind migrants' decisions, result in different types of return. Two dimensions address the length of time migrants intended to remain abroad and reasons for returning. A distinction could be made between migrants who intended their emigration to be *temporary* and those who desire it to be *permanent*. There are also intermediate positions as either desire or intention is contingent upon other individual, family, contextual and external factors. Thus, intended temporary emigration might turn out to be permanent while those who desired, or at least hoped for a permanent emigration might have their hopes and dreams dashed or short-lived. This can result from a multiplicity of factors including family circumstances, improved or faltering economic conditions in either home or host contexts, sociocultural adaptation or failure of adaptation and integration, illness, marriage etc.

The core variables of the typology of reverse migration revolve around returnees who intended temporary migration; returnees who desired permanent migration but were forced to return; and those who desired permanent but chose to return. An additional variable is that some migrants, even after several years, retain an ideology or a myth of return even though they actually never return back to their homelands on a permanent basis. Most migrants, including those who resettle in other countries, relive the idea that they will return one day.

The International Organization for Migration (IOM), the largest global provider of Assisted Voluntary Return (AVR) and Assisted Voluntary Return and Reintegration (AVRR) programs, identifies two typologies of reverse migration as voluntary and involuntary (forced) return. Voluntary return is "the assisted or independent return to the country of origin, transit or another country based on the voluntary decision of the returnee" while the forced return is "a migratory movement which, although the drivers can be diverse, involves force, compulsion, or coercion" (IOM 2019: 77, 229).

In fact, what is voluntary or involuntary is relative and depends largely on the agency, actor, benefactor, and depending on the circumstances involved in the mobility or the lack of it. Voluntary returns can be either spontaneous or assisted. Spontaneous return is "the voluntary, independent return of a migrant or a group of migrants to their country of origin, usually without the support of States or other international or national assistance". Assisted voluntary return and reintegration is the "administrative, logistical or financial support, including reintegration assistance, to migrants unable or unwilling to remain in the host country or country of transit and who decide to return to their country of origin" (IOM 2019: 205, 12). Voluntary humanitarian return is the application of assisted voluntary return and reintegration principles in humanitarian settings and "often represents a life-saving measure for migrants who are stranded or in detention" (IOM 2020). When return programs involve additional reintegration support for returnees, these programs are referred to as assisted voluntary return and reintegration (AVRR). IOM defines AVRR as "Administrative, logistical or financial support, including reintegration assistance, to migrants unable or unwilling to remain in the host country or country of transit and who decide to return to their country of origin" (IOM 2019). However, the IOM definition is limited as it mostly refers to reverse migration within the territorial boundaries of a country, as in the case of returning internally displaced persons (IDPs) or between a country of destination or transit and a country of origin, as in the case of migrant workers, refugees or asylum seekers. Reverse migration may also occur between rural-rural, urban-rural, urban-urban or rural-urban contexts.

1.2 Motives and Determinants

Migration research shows numerous motives and determinants for migration, however disproportionate attention has been adduced to reverse migration. Thus, pertinent to reverse migration processes is the question of the motives and determinants for returning. It is perhaps difficult to study reverse migration without first considering the emigration factors. Was the emigration originally intended as permanent or only as a temporary, stop-gap measure? Just as research on motives for migration can help to establish the reasons for reverse migration, research on the latter can help to deepen understanding of the former. The complexities of reverse migration warrant an elastic conceptual framework to capture the diverse reasons, motivations, determinants and contextual situations under which individuals, families and groups return or hope to return.

The motives and determinants for reverse migration are context-specific and they are not mutually exclusive. Battistella situates the main types of return on a continuum utilising two variables: the time for return and the decision to return, under four general situations that results in return of achievement, return of completion, return of setback, and return of crisis (forced return) respectively (2018: 10). Based on results of microeconomic analysis, Stark (1993) posits twelve reasons for return migration as: purchasing power; strategic accumulation of human capital; occupational status and social prestige; asymmetric information; social comparisons: ordinal preferences; relative deprivation: cardinal preferences; matching in the marriage market, and divorce; receipt of inheritance; meeting a target; shocks at home; costly separation caused by migration; and failure.

The determinants of reverse migration encapsulate the crucial role of personal factors, family situations, contextual as well as structural factors, both in the countries of origin and destination, as influencing the hope, decision, (un)preparedness to return. Earlier, Rogers (1984) proposed a matrix, based on combined events related and unrelated to migration with aggregate and individual events respectively in the home and host country. The decision to return is linked to structural and individual factors as well as policy interventions (Black et al. 2004). Each of these aspects hold significant gender dimensions.

The motivations for reverse migration are more involved than a simple dichotomy between success and failure, voluntary and imposed, planned and exogenously mandated, prevailing singly or jointly with others. Interest in the circumstances and conditions under which migrants return is obviously shared by policy-makers. Implications for new policy options can be drawn from the complex reasons for reverse migration. These reasons point to a new utility representation and to a corresponding policy regarding reverse migration. Thus, enacting policies that align with motives for return migration will be more efficient than devising measures that are independent of these motives (Stark 1993). As Stark remarks, there is a pressing need, including from the perspective of policy design, to understand what prompts migrants to return to their countries of origin. If policy formation in this sphere aligns with underlying motives for return, the need to resort to costly inducements to encourage migrants to depart could be less, and repatriation would be more effective (105).

2 Theorising Reverse Migration

The discourse on international migration has attracted huge interdisciplinary interests, from the social sciences to the humanities, with each discipline constructing a matrix for framing its own objectives and goals, specific research questions, concepts, methodologies, appropriating dominant theories and hypotheses, and generating its respective frameworks of analysis and conclusions. Rather than a shared paradigm, the study of migration benefits from a variety of competing theoretical viewpoints fragmented across disciplines, regions, and ideologies (Brettell and Hollifield 2000: 2). This cross-disciplinary approach on international migration across disciplines has tended to "ebb and flow" with differing waves and trends of emigration and immigration. Massey et al. (1993) have provided a detailed survey and appraisal of theories of international migration.

A plethora of factors also make reverse migration a multifaceted and heterogenous phenomenon. "As a subprocess of international migration, return migration has been subject to various approaches that offer contrasting sets of propositions stemming from neoclassical economics, the new economics of labour migration, structuralism, transnationalism and social network theory" (Cassarino 2004). While classical theories of migration have, in many respects, become moribund as explanation paradigms of migratory processes, this has paved the way, especially since the 1980s, to new definitional approaches and analyses of the phenomenon of return migration and its impact on origin countries (Kubat 1984; Kritz 1987). Although driven by differences in theories and methods, disciplinary orientations, interpretations, biases and pretensions, these varied explanations are not, in themselves, mutually exclusive. For instance, there are recent theories that analyse reverse migration from an economic standpoint (Sabharwal and Verma 2020) and an emotional perspective (Purkayastha and Sahoo 2020). A theory of return migration should be logically derivable as a subset of a more general theory of migration (Richmond 1983). Generally, the theorising of reverse migration has evolved around distinctive trajectories with each historical phase contributing to the dynamic process of theory-building and analysis of the reverse migration phenomenon.

2.1 The Neo-classical Economic Perspective and New Economics of Labour Migration

It was perhaps Ernest Ravenstein (1885) who first drew attention to the principle of return (reverse) migration through his renowned list of migration laws. The fourth law states: "Each main current of migration produces a compensating counter-current" (1885: 33). He recognises return migration as one component of this counter-current. The existence of this counter-current became a prominent feature in later statistical studies of migration in measuring the volume of "reverse streams" (Wen Lang Li 1970). Ravenstein concluded that migration was governed by a "push-pull" process; that is, unfavourable conditions in one place (oppressive laws, heavy taxation, etc.) "push" people out, and favourable conditions in an external location "pull" them in. These "general theories" often expressed as "push-pull" emphasise tendencies of people to move from densely to sparsely populated areas, or from low- to high-income areas, or link migrations to fluctuations in the business cycle.

The economic variable predominates in most theories that seek to explain why the decision to migrate is made in the first place. Those who migrate are not usually the poorest of the poor or simply a movement from the poor, least-developed countries to richer ones. In fact, many who engage in frequent movements are the upwardly mobile and people of intermediate social and economic status.

Contemporary migration trends suggest a multidirectionality across different contexts: from so-called poor, less-developed contexts to rich, more-developed contexts and vice versa. There is also movement from one poor to another poor, developing country, and from one rich to another rich, more-developed country, from rural to urban and vice versa. Reverse migration is sometimes perceived as the outcome of a failed migration experience which did not yield the expected benefits. This presupposes that an individual return occurs as a consequence of failed experiences abroad or because his/her human capital was stunted or unrewarding.

One contrasting feature of the neoclassical economics and the new economics of labour migration is that while the former views reverse migration as a failure, if not an anomaly, the latter view reverse migration as a success story, if not a logical outcome (Cassarino 2004: 256). Both theoretical approaches provide some insights regarding the reasons why people move abroad and return home, although have several pitfalls at the same time. One shortcoming relates to the actors themselves and their motivations for return. The perceived sole determination by financial or economic factors fail to do sufficient justice in explaining how remittances and skills are appropriated in original home contexts or countries. Returnees are viewed narrowly as "foreign-income bearers" or "financial intermediaries", and little or no attention is paid to where migrants return.

The neo-classical economic theories of migration are largely viewed as individualistic and ahistorical, and also tends to treat the role of the state as an aberration. The neglect of historical causes of movements and the undermining of the role of the state, makes this theory rather simplistic and incapable of explaining actual movements or predicting future ones. The push-pull model is also suspect in predicting unilateral demographic shifts, movements from densely populated to more sparsely-peopled contexts. The model hardly explains why certain migrants choose to go to one country rather than the other. The success/failure paradigm is grossly inadequate in unpacking the complexity of the reverse migration phenomenon against the backdrop that returnees' decisions and strategies are insulated from their social and political realities, and hardly connect with contextual factors at home.

2.2 *The Historical-Structural Approach*

In contrast, the historical-structural approach to reverse migration views return not only as personal, but also social, contextual issue conditioned by situational and structural factors. The approach stresses the unequal distribution of economic and political power in the world economy. Migration was generally perceived as a way of mobilising cheap labour for capital, in this case mass recruitment of labour by capital. Return is both shaped by individual, personal, family experiences of the migrant, as well as social and institutional factors in contexts or countries of origin. This approach to reverse migration underscores how the financial and economic resources brought back to the home origin contexts are critical to the migrant's return decision and reintegration or the lack of it. "Not only do skills and financial capital shape return experiences, but local power relations, traditions and values in home countries also have a strong bearing on the returnees' capacity to invest their migration experiences in their home countries" (Cassarino 2004: 259).

This theory has some merits but can hardly take us too far in comprehending the complexities of migration and reverse migration on grounds that it downplayed the individual/group motivations in the process, and over-prioritised the interests of capital. The theory is not mindful of migratory trends from one rich, advanced context to another, nor does it explain internal migration within a rich country between rural-rural, rural-urban, urban-rural and rural-urban contexts. Although return is no longer viewed as being exclusively affected by the migration experience of the individual, this approach focuses more on how returnees' initiatives could favour economic development when faced with local power structures than on the reverse migration phenomenon per se. In this way, the returnees' migration experience is limited mainly to skills acquisition. Moreover, this perspective undermines the intricate connectedness between the migrant's experiences in their former host and their homecoming.

Nonetheless, historicising migration theory has led to the modification of theoretical concerns and emphases, in light of changing social realities and reverse migration dynamics. The classical and neoclassical migration theories were supplanted by new theories, paradigms of explanation. A predominant question in earlier theorising on international migration is on why people migrate (intentionality), and less on the dynamics of migration, such as the actual situation of migrants in the migratory states and process, chain migration in migration networks, contextual factors that shape migration policies and trends. Also, less captured are reasons and motivations for reverse migration, how reverse migration could be either voluntary and involuntary, how the returnee negotiates the actual contextual situations in the home context. The migration systems theory (MST), the transnational theory (TNT), and social network theory (SNT) took vantage positions as new analytic frameworks for contextualising migration and reverse migration processes.

2.3 *Migration Systems Theory (MST)*

Mabogunje (1970) defines a migration system as a set of places linked by flows and counter flows of people, goods, services, and information, which tend to facilitate further exchange, including migration, between the places. Mabogunje focuses on rural-urban migration, while Kritz et al. (1992) extends this to international migration. These interdependencies give rise to sustained and sizeable bilateral migration flows such that the determination of migration and reverse migration can presumably be made on the basis of these largely exogenous characteristics. The fundamental assumption of MST is that migration alters the social, cultural, economic, and institutional conditions at both the "sending" and "receiving" ends—that is, the entire developmental space within which migration processes operate. The basic principle is that any migratory movement can be seen as a result of interacting macro- and micro-structures, processes that are intertwined and not mutually exclusive (Castles and Miller 2009). Macro-structures refer to large-scale institutional factors, while micro-structures embrace the networks, practices and beliefs of the migrants themselves. These two levels are linked by a number of intermediate mechanisms, which are often referred to as "meso-structures".

Reverse migration dynamics are also shaped by the interacting macro-, micro-, and meso-structures. Macro-structures which include the political economy of the world market, interstate relationships, and the laws, structures and practices established by the states of sending and receiving countries to control migration and reverse migration (re)settlement. Macrostructures involving informal social networks developed by the migrants themselves, helps them to cope with migration, adaptation, reverse migration and reintegration. The informal networks which include personal relationships, family and household patterns, friendship and community ties, and mutual help in economic and social matters are integral in homecoming.

The family and community (religious, social, cultural, ethnic, economic and political) are crucial in reverse migration networks. Family linkages often provide both the financial and the cultural capital which make reverse migration possible. The intermediate meso-structures referring to certain individuals, groups or institutions take up a mediating role between migrants and political or economic institutions. The "migration industry", including inter-government agencies, recruitment organisations, lawyers, agents, smugglers, NGOs, charitable bodies and other intermediaries can be both helpers and exploiters of migrants and return migrants alike (Adogame 2013: 7).

2.4 The Transnational Theory (TNT) and Social Network Theory (SNT)

Return or homecoming does not constitute a *finis* to the migration process, reverse migration is the *afterlife* of migration, a cycle which continues the migration story that could in turn stimulate remigration. Therefore, reverse migration is integral to a migration cycle of social, cultural and economic relationships, networks and exchanges facilitating the reintegration of migrants. A further consideration of a new, emerging migrant population whose networks, activities and life-patterns encompass and transcend their home and host societies has produced a new body of theory on "transnationalism" and "transnational communities" (Basch et al. 1994). The TNT in this regard capture migrants, their lives, experiences and consciousness as one that cuts across national boundaries and bring two or more societies into a single social field.

Transmigrants, a designation for those immigrants who build such social fields, develop and maintain multiple relations—familial, economic, social, organisational, religious, and political that span borders. The notion of a transnational community puts the emphasis on human agency (Portes et al. 1999). The concept of transnationalism turns into a clearly defined and measurable object of research in acknowledging that transnationalism involves individuals, their networks of social relations, their communities, and broader institutionalised structures such as local and national governments. Portes and others distinguish between transnationalism from above, that is, activities "conducted by powerful institutional actors, such as multinational corporations and states"; and transnationalism from below, that is, activities "that are the result of grass-roots initiatives by immigrants and their home country counterparts" (221).

Transnational identities derive from and are shaped by migrant's experiences and identities acquired from their original homelands and contexts of origin, the host society where they live, the sustained interactions or networks with the home contexts, and perhaps the reverse migration when it does happen. Returnees are confronted with possibilities and challenges of reintegration or resettlement in reverse migration. The transnational mobility involving regular, back-and-forth movements between home and host contexts, the remittances sent home, the deployment of investment capital may help to leverage reverse migration. Such preparedness, illusion and false perceptions of reality about home could also result in dashed hopes, disappointment and frustration on reintegration process.

Transnationalism entails the maintenance of linkages and networks between migrants and their families and friends in the original home context and other local contexts, but also how migrants feel linked to one another by their common ethnic, national, cultural origins and in-group solidarity transnationally. In reverse migration, transnationalism provides a template for cross-border linkages between the returnees and their migrant communities abroad. Transnational social fields are not pi-bolar but multi-polar; they envelop migrants' countries of origin and other host contexts (Adogame 2013: 9). Reverse migration captures this multi-polarity of the transnational social field, just as Diaspora highlights the multi-polar mobility of migrants.

The social network theory, similar to the transnational approach to reverse migration, perceives returnees as conveyors of tangible and intangible resources, and who maintain strong linkages with the migrants' former host contexts. Resources needed to secure homecoming may also depend on investments and assets in the context where they lived, but also patterns of interpersonal relationships that may derive from the returnees' past migration experiences. Returnees are social actors who negotiate their reintegration into their home contexts.

The different theoretical approaches above help to analyse and contextualise the reverse migration phenomenon, by paying closer attention to the growing diversity of returnees, a complex variety of return motivations—real and imagined, micro-, meso- and macro-variables, contextual and institutional factors, the circumstances under which they return, and the realities of reintegration and remigration. Narratives of migration are weaved around the different kinds of people that move; the intentionality and stimulus to migrate; the realities of the local context from which they move; the anxieties of the journey to the "unknown Eldorado (land)"; the dicey expectations of moving to a new cultural context; the realised or dashed hopes upon arrival at the temporary/permanent destination; contextual factors in the new abode of sojourn; preparedness and the actual reverse migration; and the global factors that shape international migratory processes in different historical epochs (Adogame 2013: 15). The narratives of one phase or within one space/time are integral to and contingent upon our understanding of the entire migration and reverse migration story.

Narratives of migration and reverse migration should capture the reality of individual life stories and institutional narratives that shape migratory processes from: the home of origin as a point of departure through the travails of the transit—the journey itself, to the destination(s), new home(s) and then possibly return migration to the point of take-off (original home of origin). Religious repertoires and experiences are implicated in these webs of mobility both in how individual and collective motive(s), decision(s) to migrate are often imbued with religious narratives; and in whether, how and to what extent the "journey", "sojourn", "return" heightens or rather dwindles religious zeal, enthusiasm, belonging, participation, praxis, expression and experience. The religious variable thus becomes a vital marker in charting the causes, courses and conduits of contemporary migration (Adogame 2013: 13). Religious attitudes and motivations of immigrants in the context of temporary or permanent sojourn is often a reflection of religious/spiritual experiences and resources employed and deployed during the decision-making, preparatory stages of the journey, the liminal trajectories of the journey, sojourn in the host context, and reverse migration to the original home context. The process of reverse migration could also lead to religious expansion and proliferation, a case in which immigrants are not simply establishing religious communities in diaspora, but also through a kind of spiritual remittance mechanism, such communities are charting their "routes" and "roots" back to their former homelands.

Technological revolution, the ease of transport, globalisation, economic opportunities and disparities are drastically transforming human mobility into an increasingly complex, multidirectional phenomenon. Reverse migration is neither a secondary nor ancillary phenomenon, but rather an integral and crucial component of international migration (Adogame 2013: 25). Migration does not always and necessarily culminate in return and where it occurs, return may not be permanent. Instead of migration being a one-time journey, migrants and their families often subsequently move to a second or third country of destination. While a migrant has the right to return to his/her own country, return is not necessarily the final stage in the migration process. Many migrants do not envision returning to their country of origin, and a multitude of factors influence their final decision whether or not to return. Thus, the very process of reverse migration or the hope of a return to the home of origin depends on a multiplicity of factors.

Reverse migration distils a number of the major challenges pertaining to migration in general, ranging from the conditions that prompt migratory movements in the first place and that determine the incentives for and sustainability of return, to the question of how to make return "work" for the development of both individuals and societies. Several migrants express a desire to return to their original home contexts at some point. Although this desire to return may just be wishful thinking and a myth of ambiguity for many migrants, the mere fact of this consciousness rekindles forlorn hopes. Return is contingent upon a number of factors such as improving their economic base, fulfilling expected dreams, completion of academic programmes and trainings.

2.5 Readjustment/Reintegration Dynamics and Impact

Migrants confront opportunities but also huge challenges and difficulties upon return. The question of adaptation and readjustment can be approached from different perspectives, emic and etic, in which case what is determined as success or failure is relative. One approach is from the migrant's own perception of his/her alienation, adaptation and adjustment or the lack of it. The migrant is in the best position to assess how and to what extent s/he hopes and feels that the homecoming has filled self-defined aspirations, needs and social vacuum, but also provide a sense of well-being, flourishing, and personal fulfilment. Reintegration and readjustment are also evaluated based on the concrete economic, social realities and personal conditions of the returnee. This may include access to gainful employment or becoming an entrepreneur, own investments, property or affordable housing, reintegrating into old and building new family, cultural, social and sometimes political networks.

Returnees are potential brokers of information, economic and social capital, innovation and international networks. Processes of alienation, reintegration and readjustments or the lack of it in reverse migration needs to be understood against the backdrop of individual circumstances, internal and exogenous factors in the respective local contexts. It also depends on contextual factors in the host society that necessitated their home going in the first instance. Some migrants return from relatively rich countries to countries with limited job opportunities and with lower standards of living. The economies and social conditions of host countries may have dwindled in contrast to their original home contexts, thus opening manifold opportunities for returnees. Others come back from differentiated social systems to relatively simple socioeconomic structures in which much of the overseas acquired skills may prove redundant. While others return from contexts with somewhat different norms, values and ideas. Thus, returnees may have dashed hopes, experienced disillusionment, and encountered enormous personal problems during homecoming. Returning migrants may also encounter envy and suspicion from families and neighbours or display superiority and sophistication over the neighbours and families they have returned to meet. Even issues of efficiency and punctuality could be irritating and thus lead to friction.

Problems of alienation and readjustments are heightened along gender, sex differences, intergenerational identities and conflicts. Difficulties encountered in the readjustments of children and women within migrant families, and the coping mechanisms they adopt vary considerably. Gender affect, positively and negatively, the sustainability of migrant's return and reintegration. During the migration process there are often changes of gender constructions and of the identity—and life drafts and in order to address these there is a need for gender-sensitive return counselling and a gendered approach to return and reintegration programs (Black et al. 2004: 14).

The impact of reverse migration is also directly dependent on the status of the migrant—skilled, unskilled, lower, middle and upper strata levels of the returning migrants and where they choose to resettle—rural, urban, peri-urban or suburban contexts. Reintegration and readjustments are functions both of the actual contextual factors but also one of the raised or dashed hopes, expectations and imagined aspirations held by returnees. The influence of reverse migration on original home societies could be assessed in terms of the introduction of newly acquired skills, capital investment through savings, remittances, introduction of new ideas and attitudes, changes in social structure resulting from reverse migration. Return migration could in turn stimulate and encourage further emigration. Therefore, the overall impact of reintegration is dependent on variables including personal, structural, internal, external and contextual factors, whether positive or negative.

3 Reverse Migration: New Research Directions and Futures

New theoretical models on the motives and determinants of reverse migration and the varied dimensions of reintegration would help to complexify both dynamics and discourses. The narratives and experiences of migrants are rather complex, just as their aspirations, hopes, reasons and preparedness to return. For many immigrants, returning home is a dream, imagination, hope, prospect they cherish and one that sustains them during their migration history, albeit a myth or reality. Homecoming and reintegration are hardly straightforward processes to be reduced simply to success or failure, life satisfaction gains and losses, subjective well-being variables. Scholarship addressing the proliferation and diversity of global reverse migration must transcend the sociology of return, mainly from economic and political perspectives, with the former assumption that return is voluntary, and the latter recognising how state and institutional actors in countries of origin shape reintegration experiences. The complexities of reverse migration require a multimethod or mixed method approach. Hagan and Wassink (2020) provide an integrated path forward for researching and theorising return migration—a synergised resource mobilisation framework.

A proper understanding of the intricate, interpersonal relationship between women and men, and the intergenerational dynamics between migrant parents and children in reverse migration is complicated, but needs further nuancing. International and internal migration processes are gendered, thus demanding careful unpacking of the increasing feminisation of migration and reverse migration. Discourses on reverse migration should pay increasing attention to the multidirectional mobilities between rural-urban, urban-rural, rural-rural, urban-urban as well as movement from the global North to the global South and global South to North, as well as south-south, north-north networks and mobilities.

The renewed interest among global policy makers on the nexus between reverse migration and development, specifically the potential impact of migration on the development of contexts/countries of origin has refocused attention on returnees. The complex role returnees play in these reverse migration processes therefore needs deeper unpacking. Policies towards reverse migration in many countries/contexts of origin are mostly lackluster and undeveloped. Developing sustainable policies and intervention strategies will require a robust grasp of the complexities of reverse migration as well as a more ingrained understanding of the multiplicity of factors that determine it. The residual concern among policymakers is monitoring of remittances through mostly formal channels. "The more policy-oriented literature on reverse migration has been sensitive to the role of the state in bringing back its overseas national to contribute to development" (Zweig et al. 2021: 2). On the other hand, countries/contexts of destination have increasingly developed reverse migration schemes mainly to diminish the number of migrants, particularly irregular migrants, as antimigrant backlash.

Reverse migration should hardly be considered the byproduct of the migration process or the termination of the migration cycle. Reverse migration is integral to the migration process as a liminal phase. Migrant's return or homecoming is an inflection point in the migration journey that could even induce remigration. Therefore, reintegration and value-added role enhancement of returning migrants towards development may depend on the nature and implementation of reverse migration policies framework and strategies in both origin and destination contexts/countries.

Literature

Adogame, A. 2013. *The African Christian Diaspora. New Currents and Emerging Trends in World Christianity*. London, New York: Bloomsbury Academic.

Alerstam, T. 1991. Ecological Causes and Consequences of Bird Orientation. In *Orientation in Birds*, ed. P. Berthold, 202–225. Basel: Birkhäuser Verlag.

Arowolo, O. O. 2000. Return Migration and the Problem of Reintegration. *International Migration* 38:59–82.

Basch, L., N. Glick-Schiller, and C. Blanc-Szanton, eds. 1994. *Nations Unbound. Transnational Projects, Postcolonial Predicaments, and Deterritorialized Nation-States*. New York: Routledge.

Battistella, G. 2018. *Return Migration. A Conceptual and Policy Framework*. New York: The Center for Migration Studies of New York (CMS).

Black R. et al. 2004. *Understanding Voluntary Return*. Sussex Centre for Migration Research, Online Report 50/04. Publication of the UK Home Office.

Blow, C. M. 2021. *The Devil You Know. A Black Power Manifesto*. New York: HarperCollins Publishers.

Bovenkerk, F. 1974. "Defining Return Migration". In F. Bovenkerk. *The Sociology of Return Migration. A Bibliographic Essay* (Publications of the Research Group for European Migration Problems 20). Dordrecht: Springer Netherlands.

Brettell, C. B., and J. F. Hollifield, eds. 2000. *Migration Theory. Talking Across Disciplines*. New York and London: Routledge.
Carling, J. 2015. Myth of Return. In *The Wiley-Blackwell Encyclopedia of Race, Ethnicity and Nationalism*, 1–2. Hoboken, NJ: Wiley-Blackwell.
Carling, J., E. B. Mortensen, and J. Wu. 2011. *A Systematic Bibliography on Return Migration*. Oslo: Peace Research Institute Oslo (PRIO).
Cassarino, J. P. 2004. Theorising Return Migration. The Conceptual Approach to Return Migrants Revisited. *International Journal on Multicultural Societies* 6:253–279.
Castles, S., and M. Miller. 2009. *The Age of Migration*. London: Macmillan Press.
Chacko, E. 2007. From Brain Drain to Brain Gain. Reverse Migration to Bangalore and Hyderabad, India's Globalizing High Tech Cities. *GeoJournal* 68:131–140.
Cunningham, M. 2019. "In Atlanta, Reverse Migration is Beginning to Skew Suburbs Blue". *ABC News*, November 20, 2019, https://abcnews.go.com/Politics/atlanta-reverse-migration-beginning-skew-suburbs-blue/story?id=67106895.
Dandekar, A., and R. Ghai. 2020. Migration and Reverse Migration in the Age of COVID-19. *Economic and Political Weekly* LV:28–31.
Devaki, V. M., and M. V. Vanaja. 2021. Migration and Reverse Migration. Gulf-Malayalees' Perceptions During the Covid-19 Pandemic. *South Asian Diaspora* 13:157–177.
Dhar, B., and R. B. Bhagat. 2021. Return Migration in India: Internal and International Dimensions. *Migration and Development* 10:107–121.
Drummond, A. 1974. "Black Return to South in a Reverse Migration". *The New York Times*, June 18, 1974, https://archive.nytimes.com/www.nytimes.com/library/national/race/061874race-ra.html.
Friborg, S. 2020. "The Reverse Migration Paradox During COVID-19". *Borgen Magazine*, October 6, 2020.
Gmelch, G. 1980. Return Migration. *Annual Review of Anthropology* 9:135–159.
Guzzetta, C. 2004. Return Migration. *Journal of Immigrant & Refugee Services* 2:109–117. https://doi.org/10.1300/J191v02n01_07.
Hagan, J. M., and J. T. Wassink. 2020. Return Migration Around the World. An Integrated Agenda for Future Research. *Annual Review of Sociology* 46:533–552.
International Organisation of Migration. 2019. *International Migration Law. Glossary on Migration* 34. Geneva: International Organisation of Migration.
International Organisation of Migration. 2020. *Return and Reintegration Key Highlights*. Geneva: International Organisation of Migration.
Iredale, R., F. Guo, and S. Rozario, eds. 2003. *Return Migration in the Asia Pacific*. Cheltenham: Edward Elgar Publishing.
King, R. 2000. "Generalizations from the History of Return Migration". In IOM (ed). *Return Migration. Journey of Hope or Despair*. Geneva: International Organization for Migration.
King, R., Strachan, A. J. and Mortimer, J. 1983. "Return Migration. A Review of the Literature". *Discussion Papers in Geography* 19. Oxford: Oxford Polytechnic.
Kritz, M. M. 1987. The Global Picture of Contemporary Immigration Patterns. In *Pacific Bridges: The New Immigration from Asia and the Pacific Islands*. New York: Center for Migration Studies.
Kritz, M. M., L. L. Lim, and H. Zlotnik, eds. 1992. *International Migration Systems. A Global Introduction*. Oxford: Oxford University Press.
Kubat, D. 1984. Book Review: Auslaenderfeindlichkeit. *International Migration Review 18(2): 330–331*. https://doi.org/10.1177/019791838401800213 (Original work published 1984)
Kuschminder, K. 2013. *Female Return Migration and Reintegration Strategies in Ethiopia* (PhD dissertation). Maastricht: Maastricht Graduate School of Governance.
Kuschminder, K. 2017. *Reintegration Strategies. Conceptualizing How Return Migrants Reintegrate*. New York: Palgrave Macmillan.
Lewis, H. F. 1939. Reverse Migration. *The Auk* 56:13–27.

Li, W. L. 1970. A Differential Approach to the Study of Migration Streams. *Rural Sociology* 35:534–538.
Mabogunje, A. L. 1970. Systems Approach to a Theory of Rural-Urban Migration. *Geographical Analysis* 2:1–18.
Massey, D., et al. 1993. Theories of International Migration. A Review and Appraisal. *Population and Development Review* 19:431–466.
Minta, K. 2007. "The Reverse Diaspora. African Immigrants and the Return Home". *Undergraduate Humanities Forum 2006–7. Travel*, https://repository.upenn.edu/uhf_2007/10.
Muzhary, F. 2017. "Resettling Nearly Half a Million Afghans in Nangrahar. The Consequences of the Mass Return of Refugees". *Afghanistan Analyst Network.*
Nilsson, C., and S. Sjöberg. 2016. Causes and Characteristics of Reverse Bird Migration. An Analysis Based on Radar, Radio Tracking and Ringing at Falsterbo, Sweden. *Journal of Avian Biology* 47:354–362.
Portes, A., L. E. Guarnizo, and P. Landolt. 1999. The Study of Transnationalism. Pitfalls and Promises of an Emergent Research Field. *Ethnic and Racial Studies* 22:217–237.
Purkayastha, B., and A. K. Sahoo. 2020. Indian Transnationalism. In *Routledge Handbook of Indian Transnationalism*, ed. A. K. Sahoo and B. Purkayastha, 1–18. Abingdon: Routledge.
Ravenstein, E. G. 1885. The Laws of Migration. *Journal of the Statistical Society of London* 48:167–227.
Richmond, A. H. 1983. Explaining Return Migration. *International Migration Review* 17:269–275.
Rogers, R. 1984. "Return Migration in Comparative Perspectives". In D. Kubat (ed). *The Politics of Return. International Return Migration in Europe.* New York: Centre for Migration Studies.
Sabharwal, M., and R. Verma. 2020. Transnationalism and Return Migration of Scientists and Engineers from the United States to India. In *Routledge Handbook of Indian Transnationalism*, ed. A. K. Sahoo and B. Purkayastha, 53–66. Abingdon: Routledge.
Saxenian, A. 2005. From Brain Drain to Brain Circulation. Transnational Communities and Regional Upgrading in India and China. *Studies in Comparative International Development* 40:35–61.
Sen, S. 2020. "Pandemic and the Reverse Migration of Labour in India. Project: Political Economy of Labour in a Globalised Economy". *Mainstream Weekly* LVIII, New Delhi, June 6, 2020.
Sengupta, J. 2020. "Reverse Migration. An Opportunity for the Government". *India Matters*, April 4, 2020, https://www.orfonline.org/expert-speak/reverse-migration-opportunity-government-64123/.
Stark, O. 1993. *The Migration of Labor.* Oxford, Cambridge, MA: Basil Blackwell.
Thorup, K. 2004. Reverse Migration as a Cause of Vagrancy. *Bird Study* 51:228–238.
Warren, R. 2020. Reverse Migration to Mexico Led to US Undocumented Population Decline. 2010 to 2018. *Journal on Migration and Human Society* 8:32–41.
Zweig, D., K. S. Tsai, and A. D. Singh. 2021. Reverse Entrepreneurial Migration in China and India. The Role of the State. *World Development* 138:1–12.

The Securitisation of Migration

Therese Herrmann

Abstract Originating with the Copenhagen School of security studies in international relations, the concept of the "securitisation of migration" refers to the social and political processes through which migration, as a public issue, acquires security character. This article reconstructs central tenets of the Copenhagen School's speech act theory of security and discusses how it has been applied to the case of migration. While the Copenhagen School understood the securitisation of migration as the discursive construction of a threat to communal identities, subsequent accounts have emphasised the prevalence and, sometimes, discourse-independence of the material practices of securitisation. In doing so, they have substantially re-interpreted central tenets of the Copenhagen School's securitisation theory, including the role and requirement of an audience for securitisation to take place and the forms of threat construction at work in the securitisation of migration. The article concludes by discussing the implications of the securitisation of migration for migration ethics. It argues that they depend on whether one understands the securitisation of migration as a theoretical framework or a social and political condition. Read as a theoretical framework, the securitisation of migration provides a way of classifying security arguments at work in migration ethics according to formal criteria. Read as a social and political condition, it makes a point about the presuppositions within which debates about migration ethics are situated in public.

The "securitisation of migration" refers to a framework of analysis that seeks to explain the politics through which migration comes to be treated as a security issue. From a securitisation perspective, specific migration policies are premised upon prior social and political processes that fix the security character of migration as a public problem. Originating with the Copenhagen School of security studies, securitisation theory initially focussed on the capacity of speech acts to constitute a problem *as* a security problem and, within this framework, interpreted the securitisation of migration as the construction of threats to collective identities. Subsequent theories of the securitisation of migration have adopted a stronger emphasis on

T. Herrmann (✉)
University of Duisburg-Essen, Essen, Germany

A. Niederberger et al. (eds.), *Handbook of Migration Ethics*,
https://doi.org/10.1007/978-3-031-89877-8_22

institutions, and a more sociological lens to privilege a focus on security technologies and security professionals' networks and practices. While securitisation theory focusses on the processes through which an object acquires security character, the term "securitisation of migration" is also used in a wider sense to refer to the fact that migration is approached from the perspective of security in specific discourses, policies and practices.

1 Origins of Securitisation Theory: The Copenhagen School and Constructivism in International Relations

The concept of "securitisation" was coined in the 1990s through the work of international relations theorists Ole Wæver, Barry Buzan and their collaborators, who came to be referred to as the Copenhagen School of security studies (Buzan et al. 1998; Wæver 1995). The Copenhagen School offered a constructivist reading of security, understanding it as a social practice constituted through speech acts that allow an actor to suspend applicable rules by framing an issue as an existential threat. Reflecting intensifying processes of transnationalisation after the end of the Cold War, this definition built on attempts in security studies to widen the concept of security beyond its traditional focus on "the study of threat, use, and control of military force" (Walt 1991: 212; see Wæver 1995). But the Copenhagen School theorists also criticised contemporary alternative definitions in critical security studies for linking security too closely to a normative understanding of individual needs, thereby missing the concept's foundation in actual social practice (Wæver 1995). They argued that both definitions treat security threats as objective entities, when they should, instead, be understood as the product of specific types of social interactions.

According to the Copenhagen School, issues become security issues "when a securitizing actor uses a rhetoric of existential threat and thereby takes an issue out of what under those conditions is 'normal politics'" (Buzan et al. 1998: 24–25). In this reading, the security character of an object is established through a specific type of interaction with it—this is what makes security a *practice*. Security is, furthermore, a *social* practice, because the type of interaction relevant here does not directly take place between agent and object, but is mediated by interaction between various agents. Thus, actors' capacity to interact with an object in a certain way is not inherent in either actor or object, but conveyed intersubjectively, through others' recognition of the actors as having this capacity. Building upon insights from the philosophy of language (Austin 1975[1962]), the Copenhagen School further argues that the agent-to-agent interaction establishing the security character of an object takes the form of *speech acts*. This means that others' recognition of an actor's capacity to break the normal rules of politics is conveyed through an argument that succeeds in framing an issue as an existential threat for a specific entity. Securitising speech acts are "self-referential" (Buzan et al. 1998: 24) or "performative" (Balzacq et al. 2016: 495) because, in succeeding to change the social practice around a specific issue, they transform the character of the issue they are referring to.

Copenhagen School securitisation theory articulated these ideas around a number of core concepts, which subsequent securitisation theorists have taken up and developed further, sometimes amidst substantial revision. These are the *securitising actor*, who presents a particular issue as an existential threat via a *securitising move*, which according to the Copenhagen School primarily takes the form of a speech act. Furthermore, securitisation happens when a securitising move succeeds in conveying to a specific *audience* that an issue poses an existential threat to a particular entity, called *referent object*. A successful securitising move authorises a securitising actor to enact *extraordinary means* to tackle the perceived threat (Buzan et al. 1998: 24–26).

Because security is understood as a social practice in the Copenhagen School's constructivist sense, a successful act of securitisation not only results in a change in how a would-be threat is treated, but, in changing the pattern of social interaction around a specific object, also has effects on the securitising community itself. In this regard, the Copenhagen School theorists distinguish securitisation from politicisation as a mode of social interaction. Buzan et al. argue that, on the one hand, when measured in terms of the state resources that can be mobilised to handle a problem, securitisation should be understood as a "more extreme version of politicization" (ibid.: 23). But, on the other hand, from the viewpoint of the type of social interaction it generates, securitisation should be seen as contrasting with politicisation:

> "Politicization means to make an issue appear to be open, a matter of choice, something that is decided upon and that therefore entails responsibility, in contrast to issues that either could not be different (laws of nature) or should not be put under political control [...]. By contrast, securitization on the international level [...] means to present an issue as urgent and existential, as so important that it should not be exposed to the normal haggling of politics [...]" (ibid.: 29).

While politicisation, therefore, implies that the way a society handles an issue remains contestable, securitisation fixes a specific mode of interaction around an issue. Though Buzan et al. emphasise that their concept of security is largely analytic—explaining security dynamics as they actually happen rather than starting from a normative idea of security—, their efforts to contrast securitisation with politicisation also reveal securitisation theory's critical intent. In analysing securitisation as a specific mode of dealing with problems that tends to suspend public argument and strengthen executive power, Buzan, Wæver and their collaborators write from a perspective of "desecuritization" (ibid.), which they understand as a state of things in which handling an issue as a security issue remains an object of public choice (see also Wæver 1995 and, critically, Floyd 2011, 2019). However, as I will discuss at the end, securitisation theorists differ about the precise normative content of the idea of desecuritisation.

Starting from the specific pattern of a social interaction based on a security claim rather than from the inherent qualities of a security threat allows the theorists of the Copenhagen School to apply the concept of security to a variety of sectors outside of its traditional focus on the state and military force, namely, politics, the economy, the environment and society. The sectors framework shows that there may be considerable variation in the rules of social interaction around specific threats, including on the constitution of a threat's referent object and the securitising actors authorised to call out and act upon a threat. While in the military sector, for example, a threat is classically a military threat to the territorial integrity of a state, a threat in the political sector would target the authority of a political order and, as such, need not necessarily employ force. Threats in the environmental sector—an idea still fairly new at the time of Wæver's and Buzan's writing—are distinct in that they often operate at a global level and threaten "human civilization" (ibid.: 79) as a whole rather than a specific collective. The Copenhagen School's discussion of the dynamics of securitisation in the environmental sector also shows that the relation between securitisation and politicisation may shift considerably within and among sectors (see Balzacq et al. 2016). For example, Buzan et al. see debates in the environmental sector particularly confronted with the "paradox" (ibid.: 76) that conserving present levels of human civilisation in the face of environmental threats may require far-reaching politicisation of contemporary ways of life.

Theorists of the Copenhagen School were among the first to study migration from the viewpoint of security practices and criticise the securitisation of migration in a newly emerging supranational Europe (Wæver et al. 1993). In the context of post-Cold War ethnic conflicts and a backlash against intensified migration movements, Wæver and his collaborators largely discussed migration within the specific dynamics of the sector of "societal security", where it is understood as a potential threat to national and other collective identities. The concept of "societal security" and its focus on collective identities raised considerable controversy in the field (see, among others, Buzan and Wæver 1997; McSweeny 1996; Williams 1998), with critics arguing that, in arguing that societal threats were primarily identity threats, the Copenhagen School implicitly sanctioned a communitarian understanding of society that essentialises collective identities and obscures their contestedness.

In response, Buzan et al. re-emphasised the constructivist nature of securitisation theory and maintained that their point was precisely to show that collective identities were not simply presupposed in, but in fact "constructed and reproduced" (1998: 219) through a specific form of securitising speech acts. What is important, in this reading, is that an actor manages to mobilise a collective identity to the point that it allows her to make plausible the idea of an existential threat requiring an extraordinary response (see also Williams 2003). Moreover, Wæver's (1993) extensive discussion of the securitisation of migration in post-Cold War Europe suggests that sectors themselves are dynamic and constructed around the mobilisation of threats. In this reading, the societal sector is not simply "out there" to take up migration as a security issue or not, but emerges and gains ground against other sectors depending on how migration is discussed and responded to (see Huysmans 1998: 489–490).

Sectors are not only distinct when it comes to securitising actors and referent objects, but, arguably, also with regard to the types of threat responses securitising actors may adapt. If the societal sector is to be sufficiently autonomous from the other sectors, according to the Copenhagen School, threats to collective identity must at least allow for the possibility of societal actors taking up extraordinary means themselves. The task of threat response may not automatically be referred to government institutions, although Buzan et al. concede that this will often be the case (Buzan et al. 1998: 122). As Wæver elaborates in an earlier text, extraordinary means in societal terms need not involve force. Societal actors may also make use of specific cultural means, such as the promotion of a unifying cultural identity: "If one's identity seems threatened by internationalization or Europeanization, the answer is a strengthening of existing identities. In this sense, consequently, culture becomes security policy" (Wæver 1995: 66).

While the Copenhagen School securitisation theory was widely received as a departure from essentialist understandings of security, its specific design has prompted methodological debate on all of its central elements. Issues have included the role and requirement of an audience in a securitising speech act, the definition of extraordinary means, the context of securitising speech—including the power relations and institutional constellations pre-structuring the likelihood of a speech act's success as well as the influence of previous securitisation processes on these—and, most importantly for the context of migration, the significance of non-discursive practices in acts of securitisation (for an overview see Balzacq et al. 2016).

2 The Securitisation of Migration: From Discourse to Practice-Based Theories

Subsequent accounts of the securitisation of migration have retained and developed elements of the Copenhagen School's constructivism, but mostly re-conceptualised the relationship between discourse and practice set out in the Copenhagen School's speech act theory. Although the analysis of public discourse continues to play a role in some of these theories, their focus—in what has been referred to as a "sociological" shift (Balzacq 2010) in securitisation theory—is on the capacity of institutions as well as security policy practices and technologies to pre-structure public debate in migration issues.

In his analysis of the role of transnational security networks in an emerging European migration policy in the 1990s, Didier Bigo (2002; Bigo and Guild 2005) has been one of the earliest proponents of re-conceptualising securitisation theory to steer it into a sociological direction, focussing on the routine practices of security professionals and their techniques of government. Bigo understands the securitisation of migration in the European Union as driven by the functional interests of security professionals, whose often informal networks—comprised of members of the police, intelligence services and the military, public administration at both Member State and supra-state level, but also private corporations and non-governmental organisations—cut across traditional public-private distinctions. He relies on the work of Pierre Bourdieu and Michel Foucault to conceptualise how security professionals engage in fixing the character of migration as a security issue through collective, largely non-discursive practices. From Bourdieu, Bigo (2013) adopts the concepts of *habitus* and *field of practices* to convey how the object of risk management integrates professionals from various backgrounds through a specific set of routines and forms of knowledge based around confidentiality, non-partisan proactivity and technological verification. As Balzacq et al. (2016: 505) note, Bigo ascribes a colonising logic to the practices of professional security networks in the context of transnationalisation. Where transnationalisation increases the competition between different policy actors, Bigo holds that security networks have been able to use the authority gained from claiming to handle security threats as comparative advantage vis-à-vis other agencies and approaches, which enabled them to secure access and funding to handle other supranational issues such as migration. At the same time, he notes that the dynamic of securitisation has in turn encouraged transnationalisation processes, as threat control and risk management follow a functional logic that surpasses territorial boundaries.

From Foucault, Bigo borrows the idea of *governmentality,* as opposed to governance or government, to describe that the logic of security practices cuts across the public-private distinction and is "less interested in [...] in the control of territory than in the control of population" (Bigo 2002: 83). Although transnational security networks may involve public officials and may convene in the national interest, they do not usually exercise power through creating and implementing laws, but via a wider array of legal, sub-legal and extra-legal practices to direct the conduct of those subjected to them (see also Foucault 2007). Where the logic of security comes to be applied to migration, according to Bigo, it does not assume the shape of an exceptional threat-response, but is part of a routine of practices evolving in the context of transnationalisation (on this point, see Léonard and Kaunert 2019).

Empirically, Bigo's emphasis on the securitising agency of transnational networks of security professionals takes the Europeanisation of migration policy from the late 1980s onwards as its object of inquiry. Jef Huysmans (2006) has built on this to offer a longer-term perspective on the securitisation of migration in the European Union. What requires explanation, in his account, is how migration—mostly seen through the lens of economic utility in post-War Western Europe—came to be treated as a security challenge around the same time that Member States chose to yield sovereign powers to an emerging supranational polity. "The assumption", Huysmans writes, "is that abolishing internal border controls and facilitating transnational flows of goods, capital, services and people will challenge public order and the rule of law. This link has been constructed so successfully that is has obtained the status of common sense" (2006: 69).

Huysmans observes that a shift took place in the European discourse on migration from an emphasis on labour migration and social welfare issues to a more politicised approach that has focused on asylum and irregular migration since the mid-1980s. This coincided with efforts to coordinate migration and security policy at a European level, starting with the incorporation of migration into the agendas of informal transnational security networks dealing with terrorism and drug-trafficking, such as the Trevi group, the Club of Bern or the STAR group, and the emergence of intergovernmental migration control groups, such as the Schengen Group and the Ad Hoc Immigration Group (Huysmans 2000; Huysmans 2006). In initialising transnational cooperation on migration via the channel of security policy, these networks, according to Huysmans and Bigo, pre-structured the Europeanisation of migration policy, even though they were often not formally part of the European integration process. Their security framing of migration would later transpire into EU migration policy—beginning with the Treaty of Maastricht in 1993, which institutionalised migration issues overwhelmingly as part of the Justice and Home Affairs sector that, as the EU's Third Pillar, remained largely a matter of intergovernmental cooperation. When this changed with the Treaty of Amsterdam, which integrated Justice and Home Affairs into European community policy as part of the plan to progressively establish a common "area of freedom, security and justice", the formal and informal migration and security networks established previously largely continued to exercise their agenda-setting capacity. Huysmans goes on to argue that the security framework of EU migration policy continues to be reflected in the design of European asylum policies, whose security component dominates humanitarian or human rights-based approaches. Thus, the Dublin Regulation, aimed at coordinating Member States' asylum policies at a European level, was "heavily overdetermined by a policy aimed at reducing the number of applications" (2000: 756).

A related set of practice-based accounts of the securitisation of migration has put an emphasis on the securitising agency of security tools and technologies. They argue that the instruments used to implement security policies—in Europe, databases like the Visa Information System (VIS) (Mitsilegas 2012), the Schengen Information System (SIS) or Eurodac (Balzacq 2008) and the biometric technologies they run on (Ceyhan 2008; see also Torpey 2000) constitute prominent examples—may become sufficiently autonomous as to generate a securitising impetus themselves. In his analysis of the changing uses of European information exchange systems set up as counter-terrorism tools, Thierry Balzacq argues that policy tools, first, "embody a specific image of the threat and, to a large extent, what ought to be done about it" and, secondly, define "a routine set of procedures that structure the interactions between individuals and organizations" (2008: 80).

Balzacq's example is the Schengen Information System (SIS), which, he claims, has been transformed from a reporting tool to an investigative tool in the course of European security policies after 9/11 (see also Brouwer 2008). It can be argued that the SIS's design securitised migration from the outset, as it combined data on immigration and criminal law offences in a single database. Thus, the SIS data contain information on persons wanted for arrest, summoned to appear in court or under secret surveillance, but also on third-country nationals to be refused entry because of a failure to comply with Member States' immigration laws. As the function of the SIS was extended in the wake of the 9/11 terrorist attacks, criminal investigation agencies such as Europol, Eurojust and national judicial institutions received access to its database. Although the European Council explicitly excluded the data on third-country nationals to be refused entry from Europol and Eurojust's access rights, national judicial authorities were given full access to the SIS database, including its entries on third country nationals. National authorities in turn face no restrictions to share these data with EU criminal investigation agencies as part of transnational security cooperation initiatives. As a result, "persons registered for immigration law purposes will be at greater risk of being affected by law enforcement measures or secret surveillance" (Brouwer 2008: 514).

Following Balzacq's argument on securitising policy tools, the securitising dynamic at work in this case is at least partly generated by the design of the SIS database itself. In linking criminal offences to immigration offences, the SIS extends the scope of persons who are processed as threats to include unwanted immigrants. Although it is ultimately political decisions and their underlying power relations which determine who will have access to which part of SIS data, the databasis's design pre-selects objects of potential relevance to security agencies, thereby presenting their security character as given to future users.

An alternative, only loosely connected set of texts on the securitisation of migration has emphasised the institutional context of securitisation moves (Bourbeau 2013; van Munster 2009; Watson 2009). Although they mostly share Bigo's assessment that the Copenhagen School's speech act theory underrated material practices, they are equally critical of Bigo's focus on transnational executives, arguing that securitisation theory still requires the identification of tangible connections between specific discourses and the policy moves they enable.

3 Criteria for the Securitisation of Migration: Discourse and Threat Construction

In emphasising the capacity of practices to pre-structure discursive environments, practice-based accounts of the securitisation of migration significantly modify the original tenets set out by the Copenhagen School—most importantly, the requirement of an audience, but also the form of the threat constructed as a result of securitising moves. The question of whether the securitisation of migration requires the discursive construction of migration as a threat to an identifiable audience has arguably raised most controversy in the field. The majority of analyses, including heavily practice-based accounts, foresee some role for discourse in securitisation practices. Bigo has maintained a loose internal connection between security practices and securitising speech acts, arguing that security professionals' practices and technologies have enabled actors in public discourse "to create a 'truth' about the link between crime, unemployment, and migration" (2002: 65). This, however, inverts the relationship between speech act and practice as set out by the Copenhagen School, and makes their connection much more indirect than originally envisaged—it may suffice, on Bigo's account, that security practices create a discursive environment in which claims about the connection between migration and security acquire self-evidence. *If* and by whom such claims are actually made is secondary for acts of securitisation to take place.

It has, therefore, remained contested among practice-based theories of the securitisation of migration how to deal with cases where discursive securitisation moves are largely absent. Huysmans (2006: 4) has argued that, while some level of threat construction in public discourse is required for migration to be securitised, this may take the form of indirect association, as when migration is listed together with terrorism as a rationale for introducing identity cards in communications by the British Home Office. In her analysis of the securitising practices of FRONTEX, Sarah Léonard has gone further, arguing that institutional practices need not be accompanied by a complementary public discourse, and may even contradict it. She suggests that FRONTEX's establishment in 2004 and its subsequent expansion could build on an already solidified consensus that migration should be understood as a security problem, which made redundant the re-establishment of migration's threat character in public discourse. Indeed, Léonard (2010: 236) claims that this even allowed FRONTEX to adopt a humanitarian discourse on migration and asylum, while its practices remained fixed on securitisation (see also Moreno-Lax 2018).

This raises wider questions about the relation between public discourse and institutions implied in securitisation theory, and how it might be made fruitful for the case of migration. A potentially inverse relation between institutionalised and discursive securitisation was already implied in the Copenhagen School's original account. Buzan et al. noted that successful securitisation moves will usually result in a degree of institutionalisation, but that "the need for drama in establishing securitization falls away" once a claim to securitisation has been institutionalised, since this usually implies that a "security argument [has been] so well established that it is taken for granted" (1998: 28). The relation between discourse and institutionalisation is further complicated where analyses move beyond the site of democratic states. Feminist (Hansen 2000) and non-Western (Kapur and Mabon 2018) critiques of Copenhagen School securitisation theory have made the broader point that securitisation theory's emphasis on speech acts is based on a liberal conception of the public sphere and its relation to the state, making it difficult to be applied to cases where public discourse is constrained or societies heavily stratified.

Questions about securitisation theory's implicit conception of the public sphere and the institutional context it is embedded in are also relevant in discussions about securitisation in transnational contexts. The European Union has received specific attention from theorists of the securitisation of migration from the Copenhagen School onwards, although similar dynamics have also been observed in other transnational settings (Mitsilegas 2012). In their accounts of the securitisation of migration in EU in the wake of the 9/11 terror attacks, both Christina Boswell (2007) and Andrew Neal (2009) have argued that no immediate link could be found between securitising speech acts in the attacks' immediate afterwards and practices that securitised migration and asylum, such as those leading to the establishment of FRONTEX in 2004. While this leads Boswell to conclude that no securitisation of migration took place in this period, Neal suggests that transnational settings like the EU pose a specific challenge to the framework of securitisation theory's speech act analysis, since neither a clearly institutionalised public sphere nor an identifiable monopoly of violence exists (see also Sperling and Webber 2019). Indeed, it is notable that those analyses of the securitisation of migration that have focussed on discourse (Bourbeau 2013; Watson 2009) have chosen nation-states as their site of inquiry, while accounts directed at the European Union and other transnational contexts remain heavily reliant on a practice-based reinterpretation of the Copenhagen School's tenets.

Neal has suggested that a way to bring in line diagnoses of securitising practices with the absence of threat construction in migration-related discourse is by modifying the concept of the threat that is constructed in securitising moves. Building on Bigo and studies on the EU's counter-terrorism policies (Aradau and van Munster 2007), he has used the case of FRONTEX to argue that securitising migration in the EU is better understood as the "quieter and more technocratic" (Neal 2009: 348) practices of managing risks rather than the mobilisation of extraordinary means to tackle threats. While threat response, according to Neal, is largely reactive and works on suspending normal procedures, risk management sets in at an earlier stage and analyses potential threats. As such, it proceeds proactively, in FRONTEX's case anticipating and directing migration movements before they reach European shores. Simultaneously, Neal asserts that FRONTEX's risk management is more inward-looking than threat response would be, focussing on expanding Member States' capacities to anticipate migration-related risks and, thus, it "has a regulatory and disciplinary function that promotes the 'improvement' of external border management through integration" (2009: 349).

4 Outlook: Securitisation and Migration Ethics

What, if anything, does the securitisation of migration imply for migration ethics? The question may be answered from various angles, depending on whether one reads the securitisation of migration primarily as a specific theoretical framework through which to analyse linkages between migration and security or, alternatively, as an empirical diagnosis, which observes that migration has been successfully securitised in specific settings and analyses what the condition of migration's securitisation means for those involved.

Where the securitisation of migration is mainly understood as a theoretical framework, it offers a formal vocabulary to analyse a variety of security dynamics involving migration, including discourses and practices not usually associated with security politics. For example, Michaela Ceccorulli (2019) has used securitisation theory to study non-typical security claims and their effects on migration when she proposed that, in the wake of the European "refugee crisis" the Schengen zone itself was securitised, meaning that European actors (re-)established it as an essential value of the European Union that had to be protected, even if that would undermine the EU's commitments to the rule of law and to migrants' rights. Writing on the securitisation of humanitarian migration, Scott Watson (2009) made a similar point when arguing that a commitment to the institution of refugeehood may itself be securitised as part of a communal identity and may then, paradoxically, be invoked to justify the deterrence of migrants. *Because* state actors construct the institution of refugeehood as an essential feature of the international order, he contends, they can frame irregular migration as a threat to it, eroding the claims of "real" refugees and states' willingness to take them in via resettlement. Although they are not explicitly normative themselves, these studies can open up new directions for normative debate, as they use securitisation theory to connect arguments about a community's essential values to an analysis of practices that erode or suspend international norms concerning migration.

Securitisation theory can also help to categorise and contrast security arguments at work in classic accounts of migration ethics. Indeed, theories of migration ethics discuss emergency situations in which applicable norms may be suspended for the sake of security. Yet what precisely counts as security and what is permitted to safeguard it strongly differs across the various accounts, and often remains undertheorised. For example, in his defence of open borders, Joseph Carens (2013: 277–279) argues that temporary border closures are permissible when it can be shown that the large-scale arrival of immigrants poses an imminent threat to a state's public order, even though he takes this to be empirically unlikely. He further qualifies such an emergency scenario by adding that, for the restrictions on immigrants' freedom of movement thus imposed to be justified, it must be shown that the public order threat in question is directly caused by the number of immigrants arriving, rather than by violent reactions to it from the state's current population. In his communitarian defence of a state's right to exercise discretionary control over its borders, David Miller (2016), on the other hand, understands the continuation of a nation's cultural identity as a precondition for upholding a democratic welfare state. If a nation takes its cultural identity to be disrupted by immigration, this may justify it in closing its borders—in certain situations, even for humanitarian immigration, which Miller finds states under a duty to permit otherwise (ibid.: 163).

Securitisation theory, with its specific vocabulary, its taxonomy of security sectors and its focus on the internal consequences of securitisation moves is well-placed to analyse and contrast these arguments further. Beyond that, specific accounts of the securitisation of migration can fulfil a critical function by situating the creation of links between migration and security in historical context. Doing so, they make concrete the wider point that migration's constitution as a security problem cannot be reduced to the presence of specific objective factors, but is the product of a distinct type of social interactions, which in turn takes place under specific historical conditions. Yet, importantly, even where studies on the securitisation of migration can lift the veil of necessity from threat scenarios involving migration, it cannot arbitrate between them. Precisely because securitisation theory is unable what a "real threat" would consist in, it is not in a position to defend—for example—Carens's narrow definition of a public order threat against Miller's account of a threat to cultural identity on the grounds that Carens's threat definition is less reliant on a political community's subjective fears and projections. For similar reasons, it has no resources to weigh Western states' perceived security interests against the more immediate human security needs of forced migrants (cf. Hammerstadt 2014). Indeed, in laying out the premises of securitisation theory, Copenhagen School theorists already noted that securitisation theory's "inability to counter securitization (say, of immigrants) with an argument that this is not really a security problem or that the environment is a bigger security problem" (206) may be its main disadvantage.

Yet this does not mean that securitisation theory has no normative dimensions at all. Wæver emphasised that, while securitisation theory does not make substantive normative claims, its conceptualisation of security as a social and political practice does lay the ground for a normative analysis of security claims and security conduct. Securitisation theory, he notes, "points to the inherently political nature of any designation of security issues and thus it puts an ethical question at the feet of analysts, decision-makers and activists alike: why do you call this a security issue? What are the implications of doing this—or not doing it?" (1999: 334) According to Wæver, the "ethical question" about when it is appropriate to treat something as a security issue should be answered by democratic publics (ibid.; Williams 2003). Securitisation theory can contribute to their deliberation by highlighting the negative consequences of security practices, but is it unable to say more.

A second, alternative angle from which to analyse the impact of the securitisation of migration for migration ethics is to ask about the implications of migration's successful securitisation for the way migration ethics is discussed in public. Most securitisation theorists share the assessment that various forms of immigration have been successfully securitised in the Global North, specifically in the European Union—even as there are disagreements about when precisely securitisation took place and to what extent it also covers humanitarian migration (cf. Léonard and Kaunert 2019). All versions of securitisation theory make the point that successful securitisation has practical implications, which do not just come in the form of individual migration control policies, but also concern the wider assumptions and practices that determine the character of migration as a public problem. Successful securitisation, therefore, brings about a change in the broader social and political conditions—or: context (Balzacq et al. 2016: 502–504; Léonard and Kaunert 2019: 30–35)—in which public discourse takes place. Once certain forms of immigration are securitised, their threat character enters public discourse as an establish truth that is not easily re-assessed through argument. This is in part because of the argumentative logic of security claims itself, which rely on an atmosphere of urgency and an emotionally charged response. But it is also due to fact that the work of the security experts authorised via successful securitisation moves tends to be removed from public accountability, and securitisation has often had the effect of taking away avenues of legal challenge for those migrants subjected to it (Mitsilegas et al. 2020).

Securitisation theorists have discussed what follows from this diagnosis in normative terms as part of the idea of desecuritisation. As outlined above, for the Copenhagen School, desecuritisation described the state in which the question of whether to treat an issue with the means of security policy remains an object of public discussion. But others have suggested that, given the colonising dynamics of securitisation practices—their propensity to make it increasingly difficult to re-assess the threat character of a given object once it is treated by the means of security policy—desecuritisation should be interpreted in a more substantial way. Rather uniquely among securitisation theorists, Rita Floyd (2019) has recently made the case that securitisation theory should use the resources of moral philosophy, especially just war theory, to lay out ways to distinguish between "just" and "unjust securitisation". Jef Huysmans (2006) has argued that accounts of the desecuritisation of migration should ultimately take an ethical turn, sketching a vision of security that runs counter to the friend-foe logic that, according to him, is inscribed in security politics. Claudia Aradau (2004), finally, has proposed that desecuritisation would need to rely on a political vision of reconfiguring access to the public sphere, with a view to continuously opening up avenues to challenge the premises and practices of security policy, especially for those immigrants who have become its objects.

Literature

Aradau, C. 2004. Security and the Democratic Scene. Desecuritization and Emancipation. *Journal of International Relations and Development* 7:388–413.

Aradau, C., and R. van Munster. 2007. Governing Terrorism Through Risk. Taking Precautions, (Un)Knowing the Future. *European Journal of International Relations* 13:89–115.

Austin, J. 1975[1962]. *How to Do Things with Words. The William James Lectures Delivered at Harvard University 1955*. 2nd ed. Cambridge, MA: Clarendon Press.

Balzacq, T. 2008. The Policy Tools of Securitization. Information Exchange, EU Foreign and Interior Policies. *Journal of Common Market Studies* 46:75–100.

Balzacq, T. 2010. A Theory of Securitization. Origins, Core Assumptions, and Variants. In *Securitization Theory. How Security Problems Emerge and Dissolve*, ed. T. Balzacq, 1–30. London: Routledge.

Balzacq, T., S. Léonard, and J. Ruzicka. 2016. 'Securitization' Revisited. Theory and Cases. *International Relations* 30:494–531.

Bigo, D. 2002. Security and Immigration. Toward a Critique of the Governmentality of Unease. *Alternatives* 27:63–92.

Bigo, D. 2013. Security. Analysing Transnational Professionals of (In)Security in Europe. In *Bourdieu in International Relations. Rethinking Key Concepts in IR*, ed. R. Adler-Nissen, 114–130. London: Routledge.

Bigo, D. and Guild, E. (eds) 2005. *Controlling Frontiers. Free Movement Into and Within Europe*. London: Routledge.

Boswell, C. 2007. Migration Control in Europe After 9/11. Explaining the Absence of Securitization. *Journal of Common Market Studies* 45:589–610.

Bourbeau, P. 2013. *The Securitization of Migration. A Study of Movement and Order*. London: Routledge.

Brouwer, E. 2008. *Digital Borders and Real Rights. Effective Remedies for Third-Country Nationals in the Schengen Information System*. Leiden: M. Nijhoff.

Buzan, B., and O. Wæver. 1997. Slippery? Contradictory? Sociologically Untenable? The Copenhagen School Replies. *Review of International Studies* 23:241–250.

Buzan, B., Wæver, O. and de Wilde, J. 1998. *Security. A New Framework for Analysis*. Boulder: Lynne Rienner Publishers.

Carens, J. H. 2013. *The Ethics of Immigration*. Oxford: Oxford University Press.

Ceccorulli, M. 2019. Back to Schengen. The Collective Securitisation of the EU Free-Border Area. *West European Politics* 42:302–322.

Floyd, R. 2011. Can Securitization Theory Be Used in Normative Analysis? Towards Just Securitization Theory. *Security Dialgoue* 42:427–439.

Floyd, R. 2019. *The Morality of Security. A Theory of Just Securitisation*. Cambridge: Cambridge University Press.
Foucault, M. 2007. *Security, Territory, Population. Lectures at the Collège de France, 1977–1978*, ed. M. Senellart. Basingstoke: Palgrave Macmillan.
Hansen, L. 2000. The Little Mermaid's Silent Security Dilemma and the Absence of Gender in the Copenhagen School. *Millenium* 29:285–306.
Hammerstadt, A. 2014. The Securitization of Forced Migration. In *The Oxford Handbook of Refugee and Forced Migration Studies*, ed. N. Sigona, K. Long, G. Loescher, and E. Fiddian-Qasmiyeh, 294–306, New York: Oxford University Press.
Huysmans, J. 1998. Revisiting Copenhagen. Or, On the Creative Development of a Security Studies Agenda in Europe. *European Journal of International Relations* 4:479–505.
Huysmans, J. 2000. The European Union and the Securitization of Migration. *Journal of Common Market Studies* 38:751–777.
Huysmans, J. 2006. *The Politics of Insecurity. Fear, Migration and Asylum in the EU*. London: Routledge.
Kapur, S., and S. Mabon. 2018. The Copenhagen School Goes Global. Securitisation in the Non-West. *Global Discourse* 8:1–4.
Léonard, S. 2010. EU Border Security and Migration into the European Union. FRONTEX and Securitisation through Practices. *European Security* 19:231–254.
Léonard, S. and Kaunert, C. 2019. *Refugees, Security and the European Union*. Abingdon: Routledge.
McSweeny, B. 1996. Identity and Security. Buzan and the Copenhagen School. *Review of International Studies* 22:81–93.
Miller, D. 2016. *Strangers in our Midst. The Political Philosophy of Immigration*. Cambridge, MA: Harvard University Press.
Mitsilegas, V. 2012. Borders, Security and Transatlantic Co-operation in the 21st Century. Identity and Privacy in an Era of Globalised Surveillance. In *Immigration Policy and Security. US, European and Commonwealth Perspectives*, ed. T. Givens, G. Freeman, and D. Leal, 148–166. New York, London: Routledge.
Mitsilegas, V., Moreno-Lax, V. and Vavoula, N. (eds) 2020. *Securitising Asylum Flows. Deflection, Criminalisation and Challenges for Human Rights*. Leiden: Brill Nijhoff.
Moreno-Lax, V. 2018. The EU Humanitarian Border and the Securitization of Human Rights. The 'Rescue-Through-Interdiction/Rescue-Without-Protection' Paradigm. *Journal of Common Market Studies* 56:119–140.
Neal, A. 2009. Securitization and Risk at the EU Border. The Origins of FRONTEX. *Journal of Common Market Studies* 47:333–356.
Sperling, J., and M. Webber. 2019. The European Union. Security Governance and Collective Securitisation. *West European Politics* 42:228–260.
Torpey, J. 2000. *The Invention of the Passport. Surveillance, Citizenship and the State*. Cambridge: Cambridge University Press.
Van Munster, R. 2009. *Securitizing Immigration. The Politics of Risk in the EU*. Basingstoke: Palgrave Macmillan.
Wæver, O. 1995. Securitization and Desecuritization. In *On Security*, ed. R.D. Lipschutz. New York: Columbia University Press.
Wæver, O. 1999. Securitizing Sectors? Reply to Eriksson. *Cooperation and Conflict* 34:334–340.
Wæver, O., Buzan, B., Kelstrup, M. and Lemaitre, P. (eds) 1993. *Identity, Migration and the New Security Agenda in Europe*. London: Pinter Publishers.
Walt, S. 1991. The Renaissance of Security Studies. *International Studies Quarterly* 35:211–239.
Watson, S.D. 2009. *The Securitization of Humanitarian Migration. Digging Moats and Sinking Boats*. London: Routledge.
Williams, M.C. 1998. Modernity, Identity and Security. A Comment on the 'Copenhagen Controversy'. *Review of International Studies* 24:435–439.
Williams, M.C. 2003. Words, Images, Enemies. Securitization and International Politics. *International Studies Quarterly* 47:511–531.

Part IV
Ethics of Refugees and Asylum

The Case of Migrants, Refugees and Tourists

Georgiana Turculet

Abstract In this chapter I advance the view that migrants, refugees and tourists are individuals who move to pursue fundamental interests. Movement across international borders is the common feature in their life projects. I suggest that distinguishing between broader categories of movement, *movement out of need*, *forced movement* and *movement out of choice*, helps us to reflectively grasp the distinction between moral and political arguments about migrants, refugees and tourists. When advanced at the level of public discourse, I call them political arguments; and when in political philosophy, moral arguments. I will suggest ways in which moral arguments are instrumentalised politically, and political arguments moralised—both rhetorical strategies which obfuscate, rather than help our understanding of the complex field of migration and movement processes.

Movement is an essential interest for human and non-human animals (Turculet 2018). While the latter undertake migratory routes to escape predators, cold seasons, to mate and find green fields, the former escape dictators and authoritarian regimes. Unlike non-human animals, we aim at fulfilling more complex interests than the basic needs ensuring survival. My aim in this chapter is to provide non-academic audiences with some modest conceptual tools to reflectively engage with typical arguments on the complex topic of human movement, presented to us daily in political discourse and media, which we are called on to comprehend and act on politically, as citizens (Turculet 2020, 2022).

G. Turculet (✉)
Department of Law, University Pompeu Fabra, Barcelona, Spain
e-mail: georgiana.turculet@upf.edu

A. Niederberger et al. (eds.), *Handbook of Migration Ethics*,
https://doi.org/10.1007/978-3-031-89877-8_23

1 Broad Distinctions Between Migrants, Refugees and Tourists

Humans have always moved in various patterns, from empire to empire, from one continent to another, from rural to urban places, and from a country or state to another, in the pursuit of many legitimate and important interests and life goals (Goldin et al. 2012). Some may move to marry somebody in a far-away land, to form political and commercial alliances, to join different communities and ways of life, to work and exchange goods, to pursue science and arts, or to share knowledge and innovative ideas with others. In so doing, those who move enrich the society as a whole and benefit us all. To illustrate this point, let us glance at the migratory trajectories of two well-known scholars.

Marie S. Curie, the only scholar awarded two Nobel Prizes in two distinct disciplines, arguably could not have become one of the greatest scientists of all times, had she not had the opportunity to move during her lifetime. As a woman in her native country Poland, she was in *need to move* as she was not allowed to dedicate her life to the pursuit of science—on grounds of her gender (Rockwell 2004). However, following her migration to France, and the subsequent opportunity to study and to become a researcher at Sorbonne University in Paris, she became the scientist whose discoveries are now well-known to the entire world.

Curie's contemporary colleague, Albert Einstein, while fearing the persecution of the Nazis, was *forced to move* for his safety. He found *safe haven* in the United States of America (USA hereafter), and like Marie Curie in France, he received the opportunity to join a scholarly community, in Princeton University. Both scholars pursued their scientific careers despite dire circumstances. Einstein and Curie, however, might have never met, cooperated and exchanged ideas, even if they were contemporary colleagues. Their cooperation was rendered possible by the opportunity to undertake *movement out of choice*—that is to take short trips for scientific conferences and alike (Pycador 1999; Rockwell 2004; Passachoff 1996). The above examples show that movement is essential to sharing knowledge. Those who engage with and create knowledge must enjoy a necessary and sufficient level of autonomy to pursue both movement and knowledge (Raz 1986; Oberman 2016; Miller 2016; Turculet 2018). Just as it is with knowledge, which cannot be obtained in isolation or under certain constraints, many human activities will also expectedly flourish under a regime of personal autonomy (Raz 1986), independently of whether those who pursue legitimate goals are migrants, refugees or tourists (Oberman 2016; Turculet 2018). Finally, just as this holds true for sedentary individuals.

However, regardless of the primary reason prompting one to move, as in Einstein and Curie's cases above, and the grounds on which one is accepted as a newcomer in a community at one given point, it is worth noting that one might hold contemporarily all three normative grounds to move: out of need, force, and choice (Turculet 2018). That is, one might fit one of the three or all three distinct categories of movement at the same time, as a migrant, a refugee and a tourist. We nevertheless distinguish between three scenarios. First, Einstein moved from Germany to the United

States because of an imminent threat to his life. In this instance, we take that he was *forced to move*, or moved as a refugee or an asylum seeker. Second, moving from the United States, where he has found safe haven, to Canada—hypothetically to take an academic position in Montreal—he moves in this second circumstance like Curie, as a migrant. In other words, given his safety already obtained in the United States, he moves *out of need*—to take a job opportunity, where the opportunity is located. He is a migrant, either or not his departure begins in his country of origin or a third country, so long as the location from which he departs is fairly safe. If his human rights are not protected counterfactually neither in the country of origin, nor in the one which promised him a safe haven, we take that he is still forced to move. Many contemporary refugees and migrants move from one place to another multiple times in search of safe haven and opportunities. We inhabit a world in which deep global inequalities between countries, and between individuals within countries exist, and in several places, there are armed conflicts, civil wars or authoritarian regimes (Donmez and Duman 2020; Vatansever 2020). Also, see present murderous armed attacks on Palestine, Gaza and Ukraine. Third, if Einstein travelled to meet colleagues and friends for short trips, his movement might be closely described as a tourist, independently of whether he travels from Germany, from the United States or Canada, and regardless of his destination. Depending on the circumstance, Einstein can plausibly be termed all at once a refugee, a migrant and a tourist. So far, I suggest that refugees are forced to move, migrants move out of need, and tourists move out of choice.

To mark the distinction between tourists from migrants and refugees, is *inter alia* the duration of their projects (Yong 2017), as well as the level of autonomy and control over one's own established timeline (Turculet 2022). Whereby only refugees and migrants move, typically take residency and labour in and for the newly adoptive country. In the case of migrants, it could be about an important amount of years, which permits one to become a member of a new society, socially and politically, which is a moral desideratum (Carens 2013). In the case of refugees, given that it is a threatening context to prompt their movement, their return will only be foreseeable, if the threat is significantly reduced or it disappears. In many cases this could mean an indefinite amount of time, and not seldomly a lifetime, which prompts many generations to get settled away from the place of origin. Both, migrants and refugees, have in common that they often do not have individual control over the very circumstances that prompted their departure and the place and conditions of admittance in a new place. Yet, their return, if ever foreseeable, depends upon these complex circumstances, which drastically limit their ability to plan their time in one or more places.

Each one of the above-mentioned circumstances of movement, which Curie and Einstein presumably undertake, indicate a distinct type of movement. Moving elsewhere to pursue a career, although it entails pursuing an important life goal, appears less urgent than escaping a murderous regime. While people move in the pursuit of several important interests, regardless of what prompts movement at one given point, how should we draw the line between who is a refugee, a migrant, or a tourist? Is the distinction between different types of movement as clear-cut as it appears considering these illustrations? If so, on which precise features does the distinction hold true, upon moral and empirical scrutiny?

2 Reflections on Distinct Types of Movement

The distinction between migrants, refugees and tourists is intuitively neat. Given the brief introduction of the scientists' life trajectories of the past century, few migration scholars would disagree that we ought to consider for important reasons, three conceptual categories: *movement out of need*, *forced movement* and *movement out of choice* (Turculet 2018). Also, few would disagree that migrant, refugee and tourist is compatible with respectively moving out of need, because one is forced, or one chooses so, at least to some degree.

Consider that Marie Curie moved from Poland to France to access a career opportunity, one which did not exist and was not likely to exist in her country of origin during her lifetime. Her migration is compatible with movement out need—given the level of patriarchy of her country, in which only men's advancement in science was permitted and promoted. Correspondingly, most migrants move to access and enjoy more or better opportunities than the limited range offered in the country of origin, independently of their specific life plans, as plumbers, nurses, doctors, professors or unspecialised workers. Plausibly, some may wish to live in a less patriarchal society, have more material means, financial stability as well as more ability for long term planning for themselves and their children—all opportunities which they count to find in the place, which they join as newcomers, often characterised by more advanced political and economic stability. Put this way, the distinction between migrants, refugees and tourists, then does not rely on which kind of legitimate life goals anyone pursues and the form in which they do so. All humans have fundamental interests, independently of whether they move or have sedentary lives. All have an essential interest in moving or not moving, as they see fit (Oberman 2016; Turculet 2018).

Both scholars have legitimate life goals, which I take to be equivalent in that both have an interest in scientific discoveries. Such life goals, we fairly assume, is of fundamental importance to each one of them, regardless of the main cause prompting their moving in the first place, at one given point. But notice, this is true, also regardless of whether they would hardly ever move. As a counterexample, when Socrates was sentenced to death under the accusation of corrupting the youth with his philosophical views, decided to stay and face the death sentence in Athens (Waterfield 2009). His decision to stay, if he had an opportunity to flee, does not set the moral ground for most who face analogous circumstances to act similarly. It however informs us about one possible set of beliefs and considerations, which I take, each person should set for themselves, even if they decide to stay and face death. He was entitled to his belief that he was more of a philosopher, if he stayed, and stood by his philosophical beliefs.

Einstein pursuing science as his life goal is compatible with undertaking all three categories of movement, including all three in a short period of time. Who do *we* conceive a migrant, a refugee or a tourist then? He might take a trip to France to meet Curie, while he waits for his transcontinental journey to be organised towards the USA, or after he finds safe haven. Just as I live in a Denmark, and while here, I travel abroad to conferences, to meet friends and family in my country of origin and other countries and enjoy months of research stay in yet another country—as a lifestyle mover. If I were in Denmark because persecuted for my scholarly work, as Socrates was, my need of safety shall not inform us on my interest to be safe as the only interest in movement I hold. By far, movement serves many more purposes, all valuable to me. When we are free to move between states, as we are in Europe today, it might feel so ordinary to be able to satisfy all the above fundamental interests, in whatever pattern movement is necessary to our ends. Nevertheless, seeing my movement such a big part of my lifestyle, shall not obscure the fact that for many contemporary refugees and migrants it might prove virtually impossible to obtain a tourist visa, even after they obtain their refugee status or permission to reside and live in a safe place, which are also extremely hard to obtain. Border regimes are set in such a manner as to reify categories of movement, as if when one is a refugee, their interest in sheer safety is all that defines their interests and life goals.

Nonetheless notice, one might apply for a tourist visa, and take the trip to satisfy a secondary scope, for example to bring oneself to safety. It is not morally condemnable that if I visit a place with the primary aim to attend a conference, I have the scope of meeting my good friend who lives in town—as both can be fundamental interests I hold. Many refugees fleeing Nazi Germany have forged travel documents to be able to move or faked their identities altogether, with the aim of finding a safe place. One might hold an interest in accessing a summer program in another country, even if the motive for signing up for it in the first place is to find safety somewhere. Even if we can and should distinguish between types of interests that each person has, assuming that it is morally relevant comprehending those narratives, we may still not define anyone as belonging to a fixed category of movement. I take that each person has a multitude of fundamental goals in life, independently from what prompts one's movement. Does then the distinction between migrants, refugees and tourists, rely instead more firmly on the background conditions that prompts one to move in the first place?

On the one hand, it is morally relevant, whether one's life is under imminent life-threatening risk, as in Einstein's case fleeing Germany, while another is attempting to live a more fulfilling life, such as by pursuing a career, as in Curie's case—to become a scientist in a place where this opportunity exists. It is morally relevant to distinguish between these two categories, especially for enabling us to prioritise who enters first, or who enters, if there is only one available place. Most would agree that the greater the need or urgency, the bigger should be the chance for them to obtain entry in a safe state.

On the other hand, however, from a strictly moral point of view, this distinction between refugee, migrant and tourist also collapses, if the background conditions prompting movement out of need, forced movement, and to some degree also chosen movement, are equally extreme. We accept no morally relevant distinction between one who moves to escape extreme poverty and direct persecution—if both lead to one fearing death (Oberman 2011; Turculet 2017). In the case of a *migrant boat* approaching European shores, many passengers are plausibly *forced* to undertake this life-threatening trip (Olsaretti 1998; Turculet 2018), even if, among those so described, some are more properly termed migrants in search of opportunities while others, refugees looking for safe haven (Carens 2013). Death caused by hunger or disease, by a gun or a bomb, by severe urban destitution, by some natural disaster such as a Tsunami or a volcano about to erupt, or because of a combination of some of the above dire conditions, amount to equivalent dire background conditions. We permissibly weigh one's fear of death equally.

This distinction collapses not only between migrants and refugees, but also between migrants, refugees and tourists. If on the migrant boat approaching the European shores, there is also a tourist, who fears arbitrary imprisonment, torture or death in the place she fled, she qualifies no less than others as forced movement as well—at least in the instance in which such background conditions obtain. Thus, if everyone who moves with fundamental life goals, also face very similar background conditions, the distinction between migrants, refugees and tourists is less clear, from a moral point of view. Let us consider two literary works inspired by true stories.

Little Bee by Chris Cleave, is the title of the novel and the nickname of the moral hero, a Nigerian teenage girl who escapes from Africa. She is on the run because she faces direct persecution by the hand of a murderous group in their mission to seize control over natural resources. Although she manages to escape and to travel to Europe hidden on a boat, she is deported to Nigeria, after being detained for two years by the United Kingdom's migration authorities—on the grounds that her country of origin was considered safe to be returned to. What the UK failed to recognise in deporting her to Nigeria is that the group she feared back in Nigeria had murdered most people in her village, including her family. She had nobody to go back to, but fear and trauma. However, as the story unfolds, the author zooms out from Little Bee's faith alone, and shows the extremely precarious life conditions, which ordinarily affected everyone in her region at the time (Cleave 2010). The complex relationships between armed conflict over the control of natural resources by various factions of the society, political corruption, absence of the rule of were only some of the causes of this extreme precariousness, which affects to some degree most members in many developing societies (Collier 2013).

This generalised perspective of dire precariousness appears even more clear, when the author describes Sarah, the UK journalist who travels to Africa to investigate the case of Little Bee. Although she travelled as a tourist, she boarded a specific flight on a specific day, which she did not choose—as she boarded the flight on which the UK authorities scheduled Little Bee to be deported. First, Sarah's mission was to ensure that Little Bee was not going to face the authorities of her country, from which she escaped, as an unaccompanied minor, which is a vulnerable position for anyone in her situation. Second, Sarah was going to attempt to obtain papers for Little Bee to return to the United Kingdom safely, as the newly adopted member of their family. Obtaining such papers entailed for Sarah to put her life at risk while, investigating the case of survivors like Little Bee, whose life was drastically affected by the ongoing fight between various groups over natural resources. Proving to the UK authorities that the risks Little Bee faced did not affect only her fate, but a destiny that was shared among many—Sarah had hoped—would have made a legal claim for Little Bee to obtain refugeehood, and it would have led to Little Bee's safe return to the UK and their family.

While in Nigeria, as they were resting on the beach in plain day, Sarah is approached by a group of armed military personnel, shouting and pointing heavy weapons at her, in front of everyone, including her own toddler. As Sarah saw them approaching on the horizon, however, she feared for Little Bee's life more than for her own. She counted on being somewhat safer in view of her European passport, the whiteness of her skin, and her tourist status, yet she feared for herself as well. The novel offers an important observation here. Being a tourist in one place, ordinarily provides fewer rights than if one were a citizen or a permanent resident. But this rule did not appear to apply. Sarah's safety was contrarily guaranteed in the highest degree possible by her tourist status instead. This explains why in view of the armed group approaching, Sarah asked the African teen, who did not benefit of such basic immunity in her own country, to run as fast as she could and hide. The scene showing Sarah facing this danger, vividly illustrates how one's status, either as a tourist or a citizen in this case, will be the decisively arbitrary factor, from a moral standpoint, influencing the range and quality of rights protection each person realistically expects.

Little Bee had many less rights than Sara in every place in the world, including her own country of origin. As a Nigerian citizen, Little Bee was subjected to a higher degree of state power and abuse in Nigeria, but without the counterbalanced justified level of rights protection, as a newcomer elsewhere, as in the UK. It appears in the novel that Little Bee was not detained by Nigerian authorities, after being deported to Nigeria by UK authorities, because Sarah was in her company, and she would have reported on the case on international press. Little Bee also enjoyed many less rights than most, as a newcomer in the UK, where she was unable to obtain a legal status permitting her to stay safe. The United Kingdom deported her, after detaining her for two years—without her committing a crime, but the one of being found by the UK authorities on the boat, which she boarded to save her life and find safe haven in Europe.

Even if Little Bee's deportation from the UK to Nigeria would not directly lead to her being still targeted by any faction in the fight over natural resources, it is still morally wrong to force her separation from Sarah—the only person she knew and who offered her refuge, a family and a home. She was an unaccompanied minor who lost her community, her village and family of origin, and who was struggling visibly in the novel to overcome the trauma of her loss, which was not less significant, after facing a long, risky travel across the ocean and prolonged detention by the UK authorities. The UK has morally failed in deporting and detaining Little Bee. Little Bee's background conditions amounted clearly to the claim that she was forced to move in the first place. Her deportation shows the failure of the protection system to recognise her situation in full light. Her deportation put her in an extremely dire situation, a possibly life threatening one, in which she is unlikely to achieve a sufficient level of stability—from the perspective of general safety and the means to fend for herself.

In other words, the above scene demonstrates that the distinction we assumed between migrants, refugees and tourists, although it appeared intuitive and sharp in the first section, collapses between all three categories, in cases in which the background conditions are equivalent—from a moral standpoint. There is no safe haven for very many in a place in which resources are fought over so violently in the open air, and secured by those who have bigger guns and who are part of stronger criminal networks—not even for tourists like Sarah. This point is brilliantly illustrated by the author, when they describe one of the safest areas of the African region, an enchanting beach resort, where Sarah took her first tourist trip to the Nigerian enchanting beach resort, for a romantic getaway with her husband. The encounter between Sarah, who was a tourist and Little Bee, a local teen on the run, is marked by a tragic event in this heavily protected area. As Little Bee was found by the armed violent men, who intended to harm her, Sarah is forced to cut her finger with an axe in exchange of Little Bee's life—a choice which can hardly be found to meet criteria of voluntariness (Olsaretti 1998; Turculet 2018). Prior to the scene, the young man who was working as a guard for their hotel, was mercilessly killed as well—leaving them all unprotected and to the mercy of the gang. Following the bloody scene, Little Bee's sister is brutally gang raped and murdered—because Sarah's husband could not bring himself to cut his own finger in exchange for Little Bee's sister's life, too. This novel, shows very clearly that living without fear in some places, is an exception, rather than the norm. This fear is deep and not merely about potentially facing deprivation of everything material one has, as it could happen to anyone making a wrong financial investment in a stable and developed democracy. But also fear of losing one's community, family and one's own life or bodily integrity on someone's whim—as indeed all these occurrences happened in front of Little Bee and Sarah's eyes, in their first encounter in Nigeria.

The juxtaposition of Sarah and Little Bee's choices brilliantly shows two important points. First, the divide between the two heroines, in terms of their rights, comfort and opportunities, was deep: Sarah co-habits our planet with Little Bee, who had nearly no rights whatsoever. And second, to the extent to which Sarah's life got intertwined with Little Bee's, she too was less safe, if safe at all. In returning to Nigeria as a tourist, who was trying to rescue Little Bee, Sarah used her presence as a shield against any group who would attempt to capture or harm Little Bee. Sarah's mission to document Little Bee's case, and others who have faced a similar tragedy exposes her to an amount of life risk, which many would qualify as dire background conditions. One might object at this point that Sarah, however, had many more choices than taking the risks she took, including the choice which the UK had opted for—to leave Little Bee abandoned to her tragic and unjust destiny. One who might consider this choice a genuine one, fails to consider that Sarah's choice was at that point between bringing one's adopted family member to safety and her own safety—a too dire circumstance for anyone to face (Olsaretti 1998). The novel counts a number of scenarios, in which Sarah is exposed to a too dire circumstance, which I will not describe for brevity. Thus, the deeper the relationship between Sarah and Little Bee, the more analogous their exposure to risks, as their movement and return to safety is constrained by choices not of their own.

Enrique's Journey, written by Sonia Nazario similarly describes how the drug cartels and their expanding control over areas of transit from South America to North America produced detrimental social dynamics to human flourishing and local economies in Central America. This book features the Honduran teen, named Enrique during his odyssey from Honduras to the USA, in the hope of finding his mother, who was a migrant. Due to the unacceptably scarce opportunities to feed her children, Enrique's mother was constrained to migrate to the USA and left Enrique behind when he was a very young child (Nazario, 2006). I assume that her migration to the USA is a typical case of movement out of need, from a dire level of poverty to a better life (Cheng 2013; Grande 2013; Kittay 2008). Enrique succeeds in uniting with his mother in the USA after several attempts to cross the USA borders, resulting in his deportation and life-threatening situations, which he encounters as a penniless teen travelling alone. Crossing heavily patrolled national borders, while avoiding corrupt police officers, dangerous civilian militias, and other ordinary risks any child would face alone in the world are dire circumstances for anyone.

This non-fiction book, currently adopted in several universities' curricula, helps us distinguish between those who migrate towards rich economies, such as the USA, yet in a new light. Some are victimisers operating within powerful networks of organised crime, which often involve the trafficking of drugs and natural resources, and others, like Enrique and Little Bee, who are the victims of these networks. It is the movement of those who traffic drugs to generate movement of those who wish to escape the effects of this market. Many migrants in today's world have claims that are paralleled to refugees, if their life is equally endangered, even if they are not directly persecuted. Enrique's story, like Little Bee and Einstein's—triggers the same moral impetus for protection. Einstein left Germany for the USA for similar reasons, which prompts most contemporary refugees and migrants to leave their home country—for safe haven—even if we accept describing the Holocaust as a unique type of evil humanity has ever experienced (Arendt 1963). However, similarly to Enrique's situation, it is not too far-fetched conceiving of some who left Nazi Germany towards other continents, active or passive collaborators with and sympathisers of Nazi crimes. Movement is undertaken by victims and victimisers. While I will not discuss further the relationship between victims and victimisers, I pointed at different background circumstances, which result in a similar moral demand for protection.

Those in need of protection should receive it in safer states, regardless of their recognition under a specific legal status as migrants or refugees. Little Bee and Enrique's deportations represent the moral failure of the protection system in place, given that their background circumstances are analogously dire to typical refugees, thus, their claims equally strong against states. Put in this light, the USA border policies, which Enrique faced are unjust because they are not set to curb down on the drug cartels, which harms transit countries in Central America, prompting many to move in the first place. And second, for denying the protection he was entitled to, as a minor in need of safe haven, and as a minor who was trying to reach his mother across the border.

Finally, the three categories of movement, broadly defined, have in common that movement appears to be an essential interest for each person. When regulated, movement in a just society, requires a level of restriction, protection, or promotion, which leads ideally to the human pursuit of important life goals, such as the pursuit of safety, knowledge, and cooperation with other like-minded individuals, regardless of the geographical location (Turculet 2018). If Einstein were restricted to live in a refugee camp or detention facility on Elis Island, distanced for several years from all communities and societies, his life pursuits would have been reduced to harsh life conditions. This holds true even if the refugee camp or detention facility would sufficiently shield him from Nazi persecution, as these arrangements are not sufficient to pursue science or a minimally meaningful life. Movement then seems essential not only to protect basic rights, but also to fulfil life goals, and live a meaningful life.

3 Desiderata for Future Research: Ethics of Migration Versus Ethics of Movement

The field of ethics of migration has exponentially grown in the last three decades as a field of inquiry, as more scholarship challenges the groundbreaking 1987 contribution of Joseph Carens *Aliens and Citizens. The case for Open Borders*. He argues that *freedom of movement* is compatible with all main moral theories, which are anchored in our liberal democratic philosophical tradition. If movement were a basic right, anyone would be able to move for whichever reason to a new place (Carens 1987). His theory poses a threat to the assumption, which was taken for granted, that states have a right to self-determination, which implies discretion regarding who joins as a newcomer and under which conditions (Miller 2016). The debate is commonly described as the "open versus closed borders" debate (Turculet 2017), although neither theorist in both sides of the debate argues that borders should be completely closed or open, as a policy recommendation. The public debate on migration, however, is heavily polarised in most contemporary democracies, alongside the two moral positions, advocating open or closed borders. An example of "closed ports" policy was recently promoted by the Italian ex Minister of Interior, Matteo Salvini, producing a split in the Italian opinion. In the increasing incendiary public debate on migration, which recently attracted heavy media attention, nearly half of the Italian public support "closed ports" policies, even if they were not supporters of Salvini's far-right political party. Many associate migration entries with a threat to the public order. Many others recently showed sympathy for Carola Rackete, the 31 years old German captain of the Sea-Watch vessel, who rescued migrants at sea. She docked the boat in Southern Italy, against the order of Italian authorities (Aljazeera 2019). Recently, the highest Italian court ruled that Rackete, docked in Italy "out of necessity". As the captain, she was responsible to evaluate and act in the given situation in view of the migrants' condition on board. Her arrest by the Italian authorities is legally unwarranted (The Local 2020). Rackete is widely admired, nationally and internationally, for her courage to defy Salvini's "close ports" orders, and for the moral ground she held in rescuing migrants and bringing them to safety.

An example of "open borders" political stance, as reported by the BBC on 30 September 2019, features Pope Francis unveiling a bronze sculpture in St Peter's Square, dedicated to the world's migrants and refugees. The BBC news reports about the sculpture titled "Angels Unaware" which significance is derived from the portrayal of the boat with on board migrants "sailing" towards St. Peter's Basilica and from the boat being welcomed by the Pope in person. While the monument represents migrants from throughout history on board of the boat, counting a Jewish man fleeing Nazi Germany, a pregnant Polish woman escaping communism, a contemporary Syrian child fleeing war and another fleeing the 1840 famine in Ireland, the Pope openly called on the faithful to welcome, protect, promote and integrate, taking the moral stance I suggested—that analogous background conditions will help with deliberating on who should receive protection, even if upon arrival some are labelled migrants and others refugees.

These political stances, advocating closed or open borders, should be distinguished from philosophical (moral and political) positions, even if they endorse valid moral arguments for open or closed borders. Most theorists agree that citizens have a right to safety and to live free of fear of invasion, as a plausible moral position—in contrast, however, with the position often adopted by far-right parties, misdescribing migrants as "invaders" (Diamanti 2017). Conversely, welcoming, offering hospitality and protection to those in need is what most theorists and philosophers believe that justice requires.

Both lines of arguments, as advanced by the far-right and Pope Francis', are political, and arguably in line with our liberal democratic philosophical tradition. Salvini advocates "closed borders", while Pope Francis "open borders". There is danger in both political positions, however, if mistaken by moral arguments aimed at justifying a policy over another. If distinguishing between movement out of need, forced movement and movement out of choice, reflectively, helps us to make sense of what justice requires, the normative implication amounts to claiming that each person fitting one of the three categories of movement should enjoy a regime of movement that leads to human rights protection as well as adequate social and political rights—a position which is endorsed by most theorists on both sides of the "open versus closed borders" philosophical debate. However, it is not guaranteed that the best course of action for migration policies would support Marie Curie's migration to France to pursue her career, even if we endorse, as I do, the view that movement is an essential interest. Some might point out that dismantling the patriarchal management of higher education, which discriminates against female scholars, in Poland and everywhere is a superior moral goal to advocating open borders for all who face patriarchal injustice. Taking the moral position that all who are victims of a particular form of injustice should move might be too hasty and altogether unjustified.

Yet, even if one might validly argue that addressing the injustice that prompts migration is a prior moral goal to advocating open borders, it might not provide guidance on the specific cases of Little Bee and Enrique. Both were *already* forced by dire background conditions to undertake dangerous migratory journeys. Their cases resemble Einstein's, whose protection would be obtained, only if granted safe haven somewhere. It could be argued that capable states facing individuals in a condition of extreme vulnerability should offer them refuge, if all cases qualify for it. Contrarily, some might argue that in these cases, the right course of action is to defeat the Nazis, and to break the criminal networks related to the drug cartels and natural resources, as a prior moral goal. It is important to notice, as we reflect on these cases, that normative positions provide *moral desiderata*, but do not generally aim at providing direct guidance on what should be done politically in specific situations. Taking such moral stances, as political, requires further moral reasoning and empirical investigation of each individual situation and of movement more generally.

Ethics of migration, as a field of political philosophy, appears to inherit a bias similar to one existing in the public discourse. Most theorists distinguish migrants and refugees, from all other individuals who move, such as tourists. Some theorists (Oberman 2011; Turculet 2017) believe that the distinction between migrants and refugees, as people who choose or are forced to move, might be less easily distinguishable, from the perspective of the background conditions of need, which is morally relevant. I have shown nonetheless that the distinction between migrants, refugees and non-migrants (e.g. tourists) is plausibly blurry, too. I purposefully illustrated in great detail the story of Sarah, the journalist who was helping the refugee, Little Bee. Reflecting on her movement as a tourist compels us to enlarge the inquiry of ethics of migration to a larger category—*ethics of movement*—one which captures in ethical focus her experience as well. Even if she travelled as a tourist, her encounters prompted her to move again and face dire background conditions. A wider approach of ethics to movement could also capture Carola Rackete's experience, who is neither a migrant, a refugee, nor a tourist—yet someone whose fundamental interests require her movement across seas be protected and promoted morally and politically. She was arrested at the Italian border for doing her job as a captain for the sea rescue organisation, Sea-Watch. Specifically, for her act of solidarity with migrants shown in bringing them to a safe country, and for her refusal to leave them in Libya, a country which she did not consider safe—all which is foreseen in her job description as a captain bound by maritime rules and regulations, as well as international and humanitarian law, within which she adequately formed her judgement, informing her course of action.

In a nutshell, it is arguably defensible to exclude non-migrants, such as tourists, from the sculpture which features 140 migrants and refugees from throughout history in Saint Peter's Church. The moral view, which Pope Francis promotes politically, is imbedded in our historical context. He sides with the struggle of the most vulnerable, those who we know as migrants or refugees (Vatican 2019). This political stance helps us appreciate, however, that the perspective of political philosophy limiting *ethics of migration* mainly to an inquiry of migrants and refugees, is not coherent in all cases. It excludes unjustifiably from the ethical inquiry some who might be described as undertaking movement out of need or forced movement. Unless we take under scrutiny a larger conceptual category, *ethics of movement,* that defaults the scope of the inquiry to including *all* those who move, we are unlikely to evaluate cases of movement reflectively and appropriately as out of need, forced movement and movement out of choice.

Finally, we risk leaving at the margin of the ethical inquiry individuals whose movement equally qualifies for serious moral scrutiny, even if they are not migrants or refugees, like Sarah and Carola. Philosophically, we should not take this risk, even if there might be justifications to do so, politically. My line of argumentation is compatible with the wider and more inclusive moral perspective I suggested advancing as a new development of the field of inquiry, beginning with renaming the field "ethics of movement", instead of "ethics of migration" (Turculet 2018) in order to work out philosophically, who we ought to coherently be included in the scope of our study and for which justified reasons.

Accordingly, I suggested that there is a moral case for Carola Rackete to be represented in the Vatican's bronze boat sculpture in the company of the 140 chosen historical migrants, not least because boats need skilled and virtuous captains not to sink, but also, they need *her* to bring passengers, whoever they are, safely to ports against political odds (Turculet 2020). In taking this political stance, however, the Pope would prompt a morally sound revival of the very field of ethics of migration, in my view as more coherent and inclusive. However, his act, if undertaken, would be a direct political affront to Salvini, an elected Italian politician, and possibly to the European refugee regime, termed as a "fortress" (Turculet and Baycan 2018). Judging by the recent historical past—one such political stance as placing Carola in the sculptured migrant boat is unlikely to be advanced by Pope Francis. The Catholic Church has not used its enormous influence to politically resist the Italian government turning Italy during the past century into a fascist country under Benito Mussolini and this is not a distant historical past. Hence, recent history cautions us that political stances are often morally suboptimal, even if motivated by the highest moral standards, as it is the case with the sculptured migrant boat recently welcomed in the Vatican.

To conclude, I suggested caution—as citizens, migration practitioners, policy makers, scholars or students—in forming our judgement about when the debate on migration is *too politicised*—that is when moral arguments are promoted politically, regardless of whether they make moral and political sense in specific circumstances. Salvini's popular or "typical" political proposal "to help migrants back home" is one such example of flawed moral and political judgement. Although this moral argument enjoys some plausibility, it does not justify the exclusion of those in need who reach our borders, as the Rackete's case demonstrated above. The risk of endorsing unreflectively politicised moral arguments is that we might support political positions, which are full blown immoral, even if they appear justified by plausible moral arguments. Finally, we should be wary when the debate is *too philosophised*, such as when we unreflectively leave out of the ethical inquiry, individuals who move in view of background conditions which, literally or figuratively, place them in the same boat as migrants and refugees, even if they are neither migrants, nor refugees themselves (Turculet 2020). Therefore, "Is migration too politicised or too philosophised?" is a pertinent question, which citizens would benefit from raising, while forming moral judgements on specific cases involving movement of people around the world, as well as on their own choices to move.

Literature

Aljazeera. 2019. "Sea-Watch hails Italian court's decision to free Carola Rackete". *Aljazeera*, July 3, 2019, https://www.aljazeera.com/news/2019/7/3/sea-watch-hails-italian-courts-decision-to-free-carola-rackete (last checked December 17, 2023).

Arendt, H. 1963. *"Eichmann in Jerusalem. 5 Parts". The New Yorker,* https://www.newyorker.com/contributors/hannah-arendt (last checked December 17, 2023).

Carens, J. H. 1987. Aliens and Citizens. The Case for Open Borders. *The Review of Politics* 49:251–273.

Carens, J. H. 2013. *The Ethics of Immigration*. Oxford: Oxford University Press.

Cheng, S. J. A. 2013. The New Maids. Transnational Women and the Care Economy. *Migration Studies* 1:245–247.

Cleave, C. 2010. *Little Bee*. New York: Simon & Schuster.

Collier, P. 2013. *Exodus. Immigration and Multiculturalism in the 21st Century*. Oxford: Oxford University Press.

Diamanti, I. 2017. "Immigrati, cresce la paura. il 46% si sente in pericolo. E' il dato più alto da dieci anni". *La Repubblica*, September 13, 2017, https://www.repubblica.it/politica/2017/09/13/news/immigrati_cresce_la_paura_il_46_si_sente_in_pericolo_e_il_dato_piu_alto_da_dieci_anni-175338065/ (last checked January 10, 2020).

Donmez, P., and A. Duman. 2020. Marketization of Academia and Authoritarian Governments. The Cases of Hungary and Turkey in Critical Perspective. *Critical Sociology* 47:1–19.

Goldin, I., Cameron, G., and Balarajan, M. 2012. *Exceptional People. How Migration Shaped Our World and Will Define Our Future*. Princeton: Princeton University Press.

Grande, R. 2013. *The Distance Between Us. A Memoir*. New York: Simon & Schuster Children's Publishing.

Kittay, E. 2008. The Global Heart Transplant and Caring Across National Boundaries. *The Southern Journal of Philosophy* 46:138–165.

Miller, D. 2016. Is there a Right to Immigrate. In *Migration in Political Theory. The Ethics of Movement and Membership*, ed. S. Fine and L. Ypi. Oxford: Oxford University Press.

Nazario, S. 2006. *Enrique's Journey*. New York: Random House.

Oberman, K. 2011. Immigration, Global Poverty and the Right to Stay. *Political Studies* 59:253–268.

Oberman, K. 2016. Immigration as a Human Right. In *Migration in Political Theory. The Ethics of Movement and Membership*, ed. S. Fine and L. Ypi. Oxford: Oxford University Press.

Olsaretti, S. 1998. Freedom, Force and Choice. Against the Rights-Based Definition of Voluntariness. *Journal of Political Philosophy* 6:53–78.

Passachoff, N. 1996. *Marie Curie and the Science of Radioactivity*. Oxford: Oxford University Press.

Pycador, S. W. 1999. Marie Sklodowska Curie and Albert Einstein. A Professional and Personal Relationship. *The Polish Review* 44:131–142.

Raz, J. 1986. *The Morality of Freedom*. Oxford: Oxford University Press.

Rockwell, S. 2004. The Life and Legacy of Marie Curie. *Yale Journal of Biology and Medicine* 76:167–189.

The Local. 2020. "Italian Court Rules It Was Wrong to Arrest Migrant Ship Captain". *The Local*, January 17, 2020, https://www.thelocal.it/20200117/italian-court-rules-it-was-wrong-to-arrest-migrant-ship-captain (last checked December 17, 2023).

Turculet, G. 2017. The Refugee 'EU-Turkey Deal'. The Ethics of Border Politics. *Boletin de Estudios Economicos* 72:541–561.

Turculet, G. 2018. "*Movement as an Essential Interest. A Case for Just Borders*". Unpublished Monograph Dissertation, http://www.etd.ceu.edu/2018/turculet_georgiana.pdf (last checked December 17, 2023).

Turculet, G. 2020. "Coronavirus and the Populist Other. May Intellectuals Cast the First Stone?". *Global-E* 13 (Covid-19 and Populism Series).

Turculet, G. 2022. Academics Stuck in Movement Amidst Precarity, Hypermobility, and Vulnerability. In *Academic Freedom and Precarity in the Global North. Free as a Bird*, ed. A. Köleman and A. Vatansever. London: Routledge.

Turculet, G., and E. Baycan. 2018. A Fair Cooperation Scheme in a de facto Uncooperative Institution? The EU and the Current Refugee Crisis. *EuroPolis Journal* 10:243–250.

Vatansever A. 2020. *At the Margins of Academia. Exile, Precariousness, and Subjectivity*. Leiden, Boston: BRILL.

Vatican. 2019. "*It is not Just about Migrants. Message of His Holiness Pope Francis for the 105th World Day of Migrants and Refugees 2019*". https://www.vatican.va/content/francesco/en/messages/migration/documents/papa-francesco_20190527_world-migrants-day-2019.html (last checked December 17, 2023).

Waterfield, R. 2009. *Why Socrates Died. Dispelling the Myths*. New York: Cambridge University Press.

Yong, C. 2017. Immigration Rights and the Justification of Immigration Restrictions. *Journal of Social Philosophy* 48:461–480.

Refugee Ethics

David Owen

Abstract This article examines the ethical issues surrounding the modern refugee, in particular the definition of a refugee and the moral obligations which follow. It also considers the allocation of refugee responsibilities among states, the link to reparations for past injustices, and concepts of burden-sharing based on state capacity and historical responsibility. Once refugees are accepted, the question arises as to whether they have the right to become members of their host countries. Refugee ethics mainly focuses on the duties of host societies, but some also discuss whether refugees themselves have any duties, including the controversial question of whether they have a duty to seek refuge in the first safe country they reach. Following the comprehensive overview of the ethical challenges associated with the global refugee regime, this article suggests further research aimed at new refugee protection policies characterised by greater international cooperation.

The modern figure of the refugee is both a product of an international order of states and raises a range of ethical questions for such an order. First and foremost—and the one that has received most attention—is that of how to specify this figure, a question that is intimately tied to the further questions of what is owed to refugees and on what grounds. These questions in turn lead to a range of additional issues. How should responsibility for refugees be shared among states? Are refugees entitled to membership of the state of refuge? Under what conditions is the repatriation of refugees justifiable? What obligations do refugees owe to host communities and to each other? This chapter provides an overview of these issues.

D. Owen (✉)
School of Economic, Social and Political Sciences, University of Southampton, Southampton, UK
e-mail: dowen@soton.ac.uk

A. Niederberger et al. (eds.), *Handbook of Migration Ethics*,
https://doi.org/10.1007/978-3-031-89877-8_24

1 Who Is a Refugee?

The ethical question of who should be entitled to refugee status acquires its contemporary significance against the background of two factors. The first is the norm of non-refoulement as the fundamental norm of the international refugee regime composed of the principles, norms, rules, and formal and informal decision making procedures in terms of which actors orient their conduct (cf. chapter "Migration Ethics and/in Democratic Theory" in this Handbook). This establishes a strict duty on states not to return (*refouler*) refugees to any place where their basic rights are at risk. More formally, the duty of non-refoulement

> encompasses any measure attributable to the State which could have the effect of returning an asylum seeker or refugee to the frontiers of territories where his or her life or freedom would be threatened, or where he or she is at risk of persecution, including interception, rejection at the frontier, or indirect refoulement. (Feller et al. 2003: 178–179)

This norm establishes an exceptional status for refugees vis-à-vis other migrants. The second is the disjuncture between the legal delineation of the figure of the refugee established in the 1951 Refugee Convention and the post-war practice of refugee protection expressed in the 1969 Organisation of African Unity Refugee Convention and the 1984 Cartagena Declaration that has developed under the aegis of the UNHCR. The 1951 Convention specifies the refugee as:

> One who owing to a well-founded fear of being persecuted for reasons of race, religion, nationality [as belonging to a 'people'], membership of a social group or political opinion is outside the country of his nationality [as membership of a state] and is unable or, owing to such fear, is unwilling to avail himself of the protection of that country; or who, not having a nationality and being outside the country of his habitual residence [...] is unable, or, owing to such fear, is unwilling to return to it. (1951 Convention Article 1 A (2))

To which the OAU Convention adds:

> The term 'refugee' shall also apply to every person who, owing to external aggression, occupation, foreign domination or events seriously disturbing public order in either part or the whole of his country of origin or nationality, is compelled to leave his place of habitual residence in order to seek refuge in another place outside his country of origin or nationality. (1967 OAU Convention Article 1 s.2)

The current philosophical debate concerning this issue can broadly be divided between those who offer justifications for conceptions of the refugee akin to that of the OAU and those who defend the restriction of this status to persons meeting criteria of the 1951 Convention.

The *locus classicus* for the expansive view is Shacknove's 1985 article "Who is a refugee?". Shacknove points out that the 1951 Convention criteria of refugeehood involves the following assumptions:

(a) a bond of trust, loyalty, protection, and assistance between citizen and state constitutes the normal basis of political society;
(b) in the case of the refugee, this bond has been severed;
(c) persecution and alienage are always the physical manifestations of this severed bond;

(d) these manifestations are necessary and sufficient conditions for determining refugeehood. (Shacknove 1985: 275)

But this definition, it is argued, is inadequate for the following reasons:

1. Persecution is a sufficient but not necessary condition for the severing of the normal political bond—it is one manifestation of a broader phenomenon: the absence of state protection of citizens' basic needs.
2. Alienage is also not a necessary condition, being one subset of the broader category: the physical access of the international community to the unprotected person. (1985: 277)

The argument for (1) is based on the claim that the social compact between state and citizens establishes minimum conditions on the treatment of citizens, the minimum conditions of their allegiance to the state, which can be expressed in terms of the fulfilment of a set of basic needs—"physical security, vital subsistence, and liberty of political participation and physical movement"—that are *basic* because they are "equally essential for survival" (1985: 281). The argument for (2) is grounded on a distinction within the class of people whose basic needs are unmet by their state in terms of "their differing positions vis-à-vis the international community", where refugees "have a well-founded fear that recourse to their own government is futile and are, in addition, within reach of the international community" (1985: 282). Taken together these arguments lead Shacknove to propose a conception of refugeehood whereby

> a refugee is, in essence, a person whose government fails to protect his basic needs, who has no remaining recourse than to seek international restitution of these needs, and who is so situated that international assistance is possible. (1985: 282)

Shacknove's analysis thus encompasses but further extends the conception of refugeehood expressed in the 1969 OAU Convention. While this argument is the most expansive, related arguments for expanding beyond the 1951 Convention conception have been developed by Gibney (2004), Carens (2013), Miller (2007, 2016) and Owen (2016a).

The rejection of this expansive move has been developed most fully by Price (2009), Lister (2013) and Cherem (2016). Each draws a sharp distinction between refugees and what Walzer (1983) calls "necessitous strangers", they argue that what distinguishes refugees is their need for, and claim to, membership in a new state. As Cherem puts it:

> Refugees are special because persecution is a special harm. Refugees 'are targeted for harm in a manner that repudiates their claim to political membership'; their 'rights go unprotected because they are unrecognized' rather than for other reasons. [...] Refugees are distinctive because their country of origin has effectively repudiated their membership and the protection it affords. The status on which almost all their other rights hinge is gone. (Cherem 2016: 191)

In similar vein, Price argues that refugees are people who

> not only face a threat to their bodily integrity or liberty; they are also effectively expelled from their political communities. They are not only victims, but also exiles. Asylum responds not only to victims' need for protection, but also to their need for political standing, by extending membership in a new political community. (Price 2009: 248)

Expansive views such as Shacknove's are, from this standpoint, seen as erasing important normative distinctions between different types of necessary flight.

In an attempt to move beyond this opposition, Owen (2020a, b) has distinguished between the scope of the principle of non-refoulement and the scope of the institution of asylum conceived in terms of entitlement to membership in a new political community. He argues that refugees as a general class should be conceived as persons entitled to non-refoulement, where this aligns with the expansive views, but that what is owed to refugees can vary according to whether they have claims to *asylum*, *sanctuary*, or *refuge*. The first is exemplified by the case of the people who have reasonable grounds to fear persecution by their home state; the second by people fleeing generalised violence and the breakdown of civil order; the third by people fleeing specific state failures such as famine or natural disasters such as floods.

2 What Grounds Obligations to Refugees?

Debates concerning who is a refugee are closely connected to debates concerning what is owed to refugees and the grounds of such obligations (Lister 2013; Owen 2020a). This is unsurprising because conception, obligation and grounds are linked as the elements that specify the point and purpose of the institution of refugeehood. Here we can distinguish three main strands.

Perhaps the most prevalent in international policy and practice is a moral humanitarian standpoint. A clear instance of this stance is provided in Betts and Collier (2017) in which they argue that "Syrians forced to flee their homes by violence" are ethically analogous to a "drowning child" and "we have an unambiguous duty of rescue towards them" (Betts and Collier 2017: 99). On this view, states have an obligation to admit refugees *where the costs of protection are low*—and refugees are pictured as persons vulnerable to serious harms. Refugee protection can be seen as a form of moral solidarity that expresses the general humanitarian duty to prevent undeserved suffering when one can do so without incurring undue burdens. This is why the current failures of the international refugee regime have led UN Deputy Secretary-General Amina Mohammed to speak of a "crisis of solidarity" (https://www.un.org/press/en/2018/dsgsm1123.doc.htm, last checked April 4, 2019). What is owed to refugees is protection from vulnerability to harm and this is to be provided through secure provision of basic needs.

The second strand encompasses a range of justice-based arguments that take refugees to be subject to wrongs as well as vulnerable to harms. Obligations to refugees derive from a general duty to protect basic human rights typically conceived in terms of basic needs (Shacknove 1985; Miller 2007). This strand distinguishes, with some internal variations, between duties of justice and humanitarian duties conceived, respectively, in terms of duties that can characteristically be enforced (in appropriate and proportionate ways) without wronging the agent (duties of justice) and duties that cannot legitimately be enforced (humanitarian duties). The political salience of this distinction is that, on Miller's (2016) account, states are obliged to comply with duties of justice and hence have no obligation of justice to consider the preferences of their citizens concerning whether to comply. By contrast, states do have a duty to consider the preferences of their citizens concerning whether, and to what extent, to comply with humanitarian duties. It is important to note here that the issue of costs enter the picture in two distinct ways. In the case of humanitarian duties, this is a direct relationship in that our obligation is conditional on costs not being unduly burdensome. In relation to obligations of justice, the issue of cost enters in virtue of the fact that the duty to protect refugees is a collective duty that falls on a multiplicity of actors who are obliged to share the costs of refugee protection fairly between themselves in terms of their relative capacities (Miller 2007, 2013, 2016).

A third approach, adopted by Carens (2013), Owen (2016a, 2020a, b) and Brock (2020), focuses on the conditions of legitimacy of the state and of international order of states to argue that 'states have a duty to admit refugees that derives from their own claim to exercise power legitimately in a world divided into states' (Carens 2013: 196). As Carens puts it:

> Because the state system assigns people to states, states collectively have a responsibility to help those for whom this assignment is disastrous. The duty to admit refugees can thus be seen as an obligation that emerges from the responsibility to make some provision to correct for the foreseeable failures of a social institution. [...] Every social institution will generate problems of one sort or another, but one of the responsibilities we have in constructing an institution is to anticipate the ways in which it might fail and to build in solutions for those failures. If people flee from the state of their birth (or citizenship) because it fails to provide them with a place where they can live safely, then other states have a duty to provide a safe haven. (2013: 196)

The legitimacy view thus identifies a collective obligation on states to secure basic human rights as a condition of legitimacy of the international state system and conceives of refugee protection as a legitimacy-repair mechanism. It also allows for distinctions between distinct types of refugees in terms of the different normative bases of their claims in the distinct wrongs to which they are subject as discussed above (Owen 2020a, b).

3 Allocating Responsibilities for Refugees Between States

Accounts of the allocation of duties to refugees between states distinguish between general and special responsibilities. The former denotes the responsibility that, other things being equal, applies to any state; the latter arises when a particular state (or set of states) acquires responsibilities that derive from their being outcome-responsible for a given refugee flow (for example, such responsibilities would apply to the states that composed the coalition for the 2003 war against Iraq with respect to refugees arising from this conflict). The focus on special responsibilities has been most clearly developed by Souter (2014a, b) who expresses this position in terms of the idea of asylum as reparations for past injustice. The question of the reach of this reparative view given the history of imperial and colonial relations between states of the Global North and states of the Global South, where the latter are the primary sources of refugee flows and the primary sites of refugee protection, is one that deserves further exploration. Our focus, however, will be on general responsibility.

On one view, refugee protection is conceived as an *ethics of rescue* and responsibility for refugees is pictured in terms of *remedial* responsibility [for a critical review see Cherem (2025)]. On this account, the key consideration for sharing responsibilities of refugee protection concerns the capacity of states: protection should be shared on the basis of the different abilities of states to protect refugees at a given level of cost. Against the background of the existing international refugee regime, it has been proposed that wealth (Shuck 1997) or GDP and population (Gibney 2015) could serve as administratively simple proxies for capacity. It is important to note, however, that the capacity of a given state is *scheme-variable*. For example, imagine two states that are proximate to a state suffering from a civil war and each of which thereby acquire a significant number of refugees. State A is wealthy but very densely populated. State B is poor and has low population density. In a scheme in which each state is expected to bear the full costs of providing for refugees, the relative capacity of each will be limited (by population density for A, by poverty for B), whereas in a scheme in which each contributes on the basis of their joint capacities (A provides most the funds, B provides most of the land), the relative capacity of each is significantly increased and so also is the overall capacity (at a given level of cost) for providing refugee protection. The relative capacity of a state and the overall capacity of the international state system is thus conditional on the overall scheme of refugee protection. Does this entail that we should adopt whatever system, in principle, maximises overall capacity for refugee protection? Before we address this question, we need to address a second issue.

Given some legitimate method for determining fair shares, the second question that arises is whether a state that has done its fair share can legitimately choose to refuse to admit further refugees or has a duty to take up the slack where others have not fully complied. This former position is defended by Miller:

> The key argument here is that because the collective responsibility to avert injustice has been fairly distributed, ex hypothesi, by doing my fair share I have discharged my obligation, and the injustice that remains, because of partial compliance, is the responsibility of the non-compliers, and only theirs. (2013: 217)

On this view, the duty to do your fair share is a duty of justice (i.e., a duty such that you would not be wronged if forced to comply with it) and your duty of justice is exhausted by doing your fair share. States who do their fair share may still have humanitarian duties to unprotected refugees but whether to admit further refugees on this basis is a matter of choice for these states. This argument has been subject to three lines of criticism. Hoesch (2018) accepts Miller's argument that a state's fair share is determined under the presumption of full compliance but argues that the principle of non-refoulement remains binding so that a state cannot reject refugees who reach its territory and advance a legitimate claim to refugee status even if it has already done its fair share. Owen (2016b, 2018) argues that where the agents bearing responsibility for refugee protection are also the agents bearing responsibility for ensuring compliance with a scheme of refugee protection that these agents have a duty to take up the slack since it is a product of their own failure to ensure compliance. On this view, to leave refugees unprotected would be unjustly to shift the costs of the failure to ensure compliance from the group who owe that duty onto the group to whom it is owed. Finally, Stemplowska (2016, 2019) offers a more general argument that all those in a position to assist those in dire need have a duty of beneficence to step in and take up the slack left behind by others because the duties that individuals have to aid others are responsive to whether noncompliance of others leaves more people in need of aid and this is so because, in contexts of dire need, effectiveness has general priority over fairness.

Against the backdrop of this debate, we can return to the question of whether we have reasons to adopt the scheme that, in principle, maximises refugee protection and to cast some scepticism on that claim. Schemes may vary in terms of the overall protection that they would make available under conditions of full compliance but, under real world conditions, they may also vary in terms of their ability to command reasonably high levels of compliance. The obligation is then to support whatever scheme of protection we have reason to believe maximises effective protection under our non-ideal conditions where we cannot prudently suppose conditions of full compliance. The importance of the norm of non-refoulement in our current refugee regime is that it provides a mechanism for ensuring at least basic protection even where compliance with fair shares is low.

For advocates of legitimacy-based views, refugee protection is not best conceived as an ethics of rescue and responsibility for refugees is reparative rather than remedial. On this view, states have a collective responsibility to support the conditions of political legitimacy of the international order of states. An international order of states in which some states are unable or unwilling to protect the basic rights of (some of) their citizens is an illegitimate condition and the responsibility for addressing this condition lies with the collectivity of states that compose the international order of states. From this standpoint, alongside the *relative capacity* of a state to protect refugees are two further criteria that are salient to the division of responsibility: (a) the degree to which a state contributes to sustaining (or, put negatively, not overcoming) this illegitimate condition and (b) the degree to which a state benefits from this illegitimate condition. A state *contributes* to sustaining this condition insofar as it fails to do its fair share in supporting the universal securing of basic rights (i.e., the conditions of political legitimacy) where its fair share is a matter of its relative capacity to contribute to this goal. A state *benefits* from this condition insofar as its citizens are comparatively advantaged by it. The share of the collective responsibility owed by a state will be a function not only of its relative capacity to protect refugees, but also of its relative contribution to, and relative benefit from, the illegitimate conditions that make the generation of refugees an all too frequent feature of our global political life. The important contrast with the ethics of rescue view is that whereas these views treat refugee protection independently of the background conditions of production of refugees, this view links responsibility for refugee protection to the (re)production of these background conditions.

What are the limits of a state's responsibility to refugees on the legitimacy-based view? Imagine a world with radically higher numbers of refugees—perhaps as a consequence of climate change acting as a catalyst for refugee-producing events (Klepp and Riemann 2025)—than our existing world such that for any state doing its fair share would impose very considerable costs on its citizens. Is there a point at which a state could refuse to admit more refugees? Legitimacy-based arguments specify the limits of the duty to provide refugee protection in terms of the conditions under which a state is unable to meet the minimal criteria of internal legitimacy with respect to its inhabitants (i.e., respecting, protecting and fulfilling their basic human rights). The logic of this view is to say, first, that under conditions of partial compliance in which some refugees are unprotected, these refugees have no obligation to respect the norms of the international order of states with respect to border controls and, second, that if some refugees are unprotected even under conditions of full compliance, there is an collective obligation on states to transform the international state order into a different political order (i.e., to change its structuring norms) where doing so better protects the basic rights of each and every human being insofar as this is a genuine possibility (Owen 2016a).

4 Refugees and Membership in the Host State

Should refugees be entitled to membership in their host state? While there is a clear presumption in the construction of the 1951 Convention that refugees who were unable to return to their home states within a relatively short period would become citizens of the host state, that presumption is closely connected to the specific and narrow construal of the figure of the refugee expressed in that convention. This is the point that Price (2009), Lister (2013) and Cherem (2016) have stressed in identifying the figure of the refugee with those forced migrants who's very standing as members of political society has been denied by their home states—and who are consequently owed rapid access to a new membership.

In response to this political argument, Miller contends that "it seems wrong to single out those who are escaping persecution and grant them permanent residence immediately on the grounds that having arrived they will all choose to identify politically with the society that takes them in" (2016: 135–136). Price's argument, he charges:

> rests on the assumption that the causes that gave rise to persecution are such that the refugee will never be in a position to return in safety, whether because these causes themselves persist or because he has been mentally scarred by the experience and would experience psychological hardship if forced to return. Although this may be true in some cases, there will be other cases in which a change of regime means that those who have fled are more than willing to go back and pick up the threads of their previous lives. (2016: 133)

Miller's response, however, misses the point of the argument that Price and Cherem are advancing. This argument does not swing on whether the person entitled to asylum will immediately politically identify with the state of new membership nor on whether they may later choose to return to their state of original nationality under transformed conditions, but on the obligation to assert, through the granting of new membership, the political standing of the person whose political standing has been denied by their home state, who has been rendered *de facto* stateless by their home state. In a world in which effective state membership is the basic condition of political standing, to be denied such standing is both a wrong (a denial of one's equality) and a harm (making one highly vulnerable to the private or public power of others), it is thus a duty of the international order of states to ensure that all persons enjoy such standing—and when it is denied through the threat of persecution to provide protection in a way that re-affirms the right to such standing—and this is what the legal status of asylum as a distinctive type of refugee status should be conceived as providing.

At the same time, Miller's response also indicates the type of view characteristic of those who endorse an expansive conception of refugees. This response draws on a general "social membership" argument which argues that, after a reasonable period of residence, immigrants become full members of society and should be granted access to citizenship of the state (Carens 2013). In the particular case of refugees, on this view, the specific issues concern whether there are reasons to grant more rapid access to citizenship than for "voluntary" migrants and whether there are reasons to exempt them from any requirements to surrender their prior nationality where these are part of conditions of ordinary naturalisation in the host state. Both of these features are prevalent in the member states of the European Union (Owen 2019). The argument for the former is that the refugee's relation to the society in which they now reside is characterised by an indeterminate time horizon, neither open-ended like the habitual resident nor with a specified endpoint like the short-term worker, in that they may be subject to (and indeed desire to) return if the relevant conditions entitling them to refugee status cease to pertain at some future point in time. This is an extremely difficult position to occupy because it means that the refugee is situated in a condition of social and civic limbo, unable to commit to building a new life because they may be returned to the old, unable to commit to the old life because they may never be able to take it up once more. Accelerated access to citizenship provides the security needed for persons in this position (Owen 2013). At the same time, being able to retain their prior nationality while naturalising in the host state enables refugees not only to keep open the option of return but to engage in return with the security of knowing that they sustain rights of re-entry into the state of new citizenship. This reduces the risks of return for refugees who are typically desirous of returning to their home state (Bradley 2013).

An attempt to overcome the opposition between these "political" and "humanitarian" positions is provided by the legitimacy view through its differentiation of refugees who are owed asylum and those who are owed sanctuary. In the case of the former, it endorses the reparative stance of rapid access to citizenship for persons threatened by persecution. In the case of the latter, it endorses the social membership view (Owen 2020a, b).

5 Obligations of Refugees

What are the duties of refugees? This topic has recently been raised by Gibney (2018, 2020) and in concluding this chapter I want to touch on one issue that arises within its scope, namely, whether refugees have a duty to seek refuge in the first safe country that they reach.

The presumption that refugees should claim refugee status in the first safe state that they reach seems to have considerable grip on the public imagination and plays a significant role in legitimating non-arrival measures to publics of the Global North. These measures have been enabled by a transformation of the image of the refugee from a person entitled to protection into an object of suspicion—a picture of asylum seekers with valid claims to refugeehood as motivated by economic reasons for seeking entry to "developed" states rather than elsewhere, say, the camps in neighbouring states to their state of origin and hence as acting in ways that are taken to be unfairly burdensome towards the citizens of these states. This is a picture of someone who is entitled to refugee status abusing the refugee system for personal advantage by seeking refuge *here* rather than where they should do so, namely, the first safe country that they reach. This argument thus presupposes that refugees have a duty to claim refuge in the first safe country that they reach. This is, however, an implausible claim.

If we adopt the narrow 1951 Convention view of refugees in which there is a clear presumption that refugees will become members of a new state, then other considerations become immediately salient. Consider, first, that, in general, it is desirable that citizens of a state (that meets at least the minimal legitimacy threshold) identify the conditions of their own autonomy and well-being with the conditions of autonomy and well-being of their political community. This identification underwrites civic solidarity and we may think of the cultivation of this identification as the process of civic integration (Bauböck 2017). This being so, reasons that predispose a refugee to identify with a state of new membership (e.g., existing family relationships or an established home state diaspora in that state, some fluency in its language, particularly salient opportunities for educational or professional development, etc.) should carry weight in the process of determining where they are resettled. These reasons, and how particular refugee's weight them, may be very heterogeneous; some may pick out a particular state (e.g., family relations), others may be satisfiable by several states (e.g., language). The important point is that such reasons matter for the purposes served by the grant of asylum and should be given due weight in any reasonable scheme of international cooperation for protecting refugees entitled to asylum (for example through a "matching" scheme). Now, consider, second that it can only be reasonable for a refugee to have a duty to seek asylum in the first safe state if there is a well-functioning and adequate resettlement scheme in place such that they can be confident (a) that they will be resettled and (b) that the decision on settlement in State A rather than State B will be responsive to the reasons of civic integration. In the absence of such a regime—a position that characterises our current condition—there are no grounds for such a duty. Given the typical time-frames of exile for refugees construed in the wider OAU sense, similar considerations support the same conclusion.

6 Conclusion

The global refugee regime is currently under considerable pressure and while some find optimism in the recent Global Refugee Compact (Brock 2020), the overall picture is bleak. The overview of the ethics of refugee protection in our imperfect world provided in this chapter aims to highlight the key issues that redressing our contemporary predicament requires. The ethical issues we have engaged require a new politics of refugee protection, one in which the international community embraces cooperation more fully than the history of refugee protection practices exhibit.

Literature

Bauböck, R. 2017. *Democratic Inclusion*. Manchester: Manchester University Press.

Betts, A., and P. Collier. 2017. *Refuge. Transforming a Broken Refugee System*. London: Penguin Random House.

Bradley, M. 2013. *Refugee Repatriation. Justice, Responsibility and Redress*. Cambridge: Cambridge University Press.

Brock, G. 2020. *Justice for People on the Move. Migration in Challenging Times*. Cambridge: Cambridge University Press.

Carens, J. 2013. *The Ethics of Immigration*. Oxford: Oxford University Press.

Cherem, M. 2016. Refugee Rights: Against Expanding the Definition of a "Refugee" and Unilateral Protection Elsewhere. *Journal of Political Philosophy*, 24(2):183–205.

Cherem, M. G. 2025. Humanitarianism and Its Problems. In *Handbook of Migration Ethics*, ed. A. Niederberger, U. Okeja, and J. Gördemann. Cham: Springer.

Feller, E., V. Türk, and F. Nicholson. 2003. *Refugee Protection in International Law. UNHCR's Global Consultations on International Protection*. Cambridge: Cambridge University Press.

Gibney, M. J. 2004. *The Ethics and Politics of Asylum. Liberal Democracy and the Response to Refugees*. Cambridge: Cambridge University Press.

Gibney, M. J. 2015. Refugees and Justice Between States. *European Journal of Political Theory* 14:448–463.

Gibney, M. J. 2018. The ethics of refugees. *Philosophy Compass* 13:e12521.

Gibney, M. J. 2020. The duties of refugees. In *The Political Philosophy of Refuge*, eds. D. Miller, and C. Straehle, 132–153. Cambridge: Cambridge University Press.

Hoesch, M. 2018. 'Taking up the Slack' in the Context of Refugee Protection. Remarks on David Owen. *ZEMO, Zeitschrift für Ethik und Moralphilosophie* 1:163–175.

Klepp, S., and M. Riemann. 2025. Climate Mobility. In *Handbook of Migration Ethics*, ed. A. Niederberger, U. Okeja, and J. Gördemann. Cham: Springer.

Lister, M. 2013. Who Are Refugees? *Law and Philosophy* 32:645–671.
Miller, D. 2007. *National Responsibility and Global Justice*. Oxford: Oxford University Press.
Miller, D. 2013. *Justice for Earthlings*. Cambridge: Cambridge University Press.
Miller, D. 2016. *Strangers in Our Midst. The Political Philosophy of Immigration*. Cambridge, MA: Harvard University Press.
Owen, D. 2013. Citizenship and the Marginalities of Migrants. *Critical Review in Social and Political Philosophy* 16:326–343.
Owen, D. 2016a. In loco civitatis. On the Normative Structure of Refugeehood and the International Refugee Regime. In *Migration in Political Theory. The Ethics of Movement and Membership*, ed. S. Fine and L. Ypi, 269–290. Oxford: Oxford University Press.
Owen, D. 2016b. Refugees, Fairness and Taking Up the Slack. *Moral Philosophy and Politics* 3:141–164.
Owen, D. 2018. 'Taking up the Slack' in the Context of Refugee Protection. A Reply to Matthias Hoesch. *ZEMO, Zeitschrift für Ethik und Moralphilosophie* 1:177–184.
Owen, D. 2019. Refugees, EU Citizenship and the Common European Asylum System. A Normative Dilemma for EU Integration. *Ethical Theory and Moral Practice* 22:347–369.
Owen, D. 2020a. *What do we owe to refugees?* Cambridge, Medford, MA: Polity.
Owen, D. 2020b. Differentiating Refugees. In *The Political Philosophy of Refuge*, ed. D. Miller and C. Straehle, 19–38. Cambridge: Cambridge University Press.
Price, M. E. 2009. *Rethinking Asylum. History, Purpose, Limits*. Cambridge: Cambridge University Press.
Shacknove, A. E. 1985. Who Is a Refugee? *Ethics* 95:274–284.
Shuck, P. 1997. Refuge Burden-Sharing: A modest proposal. *Yale Journal of International Law* 22:243–297.
Souter, J. 2014a. Durable Solutions as Reparation for the Unjust Harms of Displacement. Who Owes What to Refugees? *Journal of Refugee Studies* 27:171–190.
Souter, J. 2014b. Towards a Theory of Asylum as Reparation for Past Injustice. *Political Studies* 62:326–342.
Stemplowska, Z. 2016. Doing More than One's Fair Share. *Critical Review of International Social and Political Philosophy* 19:591–608.
Stemplowska, Z. 2019. Coercing Compliers to Do More than One's Fair Share. *ZEMO Zeitschrift für Ethik und Moralphilosophie* 2:147–160.
Walzer, M. 1983. *Spheres of Justice. A Defense of Pluralism and Equality*. New York: Basic Books.

Refugees and International Law

Chris Boom

Abstract This chapter surveys work on three core topics in the interdisciplinary literature on ethics and international refugee law (IRL). First, it provides an overview of the most discussed issue among authors evaluating IRL from an ethical standpoint: i.e., whether the United Nations' 1951 Convention Relating to the Status of Refugees justifiably limits its definition of a "refugee" to people who have fled their countries on account of fear of persecution. While some authors maintain that this definition is ethically sound, others contend that it inappropriately excludes forced migrants who have fled their countries due to threats besides persecution, people who have not left their countries, or both. A second area of dispute covered by the chapter concerns the ethical foundations of IRL. Here too, controversy centres on the Convention's definition of a refugee, with authors dividing not only on the question of whether its restrictions with regard to persecution and alienage are justified, but also with the question of what moral judgements are implicit in conceiving of refugeehood as so-restricted. Finally, the chapter discusses work on the very connection between IRL and ethics as that topic arises within the context of the broader controversy between natural law and legal positivist approaches to the relationship between law and morality.

International law has long treated instances of forced migration differently than other forms of migration. In fact, some authors trace the roots of this special treatment to well before the emergence of international law proper, such as in the norm among the city-states of Ancient Greece of extending shelter in temples and other sacred spaces to foreign fugitives (e.g. Price 2009). Meanwhile, one can find reflection on the ethical dimensions of international law's treatment of forced migration as far back in history as in the work of Hugo Grotius, who is commonly regarded as "The Father of International Law".

By the middle of the twentieth century, a body of international law governing forced migration coalesced around the concept of the *refugee*. Before World War II,

C. Boom (✉)
Equal Freedom Institute, New Orleans, LA, USA
e-mail: cboom@equalfreedom.org

A. Niederberger et al. (eds.), *Handbook of Migration Ethics*,
https://doi.org/10.1007/978-3-031-89877-8_25

the norms of international refugee law (IRL) were largely the product of the League of Nations and its treaties. In turn, the ineffectiveness of this regime was the subject of Hannah Arendt's seminal critique in *The Origins of Totalitarianism,* which continues to play a major force in the scholarship at the intersection of ethics and IRL (Arendt 1951).

Today, both IRL and inquiry into its ethical dimensions are mainly, albeit not exclusively, dominated by the United Nations' 1951 Convention Relating to the Status of Refugees. Intended to address the mass migrations in Europe due to the Holocaust and other instances of persecution surrounding WWII, the Convention itself is limited in scope just to people fleeing events occurring before 1951 and in Europe. However, these geographic and temporal limits were later abandoned in the United Nations' 1967 Protocol Relating to the Status of Refugees, effectively transforming the Convention into a general instrument for the protection of refugees. Beginning in earnest with Andrew Shacknove's 1985 article, "Who is a Refugee?", a thoroughly interdisciplinary and steadily growing literature has since emerged to examine the ethical issues around the Convention and other aspects of the post-war IRL regime.

1 General Issues at Stake in the Controversy

The variety of ways in which IRL intersects with ethics mirrors the variety of ways in which law does so generally. Like other areas of law, IRL often—indeed, most often—receives attention from philosophers, political theorists, and others interested in normative issues simply by way of being a salient departure point for thinking about forced migration and refugees, without IRL itself being the actual object of inquiry. Meanwhile, IRL also often appears as a touchstone in reflection on the ethical questions surrounding the treatment of forced migration and refugees under the laws of states, or in the laws of regional bodies like the European Union, rather than under IRL conceived as a distinct body of law separate from these (e.g. Miller 2016).

Similarly, whereas IRL often figures into the ethics literature only indirectly, ethics also often figures into the IRL literature only casually. For, as in any area of law, IRL often involves injustices and other morally unacceptable circumstances that are claimed to be legally proper but not necessarily ethically so, or are claimed to be ethical but in a clearly disingenuous manner. Although ethics is implicated under such circumstances, the technical question of how the situation can be addressed through legal interpretation or reform will and should be at the forefront of inquiry, not the questions of moral and political philosophy.

Sometimes, however, ethics and IRL also intersect both directly and deeply. Most obviously, such intersection occurs once one begins asking broader questions about how refugees and other forced migrants should be treated within international law, and about the extent to which IRL conforms to the correct answers to those questions. Thus one might ask questions like: Who should international law protect? What protections should it extend? And how should it implement those protections?

In turn, one can then take a critical stance towards IRL, considering whether its existing features are consistent with what one takes to be the right answers to these questions. At the same time, one can also proceed in the opposite direction, starting with a view towards IRL or some feature(s) of it, and then considering the extent to which the feature(s) or IRL as a whole survives ethical scrutiny.

A second point of intersection is subtly related to, yet ultimately distinct from, this last mode of inquiry. Here, the topic is not the ethics of IRL per se, but instead IRL's ethical foundations. Just as attention is often paid to explicating the normative underpinnings of contracts and criminal law, for example, a number of authors have paid an interest in the moral presuppositions which underlie IRL and its various features. In this vein, IRL is studied with the goal of rationally reconstructing it or its features from a philosophical standpoint, identifying the moral judgements implicit in them. Though calling in part for an evaluative stance, inquiry along these lines is nonetheless not purely evaluative, as it aims in crucial part to reflect IRL as it is, not necessarily IRL as it should be.

A third, still deeper point of intersection arises where one begins reflecting on the very contrast between IRL as it is and IRL as it should be. For, as with any area of law, reflection on the ethical dimensions of IRL calls into question the nature of its connection to ethics itself. Most saliently, it prompts the question: to what extent are IRL and ethics truly distinct? IRL thus becomes implicated in the wider debate over the divisibility of law and morality. On one side of this debate stands the natural law tradition, which regards the issue of what law is as intrinsically intertwined with the issue of what it ought to be; on the other is the legal positivist, who insists that these are fundamentally separate issues.

2 Topics/Positions/Arguments in the Controversy

2.1 The Relationship Between Ethics and IRL

Unlike Grotius and other early authors approaching international law from within the natural law tradition, contemporary scholarship proceeds largely along positivist lines, explicitly or implicitly taking IRL to be exhausted by positive sources like the Convention and other treaties, customs, and general principles of international law. One possible exception worthy of note here, however, is Emma Haddad, who defends the notion that the rights of refugees are natural rights, deriving ultimately from their status as human beings, with positive law entering on this picture as a means of ensuring the refugee's enjoyment of them (Haddad 2008). Another possible exception is Shacknove, whose criticism of the Convention's definition of a refugee as having "positivistic" origins will be seen later in this chapter.

Although receiving direct attention more rarely than other topics, reflection on the relationship between ethics and IRL stretches back to the formative work of Arendt, and her accompanying critique of the concept of natural rights (Arendt 1951). Grounded in interlocking normative considerations and historical observations, Arendt's critique eludes easy characterisation. Indeed, the interpretive issues surrounding that critique are themselves the subject of a rich and still-active literature. (e.g. Benhabib 2004; Parekh 2017). However, for present purposes, Arendt's central thesis can be characterised perhaps least controversially as a denial of the very meaningfulness of rights outside of recognition within a political community.

Meanwhile, James Hathaway has offered the most sustained treatment of this topic, recognising that a position on it must be taken to determine IRL's content (Hathaway 2005). For his part, Hathaway defends "a consent-based, modern positivist view of international law", which conceives of international legal norms as being the product of state consent as reflected in treaties and other generally accepted sources of international law, against the prevailing, non-positivist approach to international law: i.e., the policy-oriented "New Haven School". Specifically, Hathaway argues that the latter is inferior to the former in that it renders international law too vague for purposes of holding states effectively accountable, is anti-democratic in regarding states as bound by norms other than those to which they have agreed, and, at least on one extreme variation, provides dominant states with an ideological pretext to ignore established norms when it suits their interests.

2.2 *The Ethical Foundations of IRL*

As for IRL's theoretical foundations, inquiry on this topic has largely centred on the Convention (Cherem 2015). Although influential exceptions exist (e.g. Benhabib 2004), it has centred in particular on the implicit assumptions underlying its definition of a refugee as a person who:

> owing to a well-founded fear of being persecuted for reasons of race, religion, nationality, membership in a particular social group or political opinion, is outside the country of his nationality and is unable or, owing to such a fear, is unwilling to avail himself of the protection of that country.

Most work on the normative foundations of this definition has focused more narrowly still on its condition that the object of the fear at hand be *persecution*, thereby excluding other circumstances resulting in forced migration. In this context, it is often asked: what ethical presuppositions are implicit in the Convention definition's exclusion of those fleeing the indiscriminate yet similarly serious threats posed by armed conflict, criminal violence, and natural disasters? Secondarily, reflection on the Convention definition has often focused on the condition that refugees be people who are outside of their countries of nationality; that is, its condition of *alienage*. Thus a question arises as to the very relationship between refugeehood and migration, Most saliently, one might ask: what moral judgements underlie the Convention definition's exclusion of "internally displaced persons" ("IDPs") and others who have not left their countries of nationality?

Although influential exceptions exist here too (e.g. Zolberg et al. 1989; Dummett 2001), the majority of work on these issues has been provided within the literature arising from Shacknove's analysis of the theoretical foundations of the persecution and alienage conditions. Specifically, according to Shacknove, underlying those conditions are the following four implicit assumptions:

> (a) a bond of trust, loyalty, protection, and assistance between the citizen and the state constitutes the normal basis of society;
> (b) in the case of the refugee, this bond has been severed;
> (c) persecution and alienage are always the physical manifestations of this severed bond; and
> (d) these manifestations are the necessary and sufficient conditions for determining refugeehood. (Shacknove 1985)

As will be discussed in the next chapter, Shacknove goes on to criticise the third and fourth of these assumptions, and thus the Convention definition.

In considering the literature spawned by Shacknove on this point, it bears noting that sometimes the responses to it do not make explicit whether they are intended to address his account of the foundations of the Convention's definition, his evaluation of those foundations and the definition, or both. For example, such ambiguity can be seen in one of the earliest and most influential responses to Shacknove, Hathaway's "Is Refugee Status Really Elitist? An Answer to the Ethical Challenge" (Hathaway 1997).

Although Hathaway does not express his view in these terms, in essence he accepts that the Convention definition is grounded in each of the assumptions identified by Shacknove except the third (Hathaway 1997). That is, while agreeing that the Convention definition should be understood as flowing from the assumption that the normal bond between citizen and state has been severed, and that persecution and alienage are necessary and sufficient conditions for refugeehood, he does not contend that the Convention definition is underpinned by the view that severance of the bond between citizen and state always physically manifests itself in persecution and alienage.

Regarding persecution, Hathaway locates the rationale for the Convention's restriction on this point instead in two empirical assumptions. One is that it would be impractical for the world to extend shelter to all people fleeing serious harms. The other is that those threatened by persecution on account of the bases enumerated in the Convention definition are less likely than others to be able to obtain effective redress in their own states. In turn, Hathaway argues that the Convention's limited scope reflects a "principled compromise", seeing to it that "the most deserving among the deserving" of international protection are protected.

Regarding alienage, Hathaway likewise regards the rationale for the Convention not encompassing IDPs as being a principled compromise in response to considerations of feasibility. Specifically, though acknowledging that IDPs have a strong moral claim to protection, Hathaway emphasises that access to them is often significantly limited by state claims to sovereign authority over their territory. And because the international community's ability to assist people outside of their states is not limited in this way, he claims, "the level of obligation and comprehensiveness of protection can and should be much greater." Thus, Hathaway insists, to maintain that the protections extended to IDPs must be the same as the protections extended to the externally displaced would have the undesirable consequence of reducing the degree of protection that can be guaranteed to the latter.

Before moving on from Hathaway, however, it is worth noting that in a later article he takes a somewhat different tack with respect to the issue of alienage (Hathaway 2010). First, in that article he arguably departs from his claim that the obligation to protect the externally displaced is greater for the externally displaced than IDPs. There, he cites his 1997 article instead for the proposition that distinguishing refugees from IDPs and other forced migrants "is not to suggest that refugees are more important or more deserving, but simply to recognise the distinctiveness of their circumstances." Second, to the idea that the externally displaced are distinguishable from IDPs with regard to the international community's abilities to help them, Hathaway adds that extending the protections of the Convention to IDPs would be of no value to them anyways. For he argues that the rights extended by the Convention are all directly tied to the needs of people who are outside of their own states. In defence of this, he draws attention to the right of "*non-refoulement*" under Article 33(1) of the Convention: the right that a state not "expel or return ('*refouler*') a refugee in any manner whatsoever to the frontiers of territories where his life or freedom would be threatened on account of his race, religion, nationality, membership of a particular social group or political opinion." Meanwhile, Hathaway argues that the remaining rights conferred by the Convention merely entitle refugees to the same level of treatment it provides to other people within its population, and that IDPs are entitled to this already.

Another arguable example of such ambiguity is provided by Matthew Gibney, who accepts Shacknove's views on the persecution condition yet offers an alternative rationale for the Convention definition's alienage condition (Gibney 2004). Specifically, Gibney argues that the circumstances of people who have already left their states are distinguishable from those who remain in them in that the only reasonable prospect for the former to gain protection is within another state, whereas this is less clearly true of the latter. For, whereas no other form of assistance can likely be provided as quickly or effectively to people who have already left their state as permitting their residence in another state, often people who remain in their countries can be helped through *in situ* assistance.

By contrast, Matthew Price is explicit in rejecting Shacknove's understanding of the Convention definition's foundations (Price 2009). This is because Price denies the adequacy of approaches to understanding it which, like Shacknove's, understand it in purely "humanitarian" terms, connected only to the "palliative" function of protecting the refugee. Instead, Price defends a "political" understanding of the Convention definition's foundations, according to which it is also underpinned by the further, "expressive" function of conveying condemnation of the refugee's state. In defending this account, Price looks beyond the Convention's principle of *non-refoulement* to the practice of "asylum", in the sense of granting "surrogate" membership in a new state. Not only has asylum historically been understood to have the political dimension of functioning to convey condemnation, Price argues, it is also a uniquely fitting response to a person fleeing persecution for both palliative and expressive reasons. As to the former, the distinctively reprehensible character of persecution gives rise not just to the need for residence in another state, but also to membership in one. As to the latter, persecution's reprehensible character similarly makes it incumbent on a receiving state to communicate its condemnation of a state which perpetrates or allows persecution by extending membership to its victims. By contrast, expressing condemnation of a person's state is not appropriate where that person is fleeing a natural disaster, war, or other threats not constituting persecution, and the plight of such people can be effectively redressed in a variety of ways besides giving them membership in a new state. Instead, Price argues, they can be helped through grants of temporary or permanent residence or *in situ* assistance.

Finally, although he does not attempt to analyse the foundations of the Convention's definition of a refugee specifically, David Owen's defence of what he calls the "political legitimacy" account of the foundations of the international refugee system more broadly deserves mention in this context as well (Owen 2016). For, as will be discussed below, Owen draws from this account to address the debate over the justifiability of the Convention definition's persecution and alienage conditions.

According to the political legitimacy account, the institution of refugeehood is a means by which the international society of states addresses the threat to its legitimacy as a global regime of governance posed by its members' failures to protect their citizens' basic rights. By standing "*in loco civitatis*" for a state when it fails to protect such rights, the international community thereby limits the impact of that state's failure on its own legitimacy. À la Price, Owen notes that the international refugee system so-conceived is a political institution in that it not only has a palliative function, but an expressive one as well. However, rather than necessarily conveying condemnation for refugees' states, the institution of refugeehood is instead a medium through which the international society of states acknowledges both its responsibilities, and its failings in discharging its responsibilities, to those subject to its governance.

2.3 The Ethics of IRL

Of the topics discussed in this chapter, by far the most attention has been paid to the issues surrounding the direct ethical analysis of IRL. A comprehensive overview of the interdisciplinary literature on this topic would need to discuss the large debate over how to accommodate inequities among states in sharing the burdens of IRL (e.g. Hathaway and Neve 1997; Schuck 1997). It would also examine Megan Bradley's work responding to the lack of normative guidance on the repatriation of refugees within IRL (Bradley 2013). Additionally, it would explore influential criticisms of the Convention's definition and other aspects of IRL outside of the literature responding to Shacknove (e.g. Zolberg et al. 1989; Dummett 2001).

However, in the interest of thematic unity, and because it has not only received the most interdisciplinary attention, but indeed the most attention generally, discussion here will be confined to Shacknove's critique of the Convention's definition and the literature that has developed in its wake.

With respect to the four assumptions he identifies as underlying the Convention's definition discussed above, Shacknove endorses the first and second, but rejects the third and fourth. That is, though accepting that the normal bond between citizen and state is one of trust, loyalty, protection, and assistance, and that the refugee is distinguished by the rupture of this bond, Shacknove denies that persecution and alienage are the only indicia that this bond has been ruptured. In turn, because he holds that the rupture of this bond can be evinced otherwise, Shacknove maintains that neither persecution nor alienage are necessary conditions for the determination of refugeehood.

Regarding persecution, Shacknove objects that severance of the bond between citizen and state can manifest itself in different ways as well. On this point, Shacknove draws from Thomas Hobbes, embracing the Hobbesian claim that the state exists for the fundamental purpose of securing its members against outsiders and one another. Given this understanding of state legitimacy, Shacknove argues that the state can fail to fulfil its purpose in numerous ways not involving persecution. For instance, he argues that the state will also fail to fulfil it where it does not protect its citizens from foreign invasion, civil war, or criminal acts by their fellow citizens. More than this, Shacknove claims, because the extent of the harms of natural disasters are often significantly subject to human control, a state's actions or inactions surrounding even famines, droughts, and the like can rupture its bond to its citizens as well.

Regarding alienage, Shacknove advances several overlapping objections to this condition. For one, he criticises this idea as originating in "the archaic positivistic theory" per which states "have equal, inviolable sovereign integrity," and that intervention in any state's internal affairs will undermine the stability of all states. And whereas it is a corollary of this theory that the victims of oppressive states can only be assisted if they have left their states' territories, Shacknove says, in actuality this is often not the case. For instance, he observes that states sometimes welcome international assistance, and that such assistance often also becomes available in the presence of foreign invasions, civil wars, or other breakdowns in the public order within a state that make its victims accessible to the international community. As such, Shacknove claims, refugeehood is "conceptually unrelated to migration."

According to Shacknove, because the breaking of the bond between citizen and state need not manifest in persecution or alienation, neither of these are necessary conditions for a person to be a refugee. Instead, on Shacknove's alternative definition, refugees are people whose government has failed to protect their basic needs, who have no effective recourse other than to seek international assistance, and who are so situated that international assistance can be provided to them.

Meanwhile, in connection with reasons noted before, Gibney's assertion *contra* Shacknove that refugeehood is "conceptually related to migration" has shaped the debate over the justifiability of the alienage condition, as the discussion below reflects (Gibney 2004). Gibney's work has also guided the debate over the Convention's definition more generally via his in-depth analysis of historical instances of backlashes even to refugee programs which fall far short of meeting Shacknove's criticisms. Subsequently, Gibney responded to Hathaway's claim that those fleeing persecution are less likely to receive assistance in their own countries than those fleeing other harms, objecting that this does not guarantee that the latter could receive assistance quickly enough (Gibney 2015). Further, Gibney objects to Price's argument that persecution creates a special need for surrogate membership on the grounds that people fleeing persecution often choose to return to their countries when it becomes safe, and that war and other forms of violence can likewise effectively sever one's relationship with one's country just as persecution can.

Price, by contrast, is clear in rejecting both Shacknove's account of the ethical foundations of the Convention definition and in dismissing Shacknove's critique of that definition. (Price 2009). This is because, in contrast to Shacknove, he endorses his own alternative account of the definition's foundations as sound not only *qua* account of its actual foundations, but also *qua* account of what its foundations should be. Thus, on Price's view, the moral judgements which we saw implicitly underlie the Convention definition also suffice to justify it. However, it is worth noticing that Price, like Gibney, has similarly shaped the debate on the Convention's definition through his arguments drawing attention to the risks of backlash against more expansive definitions.

Conversely, while rejecting Shacknove's account of the foundations of the Convention definition, Owen's competing political legitimacy account of those foundations similarly leads him to the conclusion that the alienage and persecution conditions are unjustifiable (Owen 2016). In reply to Price on the persecution condition, Owen concedes that Price may be correct that only people facing persecution need asylum so-conceived as surrogate membership, but insists that the protection of the basic rights of people fleeing harms besides persecution nonetheless often requires "*refuge*" in the sense protected by the right of *non-refoulement*. Furthermore, Owen notes, though recognising people as refugees under such circumstances would not convey condemnation of their states, refugeehood nonetheless retains its expressive dimension. This is because, in providing such refuge, the international society of states thereby acknowledges its responsibilities, and its failure to fulfil those responsibilities, to refugees, and, moreover, their states. And in reply to Gibney on the alienage condition, Owen rejects Gibney's view that refugeehood is conceptually related to migration. Here, he draws attention to situations in which a civil war or other factors cause domestic governmental order to break down within a state, and the international community consequently creates "safe havens" in its borders. Although the people in those safe havens have not left their country, Owen argues that they possess the distinguishing factor of refugees notwithstanding this in that the international community effectively replaces their state, standing *in loco civitatis* to ensure the protection of their basic rights.

Turning to authors not already mentioned, Alexander Betts has provided an influential elaboration on Shacknove's critique of the persecution condition, albeit in terms of Bett's preferred label of the 'survival migrant' rather than refugeehood (Betts 2013). In particular, Bett expands on Shacknove's argument that threats to basic rights more broadly suffice to give rise to a need for international protection by drawing on Henry Shue's work on the concept of basic rights, which Shue conceptualises as those rights one must enjoy in order for one to enjoy any other rights at all. And not only is security and liberty necessary for the enjoyment of other rights, Shue argues and Bett agrees, so is subsistence. Furthermore, according to Shue and Bett, though in IRL there is often an implicit assumption that security and subsistence rights can be distinguished on the grounds that the former is a *negative* right, imposing only a duty of inaction, whereas the latter is a *positive* right, obligating action, this distinction is without moral significance. For just as one's subsistence often calls for both action and inaction on others' parts, so too does security. Yet, also notably, Betts does not embrace Shacknove's critique of the alienage condition, in part for reasons echoing Gibney and Hathaway in his 1997 article, but also in light of the existing protections IRL extends to IDPs.

By contrast, Matthew Lister has provided an influential extended defence of the Convention definition against Shacknove, rejecting his views on both the alienage and persecution conditions (Lister 2013). Lister's reply to Shacknove is wide-ranging, involving a variety of practical, theoretical, and methodological objections, including significant elaborations on objections by other authors noted above. In fact, though his express aim is to respond to Shacknove's critique of the Convention definition rather than his account of its foundations, Lister indicates scepticism of that account in, among other things, his claim that it is objectionable on the grounds that it "separates refugeehood from *non-refoulement*". For, according to Lister, Shacknove's account fails to accommodate the fact that states are uniquely obligated not to force refugees to return to their countries, as opposed to assisting them *in situ*. Still, it suffices for present purposes to consider Lister's explicit defence of the alienage and persecution conditions against Shacknove's critique, which at bottom rests on an appeal to the contrasting considerations of proportionality affecting those who meet these conditions versus those who do not. As to the persecution condition, Lister argues that intervention in a state in order to prevent persecution will often fail a test of proportionality in that it would often require those intervening to subject themselves to undue risks, threaten more harm to the state's residents than it would prevent, or both. Similarly, as to the alienage condition, he argues that though the same will often be true of intervention in cases where a state opposes it, such proportionality concerns do not preclude assistance to people who have already fled that state.

Last, in the context of a broader study of the ethics of migration, Joseph Carens has argued that "the Convention embodies a misplaced set of priorities" in restricting its definition of the refugee to people facing persecution (Carens 2013). According to Carens, what is relevant from an ethical standpoint is "the severity of the threat to basic human rights and the degree of risk rather than the source or character of that threat." In response to those who defend the Convention definition on the grounds that people fleeing threats other than persecution can be helped in their own states, Carens objects that this defence confuses the issue of how to respond to threats that cause people to flee their states with the issue of how to respond to people who have fled their states already. When faced with people who have already fled their states, he contends, the relevant question for receiving states is whether they may be sent back, not how the circumstances they fled from ought to be addressed. And with respect to the former question, Carens argues that a receiving state bears a special responsibility for people who arrive in its territory, as they will no longer be at great risk if the state does not return them. What is more, he insists, though a receiving state may initially allow people to temporarily enter their territories in accordance with its duty of *non-refoulement*, eventually it will become unreasonable to expect them to leave once the threats they fled have passed, in which case it will become incumbent on the state to recognise them as members.

3 Desiderata for Future Research

Future research on ethics and IRL would benefit from still greater interdisciplinary engagement. On the one hand, authors arriving at the subject from a legal standpoint would benefit from greater attention to major recent developments among philosophers, political theorists, and others interested in the normative dimensions of IRL. For instance, this includes Serena Parekh's work on the ethical issues surrounding encampment, Luara Ferracioli's work on the ethics of reform efforts, and Christopher Heath Wellman's pro-sovereignty and Chandran Kukathas' anti-restrictionist critiques of the value of IRL itself (Parekh 2017; Ferracioli 2014; Wellman 2011; Kukathas 2016). On the other hand, authors arriving at the subject from a normative standpoint would benefit from greater attention to the nuances of IRL as explicated by leading scholars like Hathaway, Guy Goodwin-Gill, and Jane McAdam (e.g. Hathaway 2005, 2007; Goodwin-Gill and McAdam 2007; McAdam 2007). Meanwhile, authors arriving at the subject from either perspective would benefit from greater attention to work by legal philosophers on the ethical foundations of international law and on the relationship between law and ethics as they arise in the international law context (e.g. Besson and Tasioulas 2010; Dworkin 2013).

Literature

Arendt, H. 1951. *The Origins of Totalitarianism*. New York: Harcourt, Brace, and Company.
Benhabib, S. 2004. *The Rights of Others. Aliens, Residents and Citizens*. Cambridge: Cambridge University Press.
Besson, S., and J. Tasioulas, eds. 2010. *The Philosophy of International Law*. Oxford: Oxford University Press.
Betts, A. 2013. *Survival Migration. Failed Governance and the Crisis of Displacement*. Ithaca, London: Cornell University Press.
Bradley, M. 2013. *Refugee Repatriation. Justice, Responsibility, and Redress*. Cambridge: Cambridge University Press.

Carens, J. H. 2013. *The Ethics of Immigration*. Oxford: Oxford University Press.
Cherem, M. 2015. Refugee Rights. Against Expanding the Definition of a 'Refugee' and Unilateral Protection Elsewhere. *Journal of Political Philosophy* 24:183–205.
Dummett, M. 2001. *On Immigration and Refugees*. London: Routledge.
Dworkin, R. 2013. A New Philosophy for International Law. *Philosophy and Public Affairs* 41:2–30.
Ferracioli, L. 2014. The Appeal and Danger of a New Refugee Convention. *Social Theory and Practice* 40:123–144.
Gibney, M. J. 2004. *The Ethics and Politics of Asylum. Liberal Democracy and the Response to Refugees*. Cambridge: Cambridge University Press.
Gibney, M. J. 2015. Refugees and Justice Between States. *European Journal of Political Theory* 14:448–463.
Goodwin-Gill, G., and J. McAdam. 2007. *The Refugee in International Law*. Oxford: Oxford University Press.
Haddad, E. 2008. *The Refugee in International Society*. Cambridge: Cambridge University Press.
Hathaway, J. C. 1997. Is Refugee Status Really Elitist? An Answer to the Ethical Challenge. In *Europe and Refugees. A Challenge?* ed. J. Carlier and D. Vanheule, 79–88. The Hague: Kluwer Law International.
Hathaway, J. C. 2005. *The Rights of Refugees Under International Law*. Cambridge: Cambridge University Press.
Hathaway, J. C. 2007. Forced Migration Studies. Could We Agree Just to 'Date'? *Journal of Refugee Studies* 20:349–369.
Hathaway, J. C. 2010. Leveraging Asylum. *Texas International Law Journal* 45:503–536.
Hathaway, J. C., and R. A. Neve. 1997. Making International Refugee Law Relevant Again. A Proposal for Collectivized and Solution-Oriented Protection. *Harvard Human Rights Journal* 10:115–211.
Kukathas, C. 2016. Are Refugees Special? In *Migration in Political Theory. The Ethics of Movement and Membership*, ed. S. Fine and L. Ypi, 249–268. Oxford: Oxford University Press.
Lister, M. 2013. Who Are Refugees? *Law and Philosophy* 32:645–671.
McAdam, J. 2007. *Complementary Protection in International Refugee Law*. Oxford: Oxford University Press.
Miller, D. 2016. *The Political Philosophy of Immigration*. Cambridge, MA: Harvard University Press.
Owen, D. 2016. In Loco Civitatis. On the Normative Basis of the Institution of Refugeehood and Responsibilities for Refugees. In *Migration in Political Theory. The Ethics of Movement and Membership*, ed. S. Fine and L. Ypi, 269–289. Oxford: Oxford University Press.
Parekh, S. 2017. *Refugees and Ethics of Forced Displacement*. New York: Routledge.
Price, M. E. 2009. *Rethinking Asylum. History, Purpose, and Limits*. Cambridge: Cambridge University Press.
Schuck, P. H. 1997. Refugee Burden-Sharing. A Modest Proposal. *Yale Journal of International Law* 22:243–297.
Shacknove, A. E. 1985. Who Is a Refugee? *Ethics* 95:274–284.
Wellman, C. H. 2011. Freedom of Association and the Right to Exclude. In *Debating the Ethics of Immigration. Is There a Right to Exclude?* ed. C. H. Wellman and P. Cole, 13–155. Oxford: Oxford University Press.
Zolberg, A. R., A. Suhrke, and S. Aguayo. 1989. *Escape from Violence. Conflict and the Refugee Crisis in the Developing World*. Oxford: Oxford University Press.

Humanitarianism and Its Problems

Max Gabriel Cherem

Abstract Migration is often analysed through an exclusively "humanitarian" lens that highlights the dire and widespread nature of human need to argue for urgent relief. This can be problematic for a variety of reasons. First, humanitarianism prioritises immediate charitable aid over justice-based institutional reforms. In so doing it may disempower aid recipients, subject them to domination, or encounter them in other objectionable ways—aid recipients may be treated as supplicants rather than as applicants for what they are (supposed to be) entitled to. Relatedly, if we identify human need as the sole standard for aid we can easily overlook preexisting duties and entitlements that people have as a matter of *right*. For example, while refugee protection always entails *both* "traditional protection" through diplomatic and legal representation (e.g. assessing refugee status) as well as "relief protection" through socio-economic aid (e.g. food, shelter) there is a worry that overemphasis on the latter (especially in the context of camps) can make the former less likely. Moreover, while humanitarian duties are typically discretionary, most migration crises involve specific remedial duties. When humanitarians step in where truly responsible, rights-violating actors resist discharging their remedial duties then, absent efforts to *compel* these actors to do the right thing, they are basically "let off the hook"; this can even give bad actors incentives to manipulate or intensify mass migrations. Finally, humanitarianism can make us neglect the root causes of mass migrations or solutions that are more institutional and political in nature. While this can be an unintended side effect of humanitarianism it may also be *precisely* what is intended by certain actors: by structuring discourse in terms of discretionary humanitarian charity, other more 'politically difficult' possibilities that involve structural and rights-based reforms are either downplayed or removed from discussion.

M. G. Cherem (✉)
Department of Philosophy, Kalamazoo College, Kalamazoo, MI, USA
e-mail: max.cherem@kzoo.com

A. Niederberger et al. (eds.), *Handbook of Migration Ethics*,
https://doi.org/10.1007/978-3-031-89877-8_26

This entry surveys arguments against overemphasising humanitarianism when responding to migration crises—especially responses within the so-called "new humanitarianism". The general thought is that humanitarian responses should be pursued *in parallel with* (rather than instead or to the detriment of) human rights-based responses that compel others to address structural injustices (e.g. via "naming and shaming", financial and diplomatic pressure, and so on). I focus on "survival migrants" and refugees since theorists often analyse these populations in terms of either a "new humanitarian" lens or an attenuated understanding of human rights [e.g. a "political moralism" that is not legally binding or connected to institutional realities, which fails to specify the relevant duty bearers, etc.; see Williams (2005) and Brown (2001) on this term]. I survey the general issues at stake when we over-emphasise humanitarianism in migration theory and practice. I conclude with a series of new proposals and questions that merit future research.

1 Introductory Background

In 1859 French businessman Henry Dunant witnessed the battle of Solferino. He was so disturbed that he began advocating for in-the-field aid to wounded soldiers. He helped found the International Committee of the Red Cross (ICRC) in 1863. A year later Swiss government efforts yielded the first Geneva Convention on the treatment of wounded soldiers. This became the basis for International Humanitarian Law (IHL). IHL slowly developed to regulate the means, goals, and targets of armed conflict. The ICRC is the oldest humanitarian organisation in the world. It is the "custodian" of IHL in that signatories of IHL conventions give the ICRC a mandate over victims of armed conflict (Chandler 2001).

The ICRC is guided by the principles of impartiality, neutrality, and independence. These principles direct the ICRC to prioritise cases of urgent need, assist *all* those in need regardless of side, avoid taking political stances, and to ensure that donations are not directly connected to parties in the conflict. While the articulation of these principles has changed over time (and new principles have been added) this general approach represents the *classic strain* of humanitarianism (Orchard 2010; Chandler 2001).

Since its inception leaders of the ICRC have insisted that, in practice, one agency cannot attempt to respond to a conflict by simultaneously advocating justice and charity—and that the ICRC is firmly on the side of charity. A recent ICRC president reaffirmed this commitment by noting that humanitarianism only tries to meet human needs rather than to correct wrongs or promote justice [cited in Leebaw (2014) and Forsythe (2001), respectively]. Indeed, this general outlook appears to have informed several early UN responses to refugee problems that predated the 1951 Refugee Convention. For example, the UN concurrently established parallel agencies—one political and one humanitarian—to deal with refugees from both Palestine and Korea. The UN Conciliation Commission for Palestine (UNCCP) and the UN Commission for the Unification and Rehabilitation of Korea (UNCURK) were tasked with mediating *political* solutions to refugee-producing crises. Until such a solution could be found the UN Relief and Works Administration for Palestine Refugees in the Near East (UNRWA, which still exists) and the UN Korean Reconstruction Agency (UNKRA) were tasked with administering urgently needed humanitarian "relief" aid. Little has been written on this "two-tracked" approach to refugees in the early days of the UN (although, much *has* been written on UNRWA). Even if navigating power politics was a key factor in adopting this two-track model, it *also* seems to reflect the view that one organisation cannot simultaneously pursue justice and charity.

The ICRC tries to maintain a distinction between humanitarianism and politics in order to both safeguard their principles and facilitate access to imperilled populations. Thus, the ICRC only works with government consent and ICRC workers both avoid political stances and any comment on issues outside their mandate. In a somewhat similar vein, the office of the United Nations High Commissioner on Refugees (UNHCR) has an explicitly "non-political" mandate and is sometimes characterised as a "humanitarian" agency in that it avoids taking sides or commenting on governance and tries to avoid being caught up in coercive dynamics (e.g., when the international community tries to exert financial or diplomatic pressure on a given regime). Many other UN agencies also follow this strategy to facilitate access to those in need.

The line between partisan "politics" and neutral "humanitarianism" is clear but not sharp. Since humanitarianism is tacitly committed to a social liberalism that values individual persons over groups it cannot be *literally* neutral in every situation, especially with respect to collectivist or anti-egalitarian political ideologies (Forsythe 2001). Nevertheless, the strategy of acting *as if* that line can always be upheld is one of long-term pragmatism. For example, amidst the bipolar Cold War even food aid distributed without permission risked offending not only a given state but also its superpower patron. Any perception of "political taint" would reliably choke off access to imperilled populations in the future. The thinking here was that, in order to administer relief aid to all in need of it, humanitarians had to assiduously avoid pronouncing on (political) questions of "who governs?" or issues of structural injustices, and that they also needed to avoid (real or perceived) links to even mild forms of interstate coercion.

In the late sixties Non-Governmental Organisations began to question whether humanitarian relief needed state consent and whether it had to remain impeccably neutral. This was vividly illustrated in Biafra. Biafra was a largely Christian area trying to secede from Nigeria. The government responded in what many took to be an indefensible and unjust manner: with a blanket blockade that produced a famine. The ICRC initially administered aid and briefly participated in airlifting humanitarian supplies. But, after several incidents, the ICRC shut down operations in response to Nigerian government requests (Desgrandchamps 2012).

NGO partners were angered by the ICRC's compliance and its refusal to take sides or speak up about the blockade; for them it seemed like complicity through silence for the sake of access. Various Christian NGOs under the banner of Joint Church Aid (also known as "Jesus Christ Airlines") used civilian aircraft and flew at night so as to continue the airlift without Nigerian government consent. The legacy of this airlift is mixed: it saved many from starvation but also allowed the rebels to hold out beyond when they were militarily defeated—a period in which many more civilians starved or were killed. After the conflict was over Doctors Without Borders was created in response to the ICRC's unwillingness to condemn the Nigerian government's use of famine as a weapon of war.

The splits from the ICRC after Biafra marked a shift in aid and ushered in what has been called the "new humanitarianism". The new humanitarianism tries to not only help people in need, but to *also* condemn human rights violations *and* induce structural reforms or "development" in at-risk groups. In short, under the new humanitarianism there is no longer a strict division of labour between human rights and humanitarianism. Traditionally, the two movements had worked in parallel: human rights groups (e.g. Amnesty International) would name, shame, and marshal external pressure for long-term structural reforms while humanitarians (e.g. ICRC) could step in and (consensually) administer aid that governments were unwilling or unable to provide. But, the new humanitarianism tried to pursue both projects at once, often via paradigms like "human security" or "human development".

International human rights and humanitarian law have influenced one another, as have their associated cognate approaches to aid and advocacy. Scholars disagree about how we should conceptualise their relationship. One camp focuses on how human rights have altered dominant ideas about sovereignty and the (im)permissibility of certain types of conflict under humanitarian law in order to yield dramatic progress in international justice, democracy, and human rights (see, for example, Sikkink's "norm cascades"). This approach often endorses the "new humanitarian" assumption that the traditions can and should be synthesised. A more sceptical camp argues that this mutual influence has watered the traditions down and blunted their more radical potential. This approach emphasises how the discourses in both attenuated traditions are now often framed as a type of mish-mashed *humanitarianism-cum-human rights*; a de facto "charity" wherein people only show up as "of concern" or eligible for aid insofar as they are pure victims without rights or agency [Leebaw (2014) surveys these camps and flags a possible third approach]. When discourses shift in this way they can serve questionable ideological agendas. Some in this second camp even go so far as to describe this type of humanitarianism—which blurs the lines between human rights law, humanitarian advocacy, and refugee law—as an instrument of imperial domination to consolidate global North-South inequalities (Chimni 2000).

The philosophy of migration literature has largely adopted the first perspective. The assumption is that human rights and humanitarianism (largely through aid and advocacy) are mutually supportive and lead to (the same notion of) international "progress" or "development". Among migration theorists the propensity to conflate the two traditions shows up in a variety of ways. For example, in debates about the refugee definition most theorists endorse an expanded "humanitarian" definition of a refugee [though, as Owen (2020) notes, there are recent defences of a "political" definition]. Indeed, many philosophers of migration have either so heavily rooted rights in basic human needs that more 'traditional' rights-based *entitlements* disappear altogether (Shacknove 1985), or they have gone so far as to *explicitly advocate* a "humanitarian" interpretation of human rights (Miller 2007). Ironically, once human rights are conflated with or primarily understood in terms of need or humanitarian charity these accounts may fall prey to the worries of the more sceptical second camp. For, while refugee law is distinct from human rights law, legal scholars widely regard human rights (not humanitarianism) as the proper standard for interpreting refugee rights. Frustration over this confusion has even led UNHCR officials to argue that, when it comes to aiding refugees, "the language of humanitarianism should be systematically substituted with the language of Human Rights" (Jaquemet 2002).

2 General Issues

In general, the worry about policy responses that heavily rely on humanitarianism is that they can undermine the interests of migrants while furthering those of states or other powerful actors. This dynamic exists at both micro and macro levels. At the micro level humanitarianism can disempower migrants by treating them as passive recipients of charity. This can lead to situations of domination. At the macro level an approach that is overly focused on humanitarianism may undercut the rights and entitlements of migrants while producing benefits for the temporary host state, possible resettlement states, or even the persecutory state of origin; gains for states are bought with losses for migrants. In short, responses that overemphasise humanitarianism may make institutional and political reforms much less likely. This counterintuitive outcome is a function of the different capacities and legal statuses held by states and aid organisations, as well as the way in which a humanitarian logic interacts with strategic behaviour to produce opportunities for moral blackmail and coercion. While past scholarship has focused on the negative micro level effects more recent scholarship focuses on the problematic macro level effects as well.

The negative effects that an overreliance on humanitarianism can produce at the micro level are well known. Barbara Harrell-Bond's *Imposing Aid: Emergency Assistance to Refugees* was one of the earliest accounts of this line of critique in migration theory (Harrell-Bond 1986). Her book was an ethnographic case study of the international community's response to Ugandan refugees in the border region of southern Sudan. Through thousands of interviews she found that distributing humanitarian aid through "agricultural settlements" (basically: loosely defined precursors to refugee camps) was inefficient and counterproductive. The aid was so poorly tailored that many refugees resisted it. Moreover, for those who did accept aid, it could facilitate new rights violations. For instance, such settlements often required strangers and even those who *feared* one another to live in close proximity; food rations were not only irregularly distributed but also did not meet nutritional standards; easily anticipated language barriers were a persistent problem with domino-effects; and existing governmental and economic structures were undermined by donor choices to buy relief supplies out-of-country and avoid partnering with local government.

Sadly, the observation that humanitarian assistance can itself yield new rights violations has continued to be relevant: two decades later Harrell-Bond co-authored a book which chronicled how humanitarian aid administered through refugee camps in Kenya and Uganda can very often facilitate or even directly commit new rights violations [Harrell-Bond and Verdirame (2005) discuss, among other incidents: withholding food until camp head-counts were completed, refugees being roughly treated by UNHCR staff, an inability to protect refugees from murder by government "hit-squads", camp "community leaders" meting out lashings for "crimes" that wouldn't normally be crimes in the host country, an ever-present threat of rape and sexual assault, sub-standard or non-existent education, etc.]. If we adopt a broad view of "domination" as being "subject to the will of another" who has overwhelming power, questionable legitimacy, and largely acts with impunity then it is easy to see how humanitarian aid can lead to domination. Moreover, as noted above, some theorists extend this analysis to focus on how the situation that many 'host countries' in the "developing" world find themselves in (where externally funded camps exist and where refugee resettlement is not an option for a variety of reasons) can open the door to domination that shores up global power imbalances between states (Chimni 2000). Other theorists have chronicled further negative effects that humanitarian aid can have on migrants: it can undermine self-reliance (especially when refugees are not permitted to work), be offered in an infantilising or disrespectful manner (Harrell-Bond 2002), facilitate repressive government tactics (Branch 2008), divert international attention (Orchard 2010), or fetishise some hardships over others (Ticktin 2011).

A response that focuses too much or solely on humanitarian aid can also benefit the interests of institutions or states while undermining those of migrants. Here the concern is less with naïve or short-sighted humanitarians that poorly design or implement aid, and more with savvy actors that use humanitarian concerns as either a cover or leverage for other interests. Since these dynamics are more complex and have received less attention they deserve more explanation.

Humanitarian concerns can clearly be used as a *direct* cover for other interests—as when states appeal to humanitarian concerns to justify military intervention (Cohen 2006). But the *indirect* ways in which humanitarianism can mask other interests or serve as an excuse or "alibi" for not doing more are also important (Leebaw 2014; Brown 2001). For instance, since the early 2000s the international response to forced displacement has been to deemphasise asylum and focus on peacekeeping operations or the "soft" protection that supposedly follows from the presence of humanitarian agencies. This response is bad at protecting migrants, particularly in scenarios of regime-induced internal displacement. In those contexts it often only restrains violence in some areas for a limited time (Orchard 2010). Yet, this remains the boilerplate international response. One possible reason why is that it serves the interests of neighbouring states and the international community by "containing" potential refugees.

In short, humanitarian mechanisms that meet utterly basic human needs while also either leaving legal statuses unclear or pre-empting clear legal statuses can immunise states against the duties that they would bear had these migrants reached their (normal) territory. So, when Yugoslavia collapsed in the early 1990s Western European states successfully lobbied UNHCR to pursue "right to remain" policies that would deal with people *internally*. Regardless of whether such policies can sometimes serve the interests of migrants, in Yugoslavia the "right to remain" undoubtedly amounted to a *compulsion to stay* for many (Hathaway 2007). Similarly, while Kenya has signed the refugee convention it claims that by housing refugees in camps it avoids the obligations it would otherwise incur to refugees on its (normal) territory. Many wealthy western states (Italy, Spain, Australia, etc.) have similarly attempted to contain potential refugees on irregular territory (e.g. island "reception centres") where they receive basic humanitarian services like healthcare, food, primary education, and so on. As these responses involve some humanitarian provision the international community may point to them as an excuse to not take more forceful or structural measures. These responses are all symptomatic of a shift among states that once supported refugee law from providing asylum to strategies of "containment" (Shacknove 1993).

Apart from using humanitarianism as a cover for other interests or as an excuse for not doing more, overly humanitarian responses to migration can also create points of *leverage*; they can open the door to moral blackmail or coercion. For instance, a state with many refugees on its territory that has not signed the refugee convention can threaten to either forcibly expel them or open the floodgates of onward travel unless wealthy states give financial assistance, resettlement assurances, or both. This occurred repeatedly during the "Indochinese refugee crisis" of the 1970s (Davies 2007). More recently, Russia seems to have facilitated the onward travel of African asylum seekers to the Finnish border—just when Finland was considering closer partnership with NATO (Higgins 2016). There are *countless* examples of states using refugees to "send a message" and morally coerce other states into concessions or questionable arrangements (Stedman and Tanner 2003).

The frequency with which states leverage refugees is in part due to the prevalence of refugee camps, which basically provide a stable equilibrium, "holding pattern", or "warehousing" outside the protection system envisioned by refugee law (Smith 2004). There has been little formal modelling of the bargaining scenario created by refugee camps. The models that exist suggest that the humanitarian logic of camps can intersect with self-interested states in a way that harms refugees (Backer and Moehler 2007; Zeager 1998). This is not only due to the fact that camps facilitate new violations, but also because they make the "durable solutions" of refugee law— voluntary repatriation, local integration, or resettlement—*less* likely. Modelling shows that when there are more than two "players" in a refugee bargaining scenario the probability of achieving a durable solution *decreases* since the cost of maintaining camps is certain while each actor's burden for implementing the durable solution is not. This gives states an incentive to withhold support in the hope that others will provide it. Sadly, there are almost always more than two players: the country of origin, the "temporary" host state, possible resettlement states, the international community represented by UNHCR, and refugees themselves. Modelling also indicates that *decreasing* the cost of camps delays durable solutions since states judge the initial economic and political costs of resettlement to be higher than maintaining camps.

Finally, "weaponised" migration is the clearest example of how overemphasising humanitarianism (while also refusing to pursue more forceful structural changes) can open the door to moral coercion. Migration flows are weaponised when a state threatens to induce or intensify mass migratory outflows into neighbouring states so as to get bargaining power on other issues. Kelly Greenhill's 2011 *Weapons of Mass Migration* was the first major empirical study of this dynamic. She chronicled 56 attempts to weaponize migration between 1951 and 2006. The tactic is most effective when illiberal or authoritarian regimes use it against liberal states: it largely succeeded 57% of the time and achieved some of its goals 73% of the time (Greenhill 2010). For instance, before the US neutralised the threat by "fast tracking" asylum for all Cubans who arrived, Cuba periodically used migration as a weapon against the United States. Major transit states can also weaponize migration. For instance, Libya under Gaddafi extracted numerous concessions from Italy. Gaddafi foresaw that Italy (and Europe) would be inundated with North African migrants if his regime relaxed border patrols, and for a time was able to parlay this into dual-purpose border-security and military hardware from Italy. A parallel variant of this dynamic occurs when states with densely interdependent labour migration use things like remittances or sudden migrant worker expulsions to send political messages, exert leverage on other issues, or shore up regional regime support (Tsourapas 2018).

Weaponised migration is effective for two reasons. First, there is often high asymmetry between the states. The states inducing or intensifying the flows are often illiberal or authoritarian regimes relatively unconstrained by human rights or domestic opposition, while many of the states that are targeted are liberal or liberalising democracies. Target states must therefore try to *both* respect human rights *and* mediate among constituencies that are "for" or "against" increased migration. This tactic leverages liberal states' vulnerability to human rights "hypocrisy costs", sows internal discord, and forces political capital to be spent at home. Second, in terms cost distribution, international policy responses to migration crises tend to focus on those in need (migrants) and those who can help (host or resettlement states) while ignoring the country of origin. This affords illiberal or authoritarian states leverage through the ability to reliably create an externality that will be borne by other states (Kritzman-Amir 2009). Current policy responses do not try to re-internalise the externality by finding ways to impose costs on the source country, and burden-sharing schemes that could neutralise the power of such coercion have had difficulty getting off the ground. Yet, it is hard to see how to respond to weaponized migration with a purely humanitarian logic that focuses on meeting sheer need and screening out or otherwise giving up on more political responses.

Apart from micro or macro dynamics, over-focus on humanitarianism can blind us to other possible responses. This may occur by design. For instance, countries like Japan and Australia have signed the refugee convention but have nevertheless responded to certain groups of migrants in need with discretionary "humanitarian" visas rather than formally recognising them (and potentially others in the same situation) as refugees entitled to rights. Similarly, while Malaysia is not a signatory of the refugee convention, when speaking of UNHCR's longstanding presence in the country it often appeals to humanitarianism. Leaders tout voluntary cooperation with UNHCR as evidence that they are *already* doing enough for refugees [Lego 2012/2013; see Davies (2007) for historical dynamics behind this]. The worry is that a rhetoric of "apolitical" humanitarian charity that presents itself as "simply" trying to reduce suffering is strategically deployed to forestall other discourses. The thought here is somewhat like Wendy Brown's observation: "Yes, the abuse must be stopped, but by whom, with what techniques, with what unintended side effects, and above all, unfolding what possible futures?" [Brown (2011) speaks of human rights but the point seems best understood as a critique of a certain attenuated understanding of human rights]. Using (merely) humanitarian aid and advocacy to answer these questions can displace discussion and serious consideration of coercion/pressure against systematic and structural injustices or finding ways to institute effective legal rights and the rule of law within "humanitarian spaces" (Wilde 1999; Harrell-Bond 2002).

The strategic use of humanitarianism (that falls short of human *rights*) also extends beyond states to international organisations. For example, the International Organisation for Migration (IOM) is a member-state-organisation that emphasises the management of migratory flows (rather than protecting any refugees in those flows). In 2016 it was recognised as a "related organisation" of the UN that has "a leading role" in migration. Yet IOM does not have a human-rights-based mandate like the UNHCR or even the International Labor Organization (ILO)—a natural home for non-refugee-related migration inside the UN system. Strangely, the 2016 UN-IOM agreement claims the IOM remains a "non-normative organisation". While IOM has its own internal human rights *policy* scholars have wondered whether this language releases IOM from the same type of human rights *law* scrutiny that UN agencies are subject to (Guild et al. 2017). Other scholars have noted how the IOM's work in Indonesia focuses on strengthened border controls, returning asylum seekers, and "awareness" campaigns dissuading people from moving-on. Such work serves the interests of Australia—one of IOM's largest donors. This has led to the charge that the IOM's affiliation with the UN is "blue-washing": creating the impression of a purely humanitarian organisation (that is now also vaguely UN affiliated) while in reality carrying out the interests of member-states (Hirsch and Doig 2018). Scholars have had similar worries about FRONTEX (the EU border agency) because agency directors have insisted that, since FRONTEX only *coordinates* EU border forces, it is legally incapable of violating human rights (Frenzen 2014). This may be technically true as a matter of law but the extra involvement by FRONTEX muddies the waters of what could, practically speaking, be attributed to states in order to seek legal redress [Wilde (2006) on this general dynamic]. This has become salient in light of recent reports and video evidence showing FRONTEX involvement in illegal maritime "pushbacks" from Greek to Turkish waters (Christides et al. 2020).

3 Key Terms, Topics, and Positions

Humanitarian responses to migration crises raise issues of moral coercion, who has what duties to whom, and how to structure institutional responses at the international level. These issues are only beginning to be understood. While political theorists and historians debate how we should view the relation between humanitarianism and human rights *in general*, a large number of authors who work in migration theory seem to downplay the tensions that can be created by the two traditions. Since migration theorists generally don't foreground these problems there are not yet clearly defined scholarly "camps" about how to respond to them. I accordingly only summarise key terms and positions here. The next section surveys some more concrete institutional responses.

For present purposes we can say that pure moral coercion occurs when a coercer engineers a scenario in which a target *must* choose between two options which both harm or wrong others. The target is meant to choose the lesser of two evils. Impure moral coercion may exist insofar as, for example, one of the choices violates either special moral duties or duties to oneself (Bazargan 2014). Moral blackmail is a type of moral coercion where the coercer gets a target to do something by making the alternatives morally unacceptable *to the target*. Moral blackmail can easily arise in fiduciary relationships where an agent is charged with acting in the best interests of a beneficiary and where special moral duties exist. This is because a third party or even the beneficiary can manipulate the agent's self-conception and sense of a special moral relation in order to get the agent to do things they would not otherwise do. Even though the unscrupulous actors are the ones responsible for the target's predicament, and even though the target can complain that they now have moral duties they should not have, this complaint alone does not make their duties disappear (Keller 2018).

Many examples from the last section are clear instances of migration being used to morally coerce or blackmail. Such paradigm cases shed light on less direct examples that are equally important. For instance, UNHCR is in an incredibly fraught moral position insofar as it is *both* an international organisation created by states *and* a fiduciary of refugees. At present over 90% of UNHCR's yearly operating budget is funded through *voluntary* donations from about 11 states including the United States, Japan, and other EU countries. This creates an enormous opportunity for these states to shape UNHCR policy and responses in ways that serve their interests and not necessarily those of refugees (Vayrynen 2001). Yet UNHCR also has a fiduciary relationship with refugees since it is tasked with protecting refugees and promoting durable solutions to their plight. Many special moral relations and duties come to exist between UNHCR staff and refugees. This leaves UNHCR quite vulnerable to moral blackmail.

For instance, even if UNHCR would prefer not to set up refugee camps, camps or other (de facto permanent) "temporary" solutions may be the only option the relevant states will countenance. The choice is between the lesser of two evils: camps or nothing. The need to choose is exacerbated by UNHCR's turf war with the

IOM. While camps or other facilities run by UNHCR might (generally) be better than those run by IOM or individual states (e.g. some Thai camps) they are still problematic. Not only can camps violate rights (e.g. freedom of movement) or facilitate new violations, but they can also allow states to "outsource" their duties to a non-state actor (an international organisation) that has neither the same capacity nor mechanisms of legal accountability (Wilde 1998, 2006). Indeed, to take just one example: hundreds of Roma who were sickened by lead in the water supply of a UNHCR refugee camp in Kosovo are still seeking accountability or compensation more than two decades later.

Other examples of problematic "leverage" abound: donor countries might earmark UNHCR donations to be used for Syrian or Iraqi refugees in Jordan, who then receive more goods and services than Sudanese refugees with comparable needs. The Sudanese feel this violates impartiality and protest in front of UNHCR's Amman office, only to be swept up by Jordanian security forces and refouled to Sudan in retaliation [HRW 2015; Forsythe (2001) on this general dynamic]. UNHCR might feel pressure to mute its objections since they are in the country at the discretion of Jordan, who is not a signatory of the refugee convention. In a similar vein, the IKEA Foundation (associated with but apparently independent from the retail brand) is one of largest private donors to UNHCR. The chief executive, Jens Hansegard, has made public comments that some see as "boasting" about the leverage this gives the foundation over UNHCR policy and priorities (Sazak 2015). Even setting this individual case aside, it is not difficulty to see how the increasingly common practice of *corporate giving* to UNHCR or other humanitarian and "development" organisations can easily create opportunities for moral blackmail and coercion.

Above I noted that identifying human need as the sole or primary standard for assistance may cause us to neglect attempts to enforce preexisting duties. I mentioned one example of this dynamic: when international policy responses to migration crises focus on those in need (migrants) and potential helpers (host or resettlement states) but ignore the duties of the country of origin. This is problematic. For, if a country of origin is largely responsible for a forced mass migration then it *clearly* has particularised, remedial duties to these migrants (and even neighbouring states). If it met these duties then humanitarians would not need to step in. The point here is *not* that those who can help should refrain from doing so until the offending regime takes some responsibility. Rather, the point is that any sensible policy response needs to foreclose a regime's ability to create an externality borne by other states; directly imposing costs on the state of origin is often part of any long-term solution. But, it is often politically difficult and therefore avoided. This same point also applies to many if not most mass migrations. For instance, if a mass migration can be explained by war, colonialism, or exploitation between two states, then shouldn't the relief to these particular migrants come from the parties that primarily contributed to or facilitated the relevant rights violations? On this view the United States might have particularised remedial duties towards Afghanis and Iraqis fleeing contemporary Afghanistan and Iraq, it may have had (or still has) such duties towards those migrating from the Philippines (a former colony) or Vietnam, and it may have such duties towards Haitians fleeing agricultural collapse brought about by US farm subsidies. Similarly, France might have these sorts of duties towards those fleeing Algeria or Mali, and Italy may have them towards Libya. Precisely what remedial duties are and how they should be discharged is debated (Amighetti and Nuti 2015); asylum might play a role here (Souter 2014). But these questions are separable from the question of whether these duties exist and whether we are not letting bad actors off the hook when we simply give up on pushing for their enforcement through reparations or using pressure to push for other remedial measures and instead settle for acts of humanitarianism that are perceived as discretionary charity (Wilcox 2012).

How should we respond to the fact that many international responses to mass forced migration seem to overemphasise the charity-based ethics of humanitarianism? An extreme view is to *merely* rely on market mechanisms (I shall not consider military intervention to stop migratory flows, as I know of no scholar who advocates it as a *primary* response). Economist Gary Becker once proposed the complete commodification of all migration: all people could simply pay to migrate (Becker 2011). Given that some states have begun to experiment with selling citizenship his proposal no longer seems outlandish [see Simon (1990) on giving migrants loans to purchase citizenship]. Yet, whatever (de)merits his proposal may have for truly discretionary migration, his policy is grossly inadequate to deal with refugees and survival migrants—a topic Becker devoted *one comment to* in the question-and-answer panel accompanying his proposal. His proposal also seems to neglect considerations of brain-drain. A more sensible way forward would be to find institutional innovations that pursue human rights based justice and humanitarian charity *in parallel*—much like the classic strain of humanitarianism. Moreover, since new responses need to avoid the perverse incentives that currently exist there may be *some* place (however limited) for "marketisation" mechanisms.

4 Future Research

Most future research in this area will centre on international norms, policy options, and institutions. In recent years there have been a number of proposals that try to surmount the limitations of older approaches which rely on "humanitarian measures alone"—approaches many see as inadequate, "reactive" (Dowty and Loescher 1998), or even facilitative of mass migration crises (Garry 1998). These proposals try to tweak the perverse incentive structures that have let states engage in weaponised migration or which have made refugee camps (or much the same thing in terms of status: long-term residence granted on a discretionary basis with in non-signatory states) a stable equilibrium. One common approach is to move from thinking of refugee and forced migration scenarios as a two party-relation (e.g. refugees and resettlement states) to thinking of them as a relation between three or more parties (e.g. adding in the country of origin, "temporary" host states, the UNHCR, etc.). Other common themes are the integration of market elements (like quotas with trades and bargains, "matching" schemes) or the creative use of existing legal mechanisms to *compel* those who are responsible to bear the costs of their behaviour. In this final section I survey some of these proposals and suggest avenues of research.

Luke T. Lee worked on the first *legal* proposal along these lines (Lee 1986). He proposed that states receiving mass inflows of refugees could get compensation for the costs of hosting them. A state could lodge a complaint with the state of origin and then (absent progress) refer the issue to the International Court of Justice (ICJ). If that failed or the ICJ did not have jurisdiction the international community could suspend UN aid or assistance to exert pressure. Finally, following early ICJ precedent, the UN could bring a claim against the state of origin to obtain compensation for not only those acting on behalf of the UN but also for damages to those over whom the UN serves as a "guardian". This language has been used to claim compensation for UN workers and for refugees themselves, but it could also be used to claim reimbursement for states that host refugees.

Compensating states for the costs associated with hosting refugees faces serious challenges. But, the basic framework appears sound enough such that rendering the *threat* of compensatory damages more credible may have a mild preventative effect (Garry 1998). Recent proposals have also given the idea more teeth. For instance, the international community could freeze some (targeted) assets of refugee producing states and then use them to pay for the costs of refugee camps (Sazak 2015; Goodwin-Gil and Sazak 2015; something like this approach, though not with reference to refugees in particular, appears to be taking place with respect to Russia over its illegal aggression in Ukraine). While this is an area of further research, at least in principle it seems like such frozen assets could *also* be used for the costs of initial integration during the time period when refugees are not yet citizens of the resettlement state and (the first year or so) when it could be shown that they *may* be a net economic drain (due to initial under- or unemployment coupled with reliance on early integration services—language classes or job retraining). This could yield more resettlement since it would undercut an argument put forward by restrictionists. Indeed, while I could find *no* scholarly literature on it, it bears emphasis that after WWII the Preparatory Commission for the International Refugee Organization (PCIRO) held an auction of "Nazi Loot" in order to fund refugee resettlement. There are surely moral mistakes from how that auction was structured which could then inform how the international community crafts any future scheme to freeze and disburse the assets of refugee producing states. There also seems to be no reason why a compensatory mechanism could not be used against the frozen assets of *non-state* actors that created mass refugee outflows (e.g. groups like ISIS that had large territorial control). While there are some issues to be ironed out (e.g. how to treat the proceeds of *patently* unjust enrichment and how to approach completely non-separated budgets) implementing such a mechanism against these groups could serve as a "pilot case" or "proof of concept" that could then one day be scaled-up to a refugee-producing state.

Other scholars propose punitive or forcible policy responses to mass migrations that are either natural extensions of Customary International Law (CIL) or which may have even achieved the status of CIL over time. For instance, it has been argued that there is an emerging consensus that mass cross-border refugee flows are a threat to international peace and security. As such, they can trigger intervention under Article VII of the UN Charter (Dowty and Loescher 1996). Moreover, so-called "humanitarian safe zones" may already represent an emerging CIL norm of permissible humanitarian intervention (Ní Ghráinne 2019). Such zones are increasingly part of the response to not only protect civilians amidst conflict but to also manage mass displacement "at home".

Lastly some proposals use market elements. Since the late 1990s several prominent scholars have advocated mildly market-ized systems that involve both collectivised burden sharing and tradable quotas (Hathaway and Neve 1997; Schuck 1997, 2014). These bear some resemblance to "cap and trade" policies used for global emissions. More recently, there have been proposals to match states with refugees awaiting resettlement through a system analogous to systems that match post-med-school residents with hospitals or students with charter schools. Such a system would give weight to the rank-ordered preferences of states and those of refugees, but not absolute weight (Jones and Teytelboym 2018).

A more strongly market-focused proposal would allow states to sell and buy the *services* of asylum determination and protection amongst themselves. Since only states meeting human and procedural rights standards should participate, a UN agency would need to underwrite, mediate, and perhaps even monitor such transactions. Such a system would reverse some of the perverse incentives of the current system (Himmelreich 2019). One detail to be worked out would be how such a proposal would avoid so-called "chain refoulement" and which parties would be held responsible if it occurs. Another is how procedural rights would be dealt with—perhaps via a global minimum standard (May 2011).

Yet another strongly market-based proposal starts from the observation that refugees have claims for compensation against their country of origin. Thus they could transform this financial claim into a tradable asset that could be exchanged for admission into a resettlement state. The state could then seek to collect the debt directly, apply it to existing sovereign debts, or sell it to a hedge fund specialising in sovereign debt (Blocher and Gulati 2016). Here a main worry is that, to avoid dynamics that would yield sub-par protection outcomes, refugees would need to bargain collectively. But, it may be quite difficult to justify the enforcement of collective bargaining at the international level when the only thing binding a group of *refugees* together is that they all fled a country (perhaps for very different reasons). However, this bargaining mechanism might be appropriate for other groups of displaced persons that *do* share a dense interdependency of interests or identity and who *do* generally want to stay together—e.g. climate migrants from disappearing small islands states or displaced coastal indigenous groups. But, in those scenarios the financial claim would be of a different nature [see Donhauser (2019) on using the "losses and damages" clause of the Paris Climate agreement].

Literature

Amighetti, S., and A. Nuti. 2015. David Miller's Theory of Redress and the Complexity of Colonial Injustice. *Ethics and Global Politics* 8. https://doi.org/10.3402/egp.v8.26333.

Backer, D., and D. C. Moehler. 2007. *The Puzzling Persistence of Semi-Permanent Refugee Camps. Analyzing Humanitarian Crises Using Bargaining Models*. Working Paper. Harvard Academy of International and Area Studies.

Bazargan, S. 2014. Moral Coercion. *Philosophers' Imprint* 14:1–18.

Becker, G. 2011. *The Challenge of Immigration. A Radical Solution—With a Commentary by Diane Coyle"*. Occasional Paper No. 145. Institute of Economic Affairs.

Blocher, Joseph, and M. Gulati. 2016. Competing for Refugees. A Market-Based Solution to a Humanitarian Crisis. *Columbia Human Rights Law Review* 48:53–111.

Branch, A. 2008. Against Humanitarian Impunity. Rethinking Responsibility for Displacement and Disaster in Northern Uganda. *Journal of Intervention and Statebuilding* 2:151–173.

Brown, W. 2001. *Politics Out of History*. Princeton: Princeton University Press.

Brown, W. 2011. 'The Most We Can Hope For…?'. Human Rights and the Politics of Fatalism. In *Wronging Rights? Philosophical Challenges for Human Rights*, ed. A. Cistelecan and A. S. Rathore. New Delhi: Routledge.

Chandler, D. 2001. The Road to Military Humanitarianism. How Human Rights NGOs Shaped a New Humanitarian Agenda. *Human Rights Quarterly* 23:678–700.

Chimni, B. S. 2000. Globalization, Humanitarianism, and the Erosion of Refugee Protection. *Journal of Refugee Studies* 13:243–263.

Christides, G., E. Freudenthal, S. Lüdke, and M. Popp. 2020. EU Border Agency Frontex Complicit in Greek Refugee Pushback Campaign. *Der Spiegel*, October 23, 2020.

Cohen, J. L. 2006. Sovereign Equality vs. Imperial Right. The Battle over the 'New World Order'. *Constellations* 13:485–505.

Davies, S. E. 2007. *Legitimizing Rejection. International Refuge Law in South East Asia*. Leiden: Martinus Nijhoff.

Desgrandchamps, M. L. 2012. 'Organizing the Unpredictable'. The Nigeria-Biafra War and Its Impact on the ICRC. *International Review of the Red Cross* 94:1409–1432.

Donhauser, J. 2019. Climate Refugees. Realizing Justice Through Existing Institutions. In *A Research Agenda for Climate Justice*, ed. P. G. Harris, 118–130. Cheltenham, Northampton, MA: Edward Elgar.

Dowty, A., and G. Loescher. 1996. Refugee Flows as Grounds for International Action. *International Security* 21:43–71.

Forsythe, David. 2001. *UNHCR's Mandate. The Politics of Being Non-Political*. New Issues in Refugee Research Series Working Paper 33.

Frenzen, N. 2014. Interview with FRONTEX Director Laitinen. *Migrants at Sea blog*, March 24, 2014. https://migrantsatsea.org/2014/03/24/interview-with-frontex-director-laitinen/.

Garry, H. R. 1998. The Right to Compensation and Refugee Flows. A 'Preventative Mechanism' in International Law? *International Journal of Refugee Law* 10:97–117.

Goodwin-Gil, G. S., and S. C. Sazak. 2015. Footing the Bill. Refugee Creating States' Responsibility to Pay. *Foreign Affairs*. https://www.foreignaffairs.com/articles/africa/2015-07-29/footing-bill.

Greenhill, K. M. 2010. *Weapons of Mass Migration. Forced Displacement, Coercion and Foreign Policy*. Ithaca: Cornell University Press.

Guild, E., S. Grant, and K. Groenendijk. 2017. *IOM and the UN. Unfinished Business*. Queen Mary School of Law Legal Studies Research Paper 255.

Harrell-Bond, B. 1986. *Imposing Aid. Emergency Assistance to Refugees*. Oxford: Oxford University Press.

Harrell-Bond, B. 2002. Can Humanitarian Work with Refugees Be Humane? *Human Rights Quarterly* 24:51–85.

Harrell-Bond, B., and G. Verdirame. 2005. *Rights in Exile. Janus-Faced Humanitarianism*. New York, Oxford: Berghahn Books.

Hathaway, James. 2007. Forced Migration Studies. Could We Agree to Just 'Date'? *Journal of Refugee Studies* 20:349–369.
Hathaway, J. C., and R. A. Neve. 1997. Making International Refugee Law Relevant Again. A Proposal for Collectivized and Solution-Oriented Protection. *Harvard Human Rights Journal* 10:115–211.
Higgins, Andrew. 2016. E.U. Suspects Russian Agenda in Migrants' Shifting Arctic Route. *The New York Times*, April 2, 2016.
Himmelreich, J. 2019. Asylum For Sale. A Market Between States that Is Feasible and Desirable. *Journal of Applied Philosophy* 36:217–232.
Hirsch, A. L., and C. Doig. 2018. Outsourcing Control. The International Organization for Migration in Indonesia. *The International Journal of Human Rights* 22:681–708.
Human Rights Watch. 2015. *Jordan. Deporting Sudanese Asylum Seekers*. https://www.hrw.org/news/2015/12/16/jordan-deporting-sudanese-asylum-seekers. Last checked 16 Nov 2023.
Jaquemet, S. 2002. Refugee Protection and Humanitarian Values. *Refugee Survey Quarterly* 21:111–112.
Jones, W., and A. Teytelboym. 2018. Matching Systems for Refugees. *Journal on Migration and Human Security* 5:666–680.
Keller, S. 2018. Fiduciary Duties and Moral Blackmail. *Journal of Applied Philosophy* 35:481–495.
Kritzman-Amir, T. 2009. Not in My Backyard. On the Morality of Responsibility Sharing in Refugee Law. *Brookings Journal of International Law* 34:355–393.
Lee, L. T. 1986. The Right to Compensation. Refugees and Countries of Asylum. *The American Journal of International Law* 80:532–567.
Leebaw, Bronwyn. 2014. Justice, Charity, or Alibi? Humanitarianism, Human Rights, and 'Humanity Law' (Book Review of Ruti Teitel's *Humanity's Law*). *Humanity* 5:261–276.
Lego, J. 2012/2013. Humanitarianism in the Reception of Refugees. Implications, Contradictions, Limitations. *Asian Journal of Global Studies* 5:81–93.
May, L. 2011. *Global Justice and Due Process*. Cambridge: Cambridge University Press.
Miller, David. 2007. *National Responsibility and Global Justice*. Oxford: Oxford University Press.
Ní Ghráinne, B. 2019. *Safe Zones as Humanitarian Intervention*. Presented at the University of Michigan Law School Young Scholars in the Law Conference (on file with author).
Orchard, P. 2010. The Perils of Humanitarianism. Refugee and IDP Protection in Situations of Regime-Induced Displacement. *Refugee Studies Quarterly* 29:38–60.
Sazak, S. C. 2015. An Argument for Using Frozen Assets for Humanitarian Assistance in Refugee Situations. *Journal of International Affairs* 68:305–318.
Schuck, P. 1997. Refugee Burden-Sharing. A Modest Proposal. *Yale Journal of International Law* 22:243–297.
Schuck, P. 2014. Refugee Burden-Sharing. A Modest Proposal Fifteen Years Later? In *The Nation State and Immigration. The Age of Multiculturalism*, ed. Anita Shapiro et al. Brighton: Sussex Academic Press.
Shacknove, Andrew. 1985. Who is a Refugee? *Ethics* 95:274–284.
Shacknove, A. 1993. From Asylum to Containment. *International Journal of Refugee Law* 5:516–533.
Simon, J. 1990. *The Economic Consequences of Immigration*. London: University of Michigan Press.
Smith, M. 2004. Warehousing Refugees. A Denial of Rights, a Waste of Humanity. *World Refugee Survey* 38:38–56.
Souter, J. 2014. Towards a Theory of Asylum as Reparation for Past Injustice. *Political Studies* 62:326–342.
Stedman, S. J., and F. Tanner, eds. 2003. *Refugee Manipulation. War, Politics and the Abuse of Human Suffering*. Washington, DC: Brookings Institution Press.
Ticktin, M. 2011. *Casualties of Care. Immigration and the Politics of Humanitarianism in France*. Berkeley: University of California Press.

Tsourapas, G. 2018. Labor Migrants as Political Leverage. Migration Interdependence and Coercion in the Mediterranean. *International Studies Quarterly* 62:383–395.

Vayrynen, R. 2001. Funding Dilemmas in Refugee Assistance. *International Migration Review* 35:143–167.

Wilcox, S. 2012. Do Duties to Outsiders Entail Open Borders? A Reply to Wellman. *Philosophical Studies* 169:123–132.

Wilde, R. 1998. Quis Custodiet Ipsos Custodes? Why and How UNHCR Governance of 'Development' Refugee Camps Should be Subject to International Human Rights Law. *Yale Human Rights and Development Law Journal* 1:107–128.

Wilde, R. 2006. Enhancing Accountability at the International Level. The Tension Between International Organization and Members State Responsibility and the Underlying Issues at Stake. *ILSA Journal of International and Comparative Law* 12:395–415.

Williams, B. 2005. *In the Beginning Was the Deed. Realism and Moralism in Political Argument.* Princeton: Princeton University Press.

Zeager, L. 1998. Negotiations for Refugee Repatriation or Local Settlement. A Game-Theoretic Analysis. *International Studies Quarterly* 42:367–384.

Refugee Agency (Poststructuralism/ Normality of Migration/Illegalisation)

Mareike Gebhardt

Abstract Poststructuralist approaches are often prevalent in border regime and critical migration and refugee studies when questions of illegalisation and de/normalization of migration and *refugee-ness* are discussed. The paper introduces refugee ethics from a poststructuralist perspective that understands migration as a phenomenon produced via discursive operations. It is stabilised by performative repetitions and manifests in legal regulations. Migration thus comprises a tightly knit network of speech acts, tacit performances, semantics, affects, materialities and architectures that make up the international migration regime. The paper discusses refugee agency from a white, European feminist academic position and how it is rendered im/possible in the contemporary migration regime. It concentrates on an analysis of the bio- and necropolitical production of oppression via migration politics and shows how, in reference to Autonomy of Migration, people on the move rework migratory routes and political scripts to challenge contemporary hegemonies concerning migration.

During the first wave of the SARS-CoV-2 pandemic in the spring of 2020, most national governments imposed general lockdowns and curfews to contain the virus worldwide. Simultaneously, the hashtag *#leavenoonebehind* spread through social media platforms, media outlets and urban areas, where it was sprayed on streets and walls. The hashtag soon represented further failings of the international human rights regime to protect vulnerable and marginalised groups *left behind* during crisis management. From a critical perspective, one can suppose that they are left behind because their lives do not count (socially, politically, legally and morally) or cannot be counted by statistics because they stay invisible to the calculating eye of the biopolitical regime of crisis administration.

We seem to live in a world of crises: The financial and real estate crisis was superseded by a migration crisis that was supplanted by the COVID-19 crisis. However, these crises share the amplification precarity in groups already rendered

M. Gebhardt (✉)
Department of Political Science, Münster University, Münster, Germany
e-mail: Mareike.gebhardt@uni-muenster.de

A. Niederberger et al. (eds.), *Handbook of Migration Ethics*,
https://doi.org/10.1007/978-3-031-89877-8_27

precarious and vulnerable: exposed to poverty, violence and, eventually, death. During the COVID-19 pandemic, however, some people did not want to leave othered populations behind. Instead, they protested against the *national(-ist)* policies that were introduced, ironically, to counter a *global* pandemic—as if the virus cared about borders and nationwide lockdowns. The virus did not render all equal. In contrast, it worsened and deepened the social hierarchies of precarity. One of these groups comprises people on the move who are mostly referred to legally, morally and politically primed terminology as "migrants" or "refugees". Via the hashtag *leavenoonebehind*, often followed by or connected to *#evacuatemoria*, the protest against nationalist and authoritarian governmentality became visible, especially throughout Europe. "Moria" represents one of the EU's most notorious internment camps for refugees and migrants. It is located on the Greek island of Lesbos and was designated a European migration "hot spot". Moria camp accommodated up to 20,000 people, although its infrastructure was initially designed for only 2800. In September 2020, the camp burnt down, leaving its inmates without shelter while EU member states discussed what to do with them. Most had been in the Greek camp, which was planned as a temporary transit zone, for months, if not years. Migration management failed, asylum processes were delayed and living conditions worsened. When the camp caught fire in September 2020, media and political attention rose significantly, and Moria's inmates became visible—often addressed as victims and potential criminals who had set the camp on fire themselves (Ullrich 2022: 46f).

How some voices and discourse positions are excluded while others dominate and shape debates, deliberations and discussions are addressed in this paper under the notion of refugee agency. Poststructuralism and Autonomy of Migration (AoM) ask how society and politics render different populations in/visible and how they justify leaving people behind. Their analytical focus is the discursive construction of migratory populations as "disposable" or "ungrievable" lives; i.e., lives that do not count or are only counted in death when the *United Nations High Commissioner for Refugees* (UNHCR) releases the yearly numbers of people who died while on the move: people exposed to structural, sexualised and racialised violence in home countries, transit zones and receiving countries, people deprived of fundamental rights and infrastructure—from freedom of movement to sanitation—and people starved in deserts or drowned in oceans.

The paper, first, introduces general issues at stake in the debate, focusing on the narrative production and discursive dissemination of migration as a "crisis". Second, different positions within the debate are introduced and compared, concentrating on scholarship on migration with an emphasis on poststructuralist theories. Lastly, some desiderata for future research are identified and briefly discussed.

1 A Song of Sovereignty: The Narrative Production and Discursive Dissemination of Migration as "Crisis"

In his notorious definition in *Political Theology* (1922), "Sovereign is he who decides on the exception", Carl Schmitt connects the concept of sovereignty with the decision-making power to declare a state of exception. Via the state of exception, leaving the realm of the rule of law is justified and democratic decision-making processes, such as parliamentary debate, are suspended. What was historically most often used to justify non-democratic politics can now be witnessed in recent Euro-Atlantic migration politics. Narratives of exception shape migration politics in liberal democracies. They denote situations as crises to justify the suspension of democracy. Lauren Berlant (2007: 760) describes the inner workings of this process. The notion of "crisis ordinariness" defines the misrepresentation of "duration and scale of a situation by calling a *crisis* that which is a fact of life and has been a defining fact of life for a given population that lives it as a fact in ordinary time." In this sense, the narrative distribution and discursive dissemination of migration as a "crisis" have been used by politicians, parties and media outlets from the whole political spectrum to frame human mobility as a social, political, cultural and legal exception to destabilise the status quo. Governments and their representatives have governed through crises with emergency politics and narratives of urgency justifying the suspension of parliamentary procedures, for instance, during the 2015 so-called "refugee crisis". What logics and logistics underly such a claim? Berlant suspects that the "deployment of crisis is often explicitly and intentionally a redefinitional tactic, a distorting or misdirecting gesture that aspires to make an environmental phenomenon appear suddenly as an event because as a structural or predictable condition it has not engendered the kinds of historic action we associate with the heroic agency a crisis seems already to have called for" (Berlant 2007: 760). The crisis narrative works as a socio-political tactic to deepen, amplify and circulate negative feelings towards a particular phenomenon, such as migration, and render it an exception. Consequently, it becomes an anormal and thus threatening situation that putatively destroys the status quo's achievements.

Berlant also propagates the idea of heroism that dwells within crisis narratives. Those who manage the crisis best or even "solve" it will appear as the saviours of the common good. When Donald Trump, repeatedly and beyond any financial or realistic rationale, evoked his plan to build a wall between the USA and Mexico in his 2016 election campaign and during his presidency, or when Manfred Weber, head of the conservatives in the European Parliament, compared the selection criteria of EU migration policies to the front door of a house, both respond to a perceived crisis. Both provided putative solutions that evolve around tropes of bulwark and mechanisms of lockdown: build a wall or close the door as high, tight, and impenetrable as possible. Moreover, both rhetorical tactics of bordering follow the idea that those they seek to shut out or leave behind (a wall) are rendered invisible. They are given no chance to tell their side of the story. They are neither represented medially nor politically—as if those populations that most democracies define as refugees or immigrants do not exist, as if they do not have the right and the capability to decide where, how and when to move. The crisis narrative, therefore, works via processes of denormalization, invisibilization and unrecognizability. It reduces refugees and migrants to victims, at best, and takes from them their agency or, even worse, identifies them as threats.

Within contemporary migration politics, the narrative of crisis stretches globally. It can be witnessed in both the Global North and the Global South. The threatening tone of high numbers of 'migrant influx' often instils fear and insecurity, as we saw during the 2015 migration management crisis. Furthermore, migration has been under strict surveillance by receiving countries. Surveillance technologies have become a decisive factor in the il/legalisation of human mobility. However, human mobility—the conditions and circumstances leading to leaving one's home—has always been a defining fact of some populations' lives. It has been "ordinary", in Berlant's words. People have always been on the move and will continue moving. Legal definitions and nation-state-centric policies produce and frame this mobility as migration, as approaches in AoM have argued (Genova 2017). Consequently, denoting movements as 'migrant/refugee crisis' seeks to de-normalise mobility and, in turn, normalise immobility. Additionally, the discursive and narrative processes and the politics of de-normalisation regarding migration are often soaked with racism, nationalism, culturalism and (neo)colonialism.

2 From Biopolitics to Necropolitics (and Back?): Positions in the Controversy

In border regime and critical migration studies, poststructuralist approaches are often prevalent when questions of de-/normalisation and illegalisation of migration are discussed. From the perspective of poststructuralism, migration is a phenomenon produced via discursive operations. It becomes stabilised by performative repetitions. Referring to Michel Foucault (1980), a migration regime can be defined as a "thoroughly heterogeneous ensemble" to govern populations. This ensemble consists "of discourses, institutions, architectural forms, regulatory decisions, laws, administrative measures, scientific statements, philosophical, moral and philanthropic propositions—in short, the said as much as the unsaid." Accordingly, migration and its adjacent politics and law compose a tightly knit network of speech acts, tacit performances, semantics, affects, materialities and architectures. In Foucault's perspective, political rhetoric, social semantics, legal definitions, and media outlets represent migration—its routes, logic, politics, and effects—and produce migration as a (nation) state-centric concept framing human mobility as migration. However, globalisation, understood as mobility beyond the nation-state, does not normalise transnational movements of people but only the mobility of elites and capital. Instead, it is accompanied by authoritarian and nationalist reactions that stage immobility as the normality of living in their discursive and performative practices.

At the same time, mobility is singled out and understood as an exception. Hence, its 'management' through emergency politics and states of exception.

2.1 *Border Spaces and Governmentality: The Other's Death*

Foucault's works on power, knowledge, normalisation, governmentality and biopolitics have been used to analyse migration politics and policies. His œuvre provides both a vocabulary and an understanding of how liberal and neoliberal formations produce knowledge by interlocking it with power. Those in power decide what and who is to be known. They define what is to be located *within* the normative frameworks and, hence, understood as 'normal'. In turn, what is left out and rendered invisible is considered deviant from the norm. The knowledges, practices and groups outside these normative orders and normalising processes are subjected to othering and denormalization, such as the 'sexually deviant', 'mentally ill' or 'foreign'; the latter is represented by the social and legal constructions of refugees and migrants.

Moreover, those in power develop an "art of government"—as technology, practice, and strategy—to stabilise their rule, its normative frameworks and canonised knowledge. Foucault describes this technique and mindset as *governmentality*. In his lecture series on biopolitics, sovereignty and governmentality, he argues that "one of the greatest transformations" in politics and law during the nineteenth century is the shift from sovereignty as the "old right [...] to take life or let live" to the "opposite right [...] to make live and to let die" (Foucault 2003: 239ff.). This transformation marks the "birth of biopolitics" through which people are governed as those who must live and those who must die (Foucault 2010).

Among the most well-known interpretations of Foucauldian thought transferred to questions on human mobility, we find Achille Mbembe's and Giorgio Agamben's works. Referring to Foucault and Schmitt, Agamben (1998, 2005) discusses the juridico-political construction of a state of exception as an overarching paradigm (*nomos*) to govern populations biopolitically. He uses the *homo sacer*, a legal figure originating in the Roman Empire, to show how lives are rendered disposable by pushing them beyond the rule of law. These lawless *homines sacri* represent those allowed to kill without social, political, or moral repercussions. Agamben's highly polarising theory has been critically reviewed in migration, refugee, and citizenship studies to rework concepts of citizenship and *refugee-ness* (Isin and Nyers 2014). Scholars criticise with and against Agamben that people on the move—as the *homini sacri* of our time—have been depoliticised by the humanitarian juridico-moral category of 'the refugee'. Consequently, they push for the politicisation of both 'the refugee' as a subject "beyond 'bare life'" and 'the camp' as a political space of contestation and not only oppression (Owens 2009; Rygiel 2001; cf. Ullrich 2022).

While Foucault concentrates on the category of *life* in his considerations on *bio*-power, Mbembe (2003) focuses on *death* as the primary criterion for analysing contemporary (migration) politics: its colonial residues and neo-colonial formations.

Biopower turns into *necropower*, which defines sovereignty as the capacity to decide who does not matter and is, thus, disposable. Migration politics, Mbembe argues, are linked to the politics of death since they operate on the "shattering experience of otherness". It is, therefore, shaped by racialised questions, narratives, and productions of alterity. The idea of necropower is echoed, for example, in the hashtag *blacklivesmatter*. To Mbembe, othered and disposable lives are exposed to the biopolitical administration of racialised populations and the necropolitical management of their (social or physical) death. *Necropower*, working through the contemporary migration regime, defines which groups are deemed legal or legitimate under the rule of (supra)national or international law: First, in terms of visa regulations, 'legal immigrants' can be identified. Second, through humanitarian lenses, the 'refugee' is produced by and protected under the 1951 *Geneva Refugee Convention* and the *Protocol* of 1967. Third, migratory populations who do not fit into one of these two categories are often homogenised and become illegalised. In necropolitical operations, the power-political focus changes from sustaining healthy lives of a given population to selecting and eradicating the 'foul', 'impure' and 'diseased'. In short, the 'disposable' populations who are left behind in extra-territorialised detention centres and internment camps eventually die in desert border wastelands or oceans.

Mbembe's concept of necropower uses a postcolonial epistemology to deconstruct how current discursive operations, evolving around migration as a crisis, dehumanise migratory lives and depict them as a 'menace' within a neo-colonial mindset. As the gendered and illegalised 'migrant' putatively originates from the savage tribes of 'wild', 'uncivilised' and potentially 'dangerous' (post)colonial zones, the 'migrant' is judged through (neo)colonial frames. Therefore, the possibility of a common bond between the colonised and the colonisers are racially denied. Solidarity can neither grow nor thrive (English et al. 2019). However, Mbembe's attention on disposability neglects spaces of agency that is, for example, practiced in daily encounters and local communities to counter the devastating effects of necropower. Consequently, his theorization does not account for the ambiguity between repression and resistance, which we both find in the "Black Mediterranean" (Proglio et al. 2020) or the desert US-Mexican "borderlands" (Dudziak and Volpp 2006).

2.2 Migratory Itineraries: Spaces of Contestation

In contrast to Mbembe and Agamben, a Foucauldian research perspective allows for analysing migrant and refugee lives as subjected to repressive bio- and necropolitical management. Repressive power is challenged by or co-constituent with resistance and contestation in Foucault's thinking. Accordingly, scholars and academic fields dedicated to the critical analysis of migration and human mobility use Foucault's works to show that refugees and migrant populations are not, or at least not only, helpless victims or oppressed subjects of the bio- and necropolitical migration regime. Instead, mobilities, "refugeeness" (Picozza 2017) and populations on the move need to be understood in more complex ways: as agents, political actors, and social influencers. Mobility as refugee and migratory agency has to be acknowledged by media outlets, societal debate, politics and academic scholarship. However, the "actions of migrants" are not to be understood "as subjects of pure agency or as liberal individuals with or without rights, but as complexes of practices that always exceed the ability of migration policies and state authorities to fix and control them." Referring to AoM, scholarship on migration has shifted the focus "from the apparatuses of control to the multiple and diverse ways in which migration responds to, operates independently from, and in turn shapes those apparatuses and their corresponding institutions and practices," Casas-Cortes, Cobarrubias and Pickles (2015: 896f.) argue. Refugees and migrants are not only oppressed objects or victims of the migration regime but also agents and subjects that challenge, navigate through, shape and alter the migration regime. This perspectival turn, influences research designs and publications, especially from the early 2000s onwards. Different engagements with AoM within critical migration and border studies thus share the effort "to reinterpret the effects of seeing regular, irregular, transit and other forms of migration as constitutive factors of border policies, architectures, and practices" (Casas-Cortes et al. 2015: 896f.).

More specifically, critical migration and border studies challenge three discursive practices and socio-political interpretations: first, the construction of migration as singularity or 'crisis' in the sense of Berlant's definition. Second, they contest the imagination of migrant lives as 'helpless victims' we often find within humanitarian frameworks. Third, they challenge the ideology of migrants and refugees as 'threats' to the common good, which we can witness in anti-migration discourse and nationalist and racist politics (Friese 2017). Thus, AoM seeks to rework both scholarship on migration and socio-political understandings of migration and migrant and refugee lives as agency (Genova et al. 2018; Hess et al. 2017; Mezzadra 2020). Consequently, not only the mere humanitarian ethics of receiving countries are questioned but also those of research designs and scholarly practices. Both (may) reproduce problematic colonial stereotypes that must be identified and scrutinised. Both need to draw their attention toward the intertwined spaces of ethics and politics to show how refugee agency traverses both spaces. Thus, refugee agency is part of an ethico-political dispositive in which complex entanglements of repression and resistance thrive.

With this Copernican turn, the academic perspective on borders changes too. Borders are deconstructed in their semi-metaphysical understanding as natural and/or a means for safeguarding (national) security. Instead, critical scholarship has shown that borders are contingent: that they are governed by regressive nationalism and efficient management, which are discursively produced, affectively stabilised, and socially reproduced. Moreover, borders are not understood as a neatly drawn and clearly defined lines but as spaces of contestation. They can be modified or radically changed by counter-hegemonic practices, affective atmospheres of resistance and emancipatory social climates.

Consequently, ‘the’ border becomes pluralised: more complex, geopolitically contested “borderlands” (Balibar 2009) that politicise border surveillance technologies and challenge power structures to correspond and adapt to the subversive versatility of migratory routes, refugee agency and migration networks. Contested borderlands turn into spaces of subversion and emancipatory struggles. Consequently, migration and mobile lives are politicised as well. Migration becomes an ethico-political act challenging the topology and the inner workings of power. Migratory lives are no longer or not only passively exposed to violence, discrimination, and repression. Instead, they become “border subjects”—neither objects nor victims—that draw, challenge, transcend and create new borderlands to “dwell” in them (Schulze Wessel 2016). However, the international migration and border regime still represent stately oppression and fantasies of territorial sovereignty. Consequently, contested border spaces are both “emancipatory” and “repressive” (Dudziak and Volpp 2006).

With its politicisation, refugees’ and migrants’ acts become comprehensible as protest and emancipation. Even though refugees and migrants still live under legally, socioeconomically and physically precarious conditions, they act politically—even if their practices are, at first sight, not legible as a protest or civil disobedience in the hegemonic eyes of the white, Anglo-European Global North (Topak 2016). However, the refugee protest marches from Berlin to Vienna, the sit-ins at Budapest Main Station, the protests of undocumented populations in the Southern United States and against expedited deportations, as well as protest movements such as *Lampedusa in Hamburg* and *Refugee Tent Action* and the “no borders” and *#leavenoonebehind* movements are representations of refugee agency. Their struggles for political power marked a transformative moment in the international migration regime. When invisibilised, illegalised and disposable lives intervene in power structures, they rupture the quotidian, administrative and power-laden routines of neoliberal migration management and authoritarian politics.

Besides these macro- or meso-level “migrant upheavals”—understood as “discordant practices of freedom” (Tazzioli 2015)—also thrive at the micro-level. Here, protest often consists of practices of self-harm and auto-injury. For instance, researchers, activists, and NGO workers have witnessed and documented self-mutilation practices among “Dubliners”, a group of mostly younger men who are processed, often several times, through the EU’s Dublin Regulation. They burn and mutilate their fingertips to render themselves unidentifiable by biometric data technology and management such as EURODAC (Picozza 2017; Scheel 2019). Furthermore, lip-sewing, hunger strikes or jumping barbed-wire fences at the EU’s external borders need to be seen and recognised as ethico-political acts of protest and resistance to counter repressive migration policies and practices of bordering. At the intersection of ethics and politics, both can be identified as the articulation of dissensus and disagreement to contest and challenge hegemonic positions. As small, fleeting, and undetected as those micro-level resistances might seem, they deny the de-normalisation of mobility and claim it a right to equal and free movement. Moreover, they challenge the post-democratic scripts of liberal democracy by politicising democracy and humanitarianism and thwarting the current bio- and necropowers governing migration. Hence, these protests claim a "right to have rights" (Arendt 1973: 298): to step out of the a-political shadows to be seen and heard within the bio- and necropolitical formations that render black and brown (migrant/refugee) lives disposable, or even "ungrievable" (Butler 2016).

3 "Nothing About Us Without Us!": Desiderata for Current and Future Research

Let me return to Moria Refugee Camp: EU and national officials rarely ask interned populations what they want and how they feel when forced to live under permanently precarious legal, socio-political and sanitary conditions. Europeans heard nothing about Moria's inmates for a long time, and when they finally did, they often only heard *about* them and rarely *from* them. The migration regime of Europe, and the Global North in general, often aims to manage and contain migration efficiently. If concerns regarding the internment of migratory populations are raised, it is usually done in a humanitarian, not a political fashion. Ethics and politics are neatly separated. When we focus on the entanglement of ethics and politics, however, it becomes evident that those affected most by migration and asylum policies need to be heard more often and more clearly: their opinions, viewpoints or political agendas matter. This focal shift needs to carefully differentiate between paternalistic and often neo-colonial archives of 'giving voice' and de-privileging repertoires of "listening" (Spivak 1988; Qasmiyeh 2019): It is not so much the case that refugees and migrants cannot speak—even when their lips are sewn together—but that their voices are not heard in the echo chambers of colonial, ethno-sexism, and racism. When we separate ethics and politics neatly, we reduce migrants and refugees to passive victims of violence in their home countries, transit zones, camps or receiving countries. Additionally, they are perceived as welfare and humanitarian support recipients or, at worst, as a threat to the normative orders of the so-called receiving countries. But what about their dreams, aspirations and political rights?

In disability studies, the slogan "Nothing about us without us!" communicates the idea that no policy should be decided and implemented without members of the groups affected by that policy. The slogan, moreover, emphasises the general lack of representation of so-called marginalised groups, including discrimination based on race, ethnicity, sexual orientation, gender and dis/ability, and the unwillingness of hegemonic formations to integrate 'peripheral' positions in central decision-making processes. "Nothing about us without us!" neatly sums up everything wrong with recent migration politics and migration studies—empirically and philosophically understood in both ways.

Concluding, I want to hint towards three desiderata that are yet to be studied more thoroughly: First, in migration politics as well as in the academic field studying migration, a (neo)colonial one-way street from the Global North to the Global South is often reproduced. Benevolent humanitarian paternalism and normalised racism shape the global monodirectional relations between these two 'hemispheres'. Therefore, current and future research in empirical migration studies and the field of migration philosophy need to consider that studies *on* the marginalised cannot work by dismissing them as incompetent interpreters and storytellers of their own lives. Hence, academia must be decolonised to enable researchers from the Global South to be heard and seen or to outline participatory research designs. Developing critical landscapes within migration studies, ethics and philosophies cannot stay the privilege of white researchers or researchers educated in the Global North. Furthermore, the dominance of researchers identifying as white, male, able and heteronormative must be further challenged. Consequently, migration studies and the philosophy of migration need to be more inclusive and intersectional: race, gender, sexuality, nationality, religion or beliefs cannot be played against each other but must instead be integrated into intricate research designs, academic recruitment policies, funding, and teaching schedules.

Second, empirical studies, theoretical work, philosophical conceptualisation and applied ethics must make more profound connections. They need to forge stronger bonds to strengthen the social sciences *and* the humanities in their struggle against further quantification and neoliberalisation of research. Consequently, empirical research cannot discard theoretical and philosophical work's complexity and analytical insights. However, the latter's results and premises must be communicated more clearly and translated into the vocabularies of other scientific communities, political arenas and decision-making processes.

Third, instead of increasing efficiency, migration studies and philosophically saturated research on migration and mobility should focus on ethico-political considerations within their research designs. Moreover, they should introduce more participative research approaches to their analyses to generate knowledges not only *on* and *about*, but *with* refugee and migrant groups. Therefore, scholars and study participants must achieve higher visibility and adequate representation of refugees and immigrants within academic landscapes. The paradigm of scholarly research on migration must be shifted from studies *on* and *about* othered populations to studies *with* and *by* them.

Literature

Agamben, G. 1998. *Homo Sacer: Sovereign Power and Bare Life*. Stanford: Stanford University Press.

Agamben, G. 2005. *State of Exception*. Chicago, London: University of Chicago Press.

Arendt, H. 1973 [1951]. *The Origins of Totalitarianism*. San Diego, New York, London: Harcourt Brace.

Arendt, H. 2008 [1943]. We refugees. In *The Jewish Writings: Hannah Arendt*, ed. J. Kohn and R. H. Feldman, 264–274. New York: Schocken Books.

Balibar, E. 2009. Europe as Borderland. *Environment and Planning D: Society and Space* 27:190–215.

Berlant, L. 2007. Slow Death (Sovereignty, Obesity, Lateral Agency). *Critical Inquiry* 33:754–780.
Butler, J. 2016 [2009]. *Frames of War. When Is Life Grievable?* London, New York: Verso.
Casas-Cortes, M., S. Cobarrubias, and J. Pickles. 2015. Riding Routes and Itinerant Borders. Autonomy of Migration and Border Externalization. *Antipode* 47:894–914.
De Genova, N. 2017. *The Borders of 'Europe'. Autonomy of Migration, Tactics of Bordering*. Durham: Duke University Press.
De Genova, N., G. Garelli, and M. Tazzioli. 2018. Autonomy of Asylum? The Autonomy of Migration Undoing the Refugee Crisis Script. *The South Atlantic Quarterly* 117:239–265.
Dudziak, M. L., and Leti Volpp. 2006. *Legal Borderlands: Law and the Construction of American Borders*. Baltimore: Johns Hopkins University Press.
English, C., M. Grazioli, and M. Martignoni. 2019. Migrant Solidarity in Postcolonial Europe. Challenging Borders, Creating Mobile Commons. In *Routledge Handbook of Radical Politics*, ed. R. Kinna and U. Gordon, 194–210. New York, London: Routledge.
Foucault, M. 1980. The Confession of the Flesh. In *Power/Knowledge. Selected Interviews and Other Writings*, ed. Colin Gordon, 194–228. London, New York: Pantheon.
Foucault, M. 2003 [1975/76]. *Society Must Be Defended. Lectures at the Collège de France 1975–76*. New York: Picador.
Foucault, M. 2010 [1978/79]. *Birth of Biopolitics. Lectures at the Collège de France, 1978–1979.* London, New York: Palgrave Macmillan.
Friese, H. 2017. *Flüchtlinge: Opfer – Bedrohung – Helden. Zur politischen Imagination des Fremden*. Bielefeld: Transcript.
Hess, S., B. Kasparek, S. Kron, M. Rodatz, M. Schwertl, and S. Sontowski, eds. 2017. *Der lange Sommer der Migration. Grenzregime III*. Berlin: Assoziation A.
Isin, E., and P. Nyers. 2014. *Routledge Handbook of Global Citizenship Studies*. London, New York: Routledge.
Mbembe, A. 2003. Necropolitics. *Public Culture* 15:11–40.
Mezzadra, S. 2020. Abolitionist Vista of the Human. Border Struggles, Migration, and Freedom of Movement. *Citizenship Studies* 24:424–440.
Owens, P. 2009. Reclaiming 'Bare Life'? Against Agamben on Refugees. *International Relations* 23:567–582.
Picozza, F. 2017. Dubliners. Unthinking Displacement, Illegality, and Refugeeness within Europe's Geographies of Asylum. In *The Borders of 'Europe'. Autonomy of Migration, Tactics of Bordering*, ed. N. De Genova, 233–254. Durham: Duke University Press.
Proglio, G., et al., eds. 2020. *The Black Mediterranean. Bodies, Borders and Citizenship*. London: Palgrave Macmillan.
Qasmiyeh, Y. M. 2019. To Embroider the Voice with Its Own Needle. *Berghahn Books Blog*. https://berghahnbooks.com/blog/to-embroider-the-voice-with-its-own-needle.
Rygiel, K. 2001. Bordering Solidarities. Migrant Activism and the Politics of Movement and Camps at Calais. *Citizenship Studies 15:1–19.*
Scheel, S. 2019. *Autonomy of Migration? Appropriating Mobility within Biometric Border Regimes*. Abingdon, New York: Routledge.
Schulze Wessel, J. 2016. On Border Subjects. Rethinking the Figure of the Refugee and the Undocumented Migrant. *Constellations* 23:46–57.
Spivak, G. C. 1988. Can the Subaltern Speak? In *Marxism and the Interpretation of Culture*, ed. C. Nelson and L. Grossberg, 271–313. Basingstoke: Macmillan Education.
Tazzioli, M. 2015. *Spaces of Governmentality. Autonomous Migration and the Arab Uprisings*. London, New York: Rowman & Littlefield International.
Topak, Ö. E. 2016. Migrant Protest in Times of Crisis. Politics, Ethics and the Sacred from Below. *Citizenship Studies*. https://doi.org/10.1080/13621025.2016.1191428.
Ullrich, M. 2022. Refugee Agency Through Bare Life? News Forms of Voice and Strategies of Imperceptibility at the European Borders. *Deusto Journal of Human Rights* 9:39–56.

Refugees and Family Unification

Eilidh Beaton and Matthew Lister

Abstract The topics of refugee protection and family migration have both received significant attention in the philosophical literature. However, until recently, issues at the intersection of these two subjects were rarely discussed. In this entry, we outline and explore some of the most important questions concerning these issues of refugee family unity, separation, and reunification, considering what obligations (if any) states might have to respect and protect the value of family life for refugees in light of the different sorts of situations refugees may be in. In particular, we consider how the family unification rights of resettlement refugees may differ from the rights of first-asylum refugees, and how these issues might be complicated when unaccompanied minors are involved. Along the way, we discuss a number of related questions, such as: why should states respect and protect the unity of refugees' families? which family members do refugees have rights to reunite with? and, how much weight should the importance of family life have in non-ideal circumstances when places for refugees in safe states are scarce?

The significant, enduring, and multi-faceted relationship between refugee protection and family migration raises a number of important questions. These concern, for instance, the extent of states' obligations to facilitate refugee protection, how scarce places in resettlement programs should be distributed, how to encourage and promote safe and regular travel, what demands refugees can place on receiving states, and how to set up refugee protection schemes to meet the needs of those given protection.

Although there has been an upsurge in scholarly interest in refugee issues in recent years, the topic of refugee family unification has received relatively little philosophical attention. In this entry we hope to prepare the ground for future

E. Beaton (✉)
School of Divinity, History, Philosophy, and Art History, University of Aberdeen, Aberdeen, UK
e-mail: eilidh.beaton@abdn.ac.uk

M. Lister
Faculty of Law, Bond University, Robina, QLD, Australia

A. Niederberger et al. (eds.), *Handbook of Migration Ethics*,
https://doi.org/10.1007/978-3-031-89877-8_28

research on this subject by outlining important legal distinctions, highlighting key normative questions and topics in existing research, and suggesting questions for future inquiry. While we cannot hope to give definitive accounts of any of these issues here, we will set out the main problems and provide what seem to us to be plausible courses of action. As a preliminary matter, for the sake of this entry we will assume a more or less standard account of who counts as a refugee. This is not necessarily because we accept all aspects of the standard account, or think that criticisms of it are without merit, but rather because we are here interested in seeing what the moral obligations of states are or should be in the world as we know it, even under the fairly narrow definition of a refugee found in the U.N. Refugee Convention and the 1967 Protocol.

1 The Basics: Refugee Family Reunion in Law and Practice

1.1 Refugee Family Unification and International Law

The value of family life and the right to family unity are recognised in a number of documents in international law. Both the International Convention on Economic Social, and Cultural Rights ("ICESCR") (article 10) and the International Convention on Civil and Political Rights ("ICCPR") (article 23) affirm that the family is entitled to state protection in virtue of its status as "the natural and fundamental group unit of society". The Final Act of the Conference that adopted the Refugee Convention recommends that governments "take the necessary measures for the protection of the refugee's family", including "ensuring that the unity of the refugee's family is maintained" (UN General Assembly 1951). Although not specific to refugee situations, the importance of family reunification for unaccompanied child migrants is recognised under the Convention on the Rights of the Child ("CRC"), which requires that applications for family reunification made by a child or her parents "be dealt with by State Parties in a positive, humane, and expeditious manner" (article 10). The widespread acceptance of these international legal obligations indicates an overlapping consensus on the importance of family unity, even if there is disagreement on the underlying justifications for these obligations. For a project located in political philosophy, such as this one, an overlapping consensus of this sort is what is important, rather than answering more foundational questions.

The majority of states formally observe the Conference of Plenipotentiary recommendation against separating refugee families, and, in clear cases, will grant refugee status to immediate dependents—at minimum, the spouse and minor children—when the head of a family meets the criteria for refugee status (UNHCR 2011). However, questions of refugee family unity often arise in less-straightforward circumstances. For instance, while international law clearly requires that states refrain from separating families, it does not yet demand that states actively facilitate family reunion (Hathaway and Neve 1997). So although the UNHCR insist that the principle of family unity applies not just when all family members become refugees at once, but also in cases where families are separated during flight, some states have distinguished between the duty to maintain the unity of families that arrive together and the duty to reunite families separated during flight (UNHCR 2011; Hathaway and Neve 1997). Moreover, even when states do make provisions to reunite separated families, there are often significant and sometimes prohibitively high financial costs associated with this process. Scholarly attention on these practical problems, as well as the normative values underlying the right to refugee family unity more generally, are long overdue.

1.2 Refugee Resettlement and First Asylum

In discussions of the ethics of refugee protection, it is often helpful to distinguish between the two main avenues by which refugees access safe states—refugee resettlement and "first asylum"—because different normative issues arise in each.

Refugee resettlement occurs when people have already been determined to be refugees—typically by the United Nations High Commissioner for Refugees (UNHCR)—but are in need of resettlement in a safe country. Families in such situations typically undergo refugee status determination as a group. Most often the people at issue here will be living in a refugee camp, or another setting that provides, at best, temporary and precarious safety: resettlement is specifically afforded to refugees who are in perilous situations or have specific needs that cannot be addressed in their current place of asylum. For many refugees, resettlement is the only solution that would afford long-term, stable protection. However, resettlement places are incredibly scarce. In 2017, just 0.5% of all refugees under the UNHCR's mandate were resettled, and resettlement needs were almost 12 times the number provided (UNHCR 2018). So while the circumstances of resettlement raise a number of interesting normative questions, it is important to remember that resettled refugees make up only a very small proportion of the total refugee population.

On other occasions, refugees enter safe states by seeking "first asylum" in them. In a typical situation of this sort, a refugee—possibly traveling alone but also possibly as a member of a group—travels from a country where he or she would have a well-founded fear of persecution to a state where he or she will be safe from persecution. Once in or at the border of this safe state, the refugee applies for asylum. In the typical case, the asylum claim is adjudicated by the state granting refuge. While a country of first asylum need not be the state where final resettlement—the provision of a durable solution—is granted, in a large number of cases it will be so. Many variations on the basic scenario are possible. Here, we will consider two.

In the first scenario, a refugee arrives at a safe state to request asylum and already has his or her family members with them. Of course, it is possible, even not unlikely, that when one member of a family has a valid claim to be a refugee, the other family members will also have a valid claim, making the argument for granting refuge to the whole family together straightforward. Situations where the family members of the person who has been granted first asylum do not have, or at least do not obviously have, strong refugee claims of their own are slightly more complicated.

In a second scenario, the would-be refugee arrives at the safe state alone, having left their family members behind in their home country. This may be because the initial travel was very expensive or dangerous, or because they were the only family member able to gain a visa to travel regularly to the country where asylum is sought—perhaps because of policies designed specifically for this end. Many states recognise that people granted asylum have a right to live with their family members in their host country, and have created specific immigration options for the spouses and children of refugees. These "family unification" mechanisms grant refugee families priority for admission, and remove some administrative requirements for entry. However, the right to family unification is imperfectly administered and has been attacked at times in countries of first asylum, including via so-called "increased vetting" by the Trump administration, and the use of Temporary Protection Visas, which do not allow family unification, by Australia. In some cases, these attacks on the right to bring in family members seem to be motivated by a state's desire to deter asylum seekers from applying for refugee status in the first place. Clearly, states act unjustly when they reduce the rights of people granted asylum so that they will be less likely to have to fulfil their moral or legal obligations to refugees. However, as will become clearer in Sect. 2, it is less obvious what the scope of family unification policies should be, and to what extent it is the responsibility of the hosting state to actively facilitate family reunion in circumstances of separation.

Finally, in both scenarios of first-asylum and of resettlement, we can ask which family members (if any) should count. In cases of resettlement we can ask which family members must be resettled together; in cases of families travelling together to apply for first-asylum, we can ask which non-refugee family members should have a right to remain with the refugee; and in cases where refugees travel alone to claim asylum, we can ask which family members should be eligible for family unification. In addressing these questions, we might want to consider, among other things, the vulnerability of the family members themselves, the closeness of the relationship between the refugee and the family member, and broader cultural attitudes about family life in the refugee's state of origin and/or their country of asylum.

1.3 Family Reunification for Refugee Children

An important, but often underexplored, question that arises in some cases of first asylum is whether unaccompanied child refugees should be able to sponsor their parents—or perhaps other relatives, like siblings or grandparents—to join them. There is strong prima facie reason to think that allowing such cases of family unification would be beneficial. Children are especially vulnerable and would normally be unable to live on their own without adult support. Allowing refugee children to reunite with their family should improve their emotional, social, and possibly financial position in their country of asylum. This is not even to mention the intrinsic value of being reunited with one's family. However, refugee family unification laws are generally written on the assumption that the refugee in question is an adult, and a significant number of states currently do not allow minors who have been granted asylum to sponsor their other family members to join them. For this reason, questions about the family reunification rights of unaccompanied minors merit independent consideration.

2 The Details: What Justifies Refugees' Right to Reunite—And with Whom?

2.1 Families as Unit of Resettlement

We start our discussion focusing on the case of refugee resettlement and family unity. There are a number of important normative questions here, because states, often working with the UNHCR, take an active role in determining how resettlement places should be distributed. Issues of significance for ethicists and political philosophers include how much weight states should give family ties compared to other considerations in determining the distribution of resettlement places, and when ties to a family member already settled in a safe country should be given weight in determining where a particular refugee should be resettled.

Rightly, there is philosophical consensus on the claim that states have strong moral obligations to greatly increase their resettlement programs. Still, since resettlement spaces in our non-ideal world are scarce, in high demand, and are likely to be so for the near future, we can ask how family ties should play a role in distributing these limited spaces. That is, we must see how family ties are morally relevant to the issue of distributing resettlement slots, and then ask which ties are relevant for these considerations.

If family ties are not given weight or consideration when resettlement slots are distributed, then some refugee families will be separated during resettlement to different safe states, or else forced to decide whether to accept or decline a resettlement offer that means leaving some family members behind. None of these outcomes is desirable. As outlined in Sect. 1, it is widely accepted that people have a right to live with at least their most intimate family members if they wish to do so. The overlapping consensus indicated by reference to several human rights documents recognising the value of family life is evidence of this. Refugees have already suffered a significant injustice in being forced to flee from their homes because of a well-founded fear of persecution. To then require them to either forego the opportunity for resettlement, or to be split from other family members, would be to subject them to a further harm, and arguably an injustice. In recognition of this, under current practice, families of at least spouses and children that have been displaced together typically undergo adjudication of status and resettlement as a unit.

If we think that being able to live close to one's family members is not merely something that people have an interest in, but something that people typically have a right to, then this could only be overcome when there are sufficiently weighty reasons for restricting a resettlement offer to only a subset of the family unit. But given that we are here assuming that the states in question have an obligation to do their share in refugee resettlement, it is not clear what the competing sufficiently weighty ground could be. States may have rational or prudential reasons to separate refugee communities while distributing resettlement slots, but these reasons are almost always designed to promote certain optional ends of the state. Grounds here might include trying to select refugees that one thinks will be most likely to be economic contributors, most likely to "fit in" or adapt well to the host society, or to have a certain demographic range among those resettled. It is an open question whether, other things being equal, it is permissible for states to take these sorts of considerations into account. However, the overlapping consensus on the value of the family is evidence that the existence of family ties provide one way in which "other things" may not be equal.

A more principled objection to giving weight to family ties is that doing so, given the ongoing scarcity of resettlement slots, may lead to some vulnerable people who are plausibly thought to be in greater need of resettlement to not receive a slot. A number of scholars have argued that states should prioritise certain vulnerable groups of refugees, and—at least in conditions where resettlement places are particularly scarce—some of these accounts could come in conflict with the principle of family unity (see papers in, e.g. "Refugee Crisis: The Borders of Human Mobility" and "The Ethics of Refugee Prioritization: Reframing the Debate"). Here we would have a conflict of rights or duties between helping the most vulnerable and respecting family unity, not just a conflict between a duty and prudential considerations. Conflicts of duties are naturally difficult to decide in principled ways, and we cannot here hope to provide a mechanism to solve this dilemma. However, certain considerations may weigh in favour of preserving refugee family unity—for instance, the fact that vulnerable refugees themselves are often hesitant or unwilling to resettle when this will result in separation from family members. In any case, we will have made progress if we have become clearer as to when there is an actual conflict of duties, and when there are merely prudential considerations at stake, ones which ought to give way in the face of the obligation to respect family unity. For now, it seems highly plausible that there is significant reason to give strong, if perhaps not definitive, weight to family ties when determining a distribution of resettlement slots.

This raises the question of which sorts of family ties do or should give rise to the considerations above. One of us has argued that, in the normal immigration setting, all states have an obligation to allow in at least the spouses and minor children of current citizens (Lister 2010, 2018). On that account, however, it is up to the state in question to decide whether other, arguably more distant, family ties should also provide a basis for immigration benefits. The situation of resettlement refugees is different in several important ways. First, and most importantly, because, on the approach just mentioned, the primary right to bring in one's family members resides with a current citizen, it is plausible that it is the state's conception of the family that should be the operative one in deciding whether immigration benefits apply or not. Relatedly, the account discussed depends on the idea that family migration always involves an insider to the state of migration in an essential sense. This is not necessarily so in the case of refugees seeking resettlement.

Because refugees seeking resettlement are differently situated from non-refugees applying for family-based migration in these ways, we should not be surprised if the importance of family ties work somewhat differently in the different cases. In particular, because it is not the interests or rights of host-country citizens that grounds the reunification right, it is less obvious that it should be the host-country's conception of the most important or core family relationships that should delimitate the extent of the right in question. In the case of resettlement refugees, there is significant reason to favour a definition of the family found in the society that the refugee comes from, even if it is more expansive than that found in the state of refugee. (An exception to this claim will involve same-sex couples coming from states where same-sex partnerships are not recognised. We discuss this case below.) This does not mean that just any family tie would necessarily provide a right to be resettled together. Some writers have suggested that refugee family unity be approached on a 'functional' basis, where reunification is allowed in cases where a relationship satisfies a functional understanding of the family (Hathaway and Neve 1997). For general defences of functional accounts of family ties in immigration, see Yong (2016), Song (2019) and Ferracioli (2016). Alternatively, it may be necessary to show that the family tie is one that is seen as particularly important or relevant in the society that the refugee comes from, that the family member in question had previously cohabited with the refugees to be resettled, or that the family member in question had previously taken on significant caring responsibilities for a refugee to be resettled. Any one of these approaches may not include all family relationships. However, they leave scope for the possibility that "broader" ties than those that ground a right to family immigration in most western societies would suffice.

Another reason to adopt a broader conception of the family for cases of refugee family unification is that preserving family unity tends to support valuable kinds of refugee integration. Research suggests that refugees who are not forced to undergo family separation are more likely to flourish in their new home (Canadian Council for Refugees 1995; UNHCR 2013; Beaton et al. 2018). The family provides an intimate network of support and solidarity that helps with mental and emotional well-being—which, in turn, tends to be conducive to longer-term material wellbeing. This suggests that it will often also be in the interests of host states to resettle refugees alongside their "extended" family members. Indeed, where possible, and insofar as these ties prove sufficiently valuable for successful refugee integration, states may even have reason to resettle refugees with friends and other broader social networks.

Family ties may be relevant for refugee resettlement in at least one further way. It is a widely accepted principle that, while refugees are owed the opportunity to live decent lives in safe countries, they have significantly less, if any, right to choose which country they should have access to. But, despite this general principle, the existence of family ties in a particular country may give us strong reason to favour resettling a given refugee in the country where she or he has relatives. The scenario here is distinct from the one above, where refugees making a family are resettled together, and also from one to be discussed below, where a refugee in a particular state may apply for their close relatives to "follow to join" them in the safe state. Here we are primarily considering cases where one or more family member(s) currently live in a safe country, and would not necessarily have the right to sponsor the refugee relative to come to the country through ordinary immigration means or through a family unification rule. In such a situation, while the refugee(s) may still not have a right to be resettled where she or he has family ties, there is still good reason to consider this factor. If there is no other strongly compelling reason why the refugee ought not be settled in the state in question, then there is a strong argument for allowing refugees to be resettled in states where they have relatives if they so wish, even if they otherwise would not have had the right to demand settlement in the particular state.

2.2 First Asylum and Family Unity

As outlined in Sect. 1.2, there are important differences between first asylum and resettlement cases, and these differences have implications for family reunification. In particular, resettled families will already have been all found to be refugees, while this is not so for those seeking first asylum. Additionally, in the first asylum situation, refugees are more likely to arrive alone, leaving family members behind.

Consider first cases in which first-asylum families do arrive as a unit. When all family members have strong claims to be refugees, determination of status will be relatively straightforward. However, consider a case in which an asylum seeker with a strong claim to be a refugee arrives in a safe country and is accompanied by family members who do not have strong claims to be refugees on their own. Would it be reasonable to expect the person in need of refugee protection, if she or he is granted asylum, to remain in the safe country, but to have her or his family return to the "home" country, at least until the asylees gains normal family immigration rights to bring in her or his relatives? If the discussion above about the importance of families being able to live together is correct, such a policy would be unacceptable. By definition, refugees are unable to live in their home country without being in fear of persecution. To force them to decide whether to live in a country where they face danger, or to abandon their family, would be to subject them to a serious further wrong. Here the fact that essentially all states offer family migration benefits to their own citizens is relevant. This fact shows that being able to live close to one's family, at least the immediate family, is very widely recognised as an important, even fundamental, interest, rising to the level of a right, shared by all people. To require someone seeking asylum to abandon their family members in order to be safe would be to ignore this fact without justification. Therefore, when an asylum seeker with a strong claim to be a refugee arrives in a safe country, and can establish their claim to refugee protection, the ability to enter the country, work, have access to public services, and the chance to become permanent residents and eventually citizens, should also be made available to their family members.

Potentially more difficult cases are ones in which a person seeking asylum travels alone at first, leaving their family behind with the goal of reuniting in the country of asylum at a later date. Earlier in this entry we argued that, in the case of refugees seeking resettlement, it was reasonable to use the conception of the family found in the refugee's society, rather than in the society of refuge, to delineate the bounds of the rights in question. For those who hold that the nature of a state's obligations to create immigration exceptions for family unification depend on the nature of the state's relationship to the individuals in question, it will be less clear that permitting immigration of extended family members is the right answer in this situation. In the case of refugees being resettled, the state in question already has an obligation to take in a number of refugees, and the question is simply how these scarce slots are to be allocated. In such a situation, there is no clear reason to favour the conception of the family found in the receiving country over that found in the refugee's home society. According to this line of argument, the situation would be different in the case of families following to join. Here a state is making an exception to its normal admission policies for a refugee whom it has already admitted. The refugee is now much more in the situation of a regular member of the society in question, and so, on this view, while the refugee would of course have a right to be with their immediate family, it is the conception of the family found in the state of refuge that should provide the model for the bounds of this right.

This general consideration may be subject to some conditions and adjustments. The first of these has to do with same-sex relationships. There is, we claim, good reason to think that all states should accept same-sex relationships for the purpose of family-based immigration (Lister 2007, 2018). Currently, many states that are likely to be countries of first asylum recognise same-sex relationships for immigration benefits, and so should be willing to extend these benefits to refugees in such relationships. However, there are a number of practical difficulties here to be considered. First, many people granted asylum and then seeking to have their same-sex partners join them will come from states that do not recognise same-sex marriage. Such people will not be able to rely on formal documents to establish their relationship. In many cases, they will not even be able to easily establish the sorts of facts that would be able to show the existence of a de facto partnership. Homosexual behaviour remains illegal in 72 countries and is discriminated against in even more. In such countries, same-sex couples will often not be able to live together, commingle finances, hold themselves out as a couple, and do other things that would normally be used to establish a bona fide relationship when marriage isn't possible. To demand such evidence would be, in many cases, to conclude that same-sex relationships could not qualify for family reunification in refugee cases. This conclusion is unwarranted. While it is not unreasonable to ask for evidence that a relationship is bona fide before granting family unification benefits, states and adjudicators should be sensitive to how difficult "material" evidence may be to come by in such cases, and should be willing to accept credible testimony as sufficient evidence when it would be unreasonable for the applicant to provide further evidence. Furthermore, states and adjudicators should keep in mind that cultural differences and difficult legal situations are likely to make same-sex couples from certain states act in ways that will seem unusual and different from more stereotypical cases in western countries.

An alternative position to the one described above would be to insist that family-based immigration exceptions for first-asylum refugees should be more expansive than exceptions for citizens and non-refugee migrants. Proponents of this position would point to various ways in which refugees are differently situated from these groups. For instance, refugees are unlike most other family reunion applicants in that they have been forced by persecution into circumstances of family separation. Whatever one's views about ordinary family reunion might be, one might think that the presumption that states have a duty to preserve refugee family unity is especially strong in virtue of the fact that refugees have been forced into situations of precarious family unity by circumstances completely beyond their control. This also means that refugees might not have had any choice but to live in a country with a more restrictive cultural conception of the family than their own. Moreover, there is a higher-than-average possibility that refugees will have had traumatic experiences either in their home country or during flight. Since family connections can play an important role in enabling those affected by trauma to move on with their lives, maintaining these connections may be especially important for many refugees. These considerations might lend weight to the idea that a broader concept of the family should apply when first-asylum refugees sponsor their family through reunification programs, too. The argument for this position is further strengthened by the fact that such policies would also effectively provide safer and more orderly paths of departure for the family members of refugees who are also in need of asylum and might not otherwise be able to access it.

2.3 *Family Unification Rights of Refugee Children*

Some states which host significant numbers of refugees, like the US and the UK, refuse to extend the family reunion options they already have in place for refugee adults to unaccompanied minors. If child refugees could sponsor their parents, it is said, then this might incentivise parents to send their children on dangerous journeys alone to countries of asylum. According to this argument, since we wish to prevent children from being sent into danger, we ought not extend family reunion rights to minor refugees.

However, given that these states also engage in other practices to reduce their responsibilities to host refugees, there is good reason to suspect that there might be other motivations for these policies. In particular, it seems likely that states adopt such policies out of self-interested concern that there would be an exponential increase in the number of people arriving through refugee family reunion routes if children were allowed to sponsor their families. Such an argument would be similar to those made in the US in favour of prohibiting minor US citizens from sponsoring their parents. If minors were able to sponsor their parents, it is said there would be a strong incentive for non-citizens to give birth in the US, and then, given the US's strong *jus soli* citizenship laws, have the citizen child immediately sponsor the parent. That is, a policy allowing minor citizens to sponsor their parents would create a "back door" method of gaining access to permanent residence in the US via the family migration scheme, which has been deemed undesirable from the US perspective.

However, the question of whether refugee children should be allowed to sponsor family members to join them is different from the US case in various important ways. First, whereas non-refugee minors in the US gain citizenship solely by virtue of being born on US territory, unaccompanied child asylum seekers are required to undergo a rigorous determination of status to demonstrate that they fulfil the criteria for refugee status. These criteria are identical to the criteria that apply to adult refugees. Since the majority of states already permit adult refugees to reunite with their immediate family members in their territory of asylum—and since it seems likely that this is the minimum that should be required of host states—further justification is needed to show that it is permissible to apply a different rule in the case of unaccompanied child refugees.

Perhaps it might be argued that although there is no guarantee that the asylum applications of unaccompanied minors will be granted, there is reason to think that children are much more likely to be granted refugee status than adults. If this were the case, some parents might be willing to put their children at risk in order to increase the family's chances of accessing asylum. However, this is not on its own plausible. For one, there is simply no evidence of an increase in the number of unaccompanied children arriving to apply for asylum and family reunion in countries that do permit child refugees to reunite with their family. We might make sense of this by considering that parents are very unlikely to take the risky and extreme measure of sending their child to another country alone unless they are already in desperate circumstances. In such cases, the children will almost always already be in need of protection because they cannot safely stay in their home country. There is therefore no need for further incentive for them to leave. Whatever incentive is needed is already in the asylum system as it is.

Moreover, even if it were the case that such a policy would lead to an increase in the number of unaccompanied children arriving to apply for asylum, there is no good reason to think that children are more likely to be granted asylum than adults. While officials may, in some cases, be more sympathetic to minor applicants, children are also often less articulate and less able to present a claim. They may also be less likely to successfully show that they have, say, political beliefs that would ground an asylum claim. At best, it is deeply unclear that asylum claims made by children are more likely to be granted than claims made by adults. Thus, failure to extend family reunion rights to unaccompanied minors cannot be justified by the need to eliminate a perverse incentive to send children ahead alone, or a desire to limit "abuse" of the asylum system. When we recall that this is no small matter, but rather a case where a very important right is at issue, there is good reason to err on the side of protecting the right to family unification. Given that the alternative will certainly entail many significant failures to respect the rights of child refugees, the case for granting them the right to sponsor relatives seems strong.

Finally, there is good reason to question the worry that there would be an enormous increase in the number of non-refugees entering through family migration routes if unaccompanied minors were given family unification rights. While it is quite conceivable that an adult could have a well-founded fear of persecution while her spouse or children did not, it is much less likely that a child would be a refugee in a particular case if his or her parents were not also refugees. Allowing a child to sponsor his or her parents would, in most cases, only be providing an orderly way for a person who is already in need of protection to gain that protection. So it would seem that child refugee family reunion rights would be less likely to open migration avenues for non-refugees than analogous rights for adults.

3 The Unanswered: Some Topics for Future Exploration

3.1 Family Unity as a Resettlement Priority?

In Sect. 2.1 we argued that family unification ought to be a priority, even a right, in refugee resettlement. However, family unification is not the only factor to consider when distributing scarce places for resettlement. Vulnerable groups of refugees, such as unaccompanied children, adults with children they are responsible for, and disabled refugees have often, and reasonably, been given priority in resettlement. Other factors may be relevant as well. For example, we may think that a priority should be given to LGTBQ refugees for resettlement in societies where their rights will be respected (Vitikainen 2020). As indicated in Sect. 2.1, a possible avenue for future research would be to consider how these potentially competing claims should be adjudicated. It may seem questionable that family unification should be allowed to crowd out claims based on urgent need, such as that of unaccompanied minors and, in many cases, adults caring for children or the significantly disabled. Admittedly, if all or even many states fully lived up to their resettlement obligations, it would be unlikely that the highest priority cases would fill all of resettlement places, leaving room for independent consideration of family ties. However, in the near future, where the vast majority of those in need of resettlement must wait for help, real conflicts are likely to arise. How to adjudicate these conflicts is an area ripe for further research.

3.2 Which Family Members Count?

Another area open for significant further research is the question of which family members are relevant for family unification in the refugee context and why. One area for debate is how far, when, and why the conception of the family found in the refugee's home society should form the basis for unification rights, and when the host state's conception should so serve. The related point of when, if ever, a host state should be willing to compromise its moral and political values in relation to marriages or family relations it has deemed to be against public policy—such as, for example, polygamous marriages or child marriages—is also a topic in need of careful philosophical reflection. Finally, more work is needed to show the proper relationship between philosophical arguments about family unification rights and empirical questions about the impact on refugee integration and well-being when family units of different types are given weight in resettlement or admission decisions.

Literature

Beaton, E. 2023. The Right to Family Unification for Refugees. *Social Theory and Practice* 49:1–28. https://doi.org/10.5840/soctheorpract202332178.

Beaton, E., A. Musgrave, and J. Liebl. 2018. *Safe But Not Settled. The Impact of Family Separation on Refugees in the UK.* Oxfam & Refugee Council. https://www.oxfam.org/en/research/safe-not-settled-impact-family-separation-refugees-uk.

Canadian Council for Refugees Task Force on Family Reunification. 1995. *Refugee Family Reunification.* https://ccrweb.ca/files/tffr.pdf.

Duarte, M., K. Lippert-Rasmussen, S. Parekh, and A. Vitikainen. 2016. Refugee Crisis. The Borders of Human Mobility. *Journal of Global Ethics* 12:245–251.

Ferracioli, L. 2016. Family Migration Schemes and Liberal Neutrality. A Dilemma. *Journal of Moral Philosophy* 13:553–575.

Hathaway, J. C., and R. A. Neve. 1997. Making International Refugee Law Relevant Again. A Proposal for Collectivized and Solution-Oriented Protection. *Harvard Human Rights Journal* 10:115–211.

International Organization for Migration. 2011. *Glossary on Migration.* 2nd ed. Geneva: International Organization for Migration.

Lippert-Rasmussen, K., and A. Vitikainen. 2020. The Ethics of Refugee Prioritization. Reframing the Debate. *Ethics and Global Politics* 13:1–5.

Lister, M. 2007. A Rawlsian Argument for Extending Family-Based Immigration Benefits to Same-Sex Couples. *University of Memphis Law Review* 37:745–780.

Lister, M. 2010. Immigration, Association, and the Family. *Law and Philosophy* 29:717–745.

Lister, M. 2013. Who are Refugees? *Law and Philosophy* 32:645–671.

Lister, M. 2018. The Rights of Families and Children at the Border. In *Philosophical Foundations of Children's and Family Law*, ed. E. Brake and L. Ferguson, 153–170. Oxford: Oxford University Press.

Song, S. 2019. *Democracy and Immigration.* Oxford: Oxford University Press.

UNHCR. 2011. *Handbook and Guidelines on Procedures and Criteria for Determining Refugee Status.* Geneva: UNHCR.

UNHCR. 2013. *A New Beginning. Refugee Integration in Europe. Refugee Integration in Europe.* https://www.unhcr.org/protection/operations/52403d389/new-beginning-refugee-integration-europe.html.

UNHCR. 2018. *Global Trends. Forced Displacement in 2017.* https://www.unhcr.org/globaltrends2017/.

Vitikainen, A. 2020. LGBT Rights and Refugees. A Case for Prioritizing LGBT Status in Refugee Admissions. *Ethics and Global Politics* 13:64–78.

Yong, C. 2016. Caring Relationships and Family Migration Schemes. In *The Ethics and Politics of Immigration*, ed. A. Sager, 61–83. Lanham: Rowman & Littlefield International.

Child Refugees

Gottfried Schweiger

Abstract In this article, I discuss philosophical questions related to minor refugees. In the genuinely philosophical debate, there is—in contrast to adult refugees—almost no literature that explicitly deals with child refugees. This is regrettable since some philosophical-normative questions are obvious here. Two particularly important questions for philosophy can be elaborated: What can philosophy contribute to a better understanding of the situation of minor refugees, in particular what normative and ethical questions are associated with the status of minor refugees and what rights and duties do they have? What can philosophy contribute to improving the situation of child refugees, i.e. to asking which actors should assume responsibility here and how structures should be changed in order to achieve more justice? I will then discuss five sets of issues that seem important to me for understanding the philosophical concerns associated with child refugees: who is a child and what does it mean to be a child; who is a refugee and what distinguishes refugees from other migrants; what are children's rights; what is the value of the family; what is the role of temporality in understanding the situation of child refugees (i.e. their age, the duration and timing of their flight). Finally, I will briefly mention desiderata for further philosophical research.

Minor refugees are in the focus of scientific, public, political and legal attention for several reasons. On the one hand, children are perceived as particularly vulnerable and therefore enjoy special protection (UNICEF 2016). On the other hand, there is a discourse that sees children as a vehicle for their parents' migration—in this idea, parents or other relatives take advantage of the special protection of child refugees to obtain a residence status to which they would not otherwise be entitled (Hedlund and Cederborg 2015). Unlike adult refugees (Gibney 2018), there is almost no philosophical literature that deals explicitly with child refugees (Schweiger 2019)—compared to the rich literature in refugee studies, the social and health sciences. This is regrettable since some philosophical-normative questions are obvious: What

G. Schweiger (✉)
Centre for Ethics and Poverty Research, University of Salzburg, Salzburg, Austria
e-mail: gottfried.schweiger@plus.ac.at

A. Niederberger et al. (eds.), *Handbook of Migration Ethics*,
https://doi.org/10.1007/978-3-031-89877-8_29

is the political and moral status of minor refugees in contrast to adults or other minor migrants? Who has what specific duties towards these children and how can they be protected? What role does the family and the rights of children and parents associated with it play in the context of flight? Since there is almost no philosophical literature available on these questions, this paper will attempt to provide a systematic outline of some key research desiderata.

It should be noted that I will not be able to outline the many differentiations that have been elaborated and revealed in empirical refugee research. There is no such person as the "typical" child refugee since children are highly diverse, as are their flight routes and motives for fleeing. The countries of origin, the camps and the receiving countries differ considerably. There are very many domestic refugees, and most refugees move to neighbouring countries where the situation is often precarious. This is also true for child refugees, whether they are travelling alone or with their families. The focus of this short chapter is on the philosophical systematic questions—specific questions, such as how the EU should address boat refugees in the Mediterranean or how the US should deal with minor refugees from Latin America, can, of course, only be answered in detail when philosophical reflection and empirical research are combined.

In the first section of this chapter, I will provide a very brief overview of the issues and risks that child refugees face. I then distinguish two questions. Firstly, what can philosophy contribute to a better understanding of the situation—and the moral rights and duties—of refugee minors? This question can be answered if philosophy applies the normative concepts it has at its disposal to the empirical reality of the situation of refugee minors. These are, among others, the concepts of autonomy, wellbeing or vulnerability. These not only describe the situation of refugee minors but also help to understand what is unjust or immoral about their situation. The second important philosophical question concerns the responsibility for child refugees and who should bear it. Pointing out that the situation of child refugees is precarious and that their moral rights are violated does not say enough here. To show that other states have special obligations and what responsibilities parents and other people and institutions have, it is necessary to apply philosophical theories of ethics and justice to the situation of child refugees that take into account their special status. In the second section, I will then discuss five sets of issues that seem important to me for understanding the philosophical concerns associated with child refugees: who is a child and what does it mean to be a child; who is a refugee and what distinguishes refugees from other migrants; what are children's rights; what is the value of the family; what is the role of temporality in understanding the situation of child refugees (i.e. their age, the duration and timing of their flight). Finally, I will briefly mention desiderata for further philosophical research.

1 General Issues at Stake in the Controversy

For children and adolescents, flight is always associated with great risks, and even after admission to another country not all problems have been solved. There seems to be little scientific controversy about this (UNICEF 2016; Carlson et al. 2012;

Moinolnolki and Han 2017), just as few deny that child refugees deserve special protection that should place them in a better position than adult refugees (Pobjoy 2017; Grover 2018). But what are the concrete problems faced by child refugees? Firstly, the situation of refugee minors is particularly stressful. They are often unable to protect themselves sufficiently and to take care of themselves. The dangers associated with flight are also distributed differently according to other characteristics such as race, gender, ethnicity or socio-economic status. Girls, for example, are more likely to become victims of sexual abuse and exploitation. Secondly, children are particularly vulnerable when they flee alone and without the accompaniment of adult family members. There are many reasons why children flee alone, and this can also be a conscious decision by the children or their family. In cases where children flee alone, they are forced to take responsibility, which often overwhelms them and radically tears them out of the childhood phase. Such disruptions of childhood are observable in other cases, for example child soldiers, children living alone on the streets or—in milder forms—young carers. These are all examples of children taking on adult roles, having to take care of themselves, often making traumatic experiences that they are unable to master or even understand properly. Thirdly, it should be noted that child refugees are also exposed to specific risks after flight, i.e. as soon as they have reached a first safe destination. On the one hand, it should be mentioned here that children in refugee camps often receive inadequate care. On the other hand, even if a child arrives in a safe country, care is often not satisfactorily guaranteed—unfortunately also rich countries in Europe and America often provide only insufficient care and protection for child refugees (Derluyn and Broekaert 2008). The traumas suffered before or during flight have dug deep into the soul and require intensive and long-term psychological care, which is often not available. Minor refugees are therefore exposed to procedures after flight that can be stressful and traumatic. This concerns, for example, the determination of age or interviews during their admission to the asylum procedure.

The wellbeing and development of children during and after flight is thus endangered in several ways, which reflects the specific vulnerability of children. The issues at stake here are physical and psychological integrity and medical care, but also questions of material care and child-friendly facilities in reception centres and camps with toys, open spaces, activities, etc. For children, the aspect of education and school is of central importance, which was often already insufficient during flight and in any case interrupted by it (Hart 2009). Education is, however, a time-sensitive good, i.e. it must be continuous and appropriate in developmental terms, otherwise educational deficits will occur that can only be compensated for with difficulty. Here, longer stays in camps without school education are problematic because these children then struggle with deficits in their later lives. Also key is the possibility for child refugees to process their experiences and rebuild trust and self-confidence. The resources of hope and aspiration, which are often underestimated in philosophy, are particularly endangered goods (Mullin 2019; Hart 2016), which minor refugees have lost and can only recover with difficulty and in the right environment. However, they are essential for a good life as a child to succeed and for preparing a successful adult life.

Two particularly important questions can be elaborated for philosophy: What can it contribute to a better understanding of the situation of child refugees, in particular what normative and ethical questions are associated with it and what rights and duties do child refugees have? What can philosophy contribute to improving the situation of child refugees, i.e. to asking which actors should assume responsibility here and how structures should be changed in order to achieve more justice? These two questions are not specific to child refugees but also arise for adult refugees and migrants in general. It is open to question, however, whether the answers differ when it concerns child refugees, i.e. whether they have a different (higher?) moral, legal or political status than adult refugees, whether they have special moral rights, should be given preferential treatment, whether the responsible institutions differ. In this respect, it is also the task of philosophy to reflect on moral intuitions, such as that child refugees are particularly vulnerable.

There have already been some hints to the first question. In the situation of minor refugees, it is a question of their wellbeing and of the extent to which the just rights of these children are violated. Philosophy has developed a number of normative concepts that can be helpful here. This can be posed as a question of global justice, i.e. as the question of to which goods all children are entitled and to what extent they are endangered or violated by flight (Macleod 2010). It seems obvious that underage refugees not only have less education and material resources and poorer health than other children, but that they have too little of them, i.e. an unjust shortage. It is also possible, however, to shift the perspective away from the current wellbeing of the child and focus on the question of unequal opportunities and to which extent child refugees are disadvantaged compared to other children in terms of their development opportunities. In addition to long-lasting traumas, which often impair mental health in adult life, the education sector should also be mentioned here since child refugees are massively disadvantaged by the lack of educational opportunities, and it is difficult to compensate for these disadvantages. A third alternative for understanding the situation of child refugees normatively would be to focus on the concept of vulnerability. Vulnerability is normatively relevant insofar as some groups are particularly in need of protection and entitled to care (Andresen 2014; Schweiger and Graf 2017). This is about differences between people and groups who can substantiate normative claims. The group of minor refugees is more vulnerable than the group of children and than the group of adult refugees because they are less able to protect and care for themselves. One can say that the situation of flight is a form of pathogenic vulnerability, i.e. an increase in the vulnerability of a particular group caused by social actions, processes and structures, similar to violence against minorities in a racist society.

Let us turn to the second question, namely, who should take responsibility for refugee minors; that is, who are the agents of justice here? From the perspective of social and global justice, two main actors come into focus: the state and the international institutions and structures formed by the international community. Several questions arise here. The first is what protection as a refugee is based on: Does it stem from these political agreements or is it about humanitarian protection? How should national interests be weighed against global obligations? The debate here is very wide-ranging and goes beyond the status of minor refugees (Fine and Ypi 2016; Sager 2016). Many may follow the intuition that children in the context of flight and migration have a special moral and political status that leads to their preferential treatment (Schweiger 2019). This is also seen in international law. There is the question of how far the preferential treatment of children extends: Is it sufficient within the group of refugees to especially protect and care for children or do children have the right to asylum and reception (or even a fundamental right to immigration) because they are children? It is clear that children are not able to take care of themselves and that the state and its institutions therefore have special duties towards them. These duties can be understood either as duties of justice or as moral duties of assistance, but they are broader than the duties towards adults who can be expected to take care of themselves to some extent. The duties towards children are therefore also mostly understood as one-sided since children are not involved in the exchange of services of a society: they have rights, but much less duties towards others. A distinction must also be made here between minor refugees who are at the border or already on the territory of another state and refugees who are still fleeing or staying in a camp in another country. It seems plausible that a state should take in and care for children who have already reached its territory, while the obligations towards children in other countries and on the run tend to be weaker (Schweiger 2016). Here, questions of burden sharing between countries also arise as most refugees flee to neighbouring countries, and thus only a few reach rich Western states and are admitted (Holtug 2016). For children and young people, it is particularly important to note that their ability to flee is limited: babies cannot flee on their own, even if they need refuge, and even older children often cannot flee on their own.

However, it is not only states and their institutions that are responsible for the welfare of child refugees. The second central actor discussed in the philosophical debate on justice for children is the family or the parents. Parents have a specific responsibility towards their children that results from this close relationship and which they are not allowed to cast aside without further ado. In the case of child refugees, two things must be considered: on the one hand, parents, if fleeing with them, are themselves in a weak position and often inadequately able to care for their children. This is not only a burden for the parents but also creates obligations on the part of other actors of justice to support the parents or to partially assume their responsibility. Some parental responsibilities are difficult or even impossible to substitute, although they are also central for child refugees: love, trust, emotional closeness. These goods are difficult to maintain under the conditions of flight because flight as a form of toxic stress also has a negative impact on family relationships, and parents themselves are often traumatised. Therefore, biased judgements about what parents should do for their children on the run must be viewed with caution. On the other hand, it should be noted that the family is a complex social institution that should not be idealised. The family is a central mechanism for the intergenerational transmission of prejudices and disadvantages, such as gender roles and stereotypes. This makes it a delicate task for the liberal state to find the balance between intervention and protection of the family and its privacy. This debate certainly does not only concern minor refugees and their families, but instead goes far beyond.

However, in the case of migration movements that bring new and differentiated cultural norms and practices into a society, these issues become acute. In the case of refugee families, they also affect the weakest members of a society, who in any case suffer from the trauma of the often forced uprooting and departure from their homeland. A position that considers children's rights to be central will curtail parental rights when the best interests of the child or equality of opportunity are at risk and advocate appropriate state intervention. The spectrum of interventions is wide—and many of them are ethically controversial—and includes, for example, compulsory schooling and the teaching of liberal values in schools, the prohibition of corporal punishment, or bans on religious and ideological practices (such as a ban on headscarves). In addition to the state and the family, it should not be forgotten that other agents of justice also bear responsibility for child refugees: these are, for example, collective actors of civil society who have political duties to influence the state in such a way that it pursues a humane refugee policy, or also help with obligations in the event of state failure should it deliberately neglect its duties. To what extent companies also have a social responsibility here—for example, offering young refugees training places—cannot be discussed any further here, but this seems plausible.

2 Topics, Positions, and Arguments in the Controversy

I would now like to discuss five thematic blocks that seem important to me for an understanding of the philosophical issues associated with minor refugees: who is a child and what does it mean to be a child; who is a refugee and what distinguishes them from other migrants; children's rights; family; temporality. There are some overlaps.

At first, the question of who is a child does not seem difficult to answer. It is not only a matter of a biological or medical category but also of normative questions; after all, children have special rights on the one hand, and on the other they are severely curtailed in their rights compared to adults (Archard and Macleod 2002). It therefore makes a significant moral, legal and political difference whether someone is considered a child or not. This is also relevant for child refugees because their status is derived from the fact that they are both: children and refugees. There is general agreement that childhood is a broad category, i.e. it encompasses people with very different abilities and knowledge, and that it is difficult to distinguish between them at both ends: notoriously controversial is the question of when an embryo becomes a person and thus a child, and notoriously controversial is also the question of when a person is no longer a child but an adult. Similar questions arise with finer structures such as baby, toddler, adolescent. Age limits here are always imprecise aids, and many consider normative demands to be unsatisfactory; after all, there are always people whose development does not fit into these categories (Franklin-Hall 2013). For child refugees, their own childhood is also significant in terms of legal practice, and the ethically controversial methods of age determination are also an expression of the fact that belonging to the category of children is relevant (Noll 2016). All special rights (and duties) granted to child refugees ultimately relate to the definition of childhood used and what it means in normative terms to be a child. Therefore, it is also relevant here whether special goods of childhood are understood as natural or merely socially constructed (Giesinger 2017).

Both inside and outside philosophy, it is debated who is to be considered a refugee and what significance this status should have (Ferracioli 2014; Lister 2013; Cherem 2016). While it is argued, on the one hand, that the concept of refugee should be understood broadly—at least further than the Geneva Convention relating to the Status of Refugees—and that people who leave their country due to poverty, hunger, war or climate change should be recognised as refugees, on the other hand, arguments are put forward for a narrow concept of refugee. One argument in favour of such a restriction of refugee status is, for example, that it can avoid excessive demands on the receiving states, thereby ensuring legitimacy and the willingness to consent in these states, and that it also makes ethical sense to give people persecuted by their own state, which should actually protect them, the special right that other states must assume responsibility for them. These discussions are not specifically about children and young people, but they are, of course, relevant to them.

The third thematic block asks about the specific rights (and duties) of minor refugees. Here, some distinct but connected debates come together. On the one hand, there are existing international agreements and the Convention on the Rights of the Child (CRC) as political and legal instruments (Grover 2018). The CRC deals with normative terms that are also relevant for child refugees, such as "best interest" (Zermatten 2010). On the other hand, there is the ethical discussion on the justification of children's rights (Archard 2004). Children's rights, like all human rights, have the dual character that they are both legal and politically set rights, but also moral rights. To what extent the legal rights, for example in the Convention on the Rights of the Child, coincide with philosophically justifiable moral rights is controversial, just as it is unclear what the value of moral rights can be in relation to these positively set rights. Since this article aims to deal with an ethical issue rather than a legal one, I will confine myself to discussing the moral rights of children and, as a secondary consequence, of minor refugees. Minor refugees have specific rights as children (e.g. to education and adequate care) and as refugees (e.g. to reception by another state and protection from persecution). Children's rights are characterised by the fact that they refer less to autonomy than to the need for protection and the welfare of the child. This results in a tension between the consideration of children's wishes and paternalism (Bagattini and Macleod 2014), which is justified in case of children. This is also the case for child refugees who need care according to their age and stage of development. The rights and duties of younger children are more limited than those of young people who already have semi-autonomous abilities and are able to articulate their own desires and life goals. In the area of care for underage refugees, this becomes significant when it comes to who can make decisions for these children and to what extent these children and adolescents can decide for themselves (e.g. where they live or in the case of medical interventions). Therefore, the rights and duties of minor refugees are in conflict with the rights and duties of their parents or guardians if they do not travel alone. Child refugees are doubly powerless when it comes to claiming their rights: first, as children and adolescents they are usually not in a position to claim rights and obligations since they are not part of political co-determination. They are also excluded from political co-determination and participation insofar as they are refugees and, to that extent, usually have no voice and no say in asylum procedures and their provision by the state. This is ethically problematic (Archard and Skivenes 2009).

Fourthly, the situation of child refugees cannot be discussed without reference to their families. This is particularly important in times when children are deliberately separated from their parents and relatives within the framework of migration processes (Ferguson 2018). On the one hand, children have a right to their family and depend on the protection and care of adults. Early childhood attachment to family members makes separation painful and traumatic. For child refugees, the family is an important support during flight and after admission to another state. However, it is also controversial how the right to a family should be precisely justified and limited: Is the Western image of the nuclear family of parents and children a legitimate yardstick here, what makes family relationships so special that they justify such a right, or should this also be extended to other close relationships such as friendships (Lister 2010, 2018)? It should also be borne in mind that some minor refugees flee alone and apply alone for admission and asylum. This raises not only questions of family reunification, but also, in the case of orphans or other obstacles to family reunification, questions of care by foster parents or state institutions. What are the legitimate rights of child refugees to care for their children: Are they aiming at equality with the other children in the country or is minimal care sufficient?

Fifthly, the aspect of temporality plays a major role. Time has already been mentioned in the four points discussed so far: For example, the question of when does a child become an adult or how long does childhood last? When we think of the many difficulties faced by child refugees, the importance of time becomes clear once again. Firstly, it is a question of duration. It makes an empirically well-documented difference how long certain disadvantages exist, i.e. whether a child cannot go to school for only a few weeks or months or for several years. This means that the duration of flight is important, as is the duration of stay in camps or institutions. The longer this lasts, the more serious the problems become (Vitus 2010). In addition, childhood is a particularly sensitive and vulnerable phase, and many things missed here cannot be made up for. For a child who lacks basic school knowledge, it is all the more difficult to catch up as a teenager. The same applies to a whole series of injustices associated with the status of a minor refugee: inadequate medical care, traumatic experiences and a lack of emotional stability through confidants. These things have a serious impact on children's healthy development. The normative meaning of temporality in the context of child refugees can also be understood to mean that they are deprived of a particularly important temporal phase of life, namely their childhood, and do not receive the protection and care they need. The extension of time is then also relevant to everything that happens after flight: How long do asylum procedures take, how long do children and young people stay in reception camps and when do they go to school, have the opportunity to work, have permanent residence status or apply for citizenship? It is clear that after fleeing to a

Western country, not everything automatically becomes good or even better for child refugees.

In addition to the extension of time, however, points in time also play an important role: for example, the point in time from which a minor refugee is treated as an adult, namely their 18th birthday. The setting of this arbitrary age limit, which correlates only very inaccurately with the stage of development and the needs of children and adolescents, is ethically problematic in several respects and creates consequential problems (Anderson and Claassen 2012). It is obvious that it creates a good reason for people to pretend to be younger than they are, i.e. claim to be 16 or 17 years old even though they are already 19 or 20 years old. The reasons for this are all too understandable since the status of a minor refugee, at least in Western countries and camps run by them, is associated with advantages and makes it more likely to be admitted for a longer period of time and not deported. The protection of the best interests of the child, which is in the foreground here, does not stop abruptly from one day to the next because the transition from adolescence to adulthood is a phase that is individually different, and the vast majority of these "adults" at 18, 19 or 20 years of age are just as vulnerable and needy as their younger peers. Ethically problematic, however, is not only the setting of an age limit but also to some extent its implementation. It is precisely because young people and the authorities are aware of the advantages of the status of a minor refugee that states use sometimes rigorous methods to determine the real age, and it can be assumed that in times of restrictive immigration policies a hermeneutic of suspicion spreads here, which presumes that young people tend to lie and want to "exploit" the state (Allsopp and Chase 2019). Bones are X-rayed and other medical tests are carried out to determine age, sometimes with coercion or accusation of false pretences. Ethical questions regarding the consent and the ability of children and adolescents to consent must be asked here (Alderson 2007) as well as ones concerning the legitimate interests of states to "protect" themselves from "fake" adolescents and to expel them.

3 Desiderata for Future Research

Finally, I would like to introduce two points regarding the direction in which the philosophical and ethical debate about child refugees could further develop. In general, it should be said that philosophy should begin to discover child refugees as an independent field of research. Firstly, philosophical research should critically examine to what extent statements and theories from other areas of its knowledge, such as about adult refugees, childhood goods or family obligations, can be applied here. It has already been suggested in this paper that different normative concepts can be applied to the situation of minor refugees. This also concerns the extent to which the ethics of migration and the ethics of refuge converge when it is a matter of children and adolescents. The ethics of migration, especially the ones that rely on freedom of association and movement to advocate for new border regimes, can be assumed to resolve in principle some of the questions that refugees face. Open borders are also open to refugees. But other questions remain unanswered. There is, for one, the rational agent bias, also found in most proposals on the ethics of migration: it is assumed that migrants are rational and autonomous individuals who can migrate voluntarily and participate in political processes. Children, especially younger children, are in a fundamentally different position. Even in a more liberal border regime, refugees are people who do not migrate voluntarily and who have often experienced trauma in their home countries—this, of course, also affects children, and some injuries and harm, such as the loss of education, are difficult to compensate for and require extensive support from the receiving country. Children cannot migrate on their own even with open borders, and they are tied to the place of residence of their family. They are also unable to provide for themselves or to make their political voices heard.

Secondly, it would make sense for philosophical research to be more closely interlinked with the empirical sciences, i.e. for findings on the abilities and characteristics of child refugees to be incorporated into their theory. This would probably also lead to a more precise analysis of the subtle differences that exist in this group. Sometimes in philosophy, childhood is still talked about in a very unreflected way, as if it were describing a homogeneous group of people. On the contrary, children and adolescents exhibit very different forms of agency and autonomy, which also become visible among child refugees. Otherwise, philosophy threatens to be unheard and cut off in the field of refugee research and also outside academia.

Literature

Alderson, P. 2007. Competent Children? Minors' Consent to Health Care Treatment and Research. *Social Science and Medicine* 65:2272–2283. https://doi.org/10.1016/j.socscimed.2007.08.005.

Allsopp, J., and E. Chase. 2019. Best Interests, Durable Solutions and Belonging. Policy Discourses Shaping the Futures of Unaccompanied Migrant and Refugee Minors Coming of Age in Europe. *Journal of Ethnic and Migration Studies* 45:293–311. https://doi.org/10.1080/1369183X.2017.1404265.

Anderson, J., and R. Claassen. 2012. Sailing Alone. Teenage Autonomy and Regimes of Childhood. *Law and Philosophy* 31:495–522. https://doi.org/10.1007/s10982-012-9130-9.

Andresen, S. 2014. Childhood Vulnerability. Systematic, Structural, and Individual Dimensions. *Child Indicators Research* 7:699–713. https://doi.org/10.1007/s12187-014-9248-4.

Archard, D. 2004. *Children. Rights and Childhood*. 2nd ed. London, New York: Routledge.

Archard, D., and C. Macleod, eds. 2002. *The Moral and Political Status of Children*. 1st ed. Oxford, New York: Oxford University Press.

Archard, D., and M. Skivenes. 2009. Balancing a Child's Best Interests and a Child's Views. *The International Journal of Children's Rights* 17:1–21. https://doi.org/10.1163/157181808X358276.

Bagattini, A., and C. Macleod, eds. 2014. *The Nature of Children's Well-Being. Theory and Practice*. 1st ed. New York: Springer.

Carlson, B. E., J. Cacciatore, and B. Klimek. 2012. A Risk and Resilience Perspective on Unaccompanied Refugee Minors. *Social Work* 57:259–269. https://doi.org/10.1093/sw/sws003.

Cherem, M. 2016. Refugee Rights. Against Expanding the Definition of a 'Refugee' and Unilateral Protection Elsewhere. *Journal of Political Philosophy* 24:183–205. https://doi.org/10.1111/jopp.12071.
Derluyn, I., and E. Broekaert. 2008. Unaccompanied Refugee Children and Adolescents. The Glaring Contrast between a Legal and a Psychological Perspective. *International Journal of Law and Psychiatry* 31:319–330. https://doi.org/10.1016/j.ijlp.2008.06.006.
Ferguson, S. 2018. *Separating Migrant Children from Their Families Is Wrong*. UNICEF USA. https://www.unicefusa.org/stories/separating-migrant-children-their-families-wrong/34425.
Ferracioli, L. 2014. The Appeal and Danger of a New Refugee Convention. *Social Theory and Practice* 40:123–144. https://doi.org/10.5840/soctheorpract20144016.
Fine, S., and L. Ypi, eds. 2016. *Migration in Political Theory. The Ethics of Movement and Membership*. 1st ed. Oxford: Oxford University Press. https://doi.org/10.1093/acprof:oso/9780199676606.001.0001.
Franklin-Hall, A. 2013. On Becoming an Adult. Autonomy and the Moral Relevance of Life's Stages. *The Philosophical Quarterly* 63:223–247. https://doi.org/10.1111/1467-9213.12014.
Gibney, M. J. 2018. The Ethics of Refugees. *Philosophy Compass* 13. https://doi.org/10.1111/phc3.12521.
Giesinger, J. 2017. The Special Goods of Childhood. Lessons from Social Constructionism. *Ethics and Education* 12:201–217. https://doi.org/10.1080/17449642.2017.1314168.
Grover, S. C. 2018. *Child Refugee Asylum as a Basic Human Right*. 1st ed. Cham: Springer. https://doi.org/10.1007/978-3-319-78013-9.
Hart, R. 2009. Child Refugees, Trauma and Education. Interactionist Considerations on Social and Emotional Needs and Development. *Educational Psychology in Practice* 25:351–368. https://doi.org/10.1080/02667360903315172.
Hart, C. S. 2016. How Do Aspirations Matter? *Journal of Human Development and Capabilities* 17:324–341. https://doi.org/10.1080/19452829.2016.1199540.
Hedlund, D., and A.-C. Cederborg. 2015. Legislators' Perceptions of Unaccompanied Children Seeking Asylum. *International Journal of Migration, Health and Social Care* 11:239–252. https://doi.org/10.1108/IJMHSC-08-2014-0033.
Holtug, N. 2016. A Fair Distribution of Refugees in the European Union. *Journal of Global Ethics* 12:279–288. https://doi.org/10.1080/17449626.2016.1251483.
Lister, M. J. 2010. Immigration, Association, and the Family. *Law and Philosophy* 29:717–745. https://doi.org/10.1007/s10982-010-9083-9.
Lister, M. 2013. Who Are Refugees? *Law and Philosophy* 32:645–671. https://doi.org/10.1007/s10982-012-9169-7.
Lister, M. 2018. The Rights of Families and Children at the Border. In *Philosophical Foundations of Children's and Family Law*, ed. E. Brake and L. Ferguson, 153–170. Oxford: Oxford University Press.
Macleod, C. 2010. Primary Goods, Capabilities and Children. In *Measuring Justice—Primary Goods and Capabilities*, ed. H. Brighouse and I. Robeyns, 174–192. Cambridge, New York: Cambridge University Press.
Moinolnolki, N., and M. Han. 2017. No Child Left Behind. What About Refugees? *Childhood Education* 93:3–9. https://doi.org/10.1080/00094056.2017.1275231.
Mullin, A. 2019. Children's Hope, Resilience and Autonomy. *Ethics and Social Welfare* 13:230–243. https://doi.org/10.1080/17496535.2019.1588907.
Noll, G. 2016. Junk Science? Four Arguments Against the Radiological Age Assessment of Unaccompanied Minors Seeking Asylum. *International Journal of Refugee Law* 28:234–250. https://doi.org/10.1093/ijrl/eew020.
Pobjoy, J. M. 2017. *The Child in International Refugee Law*. Cambridge: Cambridge University Press. https://doi.org/10.1017/9781316798430.
Sager, A., ed. 2016. *The Ethics and Politics of Immigration. Core Issues and Emerging Trends*. London, New York: Rowman & Littlefield.

Schweiger, G. 2016. The Duty to Bring Children Living in Conflict Zones to a Safe Haven. *Journal of Global Ethics* 12:380–397. https://doi.org/10.1080/17449626.2016.1247744.

Schweiger, G. 2019. Should States Prioritize Child Refugees? *Ethics and Global Politics* 12:46–61. https://doi.org/10.1080/16544951.2019.1649958.

Schweiger, G., and G. Graf. 2017. Ethics and the Dynamic Vulnerability of Children. *Les Ateliers de l'éthique/The Ethics Forum* 12:243–261. https://doi.org/10.7202/1051284ar.

UNICEF. 2016. *Uprooted. The Growing Crisis for Refugee and Migrant Children*. https://www.unicef.org/publications/index_92710.html.

Vitus, K. 2010. Waiting Time. The de-Subjectification of Children in Danish Asylum Centres. *Childhood* 17:26–42. https://doi.org/10.1177/0907568209351549.

Zermatten, J. 2010. The Best Interests of the Child Principle. Literal Analysis and Function. *The International Journal of Children's Rights* 18:483–499. https://doi.org/10.1163/157181810X537391.

Climate Mobility

Silja Klepp and Moritz Riemann

Abstract Climate change mobility poses questions that point to the core of global and climate justice and developments in the twenty-first century. This paper shows the main ethical aspects of the debate on climate change mobility as well as the developments since its beginnings in the 1980s and presents its most important features and pitfalls. It shows how questions such as whether and how to measure and identify direct interrelationships between environmental change and migration decisions were heatedly discussed and how some of these questions linger on into the present. The main framings and concepts to which the debate is related, such as climate justice, security issues and climate change adaptation, are presented and discussed. Furthermore, the paper illustrates the main normative aspects related to questions of voluntary or forced migration and the safeguarding of the rights of environmental migrants, as these have always been central to the discussion. Theoretical impulses for the debate are brought up, such as climate justice aspects linked to climate mobility and questions of citizenship and de- and reterritorialisation.

Debates on so-called climate refugees or on climate change mobility, which we will refer to as climate mobility in this text from now on, point to the core of two world-shaping phenomena of our times: human mobility and anthropogenic climate change. As both phenomena are highly complex and politically charged, the heterogeneous research field of climate mobility is shaped by inter- and transdisciplinarity and by grey literature, such as reports from international governmental institutions and NGOs. In most cases, migration drivers, such as environmental degradation, poverty, and political instability, intersect. Nonetheless, as the analysis of different framings and discussion strands of the debate will show, it is not empirical questions or facts alone that are most discussed. It is differing political and normative standpoints aimed at questions of how to deal with growing climate mobility today and

S. Klepp (✉)
Department of Geography, Kiel University, Kiel, Germany
e-mail: klepp@geographie.uni-kiel.de

M. Riemann
Department of Philosophy, Kiel University, Kiel, Germany

A. Niederberger et al. (eds.), *Handbook of Migration Ethics*,
https://doi.org/10.1007/978-3-031-89877-8_30

in the future (Baldwin and Bettini 2017). The debate points to crucial questions of the twenty-first century, such as human-environment relations and resource scarcity in the Anthropocene, environmental and climate justice, questions of inclusion and exclusion, and possibly new ways of thinking territory, statehood and citizenship.

Today, many environmental issues can trigger displacement. Distinctions are made between slow-onset environmental changes, such as desertification, sea level rise, or soil degradation, and rapid-onset changes, such as tropical cyclones, heavy rains and floods, and disasters such as earthquakes and volcanic eruptions. Experts expect the effects of a warming world and more frequent extreme weather events to exacerbate already existing vulnerabilities, mainly in the Global South, in the areas of food security, health issues and fresh water. Furthermore, sea level rise can make coastal areas and low-lying islands uninhabitable. However, the ways in which these environmental phenomena and environmental changes will interact with population movements and individual migration projects are debated. Although climate mobility is a rapidly expanding research field, the links between environmental change and human mobility remain under investigation.

1 Climate Mobility: Some Foundational Aspects of a Contested Research Field

Due to anthropogenic climate change, experts expect the strength and frequency of storms and floods, droughts and desertification, and sea level rise to increase and these environmental developments to cause migration and displacement. Regarding more exact predictions, there is disagreement on estimated numbers of environmental and climate migrants in the early twenty-first century and in the future. Some question the sense of such estimations altogether (Nicholson 2011). In any case, numbers are a main feature in the public debate. Despite sensational reports that see millions of "climate refugees" moving from the Global South to the Global North as a direct consequence of future climate change, findings of empirical studies are often counterintuitive. The influential 2011 Foresight report by the UK's Government Office for Science underlined that environmental change can render individuals or communities even less mobile, as they can become too poor to move. The report speaks of "trapped" populations in this regard (2011: 11). Such studies show the overall complexity of migration as well as the connection to the respective local and regional contexts. In its 2014 Fifth Assessment Report, the IPCC Working Group II underlined the highly context-specific, mostly multi-causal nature of environmental mobility. The development of socioeconomic factors, adaptation measures and climate change scenarios play a main role in this regard. Nonetheless and generally speaking, the IPCC expects an increase in the displacement of people (Nurse and McLean 2014: 20). According to IPCC 2007 (Parry et al. 2007), the regions most affected by climate change are the Arctic, which will be especially affected by global warming, Africa, due to the predicted strong effects of climate change and multi-dimensional vulnerabilities, small islands, mainly due to stronger storms and sea level rise, and Asian and African mega-deltas, which are densely populated and vulnerable to rising sea levels.

Experts predict that sea level rise and higher tides, coastal erosion and salinisation of freshwater reservoirs will affect those people living in low-elevation coastal zones and on small islands. The fact that some island states, such as Kiribati, Tuvalu and the Maldives, lie only centimetres above sea level and often lack financial resources for large coastal protection programmes shows that they are generally vulnerable to climate change. However, the fact that some nation-states could lose all or most of their territories has triggered debates on citizenship without statehood, statehood without territory and a general debate on whether and how the Westphalian order could be challenged by climate change (Risse 2009; Skillington 2015; Klepp 2018).

It is important to mention here that until now people forced to move as a result of climate change have not come under the legal definition of a refugee according to the *1951 Geneva Convention Relating to the Status of Refugees.* The *Geneva Convention* protects migrants persecuted for reasons of race, religion, nationality, membership of a particular social group or political opinion. However, it is evident that the debate has always touched on questions of categorisation and terminology that imply different political and normative standpoints. The following factors are taken into account and disputed regarding their importance and range: the (inter) dependence of migration decisions and environmental Geneva Convention causes, the more or less assumed "voluntary action" or "forced migration", temporary or permanent migration, and the crossing of international borders. Most of these factors are difficult to measure, and the possibility of measuring the environmental factor in migration decisions, for example, is itself a point of contestation. Hunter sees the various forms of environmental mobility "across a continuum from forced to voluntary" (Hunter 2013: 297). Due to the complexity of migration processes and the legal and political implications linked to this as well as the apparent clarity of the term "refugee", the question of forced or voluntary migration and its normative consequences might be the most disputed aspect in the debate on climate mobility. In this regard, the academic dispute on who is to be considered as a climate refugee or an "adaptation migrant" (Wiegel et al. 2019) does not particularly deviate from the political and legal notion of the refugee (cf. Miller and Straehle 2019). Nevertheless, the driving forces of climate mobility are less easily ascribed to intentional agents than in the case of persecution. A consistent terminology or common definition is missing. However, there is considerable debate on the ethical significance of the refugeehood of climate migrants, since one crucial aspect of the migrant-refugee dichotomy is the assessment and limitation of responsibilities towards the emerging mobilities (Wiegel et al. 2019).

There are arguments for the refugeehood of climate migrants or environmentally displaced persons aiming towards a similarly strong normative claim to protect the "territorially dispossessed", as persecuted people can rely on international law (Biermann and Boas 2010). However, the terms "environmental refugee" or "climate refugee" and "environmental migrant" or "climate migrant" are criticised for being too deterministic, that is, for suggesting environmental change will directly cause migration and for drawing attention away from the complexity and the socio-political context affecting migration decisions (Hastrup and Fog Olwig 2012). The whole debate on adequate terminology also highlights a main feature of climate mobility: the strong underlying factor of uncertainty. This refers to the uncertainty about who to count as an environmental refugee, the uncertainty about how much societies will reduce greenhouse gas emissions in the future and the uncertainty about how societies will adapt (Dwyer 2020).

The terms "environmental migrants" and "climate mobility" are used in this article to refer to the phenomenon of human climate mobility more generally and to use a less sedentary concept of and perspective on society (Sager 2018).

2 Climate Mobility and Climate Justice

As the discussion on the incoherent use of the terminology has shown, normative and ethical questions are at the core of the debate on climate mobility. What might be just protection measures? How to deal with land loss and possible compensation? What does an emerging relationship between climate change, justice, law and environmental migrants mean on different scales and for different agents? Exploring the ethical horizons of protecting environmental migrants is crucial to finding strategies for adequate and possibly self-determined ways of protecting environmental migrants in the future (Eckersley 2015).

Links to aspects of "climate justice" point to the fact that the historically high greenhouse gas emissions of industrialised countries and related responsibilities for changes in global climate have been subsumed under the notion of "shared but differentiated responsibilities", but not resulted in far-reaching concessions towards the developing world or an "expansion of distributive justice globally and assistance of the climate vulnerable" (Skillington 2015: 297). Furthermore, it is contradictory, according to Skillington (2015), that today's reasons for human mobility have often been caused by those countries that are drastically reducing their access rights, thus producing new forms of "mobility injustices". A further aspect of climate justice is the fact that the effects of climate change are becoming increasingly strong as greenhouse gases accumulate. This means that future generations will face a greater risk insofar as they will have to deal with the effects of climate change, which points to intergenerational justice (Dwyer 2020).

Climate justice aspects are especially prominent in discourses on the connections between climate change and mobility where nation-states and individuals face current or future loss of land and livelihoods and the uninhabitability of whole nation-states (Risse 2009; Klepp and Herbeck 2016). Social, political and legal orders are supposedly tied to a territory. Looking at the future possibility of the loss of whole nation-states, some authors discuss the implications of the loss of sovereignty of states (Willcox 2012; Klepp 2018) and statelessness of persons (Blitz 2011). Risse underlines a common responsibility to protect environmental migrants, referring to a concept based on human rights and on a common ownership of the Earth (Risse

2009 following Kant). Are there simple humanitarian responsibilities to protect, a redress for loss and damage or might various permutations of these responsibilities evoke a right of the displaced person to choose their host country (Eckersley 2015)? The possibility of *ex-situ* citizenship or citizenship without a state, which could be implemented through a global agreement or regional protection regimes that might protect "climate citizens" and their cultural and minority rights (Klepp and Herbeck 2016), might be a progressive way to protect climate migrants and their identities. Fladvad et al. (2020) show how inhabitants of small Pacific islands respond to the risk of losing their state territory by claiming universal and emancipatory rights "from below", asking for ex-situ citizenship and minority rights that would secure their cultural sovereignty in the receiving state. In any case, these discussions are only at the beginning and no concrete legal solutions for states without territory have so far been found. Two key questions regarding the citizenship of climate mobiles yet to be answered are: (1) Do persons move and are resettled as individuals or as communities? And (2) Is the goal naturalisation and citizenship of the host state, or should existing political communities keep their political rights, institutions and organisations, i.e. political autonomy in exile? Still, most of the normative debate on migration focuses on an emigration-immigration movement of individuals that may or may not become citizens of their host country, while the movement of whole communities remains underdetermined.

Caney (2010) emphasises the importance of dealing with climate change as a decidedly rights-based issue, contrary to a purely humanitarian approach, in order to strengthen the position of environmental migrants and to give them regular and securitised access to a protection regime. He points mainly at the right of not being deprived of one's life, the right not to have serious threats to one's health and the right not to be dispossessed of the means of subsistence.

The question of rights versus humanitarianism is crucial, as it determines the rule of the international law applied. Under humanitarian international law, disaster displacement is easily conceptualised as a matter of supererogatory aid or temporary relief (Nümann 2014), resulting in favour of temporary protection over permanent asylum or resettlement. On the other hand, an approach based on political rights exceeds the limits of humanitarian international law, legitimising claims and entitlements of the territorially dispossessed in a similar manner as in politically forced migration. The rule of rights over humanitarianism necessitates a stronger call for action and stronger entitlements for the migrants. Moellendorf and other authors connect climate justice debates to different understandings of a fair global distribution of mitigation, adaptation and compensation effects among states and individuals, but also to the crucial question of how to deal with uncertainties in climate change predictions and attribution of its effects (2015). According to Moellendorf, justice requires precautionary measures based on a "principle of vulnerability reduction" (2015: 182) in order not to make the dangers of climate change unmanageable, to prevent avoidable harm, to protect the most vulnerable and to reduce the costs of adaptation that might be much higher than mitigation or precautionary measures.

Dwyer (2020) discusses different concepts of responsibility linked to climate mobility. In his view, climate mobility is connected to a more general human responsibility and the duty to aid those in need. In a more concrete vein, he refers to the work of Iris Marion Young and the idea of a responsibility to address structural injustice. This conception emphasises political responsibility for environmental migrants and the need to change structures and conditions that are fostering global warming and ultimately increase the probability for environmentally induced migration (Dwyer 2020: 567f).

Fruh (2020) argues in favour of reparations for climate migrants to compensate for the loss of culture, history, political affiliation, and agency. These entitlements to reparations are not necessarily linked to a clear assignment of responsibilities for the loss since responsibilities for climate change are complex and a contested field of research. However, the deprivations caused by climate change are a crucial element of climate justice.

Draper (2023) offers a pluralist approach to the phenomenon of climate displacement based on five different forms of climate migration: community relocation, territorial sovereignty, labour migration, internal displacement, and refugee movement. Each of these contexts generate different normative consequences, and a mere framing of climate displacement as a refugee situation would create a concept too narrow to meet the realities and intersections of climate mobility.

3 Genealogy and Different Strands of the Climate Mobility Debate

Having presented the main ethical questions of the debate on climate mobility that underline the close connection between climate mobility and controversies about social, political, economic and ecological justice, especially in North–South relations we will now outline the development and some important strands of the discussion. We must critically analyse the development of the discussion and its most important influences and different narratives, as these factors powerfully shape the debates and define the categories in which action can occur.

Until the figure of the "environmental refugee" appeared in 1985 (El Hinnawi 1985), migration studies were shaped by a lack of consideration of environmental aspects – although migration driven by environmental factors has historically been a consistent phenomenon and played an important role in the first scientific migration theories (e.g. Ravenstein 1885). Between 1985 and 1990, three UN reports contributed to reconnecting environment and migration and shaped future debates in the media and among policymakers. They were driven by an agenda of raising awareness for climate change and environmental protection (Nicholson 2011) and (ab)used the topic of climate mobility to give a "human face to climate change" (Gemenne 2011). NGOs and humanitarian organisations often mobilised alarmist narratives that urged policymakers to "do something about the situation."

The critique that followed the alarmist voices soon set the scene for a polarised debate known as the "alarmists" versus "sceptics" or "maximalists" versus "minimalists" (Piguet 2013: 154). The "maximalist approach" focuses on the environmental variable from a cluster of causes and proclaims that migration is a direct result of environmental degradation. The agency of migrants is given little attention here, while demands for protection or compensation still played a subordinate role at this time in the debate. The "minimalist approach", in contrast, underlines that migration is a complex, multi-causal phenomenon. In the 1990s, the "minimalist approach" was mainly represented by social scientist and migration researcher Richard Black, making the debate also a discussion between natural scientists' more positivist understandings of human-environment interaction and the more constructivist, cautious readings of the phenomenon of climate mobility by social scientists.

This epistemological divide has remained relevant until today, although there seems to be more consensus recently that simplistic linkages between environmental change and migration decisions should be avoided (Hillmann et al. 2015).

The suggestion of security threats in countries of the Global North by environmental migrants and rising conflicts through climate change in the Global South were important aspects of the climate mobility debate in the 2000s. These aspects were brought up in various reports by military or national advisory boards (e.g. WBGU 2007), but also by NGOs (Kolmannskog 2008).

Various authors argue that perspectives underlining security threats triggered by climate mobility are a significant feature of a broader securitisation discourse that depicts climate change as an increasing threat to national security (White 2011; Herbeck and Flitner 2010). A variety of agents fuel racist discourses and fears of "millions of climate refugees", based on the vague assumption that an alarmistic scenario would increase the motivation of people and industries to reduce emissions. These include environmentalists, with slogans such as "if not literally flooded, [we] will most certainly be flooded by the climate refugees" (Kolmannskog 2008: 9).

Closely connected to concepts that are widely used in development studies, the impact of climate change on society is frequently framed in terms of adaptation, vulnerability and resilience. Mobility in times of changing environments has historically always been an important adaptation measure, and in general mobility is a normal societal process and not an exception. Some authors have scrutinised these concepts regarding, for example, the "agency" of climate migrants, and there are different attempts to conceptualise these terms in a more emancipatory way (Klepp and Herbeck 2016; Hastrup and Fog Olwig 2012).

Generally, since 2010, the debate on climate mobility has increasingly emphasised the individual responsibility of "more resilient" climate migrants to cope with their situation (Felli and Castree 2012). This avoids a perspective on climate change as structural violence against the world's poorest communities. The "resilient lives [sic]" (Felli and Castree 2012: 4) of individuals and communities, as discussed in Felli and Castree's criticism of the 2011 Foresight report, are discussed in terms of their ability to be absorbed by foreign labour markets. This approach sees climate-induced mobility as a successful adaptation strategy and a solution rather than (as was previously the case) a problem of securitisation within a discourse of environmental mobility (Methmann and Oels 2015). Lately, reports by international organisations have also highlighted the agency of the individual migrant, trying to reframe the securitisation of the debate in terms of successful migration management (UNHCR 2015). Emphasising aspects of more or less successful migration management could foster debates in which aspects of climate justice are excluded altogether.

In a more critical strand of literature, authors increasingly demand a consistent repoliticization of adaptation research in opposition to the depoliticising of the climate mobility debate through, for example, resilience and management approaches that shift responsibility for coping with climate change to the individual (Baldwin and Bettini 2017; Methmann and Oels 2015; Gesing et al. 2014). One standpoint that goes even further aims to supplant current vulnerability and resilience concepts with a broader "citizenship approach" that emphasises the agency of affected people and political aspects in climate mobility and climate adaptation debates and underlines aspects of climate justice (Klepp and Herbeck 2016).

Regarding the developments of legal aspects of the debate since its beginnings in the 1980s, experts have raised questions related to safeguarding the rights of environmental migrants. As mentioned above, normative and ethical aspects and questions of climate justice are central in these considerations. Recently, the debate has concentrated more on policy-oriented solutions on a regional level, creating a legal lacuna (Attapattu 2020) regarding legally binding treaties on a global level. Skillington sees a "legal violence" and a double punishment in the non-recognition of climate migrants, on the one hand by the loss of their homelands and on the other hand through a "legally sanctioned social suffering" that makes the suffering of climate migrants possible, "unavoidable, and entirely legal" (2015: 290).

All displaced persons are entitled to fundamental human rights. These rights include the right to life, health, shelter, food and movement. Regarding the development of more specific rights that guarantee the protection of environmental migrants, single governments, scientists, international organisations and NGOs aimed to create a new, legally binding treaty or an expansion of existing treaties in international law, especially in earlier stages of the debate. On the one hand, these aims were linked to the *1951 Geneva Convention Relating to the Status of Refugees*. On the other hand, there were suggestions for a treaty connected to the *United Nations Framework Convention on Climate Change* (UNFCCC) (Biermann and Boas 2010).

In the search for ways to improve the protection of environmental migrants, several new concepts and issues have been introduced to the debate, fostering aspects of environmental responsibility in the realm of environmental law at the global level. According to Michelot (2006), the figure of the environmental migrant embodies current threats facing humanity, and the acknowledgement of "environmental refugees" implements the principle of environmental or climate justice. However, overall, our article has shown that a clear distinction between environmental migrants and environmental refugees is not possible.

In order to establish more unbiased language in the realm of climate migration and to acknowledge the variety of migration realities that might be linked to environmental and climate change, some scholars have urged for a terminology shift "from climate migration to climate mobilities" or to "Anthropocene mobilities" (Boas et al. 2019: 902; Baldwin et al. 2019).

4 New Ways Forward in the Debate

This article has shown that debates on climate mobility are rarely used as an "imaginative resource" (Hastrup and Fog Olwig 2012: 2) to proactively discuss about the world in which we want to live; on the contrary, they help create conditions of fear, exclusion and discrimination. The different understandings and framings of climate mobility will shape future conditions and our collective dealing with the phenomenon. This article closes with a discussion of ways to differentiate and pluralise a debate that is often led in a way that is too narrow for times of changing climates and that could do more justice to such a vast and complex topic (Baldwin 2014). The im/mobilities perspective also calls for a pluralisation of paradigms applied to the ethics of mobility beyond the emigration-immigration nexus (Sager 2018).

To suggest alternative narratives contrary to the negative images that seem to prevail in research on climate-related movements today, Hastrup and Olwig underline the role of archaeologists, physical anthropologists, linguists and others who look at the development of human societies in a long-term perspective. Today, large-timescale perspectives are mostly missing in the debate on current environmental migration. The social effects of climate change should be studied from a particular angle that cannot be captured with the short-term analysis that dominates today (Hastrup and Fog Olwig 2012). Historical perspectives and the acknowledgement of historical responsibilities for climate change and global solidarity could also support the search for opportunities and solutions for environmental migrants that are decisively migration-friendly and take climate justice aspects into account. Negotiation processes on environmental migration in Oceania are a good example in this regard. The scientific predictions for many small island states that take into consideration the uninhabitability of whole nation-states have led to a proactive search for decisively migration-friendly solutions. They could also play a key role for other regions in the world. The government of the Republic of Kiribati has looked for migration possibilities that could secure self-determination and include South–South migration options (Klepp and Herbeck 2016). The main slogan in this regard is "migrate with dignity", and individual as well as community relocation is negotiated under this motto. A future refugee status should absolutely be avoided, as the people of Kiribati emphasise, knowing that refugees have little self-determination and often meagre living conditions (Klepp and Herbeck 2016). Oceania, with its rich transnational migration traditions (Farbotko 2012), especially before the colonisation of the region, could connect to histories of migration and find solutions for environmental migrants that are "concrete, local, social and connective" (Klepp and Herbeck 2016: 72). The negotiation processes initiated by the government of Kiribati also aim at less restricted mobility in Oceania (see also Farbotko 2012) and repoliticise the debate by directly connecting it to postcolonial questions and aspects of climate justice (Klepp and Herbeck 2016). This line of analysis is contrary to debates in the Global North, where discussions have been caught up in the widely distributed abbreviations of development and security. A decentring of the debate by closely analysing regional negotiation processes on environmental migration in the Global South and questions of climate justice can open up new perspectives for the search for accountable and responsible migration and climate policies.

These new perspectives include an acknowledgement of the manifold political aspects of climate change and its social consequences, such as questions of regional and global responsibility and solidarity as well as historical and current power relationships and dependencies.

Literature

Attapattu, S. 2020. Climate Change and Displacement. Protecting ‘Climate Refugees’ Within a Framework of Justice and Human Rights. *Journal of Human Rights and the Environment* 11:86–113.

Baldwin, A. 2014. Pluralising Climate Change and Migration. An Argument in Favour of Open Futures. *Geography Compass* 8:516–528.

Baldwin, A., and G. Bettini, eds. 2017. *Life Adrift. Climate Change, Migration, Critique*. London, New York: Rowman & Littlefield.

Baldwin, A., C. Fröhlich, and D. Rothe. 2019. From Climate Migration to Anthropocene Mobilities. Shifting the Debate. *Mobilities* 14:289–297.

Biermann, F., and I. Boas. 2010. Preparing for a Warmer World. Towards a Global Governance System to Protect Climate Refugees. *Global Environmental Politics* 10:60–88.

Blitz, B. K. 2011. Statelessness and Environmentally Induced Displacement. Future Scenarios of Deterritorialisation, Rescue and Recovery Examined. *Mobilities* 6:433–450.

Boas, I., C. Farbotko, H. Adams, H. Sterly, H. Bush, K. van der Geest, and H. Wiegel. 2019. Climate Migration Myths. *Nature Climate Change* 9:901–903.

Caney, S. 2010. Climate Change, Human Rights and Moral Thresholds. In *Human Rights and Climate Change*, ed. S. Humphreys, 69–90. Cambridge: Cambridge University Press.

Draper, J. 2023. *Climate Displacement*. Oxford: Oxford University Press.

Dwyer, J. 2020. Environmental Migrants, Structural Injustice, and Moral responsibility. *Bioethics*. SI: IAB 14th World Congress, 562–569.

Eckersley, R. 2015. The Common but Differentiated Responsibilities of States to Assist and Receive ‘Climate Refugees’. *European Journal of Political Theory* 14:481–500.

El-Hinnawi, E. 1985. *Environmental Refugees*. Nairobi: UNEP.

Farbotko, C. 2012. Skillful Seafarers, Oceanic Drifters or Climate Refugees? Pacific People, News Value and the Climate Refugee Crisis. In *Migrations and the Media*, ed. K. Moore, B. Gros, and T. Threatgold, 119–142. New York: Peter Lang.

Felli, R., and N. Castree. 2012. Neoliberalising Adaptation to Environmental Change. Foresight or Foreclosure? *Environment and Planning A* 44:1–4.

Fladvad, B., S. Klepp, and F. Dünckmann. 2020. *Struggling Against Land Loss. Environmental (in) Justice and the Geography of Emerging Rights. Geoforum* 117:80–89.

Fruh, K. 2020. Climate Change Driven Displacement and Justice. The Role of Reparations. *Essays in Philosophy* 22:102–121. https://doi.org/10.5840/eip20211292.

Gemenne, F. 2011. How They Became the Human Face of Climate Change. Research and Policy Interactions in the Birth of the ‘Environmental Migration’ Concept. In *Migration and Climate*

Change, ed. E. Piguet, A. Pécoud, and P. de Guchteneire, 225–259. Cambridge: Cambridge University Press.
Gesing, F., J. Herbeck, and S. Klepp, eds. 2014. *Denaturalizing Climate Change. Migration, Mobilities and Spaces*. ARTEC paper 200. Bremen.
Hastrup, K., and K. Fog Olwig, eds. 2012. *Climate Change and Mobility. Global Challenges to the Social Sciences*. Cambridge: Cambridge University Press.
Herbeck, J., and M. Flitner. 2010. 'A New Enemy Out There'? Der Klimawandel als Sicherheitsproblem. *Geographica Helvetica* 65:198–206.
Hillmann, F., M. Pahl, B. Rafflenbeul, and H. Sterly. 2015. *Environmental Change, Adaptation and Migration. Bringing in the Region*. New York: Palgrave Macmillan.
Hunter, L. M. 2013. Migration and Environmental Hazards. In *Migration and Climate Change. The International Library of Studies on Migration* 15, ed. G. Hugo, 273–302. Cheltenham, Northampton: Edward Elgar.
Klepp, S. 2018. Framing Climate Change Adaptation from a Pacific Island Perspective—The Anthropology of Emerging Legal Orders. *Sociologus* 68:149–170.
Klepp, S., and J. Herbeck. 2016. The Politics of Environmental Migration and Climate Justice in the Pacific Region. *Journal of Human Rights and the Environment* 7:54–73.
Kolmannskog, V. 2008. *Climate Changed. People Displaced*. Oslo: Norwegian Refugee Council.
Methmann, C., and A. Oels. 2015. From 'Fearing' to 'Empowering' Climate Refugees. Governing Climate-Induced Migratoion in the Name of Resilience. *Security Dialogue* 46:51–68.
Michelot, A. 2006. Enjeux de la reconnaissance du Statut de réfugié écologique pour la construction d'une nouvelle responsabilité environnementale. *Revue européenne de droit de l'environnement* 4:428–445.
Miller, D., and C. Straehle, eds. 2019. *The Political Philosophy of Refuge*. Cambridge: Cambridge University Press.
Moellendorf, D. 2015. Climate Change Justice. *Philosophy Compass* 10:173–186.
Nicholson, C. T. 2011. *Is the "Environmental Migration Nexus" an Analytically Meaningful Subject for Research?* COMCAD Working Paper. Bielefeld: COMCAD.
Nümann, B. 2014. *Umweltflüchtlinge? Umweltbedingte Personenbewegungen im Internationalen Flüchtlingsrecht*. Baden-Baden: Nomos Verlagsgesellschaft.
Nurse, L. A., and R. F. McLean. 2014. Small Islands. In *Climate Change 2014. Working Group II: Impacts, Adaptation and Vulnerability. Part B: Regional Aspects*, ed. V. R. Barros, C. B. Field, D. J. Dokken, M. D. Mastrandrea, K. J. Mach, T. E. Bilir, et al., 1613–1654. Cambridge: Cambridge University Press.
Parry, M. L., O. F. Canziani, J. P. Palutikof, P. J. van der Linden, and C. E. Hanson, eds. 2007. *Climate Change 2007. Impacts, Adaptation and Vulnerability*. Contribution of Working Group II to the Fourth Assessment Report of the Intergovernmental Panel on Climate Change. Cambridge: Cambridge University Press.
Piguet, E. 2013. From 'Primitive Migration' to 'Climate Refugees'. The Curious Fate of the Natural Environment in Migration Studies. *Annals of the Association of American Geographers* 103:148–162.
Ravenstein, E. G. 1885. The Laws of Migration. *Journal of the Statistical Society of London* 48:167–235.
Risse, M. 2009. The Right to Relocation. Disappearing Island Nations and Common Ownership of the Earth. *Ethics and International Affairs* 23:281–299.
Sager, A. 2018. *Toward a Cosmopolitan Ethics of Mobility. The Migrant's-Eye View of the World*. Cham: Palgrave Macmillan.
Skillington, T. 2015. Climate Justice Without Freedom. Assessing Legal and Political Responses to Climate Change and Forced Migration. *European Journal of Social Theory* 18:288–307.
UNHCR. 2015. *UNHCR, The Environment and Climate Change*. Geneva (Updated Version).
White, G. 2011. *Climate Change and Migration. Security and Borders in a Warming World*. Oxford, New York: Oxford University Press.

Wiegel, H., I. Boas, and J. Warner. 2019. A Mobilities Perspective on Migration in the Context of Environmental Change. *WIREs Climate Change* 10. https://doi.org/10.1002/wcc.610.
Willcox, S. 2012. A Rising Tide. The Implications of Climate Change Inundation for Human Rights and State Sovereignty. *Essex Human Rights Review* 9:1–19.
Wissenschaftlicher Beirat der Bundesregierung Globale Umweltveränderungen (WBGU). 2007. *Welt im Wandel. Sicherheitsrisiko Klimawandel*. Berlin, Heidelberg: Springer.

Economic Refugees

Andreas Niederberger and Uchenna Okeja

Abstract The convention in migration studies is to make a distinction between groups of immigrants, especially between the categories of economic migrants and refugees. The category of economic migrants is perhaps the most controversial in the ethics of migration because it is not clear which conclusions theorists should draw from the claims of this group. In this chapter, we discuss the positions of theorists of the ethics of migration about this category of migrants. Specifically, we explain why many theorists question the conventional political and legal distinction that is central to migration ethics, namely the distinction between economic and political refugees. Furthermore, we consider arguments that count in favour of recognising the claims to admission in destination countries by economic refugees. We conclude our discussion by highlighting some desiderata for future research on economic migrants.

Current international refugee law goes back to experiences in Europe after World War I and World War II. Already in the interwar period, after the restructuring of the European state system and due to the flight from political transformations, for example in the former Soviet Union or fascist Italy, Germany and Spain, there were numerous refugees in Europe for whom a solution was sought within the framework of the League of Nations. Due to the escape from the extermination policy of the Nazis and the persecution by other states and political organisations as well as the forced displacements in the course of the reorganisation of Europe after World War II, the situation was repeated and worsened.

In the post-World War II era, however, it became obvious that the lack of rights of the refugees in the wartime situation had fatal consequences. For example, it was clear that many Jews and political refugees had lost their lives due to a lack of

A. Niederberger (✉)
Universität Duisburg-Essen, Institut für Philosophie, Essen, Nordrhein-Westfalen, Germany
e-mail: andreas.niederberger@uni-due.de

U. Okeja
Nelson Mandela University, Gqeberha, South Africa
e-mail: uchenna.okeja@mandela.ac.za

A. Niederberger et al. (eds.), *Handbook of Migration Ethics*,
https://doi.org/10.1007/978-3-031-89877-8_31

protection. The 1951 Geneva Refugee Convention (UNHCR 1951) was therefore meant to provide a durable solution for remaining European refugees and express the obligation of European states to offer protection to those fleeing racial, ethnic, religious, or political persecution. Notwithstanding, it became evident rather quickly that the regional and temporal limitation of the Refugee Convention to Europe and to persons who had fled because of events before 1951 was not plausible in view of other relevant refugee groups that existed at the time.

Consequently, the 1967 Protocol Relating to the Status of Refugees removed the territorial and temporal limitations of the Refugee Convention, and the UN High Commissioner for Refugees (UNHCR) assumed the responsibility for the increasing numbers of refugees worldwide. Nevertheless, it is pertinent to note that the history of the 1951 Refugee Convention continued to shape the legal determination of refugees as well as the political discussion about them to this day. Flight is essentially understood as a reaction to the fear of becoming a victim of persecution or human rights violations, that is, fear of acts of other people who directly seek to significantly harm or even kill the members of the persecuted group. People fleeing poverty, hunger or other desolate situations are accused of not being genuine refugees. They are referred to as economic or poverty "refugees", which means they are thought of as fleeing mainly to improve their economic situation for perfectly understandable reasons. However, compared to those imagined to be genuine refugees, that is people fleeing persecution, the claims of economic refugees are considered to carry less weight.

1 General Issues at Stake in the Controversy

The opposition between political refugees and economic "refugees" has long dominated the political discourse of the Global North on forced migrants from the Global South—and it continues to be a guiding distinction in this discourse. For the normative discussion, however, one can observe that most positions deny that there is a clear priority of refugees fleeing from persecution over refugees whose primary goal is to escape from miserable economic or other social conditions. Specifically, there are five reasons for this development.

The first reason is that for many theorists, the distinction hardly holds anymore. Forms of clear political, racial, or religious persecution of certain groups and individuals that characterised the twentieth century still exist, but most of those who consider themselves refugees do not belong to these groups of people or are not primarily migrating because of such persecution. They are characterised by mixed motives, which include political coercion or repression, but also a lack of prospects and the search for a life worth living, as well as changes in the environment that make it difficult to remain in the previous place. It is therefore also not surprising that weighty reasons to abolish the distinction between political refugees and other merely supposed refugees are put forward regarding so-called climate refugees (cf. chapter on "Climate Mobility"). Although the persecution criterion does not apply to this group, it seems obvious that this is a form of forced migration that generates claims for protection or admission.

A second, more philosophical reason is that preferences for normative positions have changed. Most approaches in the fields of political philosophy or applied ethics now argue in a consequentialist way and, from a broad utilitarian perspective, aim in particular at welfare gains or harm reductions. However, if we compare the welfare of poor or even starving people with that of persecuted dissidents who, for instance, used to be oligarchs, it is difficult to justify why claims to admission as a refugee should be made conditional on the fact of persecution. On this view, accepting economic refugees would lead to higher welfare gains than accepting political refugees, so we should not tie the decision to accept refugees to the fact that some people want to harm others.

But even if one does not take a consequentialist approach in determining the best effects of refugee admission but focus specifically on the reasons that cause people to flee, it is, thirdly, not obvious why miserable conditions should be less of a reason for flight than persecution. If the privileging of persecution is based on the thesis that persecution is coercion and that refugees therefore have no rational choice but to leave a context, then one can ask why the same freedom of choice does not apply to poverty or hopelessness. Here, too, there is a threat of suffering, inability to act or even the possibility of death, which is why it is also rational under these circumstances to escape to better conditions. This point is reflected in the expanded concept of human rights, which now includes more than the classical human and civil rights, that is, those rights that connote the absence of arbitrary state power. If there are also human rights to social and cultural goods and humane conditions, then we must consider their absence to be just as serious as the state's suppression of political freedom.

The fourth reason has to do with the thought that the distinction between political and economic refugees is a mechanism that enables the countries of the Global North to resist entry of people from the Global South. It is claimed that the countries of the South are not comprehensively subject to state power, and it is denied that there is any state at all that could permanently persecute people in the way that total statehood did under Nazism or Stalinism. Consequently, it seems there are no real refugees from the Global South and that those who arrive in the North do not need the corresponding protection. Against this background, however, and considering a possible compensatory dimension of South-North-migration, it can be stated that colonial migration was primarily economic in nature, meaning that it is implausible to link the right to flee to the Global North today to political reasons for flight. Given the history of colonial migration, it would seem valid to argue that there are good reasons to recognise the claims of economic migrants from the Global South (Achiume 2019).

The fifth reason arises from the discussion about global justice, especially in connection to the thought that migration is a reaction to the problematic state of transnational politics. Some theories of global justice argue that global relations are unjust because the Global North has created and maintains a global structure (for example through the world-market or international organisations) that makes it difficult or even impossible for the Global South to establish just basic structures in their own countries (Pogge 2002). If this is true, then conditions in the Global South are not (only or primarily) miserable by accident or self-inflicted. Rather, these conditions are the result of more or less deliberate actions by the Global North. This implies that people in the Global South are not simply worse off or unlucky, but rather are victims of the actions of others. Lacking the power to resolve the problems of transnational politics, citizens of countries in the Global South react through migration to the Global North. The distinction between economic migrants and political refugees therefore does not apply to the movements of the latter, because they are reactions to the attempts of others to harm them, not actions to improve their own situation.

This last reason is given a new dimension by the emergence of the concept of climate refugees. For the problematisation of the discourse on economic migration is accentuated by the fact that many current so-called economic refugees are (at least also) victims of the effects of catastrophic climate change, which is largely the result of the activities of the countries of the Global North. Climate refugees therefore constitute a category of economic migrants that have both economic needs and they can claim to be forced away from their home by the Global North. Given the unequal pollution that exacerbates climate change, some theorists of migration propose that the distinction between political and economic refugees is spurious and ultimately not helpful in the discourse on the ethics of migration. It should be evident from the discussion above that there are different controversies that are central to the ethics of migration due to the various ways theorists frame the claims of economic migrants.

2 Controversies on the Extension of the Refugee Category to Economic Migrants

In this section we consider three controversies at the core of the discourse on economic refugees, some of which address the reservations about the presumably special nature of the reasons for flight recognised under international law that were presented in the previous section. The first controversy is conceptual, and it pertains to the thinking that recognising the claims of economic migrants would mean a justification of the idea that a general theory of freedom must entail endorsement of the claims of all categories of refugees. This controversy is in part traceable to David Ingram who mobilised the idea of social freedom in his *World Crisis and Underdevelopment* to argue that the coerciveness of international institutions could be considered a justification for the claim that economic migrants are owed a duty of admission by specific states.

The second controversy arises from contestations about the current framing of the migration discourse, especially the extent it is justified to argue that so-called economic refugees can stake a claim of recognition due to historical encounters such as colonialisation. Thirdly, there is the controversy around the transformation of so-called economic refugees to climate refugees. There are two layers to this controversy. On the one hand, it is argued that designating some economic refugees as climate refugees makes more evident the cogency of the claims they advance. This could be regarded as the conceptual accuracy debate. On the other hand, there is an argument about the extent it is the duty of affluent countries to take responsibility for accommodating economic refugees if the climate crisis becomes the principal cause of their economic misery prompting migration to other countries.

To understand the nuances of these controversies, it will help to contextualise them. To this end, we will provide concise summaries of the ideas of some of the prominent theorists that have contributed to these debates. We begin with the first controversy, namely, the concern about what a general theory of freedom should entail for economic refugees. David Ingram contributes to this discussion by first recognising that the current picture of "global poverty and global inequality [...] suggests that inequalities built into the global distribution of economic benefits and burdens—whatever their causes—impose greater duties beyond emergency assistance, duties that fall on everyone, more so on those living in high income countries and those who are comparatively well off" (Ingram 2018: 14).

Given the current shape of the global economy, it is untenable to hold that political refugees constitute a different category of migrants from economic refugees. For Ingram, the nature of the contemporary global political economy makes a distinction between political refugees and economic refugees difficult because of the intricate connection between political abuse and coercive economic factors. Both the self-identified political refugee and the so-called economic migrant confront the same challenge, namely, the loss that stems from their diminished agency.

The core issues at stake in this first controversy about the claims of economic migrants is a question of standpoint. The idea is that we could arrive at a plausible conclusion about the claims of economic migrants if we proceeded in a manner that does not draw an artificial line between political refugees and economic migrants and seek on this basis to decide whose needs or claims carry greater weight. Drawing on recognition theory and discourse ethics, Ingram proposes a view of migration ethics that "commands respect for the human rights of both political and economic refugees" (Ingram 2018: 127–128). He argues that "the failure of developed countries to secure the rights of refugees hinges on several factors: an outdated legal definition of who counts as a refugee accompanied by wide latitude in discharging an unspecified duty to rescue" (Ingram 2018: 128). This means that economic migrants, understood as political refugees, deserve the same level of political protection that is usually extended to political refugees. Ingram believes this is plausible because persons fleeing severe economic deprivation qualify as refugees. Accordingly, it is also appropriate that "recent changes in the legal definition of refugee imply that persons fleeing life-threatening poverty have a claim to foreign protection regardless of how they enter their chosen country" (Ingram 2018: 128).

Similarly, Michelle Foster argues that it is possible to become a refugee on the grounds of economic deprivation. People seeking admission as immigrants qualify as refugees if their economic situation is such that their chance of survival in their home countries is compromised. Foster notes that "the artificial quality of the distinction between economic refugees and political refugees" is the reason many theorists question current approaches to implementing the Refugee Convention (Foster 2007: 11). Not only are the labels "economic refugees" and "political refugees" problematic, the implications of being categorised in a particular way make the quest for conceptual clarity necessary. This is especially the case for so-called economic refugees who are not accorded the same rights as political refugees. In a situation, such as the implementation of the Refugee Convention, where conceptual categorisation could mean being granted or denied a chance to survive, we must ensure that the basis of conceptual labels is reasonable—in the sense of being aligned with the overall goal of protecting human rights.

According to Foster, the ways we try to argue for the distinctiveness of political refugees are mostly implausible. For instance, it is implausible to argue that political refugees are indeed special because the concept refers to individuals that are singled out by an oppressor. The idea is that unlike political refugees, economic refugees are not fleeing targeted oppression. By making this point, it is emphasised that what qualifies one as a refugee is the uniqueness of one's position, namely, that one is in the unique position of being singled out by an oppressor. Foster argues that "the cogency of this rationale for dismissing economic claims is open to question on the basis that the system of international refugee protection was designed initially to accommodate extremely large groups of refugees such as Jews fleeing Nazi Germany" (Foster 2007: 11). Although we find Foster's observation here to be contentious, if not inaccurate, the point we would like to note is that Foster and many other theorists insist that appeals to intent and voluntariness of the movement of the economic migrant as opposed to the political refugee do not suffice to sustain the conceptual distinctions that underlie the Refugee Convention.

One thing that emerges from the perspectives advanced by Ingram and Foster is that they contest the extent to which it makes sense under current global realities to distinguish between economic refugees and political refugees. Asking whether there are really significant differences between economic refugees and political refugees as some theorists of the ethics of migration do is not only relevant because of the normative implications but also important due to the policy implications that are at stake. For this reason, the accuracy of the distinction between political refugees and economic migrants constitutes one of the controversies in the contemporary discourse on the ethics of migration. A core issue at stake for most of the theorists interested in the debate is the implication of erasing or retaining the boundaries assumed to exist between economic refugees and political refugees. However, one can also raise the question of whether the problematic effects of the certainly observable close linkage of legal claims of refugees to (political) persecution justifies the levelling of the normative claims of all refugees. From a normative point of view, the reasons for supporting refugees fleeing the threat of being targeted for deprivation of civic rights and for supporting refugees fleeing circumstances in which a decent life is not possible remain, of course, different, at least from the perspective of theories that do not simply look at well-being but at autonomy or freedom. Normative differentiation does not necessarily mean that some claims are denied altogether. Consequently, it may be more plausible to demand the same rights and benefits for many economic migrants that political refugees enjoy, but still emphasise that the reasons for the rights and benefits in each case are different.

As noted earlier, the second controversy is whether so-called economic refugees as such can stake a claim of recognition due to historical encounters like colonialisation. One of the theorists that has considered this point extensively is Tendayi Achiume who appeals "to the history and legacy of the European colonial project to [...] argue for a different theory of sovereignty that makes clear why, in fact, economic migrants of a certain kind have compelling claims to national admission and inclusion in countries that today unethically insist on a right to exclude them" (Achiume 2019: 1510). What is at issue for Achiume is not merely a recounting of the history of colonisation to argue that former colonial powers will be doing something morally beneficent by offering so-called economic migrants a path to admission and inclusion. Achiume is rather arguing that insofar colonial structures perpetuated until this day continue to subordinate colonised populations, there is need to develop a different understanding of sovereignty. This new sense of sovereignty is one that imposes on former colonial powers an obligation to admit and accommodate so-called economic refugees. Former colonial powers ought to do this because economic refugees from formerly colonised countries are merely responding to their subordination by structures introduced through colonisation.

According to Achiume, "First World nation-states have no right to exclude Third World migrants, for reasons tied to the distributive and corrective justice implications of the legacies of colonialism" (Achiume 2019: 1517). To understand why Achiume thinks this is plausible, it is important to note her fundamental conviction, namely, that "the European colonial project [is] a mandatory invitation from European colonial powers to the rest of the world to co-depend indefinitely" (Achiume 2017: 144). The idea is that through the project of colonisation, European powers were able to secure colonial advantages through the international legal system and other structures of domination. Thus, given that these advantages have been preserved and perpetuated, it would be reasonable to recognise the sense in which migration is a project of decolonisation. For Achiume, "decolonization is the pursuit of a long overdue geopolitical reordering of benefits of a global order defined by interdependence forged in the colonial era" (Achiume 2017: 145). Against this background, migration is a "rejection" of the unfavourable political and economic conditions that result from the interdependence created by colonisation. Therefore, decolonisation should be "the pursuit of a more equitable marriage between the geopolitical center and the periphery" (ibid.). This means that so-called economic migrants are refugees whose claim to admission ought to be recognised.

The debate on climate justice is at the core of the third controversy. Climate change will inevitably lead to a situation where certain places will become either too hot or cold. It would also lead to severe flooding and other environmental disasters. Regardless of the specific challenge that will manifest in different places, one thing that is clear is that these occurrences will make it impossible for many people to just stay where they are. The notion of climate migrants therefore refers to people compelled to leave their homelands to seek admission in other places due to the effects of climate change.

In the literature on climate migrants, there is conceptual debate about what it means to regard some economic migrants as climate refugees, or environmental migrants, which is a widely used category in the discussion among theorists of climate induced migration. Whereas some argue that although the concept is not precise, it could still be applied meaningfully across a broad spectrum, some propose that the very idea of environmental migrants serves to distract attention from the ideological objective at stake in the context in which the concept is invoked. The conceptual controversy here arises due to the difficulty in answering the question: who is a climate refugee? Are they people who leave pre-emptively to avoid the coming climate catastrophe? Does the concept refer only to the people that have already been displaced due to adverse climate events that have made their homelands uninhabitable? And what about the incremental movements of people, by which they slowly adapt to shifting climatic conditions—are these people then to be denied entitlements that others receive in the case of more abrupt or spontaneous movements?

Notwithstanding the conceptual disagreements, a functional consensus in the discourse on climate migrants is that the impact of climate change will be felt much more by people in developing parts of the world, hence many of those currently regarded as economic migrants might ultimately become climate migrants. This is the case because the countries of the Global South where the majority of the current economic migrants originate from have less sophisticated infrastructure and capacity to adapt or deal with large-scale environmental disasters. Against this background, most of the so-called economic refugees can plausibly be regarded as climate refugees when their needs become even more acute due to environmental disasters, such as the flooding that occurred in Mozambique in March 2019.

Designating some economic migrants as climate refugees has normative and practical implications. Apart from implying a need to redefine the conception of refugees in international law, we will be required to reconsider the moral obligation towards this category of migrants. Although it might seem that the claim of economic migrants is given more weight once the adverse impact of climate change is taken into account, adequate notice must be taken of the complexity of this approach. This has been noted in relation to the approach of Wennersten and Robbins (2017). Both theorists have sought to suggest that the category of climate refugees comprises mainly current economic refugees, the idea being that climate refugees are often poverty stricken people from the Global South. One reason this framing of climate refugees is problematic is because it gives the impression that the "concern is not the climate refugee per se but the threat the climate refugee represents to an imagined global order overseen by the West" (Baldwin 2018: 638).

Besides the conceptual problem, the idea of climate refugees raises a question about the scope of moral obligation. Who should bear the responsibility to meet the needs of economic refugees once they become designated as climate refugees? Should the industrialised countries of the West shoulder this responsibility, given, for instance, Peter Singer's polluter pays principle? (Singer 2002: 27) Disagreements abound regarding the usefulness of this approach to thinking about the benefits and burdens of remedying the unfavourable impact of climate change. What is clear is that the concept of climate refugees carries a crucial question of responsibility, especially the scope of the obligation of countries that have contributed disproportionate emissions to the climate. This issue, like the other controversies discussed above, cannot be resolved by fiat.

3 Desiderata for Future Research

It is evident that the association of legally protected claims of migrants with their status as political refugees requires revision. Initially designed to address the unique demands of individuals who have been deprived of everything, including their right to exist, for political reasons, the category of political refugees has evolved into an arbitrary criterion in various ways (Gündoğdu 2015). However, does this mean that the differentiation between reasons for migration has become entirely irrelevant? The first desideratum is certainly a closer examination of the relationship between reasons for migration and claims to admission or residence. In many contexts, it is the case that the reasons agents have for their actions carry different weights and, corresponding to this weight, different obligations arise in other people.

Interestingly, even many of the rejections of the distinction between political refugees and economic migrants still refer to the special importance given to reasons for migration in which a more or less active harm by others is evident. Thus, despite all the criticism, the point of distinguishing between migration for the purpose of improving one's own situation and migration to escape harm stemming from others or to restore an indispensable political, legal or social status does not seem to be disputed. But what then should the distinction be based on, if the persecution criterion as found in international refugee law is too narrow, but a general global deprivation or disadvantage criterion may be too broad? The discussion on climate refugees certainly opens up important perspectives, especially in the different emphasis on the direct harm caused by climate change or the harm caused by the obstruction to a global policy of mitigation and adaptation. However, it is not immediately obvious whether and how the considerations from this context can be transferred to other groups of migrants, for example from regions with widespread violent phenomena or with other than climatic or natural transformations.

Second, the criticism of the narrow focus on political refugees should not lead us to completely disregard the special nature of this group of refugees or the claims that are particularly relevant in their case. Persons forced to flee for political reasons are certainly in some cases pure victims, i.e. they are deprived of their status and opportunities because they belong to certain groups or are accused of intentions they do not actually have. In some cases, however, they are also people who are actually fighting for certain political goals, a specific order or even for a future revolution. Many migrants have a strong interest in the development of their countries of origin and would see their future in those countries if political conditions allowed. They fight tyrannical or corrupt governments or seek entirely different forms of political organisation. The focus on political refugees has always also made visible the meaning of exile and the severe restrictions on political participation that accompany it.

The second desideratum is therefore to identify the special claims of political refugees, by which is not meant privileges in terms of residence or protection status vis-à-vis other migrants. Rather, it is about the conditions under which political refugees can remain politically active in exile and continue to participate in the transformation of their country of origin. Host countries could also be obliged to lobby for the right of return of exiles to their countries of origin. Such conditions are not always straightforward to guarantee, especially for the host countries, because they can lead to tensions with the countries of origin—one need only think of the asylum granted to activists of the militant Kurdish liberation movement PKK in Scandinavia and its consequences for Sweden's negotiations with Turkey about its accession to NATO. On the other hand, it is also not clear whether every form of political activity in exile should be supported: Should, for example, religious fundamentalists be supported in their struggle against secular dictatorial regimes, even if they themselves do not aim for (liberal) democracies? And how can we avoid the claims of political refugees becoming simply part of the foreign policy and the particular interests of receiving countries?

To evaluate the claims of migrants by referring to the political conditions in the first half of the twentieth century and totalitarian or authoritarian forms of government in that period largely misses the reality of migration in the twenty-first century. However, this should not lead us to completely relegate the relationship between politics and migration to the background. Political, legal, and social conditions and the exercise of power and domination within them remain important factors in migration—both at the places of origin and destination. The clear distinction between political refugees and economic migrants must therefore be overcome, without erasing everyone's claim to a political say in their intended place of residence or playing off this claim against economic goals.

Literature

Achiume, E. T. 2017. Reimagining International Law for Global Migration: Migration as Decolonization? *American Journal of International Law Unbound* 111:142–146.

Achiume, E. T. 2019. Migration as Decolonization. *Stanford Law Review* 71:1509–1574.

Baldwin, A. 2018. Book Review: Rising Tides. Climate Refugees in the Twenty-First Century, Bloomington: Indiana University Press. *International Migration Review* 52:637–639.

Foster, M. 2007. *International Refugee Law and Socio-Economic Rights. Refugee from Deprivation.* Cambridge: Cambridge University Press.

Gündoğdu, A. 2015. *Rightlessness in an Age of Rights. Hannah Arendt and the Contemporary Struggles of Migrants.* Oxford: Oxford University Press.

Ingram, D. 2018. *World Crisis and Underdevelopment. A Critical Theory of Poverty, Agency, and Coercion.* Cambridge: Cambridge University Press.

Pogge, T. 2002. *World Poverty and Human Rights*. Cambridge: Polity.
Singer, P. 2002. *One World. The Ethics of Globalization*. New Haven, London: Yale University Press.
UNHCR, 1951. *The UN Refugee Agency. Convention and Protocol. Relating to the Status of Refugees*. https://www.unhcr.org/3b66c2aa10. Last checked 3 Oct 2023.
Wennersten, J., and D. Robbins. 2017. *Rising Tides. Climate Refugees in the Twenty-first Century*. Bloomington: Indiana University Press.

Refugee Repatriation

Mollie Gerver

Abstract This chapter considers three central controversies relating to refugee repatriation: (1) whether states wrong refugees when forcing them to repatriate; (2) whether refugees are forced to repatriate if they face pressure to do so; and (3) whether refugees are wronged by humanitarian organisations helping them to unsafely repatriate.

This chapter addresses questions concerning refugee repatriation. Repatriation refers to refugees returning to the country from which they fled, or the country from which their parents or grandparents fled. In particular, this chapter addresses ethical dilemmas arising when refugees return either completely voluntarily, semi-voluntarily, or not voluntarily at all. An example of voluntary repatriation might involve refugees obtaining citizenship in their adopted countries who nonetheless wish to return to help rebuild the countries they fled. An example of completely involuntary repatriation might be refugees who are awoken in the middle of the night, forced into vans and driven to the airport, handcuffed to their seats as the plane takes off. Finally, an example of semi-voluntary repatriation might be refugees feeling pressure to return from deteriorating conditions in the country providing them asylum. This is very common. For example, roughly four million refugees returned from Pakistan to Afghanistan in the early 2000s, partly because in Pakistan they faced the threats of arrests and their work visas being revoked. Similarly, Somali refugees face regular pressure from the Kenyan authorities to repatriate, both because they cannot legally work and because the Kenyan government often threatens to close the Dadaab refugee camp. This would render many refugees without basic necessities, and so some have chosen to repatriate as a result (Gerver 2018).

The above types of cases, and many more, give rise to three pressing ethical controversies.

M. Gerver (✉)
Department of Political Economy, King's College London, London, UK
e-mail: Mollie.Gerver@kcl.ac.uk

A. Niederberger et al. (eds.), *Handbook of Migration Ethics*,
https://doi.org/10.1007/978-3-031-89877-8_32

1 Three Controversies

One controversy is over when governments are permitted to force refugees to repatriate. On the one hand, perhaps refugees who can live perfectly safe lives in their countries-of-origin are not wronged if required to repatriate, especially if only a short amount of time has passed. On the other hand, perhaps refugees should gain a right to remain after having fled, because it can be especially difficult to return after having faced persecution and/or extreme violence. For example, German Chancellor Angela Merkel announced in 2016 that Syrian refugees who had been given asylum would eventually be required to repatriate to Syria (Rinkel 2016). It is not clear if such a policy is just, given the challenges that would likely arise in Syria even if the war ended.

There is a second controversy. Even if forcing refugees to return is clearly wrong, it is unclear what constitutes a forced return. It is not clear if governments are forcing refugees to repatriate if denying them certain social rights, such as access to welfare benefits, but still not actively engaging in deportations.

Finally, there are controversies over whether aid organisations ought to assist refugees to repatriate. Sometimes organisations assist refugees who wish to return solely because they will be detained by a host government if they remain. For examples, many Afghan refugees in Pakistan repatriated to Afghanistan between 2002 and 2020 to avoid being arrested by Pakistan authorities. Some of these refugees could only return with the help of the UN Refugee Agency (UNHCR) (Gerver 2018). It is not clear if UNHCR acted ethically in helping them return, given that they were only returning to avoid wrongful detention.

2 Key Positions

Let us describe some key positions and arguments concerning the above three controversies.

2.1 Are Governments Permitted to Force Refugees to Repatriate?

With regards to the first controversy, some authors argue that there is an especially strong case for permitting refugees to remain in their country of asylum if they are refugees according to the 1951 Convention for the Protection of Refugees. According to this Convention, refugees are those with a well-founded fear of being persecuted in their country-of-origin because of their political views, or membership in a given ethnic, religious, or social group.

For example, Max Cherem (2016: 191–192), Matthew Lister (2013: 662, 669–670) and Michael Walzer (1983: 48) endorse the following argument, or a close variant of this argument:

1. All have a right to be accepted as a citizen by a government.
2. One is not accepted as a citizen if one is persecuted by one's own government because of one's membership in a given ethnic, religious, political, or social group.
3. When one is not accepted as a citizen, the only remedy is citizenship elsewhere.
4. Therefore, one who is persecuted because of their political, religious, ethnic, or social group membership has a right to citizenship elsewhere.
5. With citizenship comes the right to remain.
6. Therefore, those who are persecuted because of their political, religious, ethnic, or social group membership have a right to remain.

The argument's conclusion is consistent with current legal practices in most high-income countries, where those who are persecuted in their home countries because of their political, religious, ethnic, or social identity have a right to asylum, and generally have a right to apply for citizenship after 5 years of holding asylum. Those who are fleeing other sorts of life-threatening conditions, such as famine or general violence, also have a right to asylum, but obtaining citizenship is more difficult, and returning home is often required once conditions are safe. In low-income countries, those fleeing because of their political, religious, ethnic, or social identity are also given asylum, but rarely can qualify for citizenship in practice. Instead, all individuals fleeing any life-threatening conditions—whether it be persecution, general violence, famine, and so forth—have a right to general residency (often in camps), with such residency rights withdrawn when the UN deems it safe to return home (Albert 2010).

Some might claim that the above argument, while focusing on those fleeing persecution, is unnecessarily narrow: many individuals lack meaningful citizenship even though they are not persecuted. One lacks meaningful citizenship if one's vote counts for little or if one lacks basic necessities because of widespread corruption. If so, then it seems many individuals have a right to new citizenship upon obtaining asylum, and it would be wrong to require that they repatriate in the future (Owen 2016; Shacknove 1985; Gerver 2018).

If so, we may wish to endorse an argument with broader implications (for related arguments, see Owen 2016; Shacknove 1985):

1. All have a right to be accepted as a citizen by a government.
2. One is not accepted as a citizen if one lacks basic rights in their home state.
3. When one is not accepted as a citizen, the only remedy is citizenship elsewhere.
4. Therefore, one who lacks basic rights in their home state has a right to citizenship elsewhere.
5. With citizenship comes the right to remain.
6. Therefore, those who lack basic rights in their home states have a right to remain.

Some might reject this argument, claiming it has counter-intuitive implications. For example, it implies that an individual without access to healthcare in the US, and so lacking a basic right, has a right to both asylum abroad to obtain medical care, and a right to remain in this country once healthcare is no longer needed. This seems odd. More generally, if all individuals lacking basic rights in their home countries had a right to asylum in another country, and the right to remain, this might imply that hundreds of million people have a right to asylum abroad and the right to remain. Some might claim that states—or at least some states—lack the capacity to accept all such individuals, especially in the absence of any binding international agreements to allocate responsibility. States therefore do not have a duty to accept everyone arriving whose basic rights are not protected in their home countries, and certainly have no duty to continue to permit them to remain.

The above considerations can be accounted for if we slightly alter the aforementioned view and accept that, while all those without basic rights should have the possibility of gaining asylum, states are permitted to

(a) prioritise some before others when capacity is limited, and/or
(b) repatriate some to increase capacity overall.

With regards to the first claim (a), perhaps states ought to prioritise those facing more severe rights violations over those facing less severe rights violations (Gerver 2020), or those facing rights violations which undermine their home state's legitimacy, such as those fleeing violent persecution or extreme violence (Hosein 2019: 119). In some cases, it might even be justified to accept refugees via a lottery (Miller 2019). More importantly for our discussion, with regards to the second claim (b), we might think states are permitted to repatriate some refugees who were granted asylum, but who no longer require asylum, so that more refugees can be accepted in the future. For example, if Germany does ever require Syrian refugees to repatriate once the war in Syria ends, perhaps this policy is justified so that Germany can accept more refugees from other countries. Under this view, states ought to recognise that lacking basic rights in their home countries have a right to be considered for asylum abroad, and maybe even have a right to asylum abroad, but states are justified in engaging in repatriation to increase their capacity to grant asylum to more individuals.

One important criticism of the above claim is that, even if refugees no longer require asylum, they might have integrated into society and become citizens in practice, even if not citizens on paper. Many Syrian refugees in Germany would have likely lived in Germany for over 5 years, or even a decade, by the time the war in Syrian ends. Many if not most would have worked in Germany, learned German, had children born in Germany, partaken in civil society in Germany, and accepted a range of German cultural practices. To deport those who have integrated into society is wrong, even if this somehow frees up room to accept more refugees in the future.

Importantly, even if refugees have not lived in the country for very long, it is not clear if requiring refugees to eventually repatriate really does increase the state's capacity to accept more refugees in the future. The opposite may be true: states who grant refugees the right to remain might reduce costs, increasing state's capacity to accept even more refugees in total. This is because refugees with a right to remain into the future can more easily find employment, open a bank account, take out a mortgage, join a building society, integrate into society, and so forth, gaining a level of self-sufficiency difficult to gain on short-term visas. For example, Uganda provides refugees with relatively long-term visas and the right to work, often permitting refugees to remain even after returning is safe. It also accepts more refugees-per-capita compared to Kenya, a country that often pressures refugees to repatriate (UNHCR 2018; Clements et al. 2016; Lindley 2011: 7; Brankamp 2019). It is possible that Uganda's policy of avoiding repatriation is what gives Uganda the capacity to accept more refugees, rather than fewer. In other words, permitting refugees to remain need not necessarily mean fewer refugees obtain asylum.

2.2 What Is Forced Return?

Regardless of one's position concerning the first controversy, nearly all philosophers accept this: it is wrong to deport refugees back to countries where their lives will certainly be at risk from persecution. In reality, states rarely deport refugees. Instead, they *pressure* or *encourage* refugees to repatriate.

For an example of *pressuring* refugees to repatriate, some states deny refugees access to the job market, refusing them sufficient welfare provisions, and/or detaining them in an enclosed refugee camp. Sometimes states threaten to partake in these actions unless refugees agree to repatriate—as when governments say they will soon detain refugees unless they return home—and other times states partake in these actions regardless of what refugees agree to do—as when states just detain all refugees and many refugees return as a result. In the latter case, states are not necessarily detaining refugees so that they repatriate, but many refugees may still repatriate as a result of detention. Regardless, if refugees choose to repatriate because of detention, it seems their choice to return is not voluntary. Assuming refugees are wronged when forced to repatriate—and at least refugees fleeing persecution clearly fall within this category—then they have been wronged by the state which is purporting to give them asylum.

In particular, one might accept the following argument:

1. If X threatens to commit a wrong against Y or does commit a wrong against Y, and Y Φ-es as a result, then X wrongly forces Y to Φ.
2. When refugees' lives will be at risk from persecution if they return home, it is wrong for the government to deny them employment opportunities, deny them welfare provisions, or threaten to detain them.
3. Therefore, when
    (a) refugees' lives will be at risk from persecution if they return home, and
    (b) the government denies them employment opportunities, welfare provisions, and/or threatens to detain them, and
    (c) refugees repatriate as a result then
4. the government wrongly forces refugees to repatriate.

This argument is relevant for a range of governments around the world which deny refugees various rights (Human Rights Watch 2016; Corlett 2005; Kukathas 2021; Barnett and Finnemore 2004). For example, Australia detains those arriving by boat in offshore detention centres (Kukathas 2021) and Kenya denies free movement and work visas to refugees (Betts 2021). When refugees repatriate because of these conditions, these governments wrongly force refugees to repatriate.

The above focuses on refugees forced to return home. Sometimes refugees are not forced to return, but *encouraged* to return with inducements. For example, Burundian refugees in Tanzania were encouraged to return home with promises of small plots of land with houses, and refugees in Germany were promised a year's worth of rent in home countries if they repatriated in the near future. Most strikingly, dozens of countries have promised hard cash to refugees repatriating, with payments ranging from $50 in Pakistan to $3500 in Israel. Indeed, some political parties over the last decade have promised even more money. In an odd election campaign in 2010, the openly-xenophobic British Nationalist Party promised £50,000 to asylum seekers agreeing to repatriate. It was never elected, but in theory a family of six would have been able to receive over half a million pounds to return to Iran, Iraq, Afghanistan, and dozens of other countries of origin (Gerver 2018: Chap. 5).

When refugees are encouraged to return with cash, they are sometimes also forced to return. Such was the case when Afghan refugees were paid to return from Pakistan, and also faced threats of arrest and encampment if they did not (Gerver 2018: Chap. 5). However, not all refugees paid to repatriate are forced to do so and, even when they are, there is still an important philosophical question: whether adding money to the mix is distinctly wrong. More generally, it is not clear if states act wrongly when using inducements, as opposed to threats, to encourage repatriation.

There are three relevant considerations for establishing whether encouraging refugees to repatriate is wrong:

One consideration is related to messaging. When a government promises material benefits to refugees agreeing to leave, it risks implying the following message: "We want you to leave so much that we are willing to sacrifice valuable resources so that you do." The more a state is willing to pay, the more it signals how much it views refugees as holding negative value. Moreover, telling refugees that they have negative value is not only offensive towards those who are offered the inducement. It can be offensive towards other members of society who are not offered this inducement. When Germany offers some asylum seekers money to return home, it implies a general desire to have fewer outsiders in the country, and other immigrants—such as naturalised refugees, students, foreign workers, and so forth—might feel unwelcome as well (Gerver 2018: Chap. 5).

One potential way to avoid xenophobic messages is to avoid advertising the money as an inducement. Instead, money could be privately offered to those who are already making an active effort to return, and have already expressed an interest for some time. A waiting period of a year or two could signal to citizens that the government is not eager to encourage refugees to leave, and that money is to help rather than encourage return.

Even if negative messaging cannot be entirely avoided, there is a second consideration. It matters whether inducements dramatically increase the options that refugees have. When refugees are given significant resources to return to home countries, they can more easily choose where they will live in their home countries, what education they will access, what businesses they establish, and what foods, clothing, and gifts they choose to purchase in their initial days, months, and years of repatriation. While today refugees receive relatively little money to repatriate, the amount they receive has been increasing over time, and enough money could dramatically improve both the options and quality of life of those returning. If options and quality of life are improved enough, this value could outweigh the negative effects of messages (Gerver 2018: Chap. 5).

A final consideration concerns risks. Sometimes refugees accept money to repatriate, later regretting their decision because of risks experienced after return. Evidence of refugees regretting risky repatriation was found amongst returning Afghan refugees (Smith 2015), returning Iraqi refugees (UNHCR 2010), returning South Sudanese refugees (Gerver 2018: Chap. 4), and returning Somali refugees (Agence France-Presse 2017). This data on past refugees regretting their decision was also predictive of future refugees experiencing similar regret and insecurity. It is not clear if states acted ethically in continuing to pay for repatriation.

In other words, there is a conflict between the value of giving refugees access to money for easier repatriations, and the value of not encouraging unsafe repatriations. While this conflict cannot be resolved entirely, there is one obvious solution: allowing refugees to return to a safe state if they feel repatriation is unsuccessful. For example, in the 1990s Sweden, France, and the United Kingdom allowed refugees returning to Bosnia to re-enter these three countries in the event of an emergency (Black 2001). Similarly, Tanzania worked with the UN to help refugees returning to Burundi later re-enter Tanzania (UNHCR 2004) and in 2017 Lebanon informally permitted some Syrian refugees who returned home to again live in Lebanon if they were unsatisfied with their repatriation. One refugee describes such repatriation and why he chose to again live in Lebanon:

> From the Lebanese border to Aleppo, [Syria] I had to pay a $100 bribe [...] I had packed a food bag for my sister: tea, coffee, powdered milk, and so on. But once I arrived at Aleppo, my suitcase was empty because at every checkpoint on the road they took something. Our apartment could be rehabilitated with some work, but it is too expensive and there is almost no electricity. Our shop was destroyed and looted.

This refugee decided, upon seeing his looted shop, to re-enter Lebanon. Life was still very difficult for him, but at least he and his family could more easily access employment, school, and a stipend from UNHCR (Balanche 2017).

2.3 The Humanitarian Organisation Controversy

The stipend that this refugee received from UNHCR helped him survive. UNHCR has become increasingly involved in the provision of aid to refugees throughout low-income countries, and is becoming increasingly involved with repatriation itself. Nongovernmental Organisations (NGOs) have played a similar role, often working alongside UNHCR to provide transport for refugees wishing to return to home countries, in addition to money for refugees re-integrating into their home countries. A number of human rights groups and scholars have opposed UNHCR and NGOs—who I shall just call "organisations"—being involved in repatriation. These groups and scholars note that refugees often choose to return because conditions have become life-threatening in host states (Barnett 2001; Bradley 2013; Crisp and Long 2016). If refugees are returning to unsafe home countries to avoid unsafe host states, then they are not returning voluntarily. They therefore cannot give their voluntary consent to organisations to repatriate. If they cannot give their voluntary consent, they cannot give their *valid consent*: valid consent is the sort of consent a person must give before they are offered a high-risk service. For example, a patient must give valid consent to a doctor before the doctor is permitted to conduct surgery. This is because surgery is risky, and so a patient has a right to not have this risk imposed on her unless she agrees to give the doctor permission. Without valid consent, the doctor would commit a wrong. Similarly, organisations helping refugees repatriate to high-risk countries commit a wrong without refugees' valid consent, and so commit a wrong when refugees' consent is involuntary.

This argument can be summarised as follows:

1. Organisations should only help refugees repatriate if they can obtain their valid consent.
2. To give valid consent, refugees must be consenting voluntarily.
3. To consent voluntarily, they must have a reasonable alternative.
4. Refugees lack a reasonable alternative if they will lack basic rights should they remain in their country-ofasylum.
5. Therefore, when refugees will lack basic rights should they remain in their country-of-asylum, organisations should not help refugees repatriate.

Some might reject the first premise. Some might claim that organisations needn't obtain refugees' valid consent because they are merely giving refugees resources they can always return. Just like a person needn't obtain valid consent from someone they transfer money to—if I sent you $100 I don't need your valid consent, because you can always just give the money back—an organisation needn't obtain valid consent from refugees before giving them a bus ticket back to their countries-of-origin. Refugees can just give the ticket back if they don't want to repatriate.

Though organisations are often just handing over assistance, they are still engaging in high-risk assistance. When providing someone high-risk resource, valid consent is necessary even if the recipient can simply give the assistance back. For example, a doctor is morally obligated to obtain a patient's valid consent when handing out a free-but-risky pill, even if the patient could just give the pill back. They must obtain patient's valid consent because patients have a right to not be placed in a position of risk, and the moment they are given the pill they are placed in a position of risk. When organisations give refugees free flights to high-risk countries, they are placing refugees in a position of risk, and ought to first obtain refugees' valid consent.

The first premise of the argument, then, is sound. What of the second? This premise is false. It is not true that valid consent must be voluntary.

The reason for this is related to consent's function. Consent's function is to give individuals autonomy over their lives by successfully giving others permission they otherwise would not have. To give consent that successfully gives permission, one needn't necessarily be giving the consent voluntarily; it is enough that the agent obtaining consent cannot stop the coercion which leads to consent, nor provide any better option. For example, imagine a woman is running from a gang and knocks on the door of her neighbour for protection. The neighbour tells her, "I will let you into my home and give you protection, but there are risks in doing so: my roof is not secure, some of my electrical fixtures might be faulty, and the gang might still break its way into the home." If the woman consents to this risky arrangement, she is not consenting voluntarily—the alternative is death from the gang—but she is still giving her valid consent. This is because she is receiving the only option her neighbour could possibly provide, assuming he cannot provide any other option that will ensure better protection. When organisations help refugees return home, they can similarly obtain refugees' valid consent if this is the only option organisations could possibly provide, and they cannot provide any option that will better protect refugees (Gerver 2018 ibid: Chap. 2).

For an example of such valid consent, consider a case involving the Relief Society of Tigray (REST), an organisation established in the 1980s by Tigrayans in Ethiopia. In 1984 REST attempted to help refugees of the Ethiopian famine living in Sudan, appealing to the international donors, but these donors refused to provide sufficient funds. The death rate amongst refugees in Sudan was very high, and so REST offered to help them return to Ethiopia (Assefaw 1993). Though the organisation was helping with a forced return, given deteriorating conditions in Sudan, it nonetheless obtained refugees' valid consent, given that this was the only option it could provide.

Consider another case, involving similarly valid consent. In 1991 NGOs working in the mountainous region between Iraq and Turkey tried to help refugees obtain asylum in Turkey, but consistently failed, eventually agreeing to help refugees repatriate to Iraq (Long 2013: 107). These refugees were essentially coerced by the Turkish government into repatriating, but they could still give their valid consent to NGOs helping them repatriate.

In contrast to the above organisations, many do not obtain valid consent. Sometimes organisations could lobby for refugees' rights in a host country, or could work to resettle refugees to safer countries, and have a duty to provide both options. They might have such a duty because they can easily provide such options, or because as humanitarian organisations they have a duty to protect refugees in the most effective way possible. If organisations do not fulfil their duties to provide refugees with options other than repatriation, then they are denying options which refugees have a right to receive from these organisations. They therefore fail to obtain refugees' valid consent to repatriate.

3 Future and Emerging Research

The above included an overview of only some of the controversies surrounding repatriation. It did not include all. For example, there are questions of when and whether refugees have a right to return. For example, Victor Tadros (2017) has addressed some of the key arguments made in support of Palestinian refugees' right to return, and responded to arguments against such a right. More recently, there have been questions surrounding when refugees have a duty to return. Matthew Gibney (2019) has responded to popular claims that refugees have a duty to return to their home country after returning is safe. He argues that refugees do not owe the same duties to their home countries that others might owe to their home countries. This is because, while refugees may feel a sense of duty to their co-nationals, they do not owe duties to the state itself, given that the state did not ensure the protection it had a duty to ensure. Javier Hidalgo (2018) has focused on similar questions of refugees' duties. He argues that refugees have a right to evade immigration officials in an attempt to cross borders, and some might claim that refugees have a related right to bypass authorities who are attempting to force them to repatriate.

Both Gibney and Hidalgo's innovative focus on refugees' duties open broader possibilities for research. For example, it opens questions concerning refugees' duties towards other refugees and displaced persons. Refugees might have a duty to repatriate owed to those who lack the means to flee, including family members left behind. For example, a refugee who fled and left behind a parent may have a duty towards this parent to repatriate. Even if these duties ought to never be enforced, they might be felt and acted upon by refugees themselves. Exploring these possibilities requires exploring a broader range of agents beyond traditional state-agents alone.

A focus on non-state agents might include a focus on non-refugee citizens as well. Hidalgo, in addition to addressing the duties of migrants and refugees, addresses the duties of citizens towards migrants when immigration laws are unjust. He argues that citizens have a duty to refuse to comply with unjust immigration laws. If Hidalgo is right, citizens presumably have a duty to refuse to comply with particular laws concerning repatriation. For example, citizens may have a duty to break laws to help hide refugees in their homes so that refugees are not forcibly repatriated, or not forced into detention centres where they might feel compelled to repatriate. More controversially, perhaps citizens have a special duty to help refugees who have been wrongly repatriated, such as by sending them remittances, or by lobbying for the government to evacuate individuals who were wrongly repatriated and faced risks in their home countries.

Finally, there remain a range of more state-focused questions that have yet to be addressed. One question concerns the duties of states towards those who have already been forcibly repatriated. It is not clear if states owe such refugees special duties, such as the duty to evacuate them in the event of an emergency. For example, after Germany and the Netherlands increased efforts to forcibly repatriate Afghan refugees in the 2010s, many refugees who were returned likely faced displacement again, in particular after the takeover of the Taliban in 2021. There was media attention surrounding the halting of these repatriation efforts to Afghanistan (DW 2021), but little focus on the refugees who had already been forcibly returned and who could not legally re-enter Germany and the Netherlands. Perhaps both states ought re-admit such refugees, or at least help evacuate them to safer states nearby.

Addressing these broader questions requires further philosophical analysis, but for now it should be clear just how broad the topic of repatriation can be; it spans questions concerning governments, citizens, and refugees, and the duties and rights of all three.

Literature

Agence France-Presse. 2017. *Somali Refugees Regret Returning Home from Kenya*. June 27, 2017. http://m.news24.com/news24/Africa/News/somali-refugees-regret-returning-home-from-kenya-20170627. Last checked 28 July 2017.

Albert, M. 2010. *Prima Facie Determination of Refugee Status*. Working Paper Series 55. https://www.rsc.ox.ac.uk/files/files-1/wp55-prima-facie-determination-refugee-status-2010.pdf. Last checked 9 Jan 2020.

Assefaw, T. 1993. Participatory Relief Management. The Experience of the Relief Society of Tigray. *Refuge* 12:23–25.

Balanche, F. 2017. A Half-Million Syrian Returnees? A Look Behind the Numbers. The Washington Institute. *Policywatch* 2826. http://www.washingtoninstitute.org/policy-analysis/view/a-half-million-syrian-returnees-a-look-behind-the-numbers. Last checked 21 Jan 2018.

Barnett, M. 2001. UNHCR and the Ethics of Repatriation. *Forced Migration Review* 10:31–34.

Barnett, M., and M. Finnemore. 2004. *Rules for the World. International Organizations in Global Politics*. Ithaca, London: Cornell University Press.

Betts, A. 2021. *The Wealth of Refugees. How Displaced People Can Build Economies*. Oxford: Oxford University Press.

Black, R. 2001. Return and Reconstruction in Bosnia-Herzegovina. Missing Link, or Mistaken Priority? *SAIS Review* 21:177–199.

Bradley, M. 2013. *Refugee Repatriation. Justice, Responsibility and Redress*. Cambridge: Cambridge University Press.

Brankamp, H. 2019. 'Occupied Enclave'. Policing and the Underbelly of Humanitarian Governance in Kakuma Refugee Camp, Kenya. *Political Geography* 71:67–77.

Cherem, M. 2016. Refugee Rights. Against Expanding the Definition of 'Refugees' and Unilateral Protection Elsewhere. *Journal of Political Philosophy* 24:183–205.

Clements, K., T. Shoffner, and L. Zamore. 2016. Uganda's Approach to Refugees. *Forced Migration Review* 52:49–52.

Corlett, D. 2005. *Following them Home. The Fate of the Returned Asylum Seekers*. Melbourne: Black Inc.

Crisp, J., and K. Long. 2016. Safe and Voluntary Refugee Repatriation. From Principle to Practice. *Journal of Migration and Human Security* 4:141–147.

DW. 2021. *Germany and the Netherlands Suspect Deportations to Afghanistan.* August 21, 2021. https://www.dw.com/en/germany-and-netherlands-suspend-deportations-to-afghanistan/a-58830259. Last checked 20 Sept 2021.
Gerver, M. 2018. *The Ethics and Practice of Refugee Repatriation.* Edinburgh: Edinburgh University Press.
Gerver, M. 2020. Sufficiency, Priority and Selecting Refugees. *Journal of Applied Philosophy* 37:713–730.
Gibney, M. 2019. The Duties of Refugees. In *The Political Philosophy of Refuge*, ed. D. Miller and C. Straehle. Cambridge: Cambridge University Press.
Hidalgo, J. 2018. *Unjust Borders. Individuals and the Ethics of Immigration*. Abingdon: Routledge.
Hosein, A. 2019. *The Ethics of Migration. An Introduction.* Abingdon: Routledge.
Human Rights Watch. 2016. *Kenya. Involuntary Refugee Returns to Somalia.* September 14, 2016. https://www.hrw.org/news/2016/09/14/kenya-involuntary-refugee-returns-somalia. Last checked 7 Jan 2018.
Kukathas, C. 2021. *Immigration and Freedom.* Princeton: Princeton University Press.
Lindley, A. 2011. Between a Protracted and a Crisis Situation. Policy Responses to Somali Refugees in Kenya. *Refugee Survey Quarterly* 30:1–36.
Lister, M. 2013. Who Are Refugees? *Law and Philosophy* 32:645–671.
Long, K. 2013. *The Point of No Return. Refugees, Rights, and Repatriation.* Oxford: Oxford University Press.
Miller, D. 2019. Selecting Refugees. In *The Political Philosophy of Refuge*, ed. D. Miller and C. Straehle. Oxford: Cambridge University Press.
Owen, D. 2016. In Loco Civitas. On the Normative Grounds and Character of the Institution of Refugeehood and of Responsibilities Towards Refugees. In *Migration in Political Theory. The Ethics of Movement and Membership*, ed. S. Fine and L. Ypi. Oxford: Oxford University Press.
Rinkel, A. 2016. Germany's Merkel Says Refugees Must Return Home Once War Is Over. *Reuters*, January 30, 2016. https://uk.reuters.com/article/us-europe-migrants-germany-refugees/germanys-merkel-says-refugees-must-return-home-once-war-is-over-idUKKCN0V80IH. Last checked 3 Jan 2019.
Shacknove, A. 1985. Who Is a Refugee? *Ethics* 95:278–280.
Smith, J. 2015. Limited Options Leave Afghan Refugees Reluctant to Return Home. *Stars and Stripes*, January 2, 2015. https://www.stripes.com/news/middle-east/afghanistan/limited-options-leave-afghan-refugees-reluctant-to-return-home-1.322129#.WXuNBdPyvre. Last checked 28 July 2017.
Tadros, V. 2017. The Persistence of a Right of Return. *Politics, Philosophy and Economics* 16:375–399.
UNHCR. 2004. Burundi. Repatriation from Tanzania—Numbers Remaining under 300,000. *Briefing Notes*, May 18, 2004. http://www.unhcr.org/40a9e0a21.html. Last checked 19 Sept 2015.
UNHCR. 2010. *Briefing Note. Iraqi Refugees Regret Returning to Iraq, Amid Insecurity.* October 19, 2010. http://www.unhcr.org/uk/news/briefing/2010/10/4cbd6c9c9/unhcr-poll-iraqi-refugees-regret-returning-iraq-amid-insecurity.html. Last checked 26 Feb 2018.
UNHCR. 2018. *Refugee Population by Country of Territory of Asylum.* https://data.worldbank.org/indicator/SM.POP.REFG. Last checked 9 Jan 2020.
Walzer, M. 1983. *Spheres of Justice. A Defence of Pluralism and Equality*. New York: Basic Books.

Part V
Beyond Philosophy: Migration Ethics Within Other Disciplines

Political Science and Migration Ethics

Benjamin Boudou

Abstract This chapter discusses the position of migration ethics within political science. The epistemological and methodological commitments of political science seem to prevent it from including ethical considerations, or to fully acknowledge the normative grounds and framing of the empirical work. The role of ethics, morality, values or even more generally, ideas in explaining migration patterns and behaviour, appears small or irrelevant. However, various attempts to bridge the gap between political science and migration ethics have been made. I identify four strategies that have been adopted: drawing policy recommendations based on ethical standards from an empirical analysis; fact-checking of migration ethics and norm-checking of the political science of migration; considering moral values, especially as a causal determinant in policy making; studying values empirically. Finally, I argue that pragmatist approaches, bridging empirical and normative inquiries are necessary to do justice to the multifaceted nature of migration issues.

Most of the questions asked in heated public debates surrounding migration, seem factual: do Western nations welcome more migrants than before? How much does migration cost? Will the "migration crisis" impinge on the economy, which already seems to be in a difficult situation? Does the arrival of migrants speaking a different language, coming from different cultural and religious backgrounds pose a threat to maintaining social cohesion and democratic functioning? On the contrary, how to welcome more migrants in danger while maintaining a decent level of integration, so that they do not feel like second-class citizens? Who are the refugees, where to they come from and where do they want to go, why are they moving? Answering these questions means bringing facts into the public sphere. The 2018 proposal for building the equivalent of the IPCC but for migration issues—an International Experts Panel on Migration and Asylum (IEPMA)—was precisely aimed at establishing the facts: "We ask for a radical change of paradigm in dealing with migration and asylum, based on a rational, realistic, scientifically-informed and humanist

B. Boudou (✉)
Department of Public Law and Political Science, University of Rennes, Rennes, France
e-mail: benjamin.boudou@univ-rennes.fr

A. Niederberger et al. (eds.), *Handbook of Migration Ethics*,
https://doi.org/10.1007/978-3-031-89877-8_33

approach" (IEPMA 2018), is the mission of this organisation. The ambition was to produce interdisciplinary but fundamentally empirical work to provide the "missing link between research and migration policy." Making sure policies rely on correct data and that the public is properly informed, are indeed absolute necessities when such a topic is more and more politicized (Freeman et al. 2013; Van der Brug et al. 2015). However, while essentially constituted as a valorisation of factual expertise and a re-appropriation by migration scholars of "knowledge claims" (Boswell et al. 2011), this proposal was scientific, political and normative at the same time. Although strictly separated in general along the fact-value distinction, both political science (and other empirical social sciences) and migration ethics could be required here, especially if one has to develop a "humanist approach" (IEPMA 2018). The goal of the IEPMA can be interpreted as a rallying cry to bring together different disciplines around a subject in order to better analyse it, to weigh in the public debate by requiring it to be dispassionate and rationalized and to evaluate the quality and desirability of policies according to certain (ethical) standards.

In this chapter, I begin by analysing the general problem of the incompatibility of migration ethics and political science, then I present different approaches to bridging this gap, using different strategies to associate empirical and normative inquiries. Finally, I discuss the use of political science for migration ethics and more general mixed method approaches that could be fruitful for future research.

1 Migration, Political Science and the Role of Normative Considerations

Empirical facts and data, provided by political scientists, sociologists, demographers or economists, are particularly essential and sensitive, precisely when they nuance or contradict political and ideological narratives. However, to what point are these questions only empirical? On the one hand, similar to any empirical science, experts might disagree with one other, leading to different policies and political stances on migration. It can simply be a matter of methodology: depending on the variable to which we give priority or salience, or on the definitions we use to categorize a sample, the conclusions of empirical studies vary. On the other hand, more importantly, there are normative issues which should be considered ahead. For instance, the point of view from which we measure the costs and benefits of migration, the weight given to the agency and the interests of actors at different levels (individual, national, regional) and the conceptual apparatus of migration such as movement, persecution, flight, violence or poverty, all these examples rest on diverging ontological, conceptual, linguistic and ethical standards that deeply affect the way we design experimental studies, hence policies. Moreover, public policies and political debates are filled with normative questions, that are usually subsumed under ideology or political attitudes: should we open borders (Carens 1987; Sager 2020)? What is the scope of our moral, legal and political responsibility towards migrants (Miller 2016)? What are their (formal and moral) rights against a state (Oberman 2016)? What do we owe refugees (Gibney 2015; Owen 2020)? Social sciences could surely provide elements to give an informed answer to these questions, but fully-fledged ethical and political considerations are required to assess the hierarchy and meaning of values at play, the coherence of the principles we might use, or to evaluate the (dis)continuity between these principles and institutional practices.

While moral philosophers and political theorists would easily agree with the importance of acknowledging and discussing these standards seriously, especially when it comes to migration, political scientists are more reluctant. To a certain extent, political science relies on the division between facts and values, empirical and normative methodologies, what is and what ought to be (Gerring and Yesnowitz 2006). The main task of political scientists is to uncover causal explanations of socio-political phenomena, based on concepts, indicators and measurements that are as neutral as possible. Political scientists have been asking a series of factual questions regarding migration, attempting to minimize ethical considerations and value judgements, and focusing on the causal relations between people moving and political institutions. While Hollifield and Wong (2015: 227) considered that "political scientists came late to the study of migration," they certainly caught up with a vast and multifaceted research agenda. We can cite for instance the early studies regarding the reasons why people move (Bhagwati 1984), the link between immigration and political attitudes (Davis and Deole 2017), the role of the state in constraining, encouraging or more generally controlling immigration (Joppke 1998; Zolberg 2006), the making of immigration policies (Freeman 2006; Boswell 2007; De Haas et al. 2018), including labour migration policies (Panizzon et al. 2015), the political responses to forced migration (Schuster 2000; Adhikari 2013), the causes of and policies relative to irregular migration (Düvell 2008; Betts 2013; Czaika and Hobolth 2016), transnational political engagements, democratisation and the diffusion of political attitudes of migrants towards the country they left (Pérez-Armendáriz and Crow 2009; Rüland et al. 2009).

At first sight, ethics does not seem to have much of a role in these causal and explanatory factors of migration. Some authors have even made a point of excluding all considerations other than material interests, i.e., costs and benefits (Freeman 2002; Messina 2007). The role of ethics, morality, values, or even more generally, ideas in explaining migration patterns and behaviours appears small or irrelevant.

2 Four Strategies to Bring Together Migration Ethics and Political Science

Various attempts to bridge the gap between political science and migration ethics have, however, been made and current research could lead to fruitful development in this direction. We can identify at least four strategies that have been adopted: drawing policy recommendations based on ethical standards from an empirical analysis; "fact-checking" migration ethics and "norm-checking" the political science of migration; considering ethics (moral values especially) as a causal determinant in policy making and studying values empirically.

Analysing labour migration restrictions from high-income countries, Martin Ruhs (2013) brings together a strictly empirical analysis with ethical standards that could determine future immigration policies regarding migrant workers' rights. He takes into consideration both the current political economy and the reasons and justifications that should guide it. He suggests a "two-dimensional matrix of 'ethical space'" (Ruhs 2013: 154), mapping the spectrum between consequence-based and rights-based policies, and from no recognition of moral standing for noncitizens to recognition of full moral standing. In other words, to understand how labour immigration restrictions function, and whether we should accept them, one needs to map the moral justifications involved, by analysing the promotion of economic welfare by comparison with guaranteeing individual rights, as well as the balance of interests between citizens and foreigners.

Arrighi and Bauböck (2017) address the impact of migration on the composition of the *demos*, therefore the relationship between migration, citizenship and voting rights. Their work clearly has a normative dimension (what is the right shape of the *demos* according to democratic values?) but is understood through its empirical manifestation (what does a postnational *demos* mean in practice and what are the political factors at play?). By comparing 53 countries and their distribution of voting rights locally, nationally and abroad, this confronts the normative theories of demos-shaping with empirical practices. But the goal is not to invalidate a theory based on empirical findings, rather the authors design a kind of reflective equilibrium between normative theory as a "critical benchmark" and empirical facts as a fine-grained reality check to a post hoc democratic theory, which, conversely, "provides normative support to the observed differentiation of the franchise". This appears as a truly interdisciplinary proposal, equally relevant both for political science and migration ethics.

Another major contribution is the inclusion of ethics as a factor in migration decision- and policy-making. Scholarships in the 1990s promoted an important research agenda on migration and citizenship, stressing the importance of rights (citizenship rights and human rights norms) in migration movements (Soysal 1994; Hollifield 1992, 2004). A focus on rights was essential in understanding how and why borders of liberal democracies were remaining more open than the policymakers wanted. The "control gap debate" showed how pre-commitments to the individual rights of migrants were constraining policies over border controls or exclusion. More recently, Bonjour (2011) takes the example of the Netherlands since the 1960s to show how moral norms, such as equal treatment and "immaterial values" (e.g. family unity) have guided migration policies more strongly than previously perceived. In summary, because the extension of rights is based on broad moral values, expressing some social consensus on the weight of equality, fairness or freedom, the ethical dimension of the material push and pull factors needs to be taken into consideration. Since rights are ethically charged and open to interpretations in court, ethics need to come into play when the political science of migration is assessed. Similarly, Boswell (2007) shows that the "ethical force" that guided the mobilisation of marginalized groups for obtaining the extension of the franchise from the state, also logically supported the recognition of migrants' rights. Hence, we must consider this ethical force when we study the state's response to the mobilisation of migrants and migrants' advocates or the enforcement (or lack thereof) of their rights. More generally, if the state remains the primary actor governing migration and border policies, its legitimacy in liberal democracies relies on ethical standards that should be factored into any explanatory work. As Boswell (2007: 90) writes: "even within domestic debates on distribution, there is a latent ethic of equal rights that is difficult to fully disregard in public discussions about immigration or refugee policy." Favell (2001: 9) made the point clearly showing that migration policies are always part of a broader "complex framework of ideas and justifications that sustains the policies", requesting an account of the normative dimension in empirical work, which he does by studying how "philosophies of integration"—from "amateur political theory" found in the grey literature to academic normative theorisation—irrigated migration and integration policies in France and the UK.

Finally, migration ethics has an important place within political science when it is studied empirically. The goal is not to ponder on what we should do, but to determine how people develop ethical beliefs and use them (first trend), or to analyse the variations of meaning given to values by those studying migration (second trend). Pichler (2011) for instance developed the first kind of scholarship on one particular value, cosmopolitanism, and its empirical counterpart, cosmopolitan attitudes, defined by Vertovec and Cohen (2002: 13) as "an intellectual and aesthetic stance of openness toward divergent cultural experiences". Gathering data from 49 countries, Pichler shows that socioeconomic properties, political orientations, place of residence or religious denomination strongly affect, although differently, political, ethical and identity-based identifications as "citizens of the world". Schlenker (2017) relates people's transnational status (dual citizen, foreign resident) to different types of liberal, communitarian and republican aspects of cosmopolitan attitudes (adherence to liberal norms, formation of post-national identities, or participation aimed at affecting international problems), in order to determine how practical transnationalism affects these attitudes. Bauder and Juffs (2019) represent the second trend: they analyse the diffusion and circulation of values among migration scholars, in this case solidarity, as a way of conceptualising empirically ethical standards. Contrary to the philosophical habit of inheriting concepts from a tradition, the goal here is to map a conceptual domain and clarify the variations of uses of an ethical concept that defines a social practice.

3 Bridging the Empirical-Normative Gap

The role of migration ethics appears more and more relevant but remains a fringe concern for empirical and explanatory-driven studies of migration. Yet, policies raise ethical concerns and rely on ethical choices, that cannot be entirely outsourced to moral philosophers. At a more general level, Barry (2002), Gerring and Yesnowitz (2006) and Abbott (2016) have emphasised the need to acknowledge the normative assumptions on which empirical studies rely. Conversely, empirical resources are essential for migration ethics in particular. Carens (2018: 14) shows for instance why the empirical work done on institutions and practices "performs [an] important unmasking function", which is "a crucial normative task", simply by pointing at gaps between policy goals, justifications and practices. Migration ethics scholars use empirical data and facts, although sometimes stylised, to frame their normative inquiry and better determine what can be expected from individuals and what is feasible at an institutional level. Some have also argued in favour of renewing ethical debates by grounding the conflicts of justifications regarding immigration and exclusion in non-ideal and non-institutional disputes, and by moving from "right-and-principle-based arguments" abstracted from concrete experiences to an approach based on "particular experiences of migrants with named social identities" (Reed-Sandoval 2016: 26; Fine 2018).

In other words, and following Abbott's terminology (2018), ethics and political science rely on a dualist ideal, where facts and values and normative and empirical studies are distinct, and necessitate two separate inquiries, while really being "monist", i.e., "not in practice isolated from political or normative action" (Abbott 2018: 159). This pragmatist reading requires open discussion of the normative background of supposedly scientific concepts, (e.g., inequality) and acknowledging the adaptive *raison d'être* of political science (Shapiro 2016), i.e., the assumption that social research aims at bettering society. This means gathering data and descriptions based on normative conceptions in order to react and adapt in the face of dysfunctional institutions or domination more generally, which is not, of course, designing blueprints for an ideal state of affairs. As Abbott (2018: 170) suggests: the "mainstream of the monist approach is not organized around a substantive political goal […]. Rather, at present it follows a meliorist trajectory, organized more around the removal of bad things rather than around the proposal or analysis of some new general good."

This kind of methodological eclecticism runs the risk of acritically incorporating results from another field of research. David Miller cites for instance Robert Putnam's work as evidence "that cultural heterogeneity does lead to lower degrees of trust between the culturally differentiated groups", justifying migration control on the part of the state to ensure unity, despite the controversial nature of the results among social scientists (Miller 2008; Putnam 2007; Morales 2013). Designing a truly monist research protocol will be essential for future research in migration ethics, both to avoid the risk of activist science—contrasted with normative assumptions—or cherry-picking normative theory, that is, selecting empirical findings for the sole purpose of confirming or justifying an argument.

There are some suggestions in that direction such as Longo and Zacka (2019), who call for an ethnographically informed theory, notably giving the example of border studies. The goal is not simply to "test the results of our inquiries but to broaden our horizons as to which questions are worth asking" (Longo and Zacka 2019: 1070). Gerver (2018) for instance mixes field work and qualitative interviews with a moral and political philosophy to evaluate morally and practically the policies of refugee repatriation, taking into account the practical experiences and attitudes of forced migrants. Reflecting on the roles of cities in migration, especially the local rights of migrants and the urban imperative of integration, de-Shalit (2018) also embraces a mixed-method approach to migration ethics, paying attention to the use and public construction of arguments and concepts. Interviewees give him an array of possible opinions—what we could call normative repertoires, closed to the "cities of justification" of Boltanski and Thévenot (2006)—his task being to assess these, interpret them in terms of ideal-typical positions based on a set of principles and evaluate their normative coherence and appeal. The work of Taylor (2020; Taylor and Lefebvre 2020) on the concept of hospitality and how it is used and justified by humanitarian actors at the border between France and Italy follow the same trend of what Cabrera (2009) called "qualitative political theory". A robust methodological programme has recently been provided by Ackerly et al. (2021) to identify how normative theorists have been working with ethnographic material, especially in border studies, and to uncover the epistemic benefits and responsibilities that come with empirically informed theory.

On a more general level, Bader (2012) has shown that a general methodology to truly reunite empirical and normative inquiries is essential for research on migration, arguing that all "*policy options of incorporation* [...] are confronted with a veritable mine field of policy dilemmas resulting from the interplay of moral, ethico-political and realist dilemmas and conundrums" (Bader 2012: 1169). Since migration necessarily relates to practical knowledge and complex practical judgements, no moral philosophy can be self-sufficient to address it (Bader 2008). As we have seen, moral concepts, when translated into rights, are open to interpretation and specific applications, requiring socio-legal approaches to interpret their meanings. Contextualisation and interdisciplinarity—which are not division of labour—are the only possible means of making a significant contribution about migration, citizenship or border control, i.e., a contribution that has a reasonable chance of fostering a policy change in a problematic situation.

I have argued for a reconciliation of empirical and normative studies, trying to sidestep the methodological disputes by defining political research through a series of problems and functions (Boudou 2016): empirical and normative inquiries share a heuristic function (to make sense of a problem), a pedagogical function (to situate one's work in the history of the discipline), a critical function (to suggest ways to solve the problem in question) and an ethical function (to articulate institutions and individual aspirations to change, including reflectively, the researcher). Bauböck (2008) has suggested various practical ways of combining normative theory and empirical research ("the empirical study of normative attitudes and beliefs", "studying institutionally embedded norms", "qualitative case studies" and "quantitative comparative studies") and advocated exposing normative theory "to the full force of critique from explanatory theory and empirically grounded research that analyse the application context for normative ideas" (Bauböck 2008: 59). He adds that this factual critique can be done by theorists, in a similar fashion as political scientists, who should uncover their normative assumptions.

4 Conclusion

The epistemological and methodological commitments of political science seem to prevent it from including ethical considerations and engaging fully with normative debates. The paucity of literature indicates that the fact-value distinction is still prevalent. I have argued, however, that a pragmatist approach, bringing together disciplines around a problematic subject such as migration, will more aptly capture the intertwined nature of its empirical and normative components. I introduced various works to bridge this gap, either by bringing political science within migration ethics, or the other way around. Rather than relying on a binary opposition, I showed that mixed-methods that both acknowledge the normative assumptions of data-gatherings and the need for empirical grounds for normative theorising, are the best ways of doing justice to the multifaceted nature of migration issues.

Literature

Abbott, A. 2016. The Future of the Social Sciences. Between Empiricism and Normativity. *Annales. Histoire, Sciences Sociales* 71:343–360.

Abbott, A. 2018. Varieties of Normative Inquiry. Moral Alternatives to Politicization in Sociology. *The American Sociologist* 49:158–180.

Ackerly, B., L. Cabrera, F. Forman, G. Fuji Johnson, C. Tenove, and A. Wiener. 2021. Unearthing Grounded Normative Theory. Practices and Commitments of Empirical Research in Political Theory. *Critical Review of International Social and Political Philosophy.* https://doi.org/10.1080/13698230.2021.1894020.

Adhikari, P. 2013. Conflict-Induced Displacement, Understanding the Causes of Flight. *American Journal of Political Science* 57:82–89.

Arrighi, J.-T., and R. Bauböck. 2017. A Multilevel Puzzle. Migrants' Voting Rights in National and Local Elections. *European Journal of Political Research* 56:619–639.

Bader, V. 2008. Global Justice in Complex Moral Worlds. Dilemmas of Contextualized Theories. *Critical Review of International Social and Political Philosophy* 11:539–552.

Bader, V. 2012. Moral, Ethical, and Realist Dilemmas of Transnational Governance of Migration. *American Behavioral Scientist* 56:1165–1182.

Barry, B. 2002. Why Political Science Needs Political Theory. *Scandinavian Political Studies* 25:107–115.

Bauböck, R. 2008. Normative Political Theory and Empirical Research. In *Approaches and Methodologies in the Social Sciences. A Pluralist Perspective*, ed. D. Della Porta and M. Keating, 40–60. Cambridge: Cambridge University Press.

Bauder, H., and L. Juffs. 2019. 'Solidarity' in the Migration and Refugee Literature. Analysis of a Concept. *Journal of Ethnic and Migration Studies* 46:46–65.

Betts, A. 2013. *Survival Migration. Failed Governance and the Crisis of Displacement.* Ithaca, London: Cornell University Press.

Bhagwati, J. 1984. Incentives and Disincentives. International Migration. *Weltwirtschaftliches Archiv* 120:678–701.

Boltanski, L., and L. Thévenot. 2006. *On Justification. Economies of Worth.* Princeton: Princeton University Press.

Bonjour, S. 2011. The Power and Morals of Policy Makers. Reassessing the Control Gap Debate. *International Migration Review* 45:89–122.

Boswell, C. 2007. Theorizing Migration Policy. Is There a Third Way? *International Migration Review* 41:75–100.

Boswell, C., A. Geddes, and P. Scholten. 2011. The Role of Narratives in Migration Policy-Making. A Research Framework. *The British Journal of Politics and International Relations* 13:1–11.

Boudou, B. 2016. What Is Political Theory for? *Raisons politiques* 64:7–27.

Cabrera, L. 2009. An Archaeology of Borders. Qualitative Political Theory as a Tool in Addressing Moral Distance. *Journal of Global Ethics* 5:109–123.

Carens, J. 1987. Aliens and Citizens. The Case for Open Borders. *The Review of Politics* 49:251–273.

Carens, J. 2018. On the Relationship between Normative Claims and Empirical Realities in Immigration. In *Proceedings of the 2018 ZiF Workshop. Studying Migration Policies at the Interface between Empirical Research and Normative Analysis*, ed. M. Hoesch and L. Laube, 13–29. Münster: ULB Münster.

Czaika, M., and M. Hobolth. 2016. Do Restrictive Asylum and Visa Policies Increase Irregular Migration into Europe? *European Union Politics* 17:345–365.

Davis, D., and S. Deole. 2017. Immigration and the Rise of Far-Right Parties in Europe. *ifo DICE Report* 15:10–15.

De Haas, H., K. Natter, and S. Vezzoli. 2018. Growing Restrictiveness or Changing Selection? The Nature and Evolution of Migration Policies. *International Migration Review* 52:324–367.

de-Shalit, A. 2018. *Cities and Immigration. Political and Moral Dilemmas in the New Era of Migration*. Oxford: Oxford University Press.

Düvell, F. 2008. Clandestine Migration in Europe. *Social Science Information* 47:479–497.

Favell, A. 2001. *Philosophies of Integration. Immigration and the Idea of Citizenship in France and Britain*. London: Palgrave Macmillan.

Fine, S. 2018. Refugees, Safety, and a Decent Human Life. *Proceedings of the Aristotelian Society* 119:25–52.

Freeman, G. P. 2002. Winners and Losers. Politics and the Costs and Benefits of Migration. In *West European Immigration and Immigrant Policy in the New Century*, ed. A. Messina, 77–95. Westport, CT: Praeger.

Freeman, G. P. 2006. National Models, Policy Types and the Politics of Immigration in Liberal Democracies. *West European Politics* 29:227–247.

Freeman, G., R. Hansen, and D. Leal. 2013. *Immigration and Public Opinion in Liberal Democracies*. New York: Routledge.

Gerring, J., and J. Yesnowitz. 2006. A Normative Turn in Political Science? *Polity* 38:101–133.

Gerver, M. 2018. *The Ethics and Practice of Refugee Repatriation*. Edinburgh: Edinburgh University Press.

Gibney, M. 2015. Refugees and Justice between States. *European Journal of Political Theory* 14:448–463.

Hollifield, J. F. 1992. *Immigrants, Markets and States. The Political Economy of Postwar Europe*. Cambridge, MA: Harvard University Press.

Hollifield, J. F. 2004. The Emerging Migration State. *International Migration Review* 38:885–912.

Hollifield, J. F., and T. Wong. 2015. The Politics of International Migration. How Can We 'Bring the State Back in'? In *Migration Theory. Talking across Disciplines*, ed. C. Brettell and J. F. Hollifield, 227–288. Abingdon: Routledge.

IEPMA. 2018. *Call for an International Panel on Migration and Asylum*. https://giema.eu/posts/call-for-iepam. Last checked 4 Mar 2020.

Joppke, C. 1998. Why Liberal States Accept Unwanted Immigration. *World Politics* 50:266–293.

Longo, M., and B. Zacka. 2019. Political Theory in an Ethnographic Key. *American Political Science Review* 113:1066–1070.

Messina, A. 2007. *The Logics and Politics of Post-WWII migration to Western Europe*. Cambridge: Cambridge University Press.

Miller, D. 2008. Immigrants, Nations, and Citizenship. *The Journal of Political Philosophy* 16:371–390.

Miller, D. 2016. *Strangers in Our Midst. The Political Philosophy of Immigration*. Cambridge, MA: Harvard University Press.
Morales, L. 2013. Assessing the Effects of Immigration and Diversity in Europe. *Journal of Elections, Public Opinion and Parties* 23:241–254.
Oberman, K. 2016. Immigration as a Human Right. In *Migration in Political Theory. The Ethics of Movement and Membership*, ed. S. Fine and L. Ypi, 32–56. Oxford: Oxford University Press.
Owen, D. 2020. *What Do We Owe to Refugees?* Cambridge: Polity Press.
Panizzon, M., G. Zurcher, E. Fornalé, and G. Zürcher, eds. 2015. *Palgrave Handbook of International Labour Migration*. London: Palgrave Macmillan.
Pérez-Armendáriz, C., and D. Crow. 2009. Do Migrants Remit Democracy? International Migration, Political Beliefs, and Behavior in Mexico. *Comparative Political Studies* 43:119–148.
Pichler, F. 2011. Cosmopolitanism in a Global Perspective: An International Comparison of Open-Minded Orientations and Identity in Relation to Globalization. *International Sociology* 27:21–50.
Putnam, R. 2007. *E pluribus unum*. Diversity and Community in the Twenty-first Century. *Scandinavian Political Studies* 30:137–174.
Reed-Sandoval, A. 2016. The New Open Borders Debate. In *The Ethics and Politics of Immigration. Core Issues and Emerging Trends*, ed. A. Sager, 13–28. London, New York: Rowman & Littlefield.
Ruhs, M. 2013. *The Price of Rights. Regulating International Labor Migration*. Princeton: Princeton University Press.
Rüland, J., C. Kessler, and S. Rother. 2009. Democratisation Through International Migration? Explorative Thoughts on a Novel Research Agenda. *European Journal of East Asian Studies* 8:161–179.
Sager, A. 2020. *Against Borders. Why the World Needs Free Movement of People*. London: Rowman & Littlefield.
Schlenker, A. 2017. Transnational Status and Cosmopolitanism. Are Dual Citizens and Foreign Residents Cosmopolitan Vanguards? *Global Networks* 17:321–348.
Schuster, L. 2000. A Comparative Analysis of the Asylum Policy of Seven European Governments. *Journal of Refugee Studies* 13:118–132.
Shapiro, I. 2016. *Politics Against Domination*. Cambridge, MA: Harvard University Press.
Soysal, Y. N. 1994. *Limits of Citizenship. Migrants and Postnational Membership in Europe*. Chicago: University of Chicago Press.
Taylor, A. 2020. Domopolitics, Citizenship and Dissent: An Analysis of 'Crimes of Solidarity' and Hospitality in Contemporary France. *International Journal of Politics, Culture, and Society* 33:495–512.
Taylor, A., and A. Lefebvre. 2020. Three Reasons for Hospitality. Care for Others, Care for the World, and Care of the Self. *Perspectives on Politics* 20:139–153.
Van der Brug, W., G. D'Amato, D. Ruedin, and J. Berkhout. 2015. *The Politicisation of Migration*. London: Routledge.
Vertovec, S., and R. Cohen. 2002. Introduction: Conceiving Cosmopolitanism. In *Conceiving Cosmopolitanism. Theory, Context, and Practice*, ed. S. Vertovec and R. Cohen, 1–24. Oxford: Oxford University Press.
Zolberg, A. 2006. *A Nation by Design. Immigration Policy in the Fashioning of America*. Cambridge, MA: Harvard University Press.

Sociology and Migration Ethics

Anja Weiß

Abstract Sociological positions draw on ethical debates in other disciplines and the public sphere, but they import their arguments, rather than translating them into a specifically sociological agenda. It is therefore difficult to find sociological research that debates normative concerns. Nonetheless, migration research often follows normative agendas at least implicitly. The chapter outlines four research approaches in migration and refugee studies with a focus on their normative assumptions. Scholars who work (a) within rational choice theory or who (b) study the functioning of the state define sociology in a manner that avoids normative agendas. Both approaches base research questions within the context of the state. The (c) transnational approach to migration studies criticises this as methodological nationalism. Critical migration scholars (d), on the other hand, do not shy away from normative agendas. Rather, they start their research by criticising generally held assumptions and use ethical positions as yardsticks for assessing the value of specific research agendas. Options for a future collaboration between migration ethics and sociology are thus found on the fringes of migration sociology. The last part of the chapter considers how specific sociological research agendas connect at least indirectly with ethical debate.

Scholarship on migration and migrants' integration is interdisciplinary, but sociology often takes a lead role. Sociologists provide much of the applied research on migration that policymakers sponsor in search of guidance and legitimacy. Sketching a long-established division of labour, one might think that policymakers turn to political science and philosophy concerning the goals of policymaking. However, when politicians contemplate the feasibility and likely outcomes of policies, the insights of economy and sociology are in high demand as they appear to offer "hard facts" on migration and the efficiency of particular policies (Baldwin-Edwards et al. 2019).

This sketch of applied migration research may in good part be polemic. It does, however, resonate with the strengths that sociologists themselves would attribute to the core of their discipline. Sociologists often deal with politically contentious

A. Weiß (✉)
Department of Sociology, University of Duisburg-Essen, Duisburg, Germany
e-mail: anja.weiss@uni-due.de

A. Niederberger et al. (eds.), *Handbook of Migration Ethics*,
https://doi.org/10.1007/978-3-031-89877-8_34

social issues such as inequality, education, labour market and migration. In light of this, many sociologists tend to emphasise that their research is empirical and descriptive. Mainstream sociological theorising works towards a perspective on social phenomena that differs from everyday notions of the social by being analytical rather than by taking a political stance.

It is therefore difficult to find sociological research that debates normative concerns (Niederberger 2021: 100). Of course, parts of sociology contest the politically abstinent mainstream by taking a critical stance, and this is particularly true in the field of migration sociology. But taking a stance is not equivalent to *debating* ethical and normative challenges *explicitly* in *sociological* discourse. With sociology describing and analysing society "as it is now" and migration ethics considering the "ought", the two discourses are situated at opposite ends of migration scholarship.

Studying existing sociological literature (Iskander 2018), a particular division of labour is prevalent. Some established scholars among professional sociologists "dare" to voice ethical concerns, but they do so in a manner that distinguishes between the core of their work and the essays that they write as intellectuals [cf., for instance, Zolberg (2012) or Bauman (2016)]. Critical sociologists align with particular philosophical positions, but they do not contribute to the normative debate about migration ethics.

The reasons for this lack of ethical debate will become more apparent in the first part of this chapter, which explains variations in the way sociologists in general deal with the relationship between research and political agenda setting. Against this background, relevant research approaches in migration and refugee studies are introduced in the second part, with a focus on their often implicit normative assumptions and implications. Options for a future collaboration between migration ethics and sociology are thus found on the fringes of migration sociology. The last part of the chapter considers how sociology could contribute at least indirectly to ethical debate.

1 Sociologists' General Avoidance or Adoption of a Particular Political Stance

As a sociologist advocating dialogue between sociology and multiple publics and on his way to being elected as president of the International Sociological Association, Michael Burawoy (2005) recognised the value of all sociological approaches. He also distinguished between four kinds of sociology, which he labels professional, critical, policy-oriented, and public sociology. About professional sociology he says that it "supplies true and tested methods, accumulated bodies of knowledge, orienting questions, and conceptual frameworks [...]—providing both legitimacy and expertise for policy and public sociology" (2005: 10). Burawoy finds that this kind of sociology has become hegemonic in the Global North by offering instrumental knowledge to academic audiences.

Many sociologists will agree. Even though sociology is multi-paradigmatic and cleavages between paradigms are profound, especially the sociologists labelled as "professional" by Burawoy will assert that their research programmes do not respond to demand by external publics but are based on theoretical and empirical puzzles evolving within their academic discipline. As "the social" is everywhere and as people use everyday notions of the social all the time, it is vital for sociology as a scientific discipline to go beyond everyday knowledge and to offer theoretical perspectives that differ from everyday theorising (Bourdieu et al. 1991; Weber 1946). Responding to internal debates within their discipline is seen as the best and most professional way to do sociology. According to this reading, research should be guided by empirical findings, improved concepts and theoretical models rather than everyday notions or political bias. Claims of this kind are important for sociology's legitimacy as a discipline as well as for acquiring funding in policy-oriented research.

While this set of arguments is generally agreed on—even by public sociologists such as Burawoy—the particular relationship between research questions and political and ethical agendas varies between schools of sociology, and it is continually debated even within professional sociology itself. Many sociologists firmly believe that they can and should avoid political debate, but they will accept that they are also part of a social world which may impact on their research agenda. However, this insight has little impact on the way that these sociologists produce social science and on the manner in which they present their findings. In contrast, authors such as Bourdieu et al. (1991) posit that a second epistemological break must follow upon the first break with everyday notions. Sociologists are indebted materially and intellectually to schools of thought, and they are positioned unequally in societal hierarchies. Based on insights from the sociology of knowledge, on the workings of the scientific field and on his theory of class and symbolic violence, Bourdieu and others argue for methods that enable reflexivity within the discipline, but they assert that these methods must be sociological rather than political (Bohnsack 2017; Mannheim 1952).

This is different for critical sociology and the kind of scholar activism found more often in the margins of the (global) sociological field. Critical sociology, in Burawoy's words, examines "the foundations—both the explicit and the implicit, both normative and descriptive—of the research programs of professional sociology" (2005: 10). Concerning political agendas, critical sociologists take a stance, criticising domination and professional sociology's complicity in hegemonic social relations. Based on continuous reflection of perspectives, critical sociologists offer detailed and enlightening deconstructions of everyday and social scientific perspectives on social phenomena. Nevertheless, their arguments aim to address empirical phenomena, and thus they differ from philosophers who feel compelled to explicate the premises of their position, who consider alternative ethical arguments and justify their position in relation to them.

In sum, sociological positions draw on ethical debates in other disciplines and the public sphere, but they import their arguments, rather than translating them into a specifically sociological agenda. "Professional" sociologists feel that political agendas must and can be contained through the use of scientific methods. Critical sociologists, in contrast, argue that the political implications of research call for in-depth reflection and that sociology should follow a liberating and emancipatory path.

2 Hidden and Declared Normative Agendas in the Sociology of Migration

The field of migration studies and the sociology of migration are highly productive and growing, drawing on resources from other disciplines and public debate. Therefore, this section cannot offer a comprehensive review. Rather, I will group research approaches in a way that clarifies their theoretical premises with a focus on—often implicit—normative implications.

Accounting for theoretical premises is relatively easy as sociologists will usually situate themselves within a well-explained conceptual universe, such as rational action or postcolonial theory. As discussed in part one of this chapter, the normative implications that are part and parcel of social theory tend to be treated with more ambivalence. Thus, with respect to normative implications, the chapter must be seen as a first step towards clarification that might not be consensual.

2.1 Rational Choice Theory and Migrants' Access to Labour Markets

In the "professional" sociology of migration and integration, the integration and assimilation paradigm has achieved an almost hegemonic position. The normative position of this approach can be exemplified by the arguments of a leading scholar, Hartmut Esser. Esser has contributed foundational work to sociological rational choice theory and studied the integration of immigrants.

In his publications on migrants' integration, Esser (2003, 2007: 336) avoids the "melting pot" idealism prevalent in earlier teleologies of migrants' assimilation (Taft 1953) with the help of utilitarian reasoning and empirical observations. Since migrants are seen as rational actors maximising utilities, they are expected to strive towards higher income and higher social status. Thus, "structural assimilation" will likely be the most important dimension of assimilation. Yet, in another combination of rational choice and empirically descriptive arguments, Esser also shows that cultural assimilation, such as learning a new language, will usually be necessary to achieve structural assimilation (Esser 2003, 2007: 336). Also, interethnic friendships and identification with the country of arrival will often be a result of structural assimilation. In consequence, Esser agrees with the teleologies introduced by the Chicago School's race relations cycle, according to which migrants will eventually assimilate in all respects, but his explanation is deduced from rational choice theory rather than agendas more open to contestation.

Within this general argument, paradox relations between various utilities are considered. For example, some immigrants might be better off if they invest in an ethnic economy rather than structural assimilation, at least in the short term (Esser 2007; Portes and Sensenbrenner 1993; Wiley 1967). Also, the choices of individuals impact on the integration of (nation-state) societies. According to Esser (2007: 329), a structurally integrated society will often be the long-term outcome of immigration. However, Esser also considers societies in which ethnic differentiation and inequality coincide empirically, and he can explain ethnically segmented societies through methodologically individualist utilitarian logics. In a similar vein, classical economists and proponents of the human capital approach detail conditions under which split labour markets (Bonacich 1972; Piore 1979) are functional for employers, even though they expect that "taste[s] for discrimination" (Becker 1973 [1957]: 14) will be too costly in the long run.

In sum, proponents of rational choice theory in migration scholarship need not discuss matters of equality or human rights as they focus on rational action and the maximisation of utilities both for individuals and societies/markets. Their argument implies that free markets and access to institutions based on meritocracy (equal opportunities) will maximise productivity in a market or society and offer the highest utility to actors. That these premises are clear and that the goal of maximising utilities is beyond doubt leaves little room for ethical debates. This assertion is, of course, highly contested. Critical sociologists see meritocratic assumptions as ideological, and scholars of transnational migration are wary of studying migrants' integration solely in the context of one nation-state.

2.2 Functionalism and Migrants' Challenge to the National Welfare State

A closely related school of thought uses functionalist arguments to circumvent political debates. Sociologists concerned with the viability of democratic decision-making and the welfare state reference political philosophers of diverse provenance as support for the argument that legitimate (democratic) orders must constitute a lasting and reliable relationship with their citizens. In order to enable public deliberation and to provide all citizens with basic provisions that enable them to partake in collective decision-making, constituents of a state will need to close their ranks to some degree (Brubaker 1992).

Michael Bommes (1999, 2000) is a typical proponent of this position. He uses a radically constructivist Luhmannian systems theory in which functionally differentiated subsystems of society include aspects of persons depending on the content-differentiated code of the respective system. This means that even undocumented migrants are partially included by social systems, such as the economy, religion and family, just like anyone else. Within this universalist and constructivist theory, the political system constitutes an anomaly because it is territorially segmented into nation-states and it in- and excludes whole persons rather than aspects of persons (Bommes 1999, 2000). Loosely referencing Marshall's (1950) argument for social citizenship rights, functionalists like Bommes argue that starving and illiterate citizens cannot give legitimacy to a state claiming to represent a nation. Therefore, democratic national welfare states will eventually develop institutions in support of needy citizens so that they can really partake in collective decision-making. As literacy and economic independence are long-term achievements, the national welfare state must constitute a lasting relationship between citizens and state. Nationalism supports this relationship by ensuring loyalty in times of war and by providing a legitimating semantic for the state's monopoly of power. In sum, the combined provision of basic welfare and symbolic resources such as nationalism enable states to ensure full citizenship in a functionally superior manner. That this system produces the partial exclusion of migrants from national welfare states and a threshold of inequality at their borders is seen as a side effect that cannot be avoided.

This general functionalist argument does not say much about the more specific ways in which migrants are in- and excluded in the national welfare state. Functionalist arguments give migration scholars good reason to see migrants as fundamentally different from citizens, but they allow for a wide range of empirical studies that see specific migrations as challenging or supportive of the national welfare state. Depending on historical situation and normative inclinations, some migration scholars find that migration remedies a state's problems such as labour shortages or demographic ageing (Castles et al. 2020). Others show the opposite, namely that migration feeds right-wing populism, or that migrants undermine public institutions for want of cultural assimilation, language acquisition or simple poverty. Thus, the general idea that world order depends on nation-states, that democratic national welfare states are an important achievement and that inequality and cultural difference challenge nation-state institutions enables very different political assessments.

One way in which the abstract functional argument is supplemented are distinctions between necessary, needed and controlled immigration on the one hand, and problematic, culturally distant or irregular migration on the other. Even though professional sociologists prefer to use categories that can describe migrations rather than participate in their evaluation, they can hardly avoid normative implications since debates of this kind are closely related to political contentions. For example, research typically uses the categories or visa classifications states provide, rather than creating social scientific concepts. Also, by focusing on specific subgroups of migrants, migration scholars tacitly position themselves politically. Important research agendas focus on the resources and contributions of migrants and on the many cases where migrants do not only pose no problems but are either contributing or not remarkable at all. This is the case for most research on high-skilled and labour migration, on the immigration of co-ethnics, EU and leisure migration (Recchi and Favell 2009).

A research interest that is very loosely related to functionalism's justification of the nation-state system empirically studies contradictions between declared policy goals or normative commitments by states—such as the Geneva Convention relating to the Status of Refugees—and (non-)compliance or unintended side effects of specific programmes. This research responds to Hollifield's (2000) diagnosis of a liberal paradox. According to Hollifield, liberal states adhering to the rule of law will be unable to close their borders effectively. They cannot violate the human rights of (potential) immigrants or part of their constituencies will form pro-immigration coalitions. Castles (2007) discusses further reasons why migration policy rarely achieves its goals. He emphasises that the term policy failure "is not used here as a normative term" (2007: 31). Instead, he wishes to "deconstruct official goals and look for hidden agendas" (2007: 31).

In sum, a functionalist approach to the legitimacy of the national welfare state enables migration researchers to do empirical studies on state policies without openly adopting a particular political agenda. In using the policies and conventions that polities have agreed on as starting points of research, sociologists evaluate what states do empirically, and they relate their findings to declared claims of polities. That states often follow exclusionary policies and that they treat migrants with discretion are points that may be frowned upon but are not contested as long as states do not openly violate international law.

As functionalists are concerned with the ability of states to fully include citizens and provide them with the ability to act as citizens, sociologists in this research strand often connect conflicts about migration and integration with other social conflicts. By describing inequalities, closure, exploitation and discrimination, they imply that these are undesirable or problematic phenomena. Mostly, they do what Bourdieu's first epistemological break suggests: rather than following the self-description of actors, they attempt to understand social processes in a more comprehensive manner. And while migration scholars following rational choice theory tend to criticise migrants for making the wrong choices, sociologists focusing on legitimate governance and collective decision-making are more concerned with the functioning of state institutions.

2.3 *Transnational Migration Studies*

Rational action and functionalist arguments both focus on the state as a country of arrival. This is criticised by scholars of transnational migration. They concede that the state remains important for migration studies. However, they posit that the state should not be seen as a "container space" (Pries 2005) with clear-cut borders but rather as a social and potentially extraterritorial entity. Migration is seen as a pluri-local phenomenon spanning several states (Pries 2005). Migrants should be studied "beyond the ethnic lens" (Glick Schiller et al. 2006), without groupism (Brubaker 2004) and avoiding methodological nationalism (Wimmer and Glick Schiller 2003). Also, this scholarship differs from the aforementioned approaches by focusing on meso-level networks and organisations and by turning its attention towards migrants' bottom-up self-organisation and pluri-local positionings.

Scholars of transnational migration do not see themselves as "critical" in the narrow sense of the word, but rather as correcting bias induced by methodological nationalism. For example, they show that legal categories may not be helpful for empirical studies of migration (Crawley and Skleparis 2018). A migrant's visa status or nationality may have less explanatory power than her education, her transnational social embeddedness or her migration motivation. At the same time, the ways in which the state continues to matter even in transnational migration scholarship are not well understood, as the term "transnational" shows. Faist (2015) has argued that "trans-state" would be a better term, especially since transnational communities are often connected by their nationality, but his argument did not prevail.

2.4 *Critical and Reflexive Migration Studies*

Critical migration scholars agree with transnationalists' scepticism of nation-state categories and share their interest in migrants' self-organisation. However, they differ from transnationalists by putting Bourdieu's second epistemological break centre stage. Critical migration scholars start their research by reflecting on their social and epistemological position. They then go on to deconstruct common sense knowledge as hegemonic. In contrast to the other schools, which see the exclusion of immigrants as legitimate or unavoidable, critical migration scholars study the ways in which nation-states and their border and citizenship regimes create the category of migrant, i.e. they focus on processes of bordering, classification and boundary making (Scheel and Squire 2014). In their reading, migrants themselves may take an active—but not complicit—role in processes of boundary making (Tsianos and Karakayali 2010). Critical sociologists also remind us that nation-states were established through violent civil wars and that they continue to uphold relations of colonial, racist and antisemitic exploitation and ethnic cleansing (Boatcă 2015; De Genova 2016).

Both transnational and critical sociologists reflect on the positionality of researchers and research participants in the course of research (Nowicka and Cieslik 2013; Shinozaki 2012), but critical migration scholars do this in order to understand the ways in which research stabilises or challenges existing relations of power. They give voice to marginalised groups as one way in which a tacit alliance between social research and the nation-state hegemon can be challenged.

3 Contributions of Sociology to Philosophical Debates on Migration Ethics

Portraying sociological migration research in a manner that highlights its normative implications is a difficult endeavour as normativity is seen as lack of professionalism by well-established members of the sociological profession. Critical sociologists, on the other hand, are not always prepared to argue about the "ought" that they have accepted as true. Searching for normative implications under these conditions is bound to produce misperceptions. To clarify those could be a first step towards an exchange of ideas between migration sociology and migration ethics.

Yet, it might also be a step towards expanding connections that already exist. For example, many professional migration sociologists see themselves as contributing empirical "facts" to policymakers. This may also be attractive for philosophers with an interest in non-ideal positions (Sen 2006). This concerns empirical studies on justice orientations which give context and depth to theoretical debates that merely presume that immigration or diversity challenge the nation-state order. For example, van Oorschot and Uunk (2007) find that higher levels of immigrants increase the public's informal solidarity with immigrants. Burgoon (2014) shows that this kind of solidarity may be conditional and context-specific. On the level of individual attitudes toward taxation or redistribution, the "factual" degree of immigration-related diversity matters less than subjective factors and political mobilisation (Breznau and Eger 2016; Hjerm and Schnabel 2012). On this basis, thought experiments juxtaposing immigration and the welfare state may be criticised for being unrealistic.

As mentioned above, empirical research often finds gaps between policy goals and their implementation (Cornelius et al. 2004). If national and international migration policies create regimes that differ significantly from policy goals, ethical debate is challenged on a very fundamental level. Take the refugee protection regime as an example: politically and legally, it is of utmost importance to distinguish between migrants and refugees (Feller 2005) because the normative stance of people who are forced to migrate and who cannot depend on the protection of "their" state differs greatly from those who choose to go abroad. Sociological research, however, shows that it is next to impossible to distinguish between non- and semi-voluntary migrations in an empirically sound manner (Van Hear 2012). This does not contradict the normative desire to distinguish between justified and fraudulent claims for asylum. It does, however, confirm the suspicion of many sociologists that everyday notions must be replaced by theoretical concepts that can be irritated by unexpected empirical findings. In the case of the "refugee" category, we are dealing with a legal category that is helpful for those classified as refugees, but which excludes many people with similar experiences (Weiß 2018a). The latter's classification as "voluntary" migrants delegitimises their aspirations discursively and through policies targeting "illegal" migration (Erdal and Oeppen 2018).

The example of the refugee category shows that the discipline of sociology can actually deliver what is prominent in its self-description: sociologists work towards an analytical clarification of concepts that are empirically grounded (Blumer 1954). The distinction between refugees and migrants may leave little to be desired in terms of an abstract or a legally applicable classification. Yet, empirical research shows how counterfactual it really is. Sociologists thus shed light on the historical and social processes in which the categories emerge that are then taken for granted in ethical debates (Karakayali 2008; Scheel and Gutekunst 2019).

In sum, sociology is quite a heterogenous discipline in which ethical debates are confused with taking a political stance and thus with a lack of professionalism that must be frowned upon. Texts in which important scholars explain the premises of their position can be found, but we can see a notable lack of debate between positions. Scholars who draw on rational choice or functionalist explanations often portray these as non-political. Transnational migration scholarship does not engage sufficiently with the nation-state and groupism. Even critical sociologists, who start from ethical and epistemological arguments, tend to state what they see as right rather than starting a debate.

For all these reasons, the contribution of sociologists to philosophical debates on migration ethics is limited. Empirical studies of justice orientations can, however, clarify that right-wing populist mobilisation may threaten the stability of societies more than the factual level of migration-related diversity. Empirical studies of border regimes can show that their goals create unintended side effects that non-ideal philosophers would want to take into account when searching for an ethical position.

A recent development in sociological migration studies may open up new options for a collaboration between sociology and philosophy. In the founding years of sociology, classics such as Simmel or Durkheim established that sociology should be a nomothetic science in search of general concepts and "universal" causal explanations. Migration studies has always constituted a deviation from this general orientation as it focuses on a topic that is seen as an exception to the rule. If migration is a "problematic" anomaly that interrupts the "normal" relation between state and citizen, it cannot say much about society in general.

This exceptionalism in migration studies has recently drawn much criticism within the field and several leading scholars plead for a de-centring of migration (Dahinden 2016; Nieswand 2016). De-centring could be achieved by the use of more general social concepts such as mobility, capabilities, order or citizenship. These concepts must be reworked so that they take account of the fact that many people are not citizens of an idealised national welfare state. For example, inequalities should not only be studied within and between nation-states but also for non- and transnational context relations (Weiß 2017, 2021). Migration can be studied within the mobility turn (Manderscheid 2010) and as enhancing socio-spatial autonomy (Weiß 2018b). In abstract terms, this means that a marginal and exceptional topic informs and changes core concepts in the discipline. If sociologists succeed in de-centring migration, they could set an example for philosophers who urgently need to revisit what an international order of non-ideal nation-states really means.

Literature

Baldwin-Edwards, M., et al. 2019. The Politics of Evidence-Based Policy in Europe's 'Migration Crisis'. *Journal of Ethnic and Migration Studies* 45:2139–2155. https://doi.org/10.1080/1369183X.2018.1468307.

Bauman, Z. 2016. *Strangers at Our Door*. Malden, MA: Polity Press.

Becker, G. S. 1973 [1957]. *The Economics of Discrimination*. Chicago: University of Chicago Press.

Blumer, H. 1954. What Is Wrong with Social Theory? *American Sociological Review* 19:3–10. http://www.jstor.org/stable/2088165.

Boatcă, M. 2015. *Global Inequalities Beyond Occidentalism*. Aldershot: Ashgate.

Bohnsack, R. 2017. *Praxeologische Wissenssoziologie*. Opladen, Toronto: Barbara Budrich.

Bommes, M. 1999. *Migration und nationaler Wohlfahrtsstaat. Ein differenzierungstheoretischer Entwurf*. Opladen: Westdeutscher.

Bommes, M. 2000. National Welfare State, Biography and Migration. Labour Migrants, Ethnic Germans and the Re-ascription of Welfare State Membership. In *Immigration and Welfare. Challenging the Borders of the Welfare State*, ed. M. Bommes and A. Geddes, 90–108. London: Routledge.

Bonacich, E. 1972. A Theory of Ethnic Antagonism. The Split Labor Market. *American Sociological Review* 37:547–559.

Bourdieu, P., et al. 1991. *The Craft of Sociology. Epistemological Preliminaries*. Berlin, New York: Walter de Gruyter.

Breznau, N., and M. A. Eger. 2016. Immigrant Presence, Group Boundaries, and Support for the Welfare State in Western European Societies. *Acta Sociologica* 59:195–214. WOS:000381000100001.

Brubaker, R. 1992. *Citizenship and Nationhood in France and Germany*. Cambridge, MA: Harvard University Press.

Brubaker, R. 2004. *Ethnicity Without Groups*. Cambridge, MA: Harvard University Press.

Burawoy, M. 2005. For Public Sociology. *American Sociological Review* 70:4–28.

Burgoon, B. 2014. Immigration, Integration, and Support for Redistribution in Europe. *World Politics* 66:365–405. WOS:000339685200001.

Castles, S. 2007. The Factors that Make and Unmake Migration Policies. In *Rethinking Migration. New Theoretical and Empirical Perspectives*, ed. A. Portes and J. DeWind, 29–61. New York, Oxford: Bergham Books.

Castles, S., H. de Haas, and M. J. Miller. 2020. *The Age of Migration. International Population Movements in the Modern World*. New York: Guilford Press.

Cornelius, W. A., et al., eds. 2004. *Controlling Immigration. A Global Perspective*. Stanford, CA: Stanford University Press.

Crawley, H., and D. Skleparis. 2018. Refugees, Migrants, Neither, Both. Categorical Fetishism and the Politics of Bounding in Europe's 'migration Crisis'. *Journal of Ethnic and Migration Studies* 44:48–64. https://doi.org/10.1080/1369183X.2017.1348224.

Dahinden, J. 2016. A Plea for the 'De-Migranticization' of Research on Migration and Integration. *Ethnic and Racial Studies* 39:2207–2225. https://doi.org/10.1080/01419870.2015.1124129.

De Genova, N. 2016. The European Question. Migration, Race, and Postcoloniality in Europe. *Social Text* 34:75–102.

Erdal, M. B., and C. Oeppen. 2018. Forced to Leave? The Discursive and Analytical Significance of Describing Migration as Forced and Voluntary. *Journal of Ethnic and Migration Studies* 44:981–998. https://doi.org/10.1080/1369183X.2017.1384149.

Esser, H. 2003. Ist das Konzept der Assimilation überholt? *Geographische Revue* 5:5–22.

Esser, H. 2007. Does the New Immigration Require a New Theory of Intergenerational Integration? In *Rethinking Migration. New Theoretical and Empirical Perspectives*, ed. A. Portes and J. DeWind, 308–336. New York, Oxford: Berghahn Books.

Faist, T. 2015. Grenzen überschreiten. Das Konzept Transstaatliche Räume und seine Anwendungen. In *Transstaatliche Räume*, ed. T. Faist, 9–56. Bielefeld: transcript.

Feller, E. 2005. Refugees Are Not Migrants. *Refugee Survey Quarterly* 24:27–35. https://doi.org/10.1093/rsq/hdi077.
Glick Schiller, N., et al. 2006. Beyond the Ethnic Lens. Locality, Globality, and Born-Again Incorporation. *American Ethnologist* 33:612–633.
Hjerm, M., and A. Schnabel. 2012. How Much Heterogeneity Can the Welfare State Endure? The Influence of Heterogeneity on Attitudes to the Welfare State. *Nations and Nationalism* 18:346–369. https://doi.org/10.1111/j.1469-8129.2011.00523.x.
Hollifield, J. F. 2000. Migration and the 'New' International Order. The Missing Regime. In *Managing Migration. Time for a New International Regime*, ed. B. Gosh, 95–109. Oxford: Oxford University Press.
Iskander, N. 2018. How Normative Debates About Immigration Shape Analyses of the Assimilation Processes of Second-Generation Youth. Lessons From Spanish Legacies. *Ethnic and Racial Studies* 41:484–491. https://doi.org/10.1080/01419870.2018.1388923.
Karakayali, S. 2008. *Gespenster der Migration. Zur Genealogie illegaler Einwanderung in der Bundesrepublik Deutschland.* Bielefeld: Transcript.
Manderscheid, K. 2010. Mobilities as Dimensions of Social Inequalities. In *Globalization and Transformations of Social Inequality*, ed. U. Schuerkens, 31–57. Abingdon: Routledge.
Mannheim, K. 1952. On the Interpretation of Weltanschauung. In *Essays on the Sociology of Knowledge*, ed. K. Mannheim, 33–83. New York: Routledge.
Marshall, Thomas H. 1950. *Citizenship and Social Class and Other Essays.* Cambridge: Cambridge University Press
Niederberger, A. 2021. Migrationsethik in der Krise. Einige grundlegende philosophische Überlegungen. *Zeitschrift für Migrationsforschung* 1:97–123. https://doi.org/10.48439/zmf.v1i1.99.
Nieswand, B. 2016. Die Dezentrierung der Migrationsforschung. In *Migration – Religion – Identität. Aspekte transkultureller Prozesse*, ed. K. Kazzazi et al., 283–297. Wiesbaden: Springer.
Nowicka, M., and A. Cieslik. 2013. Beyond Methodological Nationalism in Insider Research with Migrants. *Migration Studies* 2:1–15. https://doi.org/10.1093/migration/mnt024.
Piore, M. J. 1979. *Birds of Passage. Migrant Labor and Industrial Societies.* Cambridge: Cambridge University Press.
Portes, A., and J. Sensenbrenner. 1993. Embeddedness and Immigration. Notes on the Social Determinants of Economic Action. *American Journal of Sociology* 98:1320–1350.
Pries, L. 2005. Configurations of Geographic and Societal Spaces. A Sociological Proposal Between 'Methodological Nationalism' and the 'Spaces of Flows'. *Global Networks* 5:167–190.
Recchi, E., and A. Favell, eds. 2009. *Pioneers of European Integration. Citizenship and Mobility in the EU.* Cheltenham, Northampton, MA: Edward Elgar.
Scheel, S., and M. Gutekunst. 2019. Studying Marriage Migration to Europe from Below. Informal Practices of Government, Border Struggles and Multiple Entanglements. *Gender, Place & Culture* 26:847–867. https://doi.org/10.1080/0966369X.2018.1489375.
Scheel, S., and V. Squire. 2014. Forced Migrants as Illegal Migrants. In *The Oxford Handbook of Refugee and Forced Migration Studies*, ed. E. Fiddian-Qasmiyeh et al., 188–199. Oxford: Oxford University Press.
Sen, A. 2006. What Do We Want from a Theory of Justice? *Journal of Philosophy* 103:215–238.
Shinozaki, K. 2012. Transnational Dynamics in Researching Migrants. Self-Reflexivity and Boundary Drawing in Fieldwork. *Ethnic and Racial Studies* 35:1810–1827.
Taft, R. 1953. The Shared Frame of Reference Concept Applied to the Assimilation of Immigrants. *Human Relations* 6:45–55.
Tsianos, V., and S. Karakayali. 2010. Transnational Migration and the Emergence of the European Border Regime. An Ethnographic Analysis. *European Journal of Social Theory* 13:373–387.
Van Hear, N. 2012. Forcing the Issue. Migration Crises and the Uneasy Dialogue Between Refugee Research and Policy. *Journal of Refugee Studies* 25:2–24. WOS:000301363200002.

van Oorschot, W., and W. Uunk. 2007. Multi-Level Determinants of the Public's Informal Solidarity towards Immigrants in European Welfare States. In *Social Justice, Legitimacy and the Welfare State*, ed. S. Mau and B. Veghte, 217–238. Aldershot: Ashgate.

Weber, M. 1946. Science as a Vocation. In *From Max Weber: Essays in Sociology*, ed. H. H. Gerth and C. Wright Mills, 129–156. New York: Oxford University Press.

Weiß, A. 2017. *Soziologie globaler Ungleichheiten*. Berlin: Suhrkamp.

Weiß, A. 2018a. Becoming a Refugee. A Life-Course Approach to Migration Under Duress. *Sociologias* 20:110–141. http://www.scielo.br/pdf/soc/v20n49/1807-0337-soc-20-49-24.pdf.

Weiß, A. 2018b. Contextualizing Global Inequalities. A Sociological Approach. In *The World-System as Unit of Analysis. Past Contributions and Future Advances*, ed. R. P. Korzeniewicz and I. Wallerstein, 75–85. New York: Routledge.

Weiß, Anja (2021), Re-thinking society. How can sociological theories help us understand global and cross-border social contexts?. *Current Sociology* 69:333–351. https://doi.org/10.1177/0011392120936314.

Wiley, N. F. 1967. The Ethnic Mobility Trap and Stratification Theory. *Social Problems* 15:147–159. http://www.jstor.org/stable/799509.

Wimmer, A., and N. Glick Schiller. 2003. Methodological Nationalism, the Social Sciences, and the Study of Migration: An Essay in Historical Epistemology. *International Migration Review* 37:576–610.

Zolberg, A. R. 2012. Why Not the Whole World? Ethical Dilemmas of Immigration Policy. *American Behavioral Scientist* 56:1204–1222. https://doi.org/10.1177/0002764212443821.

Economics and Migration Ethics

Johan Rochel

Abstract This paper deals with the ethical dimensions of the economics of migration. The first section presents the main research topics found in the economics of migration around three questions: Why do people move? Which kinds of people are moving? What are the consequences for countries of origin, transit and, above all, destination? The section summarises the main methods used and briefly addresses influential controversies. The second section deals with ethical issues in the economics of migration. It identifies normative presuppositions at work in economic research, especially in its attempt to evaluate the impact of migration. Self-reflectivity requires the consideration of at least four distinct levels: subject matter, institutional frame, temporal frame and personal frame. The third section addresses, conversely, the use of economic arguments in the ethics of migration. It highlights two main issues: the use of economic arguments for the sake of justifying the state's competence to control immigration (as part of protection arguments) and their use for the promotion of free movement in the context of reflections on global justice. The paper concludes by outlining promising synergies for ethicists in engaging more deeply with economists' work on migration. With regard to the importance of these economic insights for public discourses, ethicists have a vested interest in better anchoring their ethical expertise within economic research.

1 Migration in Economic Studies

Migration has become an important issue in economics. Due to its political and societal relevance, the economics of immigration has been an especially controversial topic among specialists. The economic analysis of cross-border mobility (setting aside mobility inside a single country) is mainly concerned with three questions: Why do people move? Which kinds of people are moving? What are the consequences for countries of origin, transit and, above all, destination? The three

J. Rochel (✉)
University of Fribourg, Fribourg, Switzerland
e-mail: johan.rochel@unifr.ch

A. Niederberger et al. (eds.), *Handbook of Migration Ethics*,
https://doi.org/10.1007/978-3-031-89877-8_35

questions represent the classical locus of migration economics [for introductions, see Bodvarsson and Van den Berg (2013); Constant and Zimmermann (2015); Fasani (2020)].

The first question focuses on explaining the rationales of why people tend to move across borders. In mainstream economics, this question is addressed on the basis of a human capital model. Migrants make the decision to migrate as an investment in their well-being. In his seminal paper, Sjaastad conceives the migration decision as a utility (profit) maximisation problem (Sjaastad 1962). In the classical migration decision model, a would-be migrant compares their wage in their country of origin with their prospective wage in the destination country; accounting for the costs of moving, they then move only if the sum of this equation is positive. As put by Bodvarsson and Van den Berg, "the migration decision, therefore, depends on how a would-be migrant with a specific set of skills and talents perceives his or her gains from migrating from a labour market where the labour force has a certain distribution of worker characteristics to a country where the labour force has a different distribution of talents, skills, and education levels" [Bodvarsson and Van den Berg 2013: 40; see the classical model developed by Borjas (1987)]. The decision by highly skilled immigrants has been an important focus of research (Kerr et al. 2016).

A strict focus on economic analysis has been criticised by social sciences as missing the complexity of the migration decision. Highly simplified uses of this economic literature in the policy debate have also been criticised, such as, for instance, the "accelerator hypothesis" defended by Collier, who argues that we are moving towards a world of mass exodus (Collier 2013). Nevertheless, the basic human capital model is able to integrate parameters of different types, clustered in categories such as "pull", "push", "stay", "stay away" factors. If the would-be migrant's preferences are not limited to better wage, the model can integrate further interests. Likewise, the family or group dimension of the decision to migrate might be accounted for.

This first classical question is strongly linked to the second classical question of migration economics: Which kinds of people are moving? To understand the connection between these two questions, it is important to recall that the decision to migrate (the migration equation) is different for would-be immigrants with different wages in either the country of origin, the destination country, or both. Wages tend to be different because of income level differences across countries and because any particular type of education might change in value between the country of origin and the destination country. Hence, the migration equation is positive for some skill groups and not for other groups, which is why those people for whom it is positive tend to migrate. Economists have used such reasoning to try to explain why, for instance, the USA attracts primarily low-skilled immigrants (Borjas 1987) while Switzerland benefits primarily from highly skilled EU workers (Beerli et al. 2017).

In addressing this second classical question, theorists distinguish two issues. Firstly, economists try to explain the self-selection induced by the decision to migrate and the impact that this has on the skills, competencies and resources that migrants tend to display. Such work on self-selection, carried out primarily by labour economists in different contexts, is drawn upon in debates about immigration. To illustrate the methodological concerns at stake, Bodvarsson and Van den Berg speak of the “Chiswick vs. Borjas debate” [for the original contributions, see Chiswick (1999) and Borjas (1987); for further references, see Bodvarsson and Van den Berg (2013: 81ff.)]. Secondly, economists focus on assimilation, defined as the capacity for participating in the labour market of the country of destination. Like for the immigration decision, the idea of investment in human capital is also an important conceptual framework for explaining how immigrants behave in the country of destination and which decisions they make (Chiswick 1978). Note that the selection issue has an important gender dimension (Triandafyllidou and Isaakyan 2016).

The third question is the most important one, itself encompassing a number of key sub-questions: What are the consequences of migration for countries of origin, transit and destination? Most research deals with the impact of immigration on countries of destination. As briefly addressed later, there are also important studies dealing with the impact on countries of origin, mainly from the perspective of those who remain (Kapur and McHale 2012; Lenard 2022).

2 Migration Ethics in the Economics of Migration

This third question about impact entails numerous normative considerations which should be linked to the ethics of migration. Above all, it requires self-reflectivity. Making one’s research focus explicit requires an explanation on at least four distinct levels: subject matter, institutional frame, temporal frame and personal frame. The subject matter encompasses all the potential areas of impact which immigration could affect: wages, employment, fiscal effects, innovation, productivity. If defined broadly, the list is potentially open-ended, including any dimension that can be expressed in economic terms (such as impact on real estate, natural resources or public trust). The most discussed issues are wages, employment level and fiscal effects. The institutional frame bears upon the level of analysis chosen. The impact might be measured on the firm, city, sub-national, national, regional or international level. Similarly, the temporal frame chosen has a deep impact on the analysis (short-, mid-, long-term). Fourthly, the personal frame raises the question of whose impact should be measured: the nationals, residents, workers, non-workers, skilled, non-skilled.

These four parameters might be combined, creating several potential objects of analysis and comparison. Current researchers increasingly take into account the connections between these different parameters and try to account for them in a holistic fashion. For instance, when investigating the impact of immigrants on natives in the labour market, researchers should not only look at the wages of natives but also at their employment. Wages might respond very flexibly to any shift in labour supply (which can be context-specific, e.g. institutional contract preventing wage flexibility for some types of workers), but the shock could be absorbed via employment. One cannot look at one particular outcome in isolation. Instead, one needs to look at all of them together, as the decisions taken by workers, firms and public authorities are all connected to each other. Furthermore, we might note that the socio-economic context of a country, not least with respect to the funding provided for specific research projects, might have an impact on the chosen focus of research (e.g. European countries with extensive unemployment systems vs. USA and its less extended welfare mechanisms) (Kerr and Kerr 2011).

Beyond the object of interest, similar considerations might also impact the method chosen to address a specific question. If we take the example of impact on wages, we could distinguish three different approaches currently in use (Dustmann et al. 2016). The first one, advocated by Borjas (2003), is to estimate the impact at the national level by proposing skill-based categorisations of immigrants. The second one, developed by Altonji and Card (1991), focuses on the regional level by exploiting variation in the total inflow of immigrants. The third one, such as, for instance, in Card (2001), also focuses on the regional level and mixes skill-based categorisations with area-specific analysis. Generally, these approaches recognise the heterogeneity of resident workers and the complementarity of the tasks performed by residents and newcomers and integrate these elements into a general equilibrium model. Following Peri, the methods now used by economists are able to take a holistic view on the multifaceted impact which immigration has on every aspect of society (Peri 2016).

To summarise, the study of the economics of migration is loaded with multiple normative presuppositions (Edo et al. 2020). In developing new approaches to measure the impact of immigration, economists need to critically assess the chosen object of investigation (defined along the four parameters identified), but also the chosen method of investigation (general equilibrium model, but also parameter values used in the simulations). This point recalls that the economics of migration is, like any scientific project, the product of the researcher's choices, but also indirectly of the broader political and social context in and for which it is produced (Hatton 2014). It also recalls that the output of scientific research on sensitive issues such as immigration plays an important role in public debates, raising important stakes about the justifications offered by economists for the focus of their interests.

3 Migration Ethics and the Economics of Migration

Having considered normative presuppositions in the economics of migration, we focus on economic arguments used in the ethics of migration. I firstly consider their relevance in arguments about the justification of the competence to control immigration or, as a follow-up, about the exercise of this competence to choose a specific immigration policy. Secondly, I will mention further ethical issues where economic arguments are used, such as arguments in favour of free movement and global justice.

Following Cassee, we could distinguish between different attempts to justify the competence to control immigration: "protection" and "control" justifications on one axis, and "institutional" and "cultural" justifications on the other axis (Cassee 2016). The "institutional/cultural" dimension relates to the putative specificity of a political community and its relevance for the justification of the competence to control immigration. The "control/protection" dimension relates to the type of normative argument used for the justification, either a community claiming a sense of self-determination or protecting its specific interests. Economic arguments are especially important for protection arguments.

In their structure, protection arguments focus on important interests to be protected through a migration competence. Following Abizadeh, the analysis of an interest has both an empirical and a normative premise (Abizadeh 2006). The normative part interrogates the legitimacy of the interest at stake. The empirical part addresses the link between this specific interest and the migration policy of a political community.

This framing of the ethical question echoes our previous discussion about state-of-the-art immigration economics. Firstly, ethicists also face the difficulty of making their definition of "economic interests" explicit. This task requires distinguishing among the four parameters identified above and addressing the ethical questions relevant to each of the combinations. Three elements remain important across diverse combinations.

The first one pertains to the structure of the protection argument. It is based on the identification of an interest and its instrumental linkage to the immigration policy of a political community. In political discourses, "collapse" arguments ("worst case" scenario) are often used on the basis of a hypothetical mass immigration scenario and its negative impact on different economic parameters. The normative component of the argument is very similar to a public order argument. The problem of "collapse" arguments does not rely on the normative component, but on the very demanding empirical premise. If found empirically plausible, the argument justifies a specific type of competence in immigration matters, comparable to an emergency brake in a train. Like we use the emergency brake to stop the train, we might exercise strict immigration control. But the argument does not justify the competence to choose specific types of migrants for professional purposes nor to choose a strict/lenient asylum policy.

By contrast to "collapse" arguments, we might speak of "strain" arguments when economic conditions come under pressure, but do not collapse. This type of argument illustrates the second transversal element of analysis. In contrast to the "collapse" arguments, the "strain" arguments appear empirically realistic. As previously shown, immigration might negatively impact specific categories of residents (if the national level and if being a resident are taken as relevant parameters). Economic advantages and burdens produced by immigration policies represent a distributive justice issue. This means that negative economic impact cannot directly be used against a specific immigration policy but should be considered as a potential argument for other distributive responses. This point is essential for the use of economic arguments within immigration ethics and, more broadly, in public debates. As proposed by Borjas (2014), it might be the case that immigration is costly to specific categories of resident workers, but this is not automatically an argument against current immigration policy. It might be an argument for stronger protection mechanisms against wage dumping, better redistribution of common resources towards those in direct competition with newcomers or insurance schemes for newcomers.

Thirdly, this distributive issue is of a specific kind because it involves members of the political community (residents) and would-be migrants, i.e. outsiders. The argument reaches the core of the problem displayed by a movement towards openness and inclusivity in immigration policy (as argued for by many immigration ethicists, for a variety of reasons) and a movement of protection of resources produced by and for the political community's members. The use of economic arguments in immigration ethics is systematically related to the complex issue of priority-giving for members of the political community. Following Scheffler's framing, this situation raises a "distribution objection" about the legitimacy of the priority-giving scheme among fellow members from the point of view of outsiders (Scheffler 2001). This objection puts the group's relationships into a larger perspective by stressing the exclusion effects of special prerogatives (such as immigration control competence). It raises the question of the (global) justice duties which the political community should fulfil in order to draw on its priority-giving scheme as an argument for immigration policy (Schlothfeldt 2002). In addressing this crucial challenge, the key insight is to acknowledge, in Scheffler's words, that "special responsibilities need to be set within the context of our overall moral outlook and constrained in suitable ways by other pertinent values" (Scheffler 2001: 108). This challenge is essential for all discussions on the fiscal effects of immigration policy and its impact on welfare provisions. In order to address this challenge, we need to specify why members of a political community have special responsibilities towards and prerogatives over each other (Abizadeh 2016), which global justice duties the political community has, and the conditionality relation between the fulfilment of these duties and the legitimacy of priority-giving schemes for members (Bader 2005). The specification of the community's global justice duties is a precondition in navigating trade-offs, such as negative impacts on specific resident workers and positive impacts on would-be migrants and their communities of origin.

To illustrate the relevance of these normative questions, imagine that we have consistent findings on displacement effects for resident unskilled workers in the country Y. Firstly, if the more general effect on the fiscal balance and productivity of Y can be said to be positive—or at least cannot be proved to be negative—the focus shifts from a "collapse" to a "strain" scenario. Specific types of workers (i.e. their wages) are put under pressure. These negative effects should be handled as an internal distributive problem. Secondly, one of the plausible positions in this distributive issue could be to ensure that the weakest members of country Y will not be harmed by the arrival of labour migrants. This shall not only be understood as financial transfers in the form of welfare provisions but also as institutional mechanisms to train affected workers and enable them to adapt to changing professional challenges. Thirdly, this position is not complete without an argument about the broader justice duties which country Y should assume and the impact of (non-)fulfilment of these duties on the legitimacy of its priority-giving scheme for members. These three elements illustrate how economic arguments (i.e. economic data, causal relations, predictive models) represent promising research questions for immigration ethics in raising new ethical challenges (e.g. specification of one's position on domestic and global justice, burden allocation questions for diverse stakeholders).

If economic arguments have been used as protection arguments to justify relatively strict immigration policy, they have also been used in promoting mobility across borders. Different types of arguments might be distinguished. The first category of arguments connects the value of freedom to move with economic research on explaining why and which people move (the first and second focuses of research identified previously). Economic data might usefully be integrated into ethical arguments to refine the analysis from the perspective of justifying a right to move, as e.g. Shachar (2009).

The second category links economic arguments to global justice arguments. In addition to the function of justice duties for a political community claiming to control its immigration policy (as previously mentioned), global justice duties might be used as a positive argument for the liberalisation of immigration policy. Among different possibilities, the focus might be placed on the putative positive impact on the aggregate global wealth. According to the concept used by Clemens, there are "trillion-dollar bills on the sidewalk" [Clemens 2011; for a critique see Borjas (2014)]. This line of argument is often proposed as an analogy to the liberalisation of trade and its positive economic impact. However, as mentioned, when dealing with the parameters of the economic analysis, the argument requires a specification of the normative presuppositions which lead to the definition of what is considered to be a relevant unit. The following questions should be answered: Which unit of analysis is chosen to measure global wealth [e.g. GDP (Ortega and Peri 2014)]? Is the positive result aggregated? If yes, how do we deal with distributive justice issues? Shall we consider a global, regional or national level of impact? Though positive results might appear at certain levels, important distributive justice issues remain about the repartition of alleged gains among countries, but also within countries (Bauböck and Ruhs 2022).

Global justice duties are also relevant in assessing the impact of immigration policy on the country of origin. This discussion is often led in the form of a "summa summarum" exercise between positive and negative effects on countries of origin. Negative impacts mainly relate to brain drain discussions. In a strict sense, the concept refers to the negative effects bound to the departure of (skilled) migrants in terms of human capital for their home country (Beine et al. 2008). More generally, the concept has come to encompass the negative effects endured by different stakeholders in the process of managing international mobility. The empirical data are complex and highly specific for every country or even regions. The integration of both economic and ethical arguments should at least entail a reflection on the legitimate interests of the would-be emigrant and an account of the relation between would-be emigrants and their countries of origin (Oberman 2013). Brain drain arguments seem to rely on an instrumental understanding of the relation between citizens and their state (Teson 2008). This understanding is in strong tension with the human right to leave one's state but also, and more fundamentally, with the prohibition of the state to instrumentalise its citizens.

The positive side of the impact on countries of origin might be exemplarily addressed with the case of "remittances". Billions of dollars are sent by immigrants back to their relatives in their countries of origin, arguably contributing to their economic development. Like in the case of brain drain, the empirical evidence prepared by economists should be integrated into a larger ethical approach. Firstly, the assessment of the empirical relevance available on remittances raises the general issue of the nexus between development and migration. Faist et al. distinguish between different historical stages of interests on the nexus between development and migration (Faist et al. 2011). After a period of scepticism, they diagnose the return of an optimistic view since the 1990s, highlighting alleged co-advantages for all stakeholders. Beyond this required self-reflexivity on the general development-migration nexus, remittances might be integrated into a global justice argument. If remittances can be shown to positively impact on countries of origin, especially their weak members, then there is a strong multiplying argument to be made for the liberalisation of immigration policy (Cheneval and Rochel 2012). Such a policy will improve life quality of the migrants themselves, but also of individuals in their countries of origin. This line of argument illustrates how economic data (on remittances in this case) might make new ethical questions appear and fuel new arguments.

"Law and economics" applied to immigration deserves a specific note. "Law and economics" is used as a shorthand to grasp several methodological approaches which try to account for the functioning of law as an institution with the help of economic concepts. It distinguishes itself from neoclassical economics in two main respects. It firstly modifies the rationality postulate of neoclassical economics, admitting that individuals act under constrained rationality with limited information and with limited capacity to process information (thereby paving the way for behavioural economics). Secondly, it admits that transactions are costly in almost all human interactions. It emphasises the role of institutions for the process of human exchange.

This methodology has been applied to immigration, with interesting ethical implications. Following the influential work by Trebilcock (2003), theorists have tried to reconstruct the debate on immigration as a debate about property rights (Brown 2011; Casella and Cox 2018). As proposed by Schlegel (2017) "control over migration" is a valuable asset of which—theoretically—either the country of destination or migrants themselves can dispose. The normative discussion echoing much of the immigration ethics debate addresses the degree to which competing potential rights-holders should be allowed to migrate or limit immigration respectively. These works investigate the primary allocation of one's right to migration and the institutional transaction mechanisms that allow for the further exchange of this right (Cox and Posner 2006).

4 Future Research and Dialogue

Stronger dialogue between the ethics of migration and the economics of migration could mean promising research opportunities, but also a better ability to raise relevant arguments in public debates on immigration policy. As with other disciplines of the social sciences (such as law; see Trachtman 2009), ethicists should actively look for opportunities to engage with economists about the normative presuppositions of the economics of immigration. The goal is to draw upon ethical expertise in making these presuppositions explicit.

With respect to the "institutional" parameter identified above, the economics of migration might gather data and explore causal relations at levels from which the ethics of immigration is currently absent. As an example, economic data at firm level could represent a new field of research for immigration ethics. As highlighted by Kerr et al. (2015), economic analysis of US firms brings new actors to light (the firm themselves, unions, employees). Similarly, as highlighted by Lewis and Peri, the impact of immigration depends upon small-scale responses, such as, for instance, by resident workers, firms, sectors and potentially by local consumers (Lewis and Peri 2014). Ethicists should be keen to investigate new ethical challenges in this specific context. More generally, these synergies might be considered from the perspective of the ambition to question the focus placed on the nation-state as a key level of analysis. As argued by Sager (2014), current ethical approaches largely reflect a widely shared methodological nationalism. By working together and by mutually challenging their normative presuppositions, both the ethics and the economics of migration might expand their fields of investigation.

Similarly, parameters considered by economists (such as the object of impact) might be handled as new ethical arguments for a specific immigration policy. For instance, the impact of immigration on innovation and creativity could be framed as a new argument in favour of the liberalisation of immigration policy (e.g. for a focus on inventors; see Moser et al. 2014).

Literature

Abizadeh, A. 2006. Liberal Egalitarian Arguments for Closed Borders. Some Preliminary Critical Reflections. *Ethique et économiques/Ethics and Economics* 4:1–8.

Abizadeh, A. 2016. The Special-Obligations Challenge to More Open Borders. In *Migration in Political Theory. The Ethics of Movement and Membership*, ed. L. Ypi and S. Fine, 105–124. Oxford: Oxford University Press.

Altonji, J. G., and D. Card. 1991. The Effects of Immigration on the Labor Market Outcomes of Less-skilled Natives. In *Immigration, Trade and the Labor Market*, ed. J. M. Abowd and R. B. Freeman, 201–234. Chicago: University of Chicago Press.

Bader, V. 2005. The Ethics of Immigration. *Constellations* 12:331–361.

Bauböck, R., and M. Ruhs. 2022. The Elusive Triple Win: Addressing Temporary Labour Migration Dilemmas through Fair Representation. *Migration Studies 10 (3): 528–552.*

Beerli, A., R. Indergand, and J. Kunz. 2017. *The Supply of Foreign Talent. How Skill-Biased Technology Drives the Skill Mix of Immigrants*. KOF Working Papers 436.

Beine, M., M. Docquier, and H. Rapoport. 2008. Brain Drain and Human Capital Formation in Developing Countries. Winners and Losers. *The Economic Journal* 118:631–652.
Bodvarsson, O. B., and H. Van den Berg. 2013. *The Economics of Immigration. Theory and Policy*. New York: Springer.
Borjas, G. J. 1987. Self-Selection and the Earnings of Immigrants. *American Economic Review* 77:531–553.
Borjas, G. J. 2003. The Labor Demand Curve Is Downward Sloping. Reexamining the Impact of Immigration on the Labor Market. *The Quarterly Journal of Economics* 118:1335–1374.
Borjas, G. J. 2014. *Immigration Economics*. Cambridge, MA: Harvard University Press.
Brown, E. M. L. 2011. Visa as Property, Visa as Collateral. *Vanderbilt Law Review* 64:1047–1105.
Card, D. 2001. Immigrant Inflows, Native Outflows, and the Local Labor Market Impacts of Higher Immigration. *Journal of Labor Economics* 19:22–64.
Casella, A., and A. B. Cox. 2018. A Property Rights Approach to Temporary Work Visas. *The Journal of Legal Studies* 47:195–227.
Cassee, A. 2016. *Globale Bewegungsfreiheit – Ein philosophisches Plädoyer für offene Grenzen*. Berlin: Suhrkamp.
Cheneval, F., and J. Rochel. 2012. An Ethical View on Remittances and Labor Migration. *Global Justice. Theory Practice Rhetoric* 5:16–30.
Chiswick, B. 1978. The Effects of Americanization on the Earnings of Foreign-Born Men. *Journal of Political Economy* 86:897–921.
Chiswick, B. 1999. Are Immigrants Favorably Selected? *The American Economic Review* 89:181–185.
Clemens, M. A. 2011. Economics and Emigration. Trillion-Dollar Bills on the Sidewalk? *Journal of Economic Perspectives* 25:83–106.
Collier, P. 2013. *Exodus. How Migration Is Changing our World*. Oxford: Oxford University Press.
Constant, Amelie F., and Klaus F. Zimmermann, eds. 2015. *International Handbook on the Economics of Migration*. Cheltenham: Edward Elgar.
Cox, A., and E. Posner. 2006. The Second Order Structure of Immigration Law. *Stanford Law Review* 59:809–856.
Dustmann, C., U. Schönberg, and J. Stuhler. 2016. The Impact of Immigration. Why Do Studies Reach Such Different Results? *The Journal of Economic Perspectives* 30:31–56.
Edo, A., et al. 2020. *An Introduction to the Economics of Immigration in OECD Countries, IZA Discussion Paper Series*. Bonn: IZA – Institute of Labor Economics.
Faist, T., M. Fauser, and P. Kivisto, eds. 2011. *The Migration-Development Nexus. A Transnational Perspective*. London: Palgrave Macmillan.
Fasani, Francesco, Joan Llull, and Cristina Tealdi. 2020. The Economics of Migration: Labour Market Impacts and Migration Policies. *Labour Economics 67:101929.*
Hatton, T. J. 2014. The Economics of International Migration. A Short History of the Debate. *Labour Economics* 30:43–50.
Kapur, D., and J. McHale. 2012. Economic Effects of Emigration on Sending Countries. In *The Oxford Handbook of Politics of International Migration*, ed. M. Rosenblum and D. J. Tichenor, 131–152. Oxford: Oxford University Press.
Kerr, S. P., and W. R. Kerr. 2011. *Economic Impacts of Immigration. A Survey*. NBER Working Paper 09–13, 1–48.
Kerr, S. P., W. R. Kerr, and W. F. Lincoln. 2015. Skilled Immigration and the Employment Structures of US Firms. *Journal of Labor Economics* 33:147–186.
Kerr, S. P., W. Kerr, C. Özden, and C. Parsons. 2016. Global Talent Flows. *Journal of Economic Perspectives* 30:83–106.
Lenard, P. T. 2022. Restricting emigration for their protection? Exit controls and the protection of (women) migrant workers. *Migration Studies* 10 (3): 510–527.
Lewis, E., and G. Peri. 2014. *Immigration and the Economy of Cities and Regions* NBER Working Paper 20428, 2–86.

Moser, P., A. Voena, and F. Waldinger. 2014. German Jewish Emigrés and US invention. *American Economic Review* 104:3222–3255.
Oberman, K. 2013. Can Brain Drain Justify Immigration Restrictions? *Ethics* 123:427–455.
Ortega, F., and G. Peri. 2014. Openness and Income. The Roles of Trade and Migration. *Journal of International Economics* 92:231–251.
Peri, G. 2016. Immigrants, Productivity, and Labor Markets. *Journal of Economic Perspectives* 30:3–30.
Sager, A. 2014. Methodological Nationalism, Migration and Political Theory. *Political Studies* 64 (1): https://doi.org/10.1111/1467-9248.12167
Scheffler, S. 2001. *Boundaries and Allegiances. Problems of Justice and Responsibility in Liberal Thought*. Oxford: Oxford University Press.
Schlegel, S. 2017. *Der Entscheid über Migration als Verfügungsrecht – Eine Anwendung der Ökonomischen Analyse des Rechts auf das Migrationsrecht am Beispiel der Schweiz*. Tübingen: Mohr Siebeck.
Schlothfeldt, S. 2002. Ökonomische Migration und globale Verteilungsgerechtigkeit. In *Was schulden wir Flüchtlingen und Migranten? Grundlagen einer gerechten Zuwanderungspolitik*, ed. A. Märker and S. Schlothfeldt, 93–109. Wiesbaden: Westdeutscher.
Shachar, A. 2009. *The Birthright Lottery. Citizenship and Global Inequality*. Cambridge, MA: Harvard University Press.
Sjaastad, L. 1962. The Costs and Returns of Human Migration. *Journal of Political Economy* 70:80–93.
Teson, F. R. 2008. Brain Drain. *San Diego Law Review* 45:899–932.
Trachtman, J. P. 2009. *The International Law of Economic Migration. Toward the Fourth Freedom*. Kalamazoo: WE Upjohn Institute for Employment Research.
Trebilcock, M. J. 2003. The Law and Economics of Immigration Policy. *American Law and Economics Review* 5:271–317.
Triandafyllidou, A., and I. Isaakyan. 2016. Re-Thinking the Gender Dimension of High-Skill Migration. In *High-Skill Migration and Recession*, ed. A. Triandafyllidou and I. Isaakyan, 293–305. London, New York: Palgrave Macmillan.

Migration Studies and Migration Ethics

Brigitte Suter

Abstract Migration studies is the multi- and interdisciplinary field devoted to examining and explaining causes, experiences, and impacts of human mobility. Researchers studying migration face a plenitude of ethical challenges, especially in studies on or with "vulnerable populations"—people that are in dire legal, economic, or psychological circumstances, or of young age. However, there are more general normative imperatives deriving from the field's traditionally close relationship to the nation-state system and the international refugee regime. Because migration studies is an academic field with direct societal relevance, its research findings often have the potential to influence policy decision-making, which renders migration scholars in a powerful position to exert influence on how societal resources are distributed. Although these circumstances clearly call for a strong ethical framework throughout the research process, from proposal writing to dissemination, normative considerations have only been observed in some parts of the field. Ethical considerations are most pronounced in methodological inquiries, and issues of representation and labelling as well as the researcher's positionality have also received significant attention. Nonetheless, many ethical considerations remain generally implicit rather than explicit. However, a growing "critical" and "reflexive" migration scholarship has been increasingly calling into question the close relationship between academia and the nation-state and pointing at the pitfalls of unreflexive scholarship. The focus of these migration scholars on power inequalities in the production and dissemination of knowledge, as well as on the effects of policy-oriented research, can be read as an ethical imperative for good scholarship.

Migration studies, an inherently inter- and multidisciplinary field, is concerned with the study of human mobility. While migration studies have emerged as degree courses at many universities, the field generally includes all research on migration in social sciences and humanities. It observes social, political, and economic transformations on a micro-, meso-, and macro level and studies how the resulting

B. Suter (✉)
Department of Global Political Studies, Malmö University, Malmö, Sweden
e-mail: brigitte.suter@mau.se

A. Niederberger et al. (eds.), *Handbook of Migration Ethics*,
https://doi.org/10.1007/978-3-031-89877-8_36

geographical differences in economic, educational, and lifestyle opportunities and physical safety may cause various forms of human mobility and vice versa. As an academic field of inquiry, migration studies thus cover a very broad range of human movements, on a continuum of coercion and voluntariness. The boundaries of the field are changing, but the field generally responds to an interest in understanding, managing, and controlling migration. It commonly addresses issues of refugee, labour, and family migration; citizenship and naturalisation; integration; belonging; ethnic relations; and categories of and access to protection. More recently, other forms of migration have received scholarly attention, such as retirement and lifestyle migration and the migration of the highly-skilled and students. In some parts of the field, the mobility turn has been influential; therefore, practices, representations, and imaginations of mobility and immobility have received considerable attention on the research agenda. Further, from being largely ignored at the onset of migration research, gender has transformed into a vital consideration throughout the field. Owing to the perspective of intersectionality in black feminist studies, other attributes of social stratification have also received attention, albeit to a lesser extent. Among these are age, sexual orientation, religion, marital status, class, and race. Thus, migration studies have been able to study the social and cultural effects of such social stratifiers on the migration experience.

Since the inception of migration studies as an academic field of inquiry in the 1980s, there has been a close relationship between academia and the nation-state system and the international refugee regime. Over time, this has sparked rigorous disgruntlement within several parts of the field. The field's long tradition of conducting research from a positivist epistemologist stance has been criticised for depoliticising refugee movements and their origins in capitalist and imperialist efforts. Some of its strongest critics describe the field as a tool of neocolonial continuity, built on the false notion of difference between Europeans and non-Europeans (Chimni 1998; Schinkel 2017). Other critical work has increasingly pointed at severe challenges for academic inquiry—such as the blurring of conceptual, epistemological, and ontological distinctions—and the subsequent shortcomings in eliciting understanding for the migratory phenomenon and alleviating human suffering. It is in these critical commentaries that the ethical pleas for the field are most vocal.

1 The General Approach to Migration/Refugees and the Role of Normative Considerations

Questions regarding human mobility in general and migration and refugeehood in particular lie at the core of the field's orientation. Research on migration often includes research with humans through interviews, (participant) observation, and surveys, leading to a number of questions of ethical nature concerning, for example, data collection, data protection, and data retention. Due to the difficult and sensitive political-legal circumstances in which migration often unfolds and is studied, many research participants are in "vulnerable" or severely disadvantaged situations.

Further, the categories and labels in use in the migration field (migrant, refugee, asylum seeker, etc.) often have tremendous legal, social, economic, and material consequences for the receivers of these labels; while some receive protection, rights, and resources, others are excluded, and for many this can be a matter of life and death. This demands particular ethical caution in migration research, for instance, with regards to the participants' safety, the relationship between the researcher and the participants, or the practice of labelling. Lastly, migration research often has an impact on policy, either directly or indirectly, and researchers thus find themselves in a situation where their findings potentially influence decisions on how societal resources are distributed. Thus, it is evident that migration research calls for strong ethical guidelines, considerations, and ongoing reflections on how to design and conduct research and how to disseminate the findings. Surprisingly, the role of normative considerations in the field as a whole is not very strong.

With the exception of studies in political philosophy on the duty and rights of refugees, migrants, and states (with which other chapters of this handbook have dealt extensively), *explicit* normative considerations are generally limited to methodological aspects in migration studies. These considerations direct attention to the many ways in which refugees and migrants may experience vulnerability, and importantly, they address the complexities of obtaining and guaranteeing informed consent, privacy, and the principles of do-no-harm and reciprocal benefit. However, there is a substantial amount of *implicit* moral considerations, mostly from critical or reflexive scholarship, that can be read as standpoints and tutelages of ethical research conduct. Influenced largely by the reflexive turn in social anthropology, feminist standpoint theory, and postcolonial and indigenous scholarship, scholars of the reflexive tradition are mainly concerned with unearthing power relations in the production of knowledge; questioning taken-for-granted, hidden, and common-sense assumptions (for example, in labelling); and foregrounding aspects of class, race, and gender in the production of power in migration. In research practice, this means, for example, including marginal voices to disrupt dominant narratives, understanding categories as performative rather than fixed, or making the conscious distinction between common-sense and analytical categories.

2 Migration Ethics in the Field

The area where ethical considerations are most prominently and most explicitly dealt with in migration studies is methodology, including epistemology. As stated above, migration researchers often study (in) contexts that involve populations in vulnerable situations. This includes people that have been irregularised, are in jail, have been smuggled, or have migrated as children without parents. It may also involve people with an insecure legal status or individuals that are distressed and traumatised. For a long time, ethical guidelines used in the field have been rather general, stemming from other disciplines and fields dealing with research on humans in general, such as the medical field, social work, or social anthropology. In the past decade, however, migration scholars have realised the need to adapt the discussion more closely to the field and to draft more field-specific guidelines and codes of conduct. Bilger and Liempt (2009) introduced the topic into migration studies a bit more than a decade ago, highlighting the methodological and ethical challenges in conducting research with vulnerable migrants in a variety of situations. The most recent addition has been developed by the International Association of Forced Migration Studies (IASFM) in 2018 in the form of a code of ethics.

Nonetheless, the field-specific guidelines still mostly consider the fundamental ethical principle to avoid harm—along with voluntary and informed consent, confidentiality, and privacy. This principle applies throughout the research process, from designing to conducting the research to disseminating the findings, including representation and policy recommendations, and it may generate serious dilemmas. For example, being interviewed can trigger painful memories that can deter the psychological well-being of the interviewee. Further, the circumstances in which the researcher and the vulnerable informant meet often easily incline a situation of dependency with the risk to harm both the research participant and the researcher and to compromise the academic quality of the study. While this is a concern for research with migrants in general, it often takes a whole new dimension in protracted-refugee and conflict situations or in research with severely disadvantaged populations. Researchers also have to be aware that harmful consequences are not always easily foreseen and that informed consent alone is not enough to guarantee no harm. Perhaps the biggest dilemma researchers may face is when the most ethical thing to do appears to be to refrain from conducting a study on a certain topic. Recent scholarly contributions stress the need to avoid treating ethical considerations as a static occurrence and propose instead a more iterative form when considering ethical challenges. For example, being ethically sound regarding the aspect of informed consent requires understanding the concept as a process, ongoing throughout the different stages of the research project and the various (sometimes unexpected) situations that may arise.

The second fundamental principle is reciprocal benefits for the participants. The conviction of Turton (1996: 96), the former director of the Refugee Studies Centre at Oxford University, is widely shared in the field:

> I cannot see any justification for conducting research into situations of extreme human suffering if one does not have the alleviation of suffering as an explicit objective of one's research. For the academic this means attempting to influence the behaviour and thinking of policy-makers and practitioners so that their interventions are more likely to improve than worsen the situation of those whom they wish to help.

In the most recent document of the code of ethics developed in 2018 by the International Association of Forced Migration Studies (IASFM), this principle is somewhat extenuated, and reciprocal benefits are stated to be a goal "whenever possible". Indeed, while reciprocal benefits are often understood in form of policy influence to the advantage of the researched population, other example from the literature mention the provision of a service valued by the participants, such as possibilities to influence the research project or training in interview and other participatory techniques of research methods. The latter can constitute both a tool to empowerment and a beneficial prerequisite for refugees eager to enter the labour market in their country of asylum.

2.1 *Biases in Migration Research: Positionalities, Representations, and the Ontology of Sedentariness*

Brought forward by postcolonial and indigenous scholarship, discussions on ethical conduct have come to necessitate reflections on the researcher's positionality or bias. The fact that research is designed by researchers who seldom share the same experiences as their migrant or refugee participants has caused intense discussions on power inequalities or specific biases in the production of knowledge. However, much more than solely a privileged outsider, the researcher often simultaneously upholds several other positions, for example, as a potential support person or a political activist. Therefore, ethical considerations on the role of the researcher among all of these positions are crucial. Many scholars advocate a participatory approach. Here the ethical imperative is to establish conditions for a more equal relationship between researcher and participants. In allowing the participants to be actively engaged in the research design and practice at several stages (for example, the co-construction of the field, the analysis, or the dissemination of the research), the ethical objective is (a) to ensure that participants can actively voice their concerns for harm or benefit in an ongoing way and (b) to avoid the instrumental use of the participants' experiences for the purpose of the researcher's career advancement (Hugman et al. 2011). Such a participatory approach implies an obvious epistemological break from the classic view of the researcher as the expert and the researched as participants; instead, it elevates participants to a co-active position in the production of knowledge.

Another important debate concerns the field as a whole and the ontology of sedentariness it is based on. Malkki (1995) was one of the first and most prominent scholars in the field to point out how the assumptions inherent in the logics of the nation-state system impact how we as humans, citizens, and more importantly, researchers see the world and how, through these taken-for-granted assumptions, distinctions between normality and problems are created. The assumptions inherent in a world view that normalises sedentarism and pathologizes uprootedness or forced mobility have deep-felt consequences for people unable to conform with these norms, such as refugees and migrants. Following these premises, many scholars have called for a more cautious research design in order to refrain from representing refugees and migrants as the anomaly. Instead, migration scholars ought to study the environments in which people settle and the underlying assumptions and processes that link people and places. Even if it remains implicit, the link to normative considerations is clear. In one of the most recent contributions on the topic, Dahinden (2016) proposes to "de-migranticize" migration studies. Other reflexive migration scholarship of the past decade urge scholars to reflect more strongly on the field's inherent problematic features responsible for the shortcomings of many taken-for-granted conceptual tools in migration studies. Apart from the aforementioned entanglement with the logic of the nation-state, these features include the status of ethnicity as the most important difference and the entanglement of policy and research. Examples of shortcomings in the field's conceptual tools are methodological nationalism, methodological ethnicity, methodological whiteness, or ethnic groupism.

2.2 On Labelling and Policy-Making

A relevant area in migration studies concerns labelling and representation, both of migration as a spatial occurrence and of the migrating subject (especially the refugee). This ethical concern has been slightly less explicit in its normative exigency, but its call to critically engage with the concepts and labels used in the field has received significant attention in migration scholarship and beyond. Since the dawn of migration studies in academia, scholars have pointed at the danger of importing labels that have been developed outside of academia—namely, those based on a nation-state logic in which labels and categories are tools to govern inclusion and exclusion. One label that has been discussed prominently is "the refugee" (Zetter 1991). The labelling of refugees takes place in a situation of severe power inequalities, at the hand of others, and it effectively leads to severe material and legal consequences for those who are being labelled, for example, in protracted refugee situations or in asylum hearings. Critical scholarly contributions have argued that research has blended common-sense categories with those of the law and social sciences, and the use of the label "refugee" has contributed to homogenise, dehumanise, infantilise, de-politicise, and even idealise or sentimentalise refugees and refugeehood. Further, the debate on the distinction between the labels of "migrant" and "refugee", which amount to administrative categories in state practices, is prevalent in the field. Along with nation-states, the international refugee regime, and often the media, strong voices from within the migration scholars' community are adamant about refugees being distinguishable from migrants—not only because national laws often provide exemptions for individuals who flee due to political rather than economic reasons but also because the refugee's geographical movement is seen as forced rather than voluntary. On the other hand, an equally loud camp of migration scholars highlights the complexities of human movement and points at the shifting meanings of categories, their location-specific component, and the migrants' consequential shifting between categories over time. As stated earlier, for mobile people outside their country of citizenship, migration categorisations have direct impacts on their access to rights and resources.

This critical discussion of operating with labels developed outside academia has to be understood against the backdrop of the close relationship between migration studies and the modern nation-state system. Many lament the widespread noncritical use of labels, categories, and concepts in migration studies. Here, scholars point at a crucial ethical dilemma. Based on good ethical practice, many scholars conduct research with the explicit aim to impact policy in a bid to improve the life and conditions of the population they study. In order to do so, researchers often operate with terms and concepts developed by policy makers of states and humanitarian organisations. Thus, in the course of research, concepts and terminology developed by the world views and agendas of these policy makers become taken for granted and normalised. As stated above, policy categories are obviously important as they grant access to rights and resources; nevertheless, scholarship building on concepts and tools developed for non-academic purposes likely ends up with weak results. Moreover, such scholarship often ultimately fails to fulfil the aim of improving the conditions of the population under study and, thus, to provide the reciprocal benefit intended (Bakewell 2008; Dahinden 2016). For example, one concept that has recently received renewed attention is "immigrant integration". As critical scholars point out, the concept has been used in a depoliticising way, thereby contributing not only to the normalisation of the term but also to the normalisation of problematising some groups of society but not others. In a recent intervention, Schinkel (2017) underlines the neo-colonial assumptions inherent in the practice and measurement of immigrant integration. He calls on scholars to study the current and historical entanglements of politics, economic production, and race rather than confining themselves to an analysis of the nation. Ultimately, the concern of critical scholars is that research which fails to challenge the common-sense ideas behind the creation of a concept ends up reiterating the logic of the state to the detriment of the individuals they study. An ethical approach to research endeavours would thus be aware of this and be better equipped to refrain from mistaking the symptoms of a concern for its causes. Accordingly, Bakewell (2008) calls for more "policy-irrelevant" research and argues that studies that are not primarily designed to impact policy often have more potential to eventually influence policy-making and, therefore, the chance to alleviate the circumstances of the studied group.

3 On the Future of Migration Ethics and/in Migration Studies

As outlined in this chapter, there is a clear need for ethics awareness amongst migration scholars given that they often conduct research in humanly sensitive and politically delicate environments. Nevertheless, *explicit* ethical considerations are limited to some parts of the academic field: the methodology and work in political philosophy. Despite its more *implicit* normative imperatives, the discussion brought forward by migration scholarship that subscribes to debunking and challenging power relations, exposing biases, and probing taken-for-granted assumptions—largely through an intersectional lens and a reflexive practice—is likely to contribute to raising awareness on ethics that goes beyond migration studies in the coming years. As an academic field of inquiry that has to reflexively probe its relationship not only to the nation-state, the international humanitarian regime, and other humanitarian actors but also to other fields and disciplines with very different historical becomings, scholars with an interest in ethics in migration studies are well advised to pick up on this discussion. A fruitful engagement with the basic tenets brought forward in this discussion promises to produce studies on migration that avoid feeding into the dominant biases in the study of human mobility and, instead, consciously navigate the various ontologically premises of human mobility. Furthermore, through a more reflexive and sophisticated epistemology from which to study migration, ethics in migration studies also provide a major contribution to other fields of inquiry. For example, in the field of philosophy, they highlight the role of context-dependent social stratifiers and their impact on the mobility of humans and their experience of it.

The modern world is characterised by an increasing rate of mobility and displacement due to climate crisis, conflicts, globalised labour markets, and cultures of mobility. Migration experts anticipate a continuation of this situation and largely foresee three main trends: first, the increased participation of women in these migratory scenarios; second, the increased emergence of multiple belongings, including various forms of ethno-racial mixedness and diverse frames of identification beyond the national context; and lastly, the need to protect refugees outside the scope of nation-states. Thematically and methodologically, migration studies as a multi- and interdisciplinary field is well suited to be on the forefront of these scholarly inquiries. Therefore, we can expect an increase of studies that investigate the impact of migration from the complexity allowed or encouraged by an intersectional (and transnational) lens. Community building, labour market participation, policy impact, or other classic tropes of the field can be studied in this light, and they can produce insights of interest to philosophy. To tie in with the discussion above on the ethical challenges in categorising and labelling, intersectional considerations can potentially provide a valid approach in this intrinsic practice of conducting research. Finally, the increased reflexivity of academia amid powerful actors of nation-states and the international refugee regime will be useful in studying protection issues in a situation of continued state withdrawal from the role of refugee protection. A stronger foothold of reflexive scholarship in the study of migration will likely also strengthen the discussion and insights on one of the main ethical challenges for migration scholars in today's climate—namely, the question on how to conduct research in the field (preferably benefitting the target group) without becoming complicit in the states' and international refugee regime's politics of labelling or problem definition. Here, there is ample room for the fields of migration studies and philosophy to cross-fertilise each other.

Literature

Bakewell, O. 2008. Research Beyond the Categories. The Importance of Policy Irrelevant Research into Forced Migration. *Journal of Refugee Studies* 21:432–453.

Bilger, V. and I. van Liempt, eds. 2009. *The Ethics of Migration Research Methodology. Dealing with Vulnerable Migrants*. Brighton: Sussex Academic Press.
Chimni, B. S. 1998. The Geopolitics of Refugee Studies. A View from the South. *Journal of Refugee Studies* 11:350–374.
Dahinden, J. 2016. A Plea for the 'De-Migranticization' of Research on Migration and Integration. *Ethnic and Racial Studies* 39:2207–2225.
Hugman, R., E. Pittaway, and L. Bartolomei. 2011. When 'Do No Harm' Is Not Enough. The Ethics of Research with Refugees and Other Vulnerable Groups. *British Journal of Social Work* 41:1271–1287.
IASFM. 2018. *IASFM Code of Ethics*. International Association for the Study of Forced Migration (IASFM). http://iasfm.org/wp-content/uploads/2018/11/IASFM-Research-Code-of-Ethics-2018.pdf.
Malkki, L. 1995. Refugees and Exile. From 'Refugee Studies' to the National Order of Things. *Annual Review of Anthropology* 24:495–523.
Schinkel, W. 2017. *Imagined Societies. A Critique of Immigrant Integration in Western Europe*. Cambridge: Cambridge University Press.
Turton, D. 1996. Migrants and Refugees. A Mursi Case Study. In *Search of Cool Ground. War, Flight and Homecoming in Northeast Africa*, ed. T. Allen, 96–110. Oxford: James Currey.
Zetter, R. 1991. Labelling Refugees. Forming and Transforming a Bureaucratic Identity. *Journal of Refugee Studies* 4:39–62.

Cultural Studies and Migration Ethics

Nausikaa Schirilla

Abstract Cultural studies rarely addresses ethical questions directly, but it does very often imply ethical questions. In this paper, after a concise introduction to cultural studies, normative issues and their relation to migration ethics will be discussed. Cultural studies is presented as an approach and as an interdisciplinary research field that has been extremely important for migration studies in recent years. It presents a field of interdisciplinary critical studies on culture and society, focusing on how culture(s) is or are interwoven with nearly every aspect of social life and how these are related to power relations. This explains why cultural studies is fruitful for discussing various ethical questions about migration. This is especially the case with regard to the definition of culture, the claim to present critical social theory, the importance of empirical research and research approaches. I will argue for stronger integration of cultural studies and postcolonial theories into migration ethics.

This paper argues for raising ethical questions in cultural studies and relating cultural studies' approaches to migration ethics. Cultural studies is both an approach and an interdisciplinary research field, and has been extremely important for migration studies in recent years because of the notion of culture and power, as will be shown later. Cultural studies presents a field of interdisciplinary critical studies on culture and society, focusing on how culture(s) is/are intertwined with nearly every aspect of social life and how culture is permeated by power relations. Cultural studies rarely addresses ethical questions directly, but it does very often imply ethical questions. Furthermore, cultural studies should be referred to when discussing migration ethics questions in the following sense: firstly, this concerns the definition of culture—some debates in migration ethics presuppose moral conflicts because of cultural clashes. Cultural studies provides a notion of culture that questions homogeneous and fixed ideas of culture. Secondly, cultural studies questions power relations within and among cultures, and ethical questions concerning migration cannot be asked without considering these aspects. Thirdly, cultural studies claims to give a voice and a symbolic space to marginalised groups, and research is often focused

N. Schirilla (✉)
Catholic University of Applied Sciences Freiburg, Freiburg, Germany

A. Niederberger et al. (eds.), *Handbook of Migration Ethics*,
https://doi.org/10.1007/978-3-031-89877-8_37

on questions of the agency of these groups. These aspects should be strengthened in migration ethics as cultural issues are often related to differences between hegemonic and suppressed social groups or cultures. Furthermore, as an empirical science, cultural studies also raises questions about research and research ethics.

The arguments mentioned above will be unfolded in three steps. First, the discipline of cultural studies will be presented, and the relation to migration and refugees will be discussed. The second part will focus on ethical questions and normative issues in cultural studies in general. The third part will discuss cultural studies' contribution to migration ethics and argue that cultural studies should be related to migration ethics.

1 Cultural Studies, Migration and Normative Considerations

Cultural studies has been very important for migration research in recent years (Hörter et al. 2018). The term "cultural studies" is used for a variety of approaches and research fields that go back to the work of the Centre for Contemporary Cultural Studies (CCCS) in Birmingham, England, in the 1960s and 1970s, referred to as British cultural studies. Authors such as Richard Hoggart and Stuart Hall analysed "low" or popular culture and their importance for identity, social position, life world and agency of social groups (Storey 1996; Winter 2014). The research following the lines of cultural studies focuses on experiences and practices in everyday cultural life and shows how subjects make sense of their life and the circumstances in which they live, to what extent they are satisfied with, or how they question, subvert or confirm given circumstances (Grossberg 2010). Concurrently, domination and power relations are detected in the cultural field, and the claim to promote resistance and change was always present in cultural studies. This approach changed both the view of culture and critical theory (Kellner 2006).

Cultural studies developed further as an interdisciplinary and critical strand of research and theory within and beyond academia, but established in the academic field mostly in communication and media sciences, social sciences, anthropology and literary criticism, and contributed significantly to the "cultural turn" in human sciences. It is important to note that cultural studies does not present a discipline but rather an approach or an intellectual activity. This activity is characterised by a critical attempt to study domination or power relations via the cultural practices of subjects or study social and cultural practices in order to see how they are interwoven with power relations, but—as Bryan Behrenshausen puts it: "Cultural Studies refuses, especially at the outset of any analysis, the overmastering influence of any immediate and singular force or factor to which one might reduce any and all operations of power; instead, it entertains (relishes, really) a sustained scepticism of any narrative claiming to identify the motor force of activities ongoing in a social formation" (Behrenshausen 2019: 69). Culture or cultural practices are therefore hard to grasp: "They're never fixed, never stable, never final—always in motion, the ongoing work of actors at innumerable sites and scales, doing the work they do, having effects both intended and unintended, struggling to navigate the conjuncture's contours, maybe even winning a bit of space in it for themselves" (ibid).

Cultural studies by no means presents a homogeneous field and has undergone various theoretical shifts—mostly by integrating poststructuralist theories and opening new fields of research, for example, focusing on popular culture and media. Different strands relating to approaches, methodology, research fields and underlying epistemologies can be distinguished (ibid). One of the most important developments was the adoption of the idea of sending and receiving or encoding and decoding processes in the production and reception of culture. Culture was to be understood as a text that provides a variety of meaning and offers manifold possibilities of being read. Polyvalence and the open nature of culture present both a theoretical background and a research approach.

If we were to identify common ideas in cultural studies, I would follow Rainer Winter, who argues that if there were common ideas in cultural studies that it would be the concern with culture, the focus on the singular or individual and the claim to present critical theory of society (Winter 2011: 13).

Concerning migration ethics, the most important aspect of cultural studies is the concept of culture. Raymond Williams strongly influenced the notion of culture in cultural studies, and here culture is seen as a way of life related to practices in different fields; it is complex, multi-layered, contextual and polyvalent (Williams 1996). As in cultural studies, culture research departs from subjective experiences and practices, research into culture offers a route to understanding social and political structures from "below". Stuart Hall and his colleagues presented research on very individual subjective cultural phenomena in relation to power relations and emancipatory theories of resistance. The space of culture in the sense of practices, articulations and representations presents a field where power relations are established but can also be questioned and overcome. Culture is conceived as a perspective from where to describe polyvalent, diverse, complex and contested processes of construction of identities and social differences (Mecheril and Witsch 2006). As Hall argues, there are two paradigms (among others) in cultural studies—the culturalist paradigm and the structuralist paradigm (Hall 1996a). The cultural paradigm conceptualises culture as interwoven with all social practices and everyday life, but Hall stresses that from this perspective the culturalist paradigm presents the activity through which men and women make history (ibid). On the other hand, the structuralist paradigm conceives human beings as being subjected to certain structures that they cannot easily change. Neither 'culturalism' nor 'structuralism' are adequate for understanding subjectivity and society, and it is the inadequacy of both approaches that provides a space for analysis from which cultural studies starts.

In the course of time, cultural studies and the concept of culture have changed, incorporating poststructuralist notions such as contextuality, the interaction of sender and audience, textuality, etc., but these innovations were ultimately adaptions of a very open, flexible and resorbing concept of culture (and also of subjectivity).

Even though a poststructuralist notion of the subject as being constituted by surrounding cultural texts and acting non-intentionally was also adopted in newer works of Hall, traces of agency will always be found. Subjects are presented as being constituted by the textual material they absorb and thus are capable of intentional acting and of changing structures (Hall 1997).

1.1 Migration Issues

As I am raising questions of migration ethics, I will mostly refer to cultural studies as it was shaped by Stuart Hall and Lawrence Grossberg, but mostly by Hall.

Since the arrival of Stuart Hall—former head of the CCCS—in the field of cultural studies, research and theory not only on popular cultures but also on ethnic cultures, media perception of migrants, and racism have been produced (Hall 1996b). In Hall's works, racism is seen as a social power structure that can be analysed in almost all cultural fields. In a certain way, Hall showed racism to be normal—devaluation and disadvantaging of groups that are constructed as "other" is to be found in everyday cultural practices, media, communication as a hidden consensus. Media coverage of migration led Hall to stress the theoretical importance of the concept of representation (Hall 1997). Furthermore, Hall developed innovative ways to conceptualise migrants' cultures and identities as complex mixtures of various influences and layers of practices related to the concepts of diaspora and narratives. Hall's famous text "New Ethnicities" (1996c) argues that migrant communities are not constituted by home cultures but formed by social practices, cultural meanings from various heritages, experiences of marginalisation and racism, and thus we can speak of migrant communities without referring to homogeneous cultures of origin.

British cultural studies thus appropriated critical race theory and explored how cultural texts promoted sexism, racism, homophobia and other forms of oppression and how cultural texts also promoted resistance and struggle against these phenomena. Instead of focusing on ethnic cultures, intersectional approaches were adopted—the crossing of distinct differences like race, class, gender, and others—and research analysed how migrants (as any other subjects) negotiate, practise and narrate different forms of belonging, identities and self-presentation.

Migration research often departed from the notion of migration as a linear process—one-directional movement from point A to B connected with gradual processes of acculturation. This idea has been replaced by transnational approaches to migration as multidirectional mobility (Glick Schiller 2010). The reception of cultural studies by migration studies also strengthened the concept of migration as a continuous dynamic process of movement and development and not as a separate social space. Consequently, migration is seen to be constitutive of society as a whole.

Cultural studies approaches were thus suitable for analysing the different forms of belonging, of narrating identities, of practices in daily life, of changing and switching cultures and languages, of rejection, marginalisation and self-assertion that can be observed in most immigrant societies (ibid). Cultural studies as an approach or activity, as mentioned above, helped to reconceptualise the manifold and complex cultural orientations and their relation to power, hierarchies and social inequality. Communities and societies are conceived as both heterogeneous and as developing and formed in processes where belonging and identities are constantly negotiated and renegotiated.

1.2 Normative Issues

As mentioned above, cultural studies rarely addresses ethical questions directly, but it does always imply ethical questions as a critical theory. Grossberg, for example, points out that cultural studies as an intellectual activity aims to give an account of how power is penetrating everyday life and hindering people from living in dignity and realising their potentialities (2010). This conception of critique is very close to subjective aspirations and practices and their interplay with power structures. Nevertheless, there is a methodological tension between what is analysed in subjective practices as hindering or liberating. Cultural studies focuses on the intersections of culture and power relations or ideologies and conceives cultural practices as both reproductions of domination AND as a site of resistance and fight for more just and appropriate circumstances. But as Douglas Kellner rightly argues, concepts such as resistance, domination and justice are highly normative and imply an ethical critique of oppression and hierarchical orders, and they promote ethical values such as justice and equality (Kellner 2006; see also Winter 2015).

Kellner argues that instead of raising ethical questions, cultural studies referred to Antonio Gramsci's model of hegemony and counterhegemony and thus sought to analyse "hegemonic" or oppressing cultural patterns and to seek "counterhegemonic" articulations of resistance and struggle. The concept of representation thus came to the fore, and research aimed to find representations that promoted equality, social justice and emancipation. According to Kellner, ethics tends to be subordinated to politics, and the moral dimension of culture tends to be underemphasised or downplayed. He rightly demands that cultural studies should openly engage in the importance of ethical analysis, examining underlying ethical norms of cultural texts and practices and "thus, concern with ethics, with the moral aspects of cultural texts, should be a central and fundamental focus of cultural studies, [...]" (Kellner 2006: 139). The concern with ethics dwells on the unsolved question of the normative standpoint of critique.

2 Migration Ethics in the Discipline in Question

Cultural studies contributes to migration ethics in various ways. I do not argue that cultural studies poses and solves ethical questions; it does not, as I discussed above. However, I argue that reference to cultural studies leads to shaping or applying ethical questions in new ways. As mentioned before, this concerns the definition of culture. Cultural studies stands for a concept of culture as flexible, open, shifting and polyvalent. Therefore, any notion of cultural difference based on an essentialist understanding of culture and on the assumption of radical differences between two cultures must be questioned.

Many debates in migration ethics presuppose moral conflicts arising out of cultural clashes due to migration (Miller 2016). Immigrant cultures are very often presented as threats to national culture, and in this setting ethical questions are discussed regarding to what extent these developments are just and have to be redirected or to what extent this supposed situation is a substantial argument for strict migration control or at least stricter control (Walzer 2012). From a cultural studies perspective, both the idea of immigrant and of national cultures is to be questioned. The construction of a juxtaposition of immigrant and national cultures is criticised as culturalist—conceiving cultures as reified, static and homogeneous across bounded groups. According to the concept of culture unfolded above, representations of immigrant and national cultures are mutually constituted in practices, policies, the media, and everyday perceptions surrounding narratives and constructions of "we" and the "other". Furthermore, the research could show that all these cultural values and practices are inherently interlinked and changing, so there is no apparent contradiction between allochthon and immigrant cultures [see, for example, the studies by Foroutan (2013)]. According to cultural studies, the same would apply to identities: identities are always mixed, changing, volatile and polyvalent.

Furthermore, fears and anxieties about cultural loss can be deconstructed as media constructions (Hall 1996b). Certain representations gain powerful roles in cultural politics and representation, and they dominate public debates when specific symbols stand for an integrated set of cultural attributes, for example Muslim headscarves. Cultural studies can show the restraints of culturalism and contribute to public debates by promoting more processual and flexible understandings of culture. Furthermore, in exploring the intersection of public cultural performance, media, art and academic research on transnational mobilities, cultural studies proposes a new understanding of "territories", one that emphasises the performative and active practices of people like migrants before, during and after their migration processes. This approach would explore territories or symbolic spaces instead of nation states or ethnic culture.

2.1 *Reference to Different Notions of Culture Does Not Replace Ethical Debates*

Supposedly ethical problems arise from asking questions about the legitimacy of migration restrictions because of fear of losing national culture. If it is legitimate to restrict migration in order to protect national culture, or if it is necessary to tolerate different and foreign cultures—all these controversies have to be reshaped or shifted as neither national nor immigrant cultures can be detected as homogeneous entities.

Following cultural studies, one could even go further and argue that the fact that such an ethical problem is constructed in this way poses an ethical problem because the question presupposes privileges of (as homogenously constructed) allochthon cultures and attaches less value to (as homogeneously constructed) immigrant cultures. This question leads to the second point mentioned above. Cultural studies detects hierarchies of cultures and questions cultural power relations.

According to Hall, cultures are never constructed in a neutral way; cultures always imply hierarchical settings, above and low, superiority and devalorisation, "us" and the "others" (Hall 1996b).

Construction of an "other" migrant culture does not only present a homogenisation of a variety but is also marked by patterns of exclusion, construction of one and other in a binary or even dichotomous mode. On the one hand, therefore, the cultural field is open, shifting, manifold, etc., but on the other hand, it is not neutral, hierarchies and power relations can be detected. Culture can be analysed as a text, textual signs are always polyvalent, and they always entail a surplus of meaning. Nevertheless, hierarchical structuring patterns can be detected that stem from institutional, political, social power relations—one aspect is the valorisation of certain cultures as high or developed and others as different (Hall 1997).

In this sense, racialist patterns are reproduced in the cultural field on many levels. Cultural studies addresses and questions power relations within and among cultures, and ethical questions on migration cannot be asked without taking these phenomena into account. Cultural issues are often related to cultural differences between hegemonic and suppressed cultures, and understanding these patterns raises normative questions.

It is essential to point out that these hierarchical relations of cultures represent a colonialist legacy. A presentation of migrants as deficient, dangerous and threatening (mostly threatening democratic values) can be traced back to Eurocentric and ethnocentric constructions related to images of the colonial period. In his famous text "The West and the Rest" (Hall 1992; Yildiz 2016), Hall argued that in the colonial period an idea of "the West" emerged despite strong differences among the colonial countries and that it was based on the self-understanding of belonging to the same civilisation—the "West"—in contrast to the non-civilised "others". One of the results of this process is that Western Europe and America's historical development is regarded as normative, ideal or a goal that all other societies have to attain. Any differences that members of the vast and manifold types of other societies exhibit are regarded as deficient and are denigrated. Hall reconstructs the epistemic sides of the colonial process as a period of discourse formation. The idea of "The West and the Rest" presents a powerful episteme that permeated political, social, cultural and individual perceptions of the world and the self-perception of subjects (Hall 1992).

Colonial patterns are underlying hierarchical constructions of cultures and inherent in many constructions of ethical questions. The presentation of migrants as the "other" always means construction as the deficient other, less civilised and less developed.

Thus, two perspectives of ethical questions must be taken into account. The first concerns equality or justice in representation and what kind of violence and suffering is related to the representations mentioned above. From this perspective, ethnocentric images have to be overcome and the question has to be asked about what non-ethnocentric constructions would entail. The ethical question implies normative grounds of equality and difference and alternative—non-hierarchical—accounts of variety and diversity.

The other question—and this one is much more difficult to discuss—is to what extent or in which ways issues in migration ethics reproduce colonial patterns, for example present (constructed) immigrants' attitudes or value orientations as a threat to democracy or gender equality. It is difficult to operationalise this issue, but I would argue that it means developing a self-reflexive attitude concerning how ethical questions related to migration are formulated and what kind of hidden messages they might imply.

Furthermore, cultural studies claims to give a voice and a symbolic space to marginalised groups within a society and on a global level. As we have seen, cultural studies often focuses on questions of the agency of these groups. Although this claim relates to theoretical questions concerning critical analytical and normative realms of social theory, this theoretical approach should be strengthened in migration ethics.

However, it is essential to point out that there is no way of defining clear identities or communities. As mentioned above, Hall talks of a *black subject* (1996c) without any reference to ethnic culture. The black subject or an immigrant subject according to Hall can be described by the self-understanding of different individual subjects. In their narratives, difference and elements of their home cultures may play a role. But the black or immigrant subject is also constituted by discrimination, experiences of negation, being devalued, and fights. Although in Hall's account the black subject is characterised by common experiences, it is neither homogeneous nor separated from society. If we talk of a black subject, we always have to consider that there are differences according to gender, class, education, etc. and that common experiences do not stem from an authentic origin but came about in social practices of society. Thus, any black or immigrant subject is interwoven and entangled with history and society.

Cultural studies makes strong theoretical claims, but basically it presents approaches for empirical research. In his publication "The Future of Cultural Studies", Rainer Winter raises the question of how qualitative social research can remain committed to an ideal of social justice and incorporate a poststructuralist critique of simple equations of experiences and science and still present scientific research (Winter 2011: 80f, 2014). He argues for reframing the concept of validity. According to Winter, validity does not mean that scientific findings provide a correct image of an existing reality. Validity would rather mean that scientific findings open up new ways of seeing reality, new ways of thinking and alternative accounts of experiences. Research should present different constructions of reality and different truths—this is not a contradiction as they stem from different positions in society, different places on the globe, etc. In this sense, cultural studies gives a voice to marginalised individuals, groups and people by giving an account of their experiences, worldviews, etc. This approach does not imply arbitrariness and does not stand for an "anything goes" position. It is committed to social justice and radical democracy and develops new criteria for validity orientated towards these values. Naturally, these perspectives are not limited to migration questions, but they imply challenges both for migration research and migration ethics. Consequently, I want to point out that ethics is about a set of normatively oriented questions, but there are ethical questions where it might be necessary to incorporate empirical research—not for applying ethics but for clarification of normative claims that result from different contexts, for example different contexts of arrival in a country due to legal regulations, etc. or effects of the management and control of migration for migrants themselves (see also Niederberger 2021).

3 Desiderata for Future Research

Many strands of cultural studies are intertwined with critical race theory and postcolonial studies. Thus, they raise questions concerning ideas of the nation-states, privileges of "born" citizens, superiority claims, racism, deconstructions of borders. Critical race theory and postcolonial studies contribute to questioning established concepts of political philosophy and migration ethics, for example concepts of cultural identity or citizen's rights (see Fine and Ypi 2016; Gündoğdu 2015).

It was Hall himself who established a link to postcolonial studies (Hall 1996d), but other postcolonial authors tend to be mentioned in connection with cultural studies, too. Like cultural studies, postcolonial theorists have established that the history of the West was and is intertwined with the former colonies. The juxtaposition of "us" and "the other" originated from that period, and this pattern was enacted together with many different forms of norms and led to domination in various contexts. We still have to analyse how colonial justification narratives such as the civilisation mission, human rights, education, etc. operated and still operate. The strategies of justification of various colonial and neocolonial discourses must therefore be questioned, and in addition the normative violence, which was exerted in the name of civilisation and humanity, must be critiqued (Spivak 2004). As a consequence, patterns of this moral, intellectual and manifold superiority have to be detected, encoded and deconstructed in all the fields of intellectual production. This form of decolonisation of thought also has to be applied to migration ethics. This quest does not solve ethical problems, but it is by nature profoundly ethical.

Literature

Behrenshausen, B. G. 2019. Cultural Studies in the Present Tense. *Cultural Studies* 33:68–74. https://doi.org/10.1080/09502386.2018.1543337.

Fine, S., and L. Ypi, eds. 2016. *Migration in Political Theory. The Ethics of Movement and Membership*. Oxford: Oxford University Press.

Foroutan, N. 2013. Hybride Identitäten. In *Dabeisein und Dazugehören*, ed. H. Brinkmann and H.-H. Uslucan. Wiesbaden: Springer. https://doi.org/10.1007/978-3-531-19010-5_5.
Glick Schiller, N. 2010. A Global Perspective on Transnational Migration. Theorising Migration Without Methodological Nationalism. In *Diaspora and Transnationalism*, ed. R. Bauböck and T. Faist, 109–130. Amsterdam: Amsterdam University Press.
Grossberg, L. 2010. *Cultural Studies in the Future Tense*. Durham, NC: Duke University Press.
Gündoğdu, A. 2015. *Rightlessness in an Age of Rights. Hannah Arendt and the Contemporary Struggle of Migrants*. Oxford, New York: Oxford University Press.
Hall, S. 1992. The West and the Rest. Discourse and Power. In *Formations of Modernity*, ed. S. Hall and B. Gieben, 275–331. Cambridge: Polity Press.
Hall, S. 1996a. Cultural Studies, Two Paradigms. In *What Is Cultural Studies?* ed. J. Storey, 31–48. London, New York, Sydney: Edward Arnold.
Hall, S. 1996b. Race, Culture, and Communications. Looking Backward and Forward at Cultural Studies. In *What Is Cultural Studies?* ed. J. Storey, 336–343. London, New York, Sydney: Edward Arnold.
Hall, S. 1996c. New Ethnicities. In *Stuart Hall. Critical Dialogues*, ed. D. Morley and K.-H. Chen, 442–452. London, New York: Routledge.
Hall, S. 1996d. When Was the Postcolonial? Thinking at the Limit. In *The Postcolonial Question*, ed. I. Chambers and L. Curti, 242–259. London, New York: Routledge.
Hall, S. 1997. *Representation. Cultural Representation and Signifying Practices*. London, Thousand Oaks, New Delhi: Sage.
Hörter, K., et al. 2018. Migration und Migrationsforschung. In *Sozialpsychologie und Sozialtheorie*, ed. O. Decker. Wiesbaden: Springer. https://doi.org/10.1007/978-3-531-19582-7_13.
Kellner, D. 2006. Cultural Studies and Philosophy. In *A Companion to Cultural Studies*, ed. T. Miller, 139–150. Malden, MA: Blackwell.
Mecheril, P., and M. Witsch, eds. 2006. *Cultural Studies und Pädagogik*. Bielefeld: Transcript.
Miller, D. 2016. *Strangers in Our Midst. The Political Philosophy of Immigration*. Cambridge, MA: Harvard University Press.
Niederberger, A. 2021. Migrationsethik in der Krise. Einige grundlegende philosophische Überlegungen. *Zeitschrift für Migrationsforschung* 1:97–123.
Spivak, G. C. 2004. Righting Wrongs. *South Atlantic Quarterly* 103:523–581.
Storey, J., ed. 1996. *What Is Cultural Studies?* London, New York: Edward Arnold.
Walzer, M. 2012. Mitgliedschaft und Zugehörigkeit. In *Migration und Ethik*, ed. A. Cassee and A. Goppel, 107–146. Münster: Mentis.
Williams, R. 1996. The Future of Cultural Studies. In *What Is Cultural Studies?* ed. J. Storey, 168–177. London, New York, Sydney: Edward Arnold.
Winter, R. 2011. Kritik, soziale Gerechtigkeit und Intervention. Qualitative Sozialforschung in der amerikanischen Tradition. In *Die Zukunft der Cultural Studies*, ed. R. Winter, 75–92. Bielefeld: Transcript.
Winter, R. 2014. Cultural Studies. In *The Sage Handbook of Qualitative Data Analysis*, ed. U. Flick, 247–261. Los Angeles, London, New Delhi: Sage.
Winter, R. 2015. Kultur, Reflexivität und das Projekt einer kritischen Pädagogik. In *Cultural Studies und Pädagogik*, ed. P. Mecheril and M. Witsch, 21–50. Bielefeld: Transcript [English: 2004. Critical Pedagogy. In *Encyclopedia of Social Theory*, ed. G. Ritzer, 163–167. Los Angeles, London, New Delhi: Sage].
Yildiz, E. 2016. Das strategische Geflecht aus Ethnizität, Migration und Geschlecht. *Österreichische Zeitschrift für Soziologie* 41:29–41.

Index

A. Niederberger et al. (eds.), *Handbook of Migration Ethics*,
https://doi.org/10.1007/978-3-031-89877-8

MIX
Papier aus verantwortungsvollen Quellen
Paper from responsible sources
FSC® C105338

If you have any concerns about our products,
you can contact us on
ProductSafety@springernature.com

In case Publisher is established outside the EU,
the EU authorized representative is:
Springer Nature Customer Service Center GmbH
Europaplatz 3, 69115 Heidelberg, Germany

Printed by Libri Plureos GmbH
in Hamburg, Germany